BAILEY AND SCOTT'S Diagnostic microbiology

BAILEY AND SCOTT'S
Diagnostic microbiology

A textbook for the isolation and identification of pathogenic microorganisms

SYDNEY M. FINEGOLD, M.D.

Chief, Infectious Disease Section,
Wadsworth Veterans Hospital;
Professor of Medicine, UCLA School of Medicine,
Los Angeles, California

WILLIAM J. MARTIN, Ph.D. (ABMM, SM)

Head, Microbiology Section, Clinical Laboratories,
UCLA Hospital and Clinics;
Professor of Pathology and Microbiology and
Immunology, UCLA School of Medicine,
Los Angeles, California

ELVYN G. SCOTT, M.S., M.T. (ASCP)

Chief Microbiologist,
Section of Microbiology, Department of Pathology,
Wilmington Medical Center,
Wilmington, Delaware

FIFTH EDITION

with 219 *illustrations and* 7 *full-color plates*

The C. V. Mosby Company

Saint Louis   1978

The C. V. Mosby Company
11830 Westline Industrial Drive, St. Louis, Missouri 63141

Library of Congress Cataloging in Publication Data

Bailey, William Robert, 1917-1974.
  Bailey and Scott's diagnostic microbiology.

  Bibliography: p.
  Includes index.
  1. Medical microbiology—Laboratory manuals.
2. Micro-organisms, Pathogenic—Identification.
I. Finegold, Sydney M., 1921-   II. Martin, William
Jeffrey, 1932-   III. Scott, Elvyn G.
IV. Title: Diagnostic microbiology. [DNLM:
1. Microbiology—Laboratory manuals. OW25 B156d]
QR46.B26 1978   616.01′028   77-28251
ISBN 0-8016-0421-4

CB/CB/B 9 8 7 6 5 4 3 2 1

We wish to honor the memory of

Dr. W. Robert Bailey

who was the coauthor of the four previous editions of this text.
His passion for scientific integrity remains
a constant guide and inspiration.

S. M. F.
W. J. M.
E. G. S.

Preface to fifth edition

The two of us new to this edition were honored and delighted when Elvyn G. Scott invited us to work with him on this book, which we have long regarded as a classic in its field. We hope to maintain the high standards of this important work so that it will continue to serve microbiologists and students of medical microbiology. The three of us have made it our aim also to make the book useful to infectious disease clinicians and trainees, to clinical pathologists, to public health workers, and to nurses and nursing students.

The text is changed significantly from the last edition. There is some reorganization, much new material is added, and everything is updated. Only medically relevant material is included, and clinical correlations are presented. A new chapter has been added on our philosophy and general approach to clinical specimens, and a separate chapter has been created for the anaerobic cocci. Many other chapters have been completely rewritten, including the chapter on parasites by Lynne Shore Garcia. Among the new topics treated in this edition are evaluation of sputum for culture; enterotoxigenic versus enteropathogenic *Escherichia coli*; new methods for detection of bac-teriuria; role of *Chlamydia* in nonspecific urethritis; new classification of gram-positive cocci; new descriptive material on certain nonfermentative gram-negative bacilli not previously discussed; *Acidaminococcus*; *Megasphaera*; newly described mycobacteria; *Plesiomonas, Cardiobacterium,* and other uncommonly encountered gram-negative bacilli; new methods for susceptibility testing; several new tests for identification; and updated procedures in serology and clinical immunology.

We acknowledge, with gratitude, the assistance of numerous individuals. Particularly, we appreciate the help of Richard Addison, Pamela Byatt, Dorothy Citronbaum, Rick Greenwood, Myrna James, Teri Landis, Gene Marso, Vera Sutter, Donna Taratoot, and Palma Wideman. We would like to thank Nobuko Kitamura for the preparation of a number of excellent new illustrations for Chapter 35.

To our wives, Mary, Marcia, and Lois, we express our thanks for their patience, indulgence, and assistance.

Sydney M. Finegold
William J. Martin
Elvyn G. Scott

Preface to first edition

Diagnostic Microbiology is the first edition of a new series and not a revision of the former publication *Diagnostic Bacteriology*, the latest edition of which we revised (1958). This new title derives in part from the fact that the new volume includes microorganisms other than bacteria. The reader will note, for example, that the former Society of American Bacteriologists has now become the American Society for Microbiology.

Since this book is designed to be used as a reference text in medical bacteriology laboratories and as a textbook for courses in diagnostic bacteriology at the college level, the material has been consolidated and placed in separate parts and chapters. The selected sequence will be commensurate with the needs of both the diagnostician and the student.

For purposes of orientation in taxonomy and ready reference, an outline of bacterial classification has been included. For the student beginning diagnostic work, some pertinent background information is presented on the cultivation of microorganisms, the microscopic examination of microorganisms, and the proper methods for collecting and handling specimens.

A number of chapters include recommended procedures for the cultivation of both the common and the rare pathogens isolated from clinical material and should serve to familiarize the microbiologist with the wide variety of pathogens that may be encountered. An additional chapter has been devoted to the methods employed in the microbiological examination of surgical tissue and autopsy material.

To effect further consolidation of the book's content, one part has been devoted to a series of chapters which cover the various groups of bacteria of medical importance—their taxonomic position, general characteristics, and procedures for their identification. The chapter on the enteric bacteria introduces the new classification of the family Enterobacteriaceae, outlines the group biochemical characteristics, and discusses the serological aspects. The chapter on the mycobacteria includes a discussion of the increasingly important unclassified (anonymous) acid-fast bacilli, giving the methods for their identification, certain cytochemical tests, and animal inoculation procedures.

The chapter on laboratory diagnosis of viral and rickettsial diseases includes a guide to the collection of specimens and offers recommendations for the appropriate time of collection. In the chapter on laboratory diagnosis of systemic mycotic infections, the biochemical approach in identifying the pathogenic fungi is brought to the reader's attention.

The remainder of the book includes prescribed tests for the susceptibility of bacteria to antibiotics, serological procedures on microorganisms and patients' sera, and a technical section on culture media, stains, reagents, and tests, each in alphabetical sequence.

We would like to express our gratitude to Mrs. Isabelle Schaub and to Sister Marie Judith for committing the continuation of the original publication, *Diagnostic Bacteriology,* to our care and responsibility. We also acknowledge the many kindnesses extended by a number of microbiologists and clinicians in permitting the use of published and unpublished materials.

W. Robert Bailey
Elvyn G. Scott

Contents

PART IV

METHODS FOR IDENTIFICATION OF PATHOGENIC MICROORGANISMS

PART I

Laboratory methods

CHAPTER 1

General requirements for cultivation of microorganisms

Great advances relative to the nutritional requirements of bacteria and other microorganisms have been realized during the current century. Consequently, with the exception of a few fastidious forms, most pathogens can be cultivated in the laboratory on artificial media. Emphasis should be placed, therefore, on the proper preparation and selection of culture media in order to isolate and grow the various microbes. The need for quality control should be stressed. A good **culture medium** contains the essential nutrients in the proper concentration, an adequate amount of salt, and an adequate supply of water; is free of inhibitory substances for the organism to be cultivated; is of the desired consistency; has the proper reaction (pH) for the metabolism of that organism; and is sterile. It is obvious that such a definition applies only to the preparation of inanimate media because different requirements exist for obligate parasites.

Additional requirements, such as those generated by the temperature and oxygen relationships of the culture and by the lack of synthetic ability of certain fastidious pathogens, must also be met.

Heterotrophic microorganisms, the group to which the pathogens belong, exhibit a wide variety of needs. Despite this, however, numerous dehydrated media that are stable and eminently suitable for use in a diagnostic laboratory are available commercially. This ready supply minimizes concern regarding deterioration of prepared media and reduces storage needs. It is strongly recommended that media makers follow the directions given on the container labels.

PREPARATION OF MEDIA

Clean, detergent-free glassware and equipment are essential to good media preparation. The screw cap, the metal closure, and the plastic plug have virtually replaced the cotton plug in tubes of media, but personnel are cautioned concerning the exclusive use of the push-on metal closure. When tubes of media are stored over a long period, air contamination is likely to occur. Tubes cooling after sterilization take in air—the most likely means of initial contamination.

If a culture medium is to be prepared from its ingredients, the latter should be weighed out accurately in a suitable container. Approximately half of the required amount of water is added first to dissolve the ingredients; then the remainder is added. If it is an agar-containing medium, heat will be necessary to dissolve the agar. Heating on an open flame is **not recommended,** although it is often practiced. The use of a boiling water bath or steam bath is preferred. Complete dissolution of the ingredients of the medium in the required amount of water before sterilization gives a consistent and homogeneous product. Flasks of agar-containing media should be mixed after sterilization to promote a uniform consistency for pouring plates.

Media may be **liquid** (broth) or **solid** (containing agar). Sometimes a semisolid consistency is desired, in which case a low concentration of agar is used. Culture media may be also **synthetic,** in which all ingredients are known, or **nonsynthetic,** in which the exact chemical composition is unknown.

The inclusion of small amounts of a carbohydrate, such as glucose, is often recommended for the enhancement of growth in routine plating or broth media. The presence of such a carbohydrate in some instances can lead to a product of fermentation that may alter the characteristic appearance of a reaction produced by a microorganism. Examples are cited later.

Dehydrated media or media prepared according to accepted formulas are now available for the cultivation of anaerobic bacteria, which often precludes the necessity for cultivation in an anaerobic jar. Thioglycollate broth or agar is an excellent example of this type. The major use of such a medium is for biochemical tests on strains of anaerobes that have been previously isolated in pure culture. In the preparation of this medium (which contains peptone, an amino acid, and other ingredients), sodium thioglycollate is added as a reducing agent. This substance possesses sulfhydryl groups (SH–), which tie up molecular oxygen and prevent the formation and accumulation of hydrogen peroxide in the medium.

The cultivation of catalase-negative microorganisms, such as the clostridia, which are unable to break down this toxic substance, becomes possible under these conditions. The addition of thioglycollate or other reducing agents, for example, cysteine or glutathione, to culture media such as nutrient gelatin, milk, and others renders them more suitable for anaerobic or microaerophilic cultivation.

STERILIZATION OF MEDIA

Sterilization may be effected by **heat, filtration,** or **chemical** methods. The method of choice will depend on the medium, its consistency, and its labile constituents. Moist heat rather than dry heat is employed in the sterilization of culture media, and the method of application will vary with the type of medium.

Moist heat
Steam under pressure

The usual application for most media is steam under pressure in an autoclave where temperatures in excess of 100 C are obtained. Although the principle of the autoclave need not be discussed here at length, one important rule about autoclave sterilization must be stressed. In the normal procedure, using 15 pounds of steam for 15 minutes, the autoclave chamber should be flushed **free of air** before the outlet valve is closed. A temperature of 105 C on the chamber thermometer is used as an index of complete live steam content. When this temperature is reached, the outlet valve may be closed and the pressure allowed to build up to the required level of 15 pounds to attain generally a temperature of 121 C. Automation, however, takes care of such operational needs.

The time of exposure to this temperature and pressure may be allowed to exceed 15 minutes if large volumes of material are being sterilized. It is not recom-

Table 1-1. Autoclave temperatures corresponding to steam pressures in the chamber

Pressure (pounds)	Temperature (° C)	Pressure (pounds)	Temperature (° C)	Pressure (pounds)	Temperature (° C)
1	102.3	10	115.6	15	121.3
3	105.7	11	116.8	16	122.4
5	108.8	12	118.0	17	123.3
7	111.7	13	119.1	18	124.3
9	114.3	14	120.2	20	126.2

mended that an exposure of more than 30 minutes be used, because overheating can cause a breakdown of nutrient constituents.

A lower pressure may be necessary at times, such as in the heat sterilization of certain carbohydrate solutions. A pressure of 10 to 12 pounds for 10 to 15 minutes will reduce the possibility of hydrolysis.

The temperatures of the autoclave that correspond to the various live steam pressures (above atmospheric) are given in Table 1-1.

It is strongly recommended that the efficiency of the autoclave be checked at regular intervals. This may be conveniently done with the use of filter paper strips impregnated with spores of the thermophile *Bacillus stearothermophilus* and an appropriate amount of dehydrated culture medium with an indicator.[3]

The paper strips,* contained in small envelopes, are inserted in the center of a basket of tubes of media or other articles to be tested. The basket or article is then placed near the bottom at the front of the autoclave chamber and the usual sterilizing cycle is carried out. When the cycle is complete, the load is removed from the sterilizer and the sporestrip envelope is sent to the laboratory.

In the microbiology laboratory the sporestrips are individually removed, using aseptic technique, and placed directly into tubes containing 12 to 15 ml of sterile distilled water. After dissolution of the medium on the strips, the tubes are gently shaken and placed at 55 C and then observed daily for several days. An unexposed sporestrip is processed in like manner as a control with each sterilization check.

The control is examined after appropriate incubation. This should reveal a change in the color of the indicator, showing that acid (yellow if bromcresol purple is used) has been produced through fer-

mentation of the glucose. If the remainder of the tubes are of the same appearance, sterilization has not been effected.

Successful sterilization is indicated by the unchanged appearance of the heated tubes after 7 days' incubation at 55 C. The spores of *B. stearothermophilus* are destroyed when exposed to 121 C for 15 minutes.

A test kit more convenient for use in the microbiology laboratory consists of a sealed glass ampule containing a standardized spore suspension of *B. stearothermophilus,* culture medium and indicator.* The ampule is exposed, incubated, and read as previously described for the sporestrip; an unheated ampule is also included as a positive control.

These ampules are for professional use only and are used just once. Because they contain live cultures, they should be handled with care to prevent breakage. Each ampule is destroyed after using, preferably by incineration; unused ampules are stored in a refrigerator at 2 to 10 C.

A third but less reliable alternative is to use adhesive tape on which the word "sterile" is printed invisibly. The word becomes visible if the autoclaving is efficient. The tape may be placed on any suitable container to be autoclaved.

Flowing steam

The flowing steam procedure represents another application of moist heat and may be employed in the sterilization of materials that cannot withstand the elevated temperatures of an autoclave. **Fractional sterilization,** or **tyndallization,** introduced by John Tyndall in 1877, is a procedure involving the use of flowing steam in an Arnold sterilizer.

Material to be so sterilized is exposed for 30 minutes on 3 successive days. After the first and second days the material is placed at room temperature to permit any viable spores present to germinate. Vegetative bacteria are destroyed at flowing

---

*Kilit Sporestrips No. 1, Baltimore Biological Laboratory, Cockeysville, Md.

*Kilit ampule, Baltimore Biological Laboratory, Cockeysville, Md.

steam temperature, whereas their endospores are resistant. This procedure is used for media such as milk containing an indicator and others that may be precipitated or changed chemically by the normal autoclave treatment.

Inspissation

A third type of moist heat application is the process known as **inspissation,** or thickening through evaporation. This is used in the sterilization of high-protein-containing media that cannot withstand the high temperatures of the autoclave. The procedure causes coagulation of the material without greatly altering the substance and appearance. Materials such as the Lowenstein-Jensen egg medium, the Loeffler serum medium, and the Dorset egg medium are inspissated.

Modern autoclaves are equipped to allow inspissation procedures. If the Arnold sterilizer or a regular inspissator is used, the tubes containing the medium are placed in a slanted position and exposed to a temperature of 75 to 80 C for 2 hours on 3 successive days. Precautions should be taken during such treatment to prevent excessive dehydration of the medium.

Manually operated autoclaves may also be used successfully. Tubes of medium are placed in the autoclave chamber in a slanted position in a rack with adequate spacing. The tubes may be closed with a screw cap, loosely fitted initially, when Lowenstein, Loeffler, and Dorset egg media are being prepared. For operation, the exhaust valve of the autoclave is closed and the door is shut tightly. This traps air in the chamber. The steam is turned on, and the chamber contains an air-steam mixture. The pressure is then raised to 15 pounds and should be rigidly maintained for 10 minutes. The temperature ranges between 85 and 90 C. After this period, through manipulation of the steam valve and the exhaust valve, the air-steam mixture is replaced with live steam while the pressure is kept **constant**

at 15 pounds. When the temperature reaches 105 C, the chamber contains only live steam. The outlet valve is then closed, and an additional 15 minutes at 15 pounds is allowed to effect complete sterilization. During the final phase the chamber temperature rises to 121 C. At the end of this period the pressure should be permitted to subside very slowly. This is achieved by closing the steam valve and keeping the outlet valve tightly closed. The chamber temperature should drop below 60 C before the door is opened. When the tubes of media are cool, their caps should be tightened.

Filtration

Certain materials cannot tolerate the high temperatures used in heat sterilization procedures without deterioration; thus other methods must be devised. Materials such as urea, certain carbohydrate solutions, serum, plasma, ascitic fluid, and some others must be filter sterilized. Filters made of sintered Pyrex, compressed asbestos, or membranes are employed. These are placed in sidearm flasks, and the entire assembly of filter and flask is sterilized in the autoclave. A test tube of appropriate length may be placed around and under the delivery tube of the filter and sterilized with the unit when only a small quantity of filtrate is to be required. Negative pressure (suction) is applied to draw or positive pressure is applied to force the materials through the filter. Membrane filters are used in preference to the Seitz asbestos filter for some materials because of the high adsorption capacity of the latter type. The Swinney filter,* consisting of a filter attachment affixed to a hypodermic syringe and needle, is fast and simple for small quantities of material. The nonsterile material is taken up in the syringe, the sterile filter attachment is affixed, and the material is expressed by positive pressure (exerted by the plunger) into a ster-

---

*Millipore Filter Corp., Bedford, Mass.

ile container. Disposable sterile plastic holders and membrane filters are now being widely used.

Chemical methods

In addition to or in lieu of some of the foregoing, chemical methods may be used. These normally comprise two main types: (1) the use of chemical additives to solutions or (2) treatment with gases. The latter applies primarily to the sterilization of thermolabile plastic ware, such as Petri dishes, pipets, and syringes, and ethylene oxide may be used. The efficiency of such sterilization may be determined by the use of commercially prepared spore-strips.* As this is not directly concerned with sterilization of media, it will not be discussed here. The use of chemical compounds, such as thymol, a crystalline phenol, as additives to concentrated thermolabile solutions is quite appropriate, however. For example, to sterilize a 20-times normal concentration of urea or a 20% solution of carbohydrate, approximately 1 g of thymol per 100 ml of medium is added and allowed to stand at room temperature for 24 hours. The dilution subsequently made for use serves to nullify any bactericidal effects of the thymol.

SELECTION OF PROPER MEDIA

The multiplicity of available media increases annually. Selection of the proper media for specific purposes requires judgment by an experienced person, but for the most part, commercially available media, used in nearly all laboratories today, carry a recommendation based on the broad experience of others in the field. A very important facet in medium selection is the purpose for which the medium is intended. Laboratory personnel are advised to keep their selections to a minimum to avoid duplication of purpose.

Culture media, by virtue of their ingre-

*Attest Biological Indicators, 3M Co., St. Paul, Minn.

dients, may be selected for general and for specific purposes. The reader will find various terms ascribed to media throughout the literature. For example, terms such as **enrichment, enriched, selective,** and **differential** are very common as applied to media and are found in various sections of this book. These terms are indicative of purpose. Usually the media are special, but for general use a good basic nutrient medium is needed in the preparation of broth and agar media. The choice of such a medium is governed partly by tradition and partly by the experience of the users. In the selection of a basic medium, however, it is generally understood that one that is glucose free tends to give more consistent and reliable results. The presence of small amounts of this carbohydrate in a medium tends to enhance growth; yet fermentation of glucose can result in a pH that is harmful to acid-sensitive organisms. For example, this has been recognized in the cultivation of pneumococci and beta-hemolytic streptococci. The presence of the carbohydrate in a blood agar base medium can lead to the appearance of a green zone around colonies of streptococci, making differentiation between alpha and beta hemolysis extremely difficult.

Trypticase or tryptic soy broth, a pancreatic digest of casein and soybean peptone, is a widely used general medium that will support the growth of many fastidious organisms without further enrichment. Because of its glucose content, however, one should be aware of its limitations in certain areas of application.

Thioglycollate medium, now available in a modified form containing 0.06% glucose, will support the growth of many facultatives, microaerophiles, and anaerobes. In the ensuing chapters of this book, which discuss the isolation and cultivation of various pathogenic microorganisms, recommendations relative to the selection and use of media are made. Chapter 41 gives the formulas and methods of preparation for many of the media in use at the present time.

Valuable information on the choice of media and the nutritional requirements of microorganisms may be gained from references 2, 4, and 8 at the end of this chapter.

STORAGE OF MEDIA

Because of the heavy work schedule of most diagnostic laboratories today, bulk preparation of media is often necessary. It should be emphasized that infrequently used media that are subject to deterioration on prolonged storage should be purchased in the smallest available quantity. For example, ¼-pound bottles or smaller packages should be specified in these situations. The average medium should be stored in the refrigerator to avoid deterioration and dehydration. Plating media used for the cultivation of certain fastidious organisms that are sensitive to drying should be kept airtight and refrigerated. Some .tubed media, especially those with tightly fitting screw caps, paraffined corks, or rubber stoppers, may be stored for relatively long periods at room temperature. Plates of media that can be readily sealed are available commercially* from several companies and have certain advantages. Such plated

media are in wide use in many laboratories and physicians' offices.

Refrigeration facilities obviously determine the storage capacity. It is recommended that plated and tubed media be stored in sealed plastic bags. Depending on the media and whether it is plated or tubed, bagged media can be stored for from 2 weeks to 4 months.[1] The laboratory worker is cautioned about the use of media that have been just removed from refrigeration. Plated and tubed media should be allowed to warm to room temperature before use.

REFERENCES

1. Bartlett, R. C.: Medical microbiology. Quality, cost and clinical relevance, New York, 1974, John Wiley & Sons, Inc., p. 192.
2. BBL manual of products and laboratory procedures, ed. 5, Cockeysville, Md., 1973, BioQuest, Division of Becton, Dickinson & Co.
3. Brewer, J. H., and McLaughlin, C. B.: Dehydrated sterilizer controls containing bacterial spores and culture media, J. Pharm. Sci. **50:**171-172, 1961.
4. Difco manual of dehydrated culture media and reagents for microbiological and clinical laboratory procedures, ed. 9, Detroit, 1953, Difco Laboratories.
5. Difco supplementary literature, Detroit, May 1972, Difco Laboratories.
6. Porter, J. R.: Bacterial chemistry and physiology, New York, 1946, John Wiley & Sons, Inc.
7. Snell, E. E.: Bacterial nutrition-chemical factors. In Werkman, C. H., and Wilson, P. W., editors: Bacterial physiology, New York, 1951, Acacemic Press, Inc.
8. Vera, H. D., and Dumoff, M.: Culture media. In Lennette, E. H., Spaulding, E. H., and Truant, J. P., editors: Manual of clinical microbiology, Washington, D. C., 1974, American Society for Microbiology.

---

*Baltimore Biological Laboratory, Cockeysville, Md.; Gibco, Grand Island, N. Y.; Pfizer Laboratories, Inc., New York; Scott Laboratories, Fiskeville, R. I.; and others.

CHAPTER 2

Optical methods in specimen examination

SPECIMEN PREPARATION

Direct microscopic examination of clinical specimens for microorganisms is extremely important. It provides immediate information on the nature and relative numbers of various microbial forms, inflammatory and other host cells, and certain other materials, such as crystals. Data of this type provide information on suitability of the specimen, the likelihood of infection being present, the likely infecting organism(s), and the predominant organism(s) in mixed infections. This information may suggest the desirability of using special media and is very important in quality control in the laboratory.

Specimens received in a laboratory for microbiologic examination can vary widely. They may consist of clinical material, such as blood, urine, pus, cerebrospinal fluid, or microbial cultures isolated from such material. In examination the laboratory worker or student should consider the nature and possible content of the specimen and be cognizant of the difficulty in cultivating and staining certain microorganisms. Special methods are often needed.

There are two ways in which microbial isolates or microorganisms in certain clinical material may be examined under a microscope: in the **living** state or in the **fixed** state.

Living state

It is often necessary to examine certain microorganisms in the living state because they are not readily stained or because they cannot be easily cultivated. Also, when they are examined in the living state, morphology is less distorted and motility and other characteristics may be observed readily.

Two methods (the wet-mount and the hanging-drop techniques) are generally used for such examination.

Wet-mount method

In the wet-mount procedure there are two approaches, the choice being determined by the nature of the specimen: (1) place a loopful of the liquid clinical specimen or culture on a clean glass slide and cover with a coverglass; (2) place a loopful of clean water on the slide and in it emulsify some nonliquid clinical material or portion of a bacterial colony and add a coverglass. In either case, to reduce evaporation of the liquid, the coverglass may be ringed with petrolatum. The preparation may then be examined, either by brightfield or by darkfield microscopy. The latter method may be essential with such microorganisms as the treponeme of syphilis, which is very difficult to stain. This procedure may also be adapted to a wet-mount staining procedure for certain organisms.

Hanging-drop method

A second procedure used in the examination of organisms in the living state is the hanging-drop method. Loopfuls of the specimen may be prepared as with the wet-mount method but on a thin coverglass rather than the slide. The coverglass is then inverted over the concave area of a hollow-ground or well slide to provide the hanging drop. The coverglass can be sealed as in the wet-mount method with petrolatum or a small amount of immersion oil to reduce evaporation. Such a preparation is examined by brightfield microscopy.

Fixed state

Bacteria and other microbes are generally examined in the fixed state and are more readily observed when they are stained. By drying and fixing the organisms on a glass slide, various staining procedures may be carried out. The choice of procedure is determined by the nature of the specimen or the desired result. Staining procedures for bacteria and other microorganisms are discussed in Chapters 3 and 42.

MICROSCOPES—THEIR USE AND CARE

It is assumed that the reader understands the elements of microscopy and is generally familiar with the parts of a microscope. For this reason, discussion of the instrument is restricted to a few points that bear emphasis. The lenses of a microscope should be kept clean and free of grease and dust. In using the average compound light microscope equipped with a condenser, only the flat plane or face of the mirror is used. The light source selected should give uniform illumination, such as a frosted bulb, a fluorescent light, or a special microscope lamp. Most of the new light microscopes have built-in lamps. For routine bacteriologic examination, the condenser of the microscope should be kept almost fully racked up. This position usually provides optimal and uniform illumination of the field. The user of a monocular microscope is strongly advised to keep both eyes open in order to reduce eyestrain.

Since many laboratories today are equipped with a variety of microscopes, several instruments are briefly discussed here with some specific advice and information. All light microscopes should be kept covered when not in use, and full attention should be given to protection of the objective lenses. Immersion oil

should never be left on the objective when the instrument is not to be used for some time. The nosepiece should be rotated so that the low-power objective is in position before the microscope is put away. Repeated use of a solvent such as xylol to remove oil from a lens is discouraged, since this will tend to loosen the mounting cement around the lens. **Clean lens paper only** should be used for wiping the objective free of oil. Other commercial tissue is too rough and will scratch the lens.

Brightfield microscope

The brightfield microscope is used for the majority of routine examinations where stainable microorganisms are to be observed. In brightfield microscopy the organisms appear dark against a bright background.

Darkfield microscope

In contrast to the effect in brightfield microscopy, organisms appear bright against a dark background under the darkfield microscope. This appearance may be produced either by an opaque stop inserted below the condenser or by a stop built into the condenser. This stop permits only peripheral rays of light to enter the condenser. These rays pass through the specimen at an angle such that the field appears unilluminated. Any particles, such as microorganisms in the field, reflect the light and appear bright in the optical system. This type of microscope is particularly adaptable to the examination of microorganisms that are difficult to stain. The treponeme of syphilis in the exudate from a lesion or in other material may be thus examined. Examination is carried out with the living organisms in a wet-mount preparation. Motility is readily determined for other bacteria as well.

In using the conventional darkfield microscope it is essential to use immersion oil on the top of the condenser as well as on the coverglass of the wet-mount preparation to minimize light refraction.

Phase microscope

The phase microscope permits relatively accurate observation of (1) bacteria in tissue sections, (2) parasites in various types of clinical material, (3) Negri or inclusion bodies in virus-infected material, and (4) histologic preparations. A halo of light produced by light passing from the source through an annular diaphragm initiates the effect obtained. Rays of light passing through an area composed of materials varying in refractive index emerge out of phase and give a pattern of bright and dark relief. Objects such as bacteria and inclusion bodies, being of a different refractive index from surrounding tissue, consequently show up in such a microscope, whereas in normal brightfield microscopy these objects must be stained to be observed.

Ultraviolet microscope

The use of ultraviolet light rather than visible white light in microscopy allows for greater resolution. This is due to the shorter wavelength of ultraviolet light. As a result, a magnification can be obtained that is two to three times higher than that which is possible with visible light. Because of the invisibility of ultraviolet light, a photographic plate is used to record the image, and since glass is impervious to such light, quartz rather than glass lenses are employed.

Fluorescence microscope

Fluorescence microscopy has become very popular in a number of laboratories because of its application to the diagnosis of disease. The light source is ultraviolet. The principle involved here is that certain fluorescent chemical complexes have the property of absorbing ultraviolet light and emitting rays of visible light. Microorganisms can be treated with a fluorescent dye or a dye-antibody complex that causes them to fluoresce and become readily distinguishable in a mixed population.

The fluorescent-antibody technique has both advantages and disadvantages in diagnostic microbiology. Among the former are the time saved by eliminating lengthy cultural study, the sensitivity of the test, the lack of interference from contaminating microorganisms if their staining reactions are known, and the detection of nonviable pathogenic organisms. The disadvantages of the technique are the same that are experienced with most serologic tests: the cross-reactions that inevitably develop between species. Anyone, therefore, planning to use the fluorescent-antibody technique should be aware of such limitations. Successful results may now be obtained with group A streptococci and other bacteria and in the diagnosis of rabies. The technique has further application in the detection of treponemal antibody (FTA). The reader's attention is directed to Chapter 39, where immunofluorescence is discussed in further detail. The value of this procedure in diagnostic parasitology and mycology is discussed by Cherry.[1]

Electron microscope

The electron microscope has become essential in various fields of research. In the biologic sciences, virology presently makes the greatest use of this instrument, but it has also become an invaluable piece of equipment to cytologists.

An electron source is used in lieu of a light source, and magnets are used instead of lenses. Only nonliving material may be examined, and specimens require special preparation. In the more modern instruments a direct magnification of 200,000 diameters may be obtained.

The image may be observed on a fluorescent screen through a window at the base of the evacuated column, or a photographic plate may be inserted to record the image. The result, however, is not a photograph, since photons are not used. The coined term electron micrograph has been in use for some years.

Great advances in the refinement of techniques used in specimen preparation have been made in recent years, and much knowledge has now been amassed relative to the interpretation of electron micrographic images.

REFERENCE

1. Cherry, W. B.: Immunofluorescence techniques. In Lennette, E. H., Spaulding, E. H., and Truant, J. P., editors: Manual of clinical microbiology, ed. 2, Washington, D. C., 1974, American Society for Microbiology.

CHAPTER 3

General principles in staining procedures

Bacteria and other microorganisms are usually transparent, which makes the study of morphologic detail difficult when they are examined in the natural state. The early methods of fixing and staining initiated by Paul Ehrlich and Robert Koch allowed the microbiologist to distinguish many structural features not formerly seen.

Unless some specific morphologic feature, dependent on the age of the culture, is to be demonstrated, the microbiologist is advised to use a young culture for routine staining procedures, as old cells lose their affinity for most dyes. Unless one is dealing with organisms that have an unusually long generation time, a 24-hour culture is expected to yield favorable results.

PREPARATION OF A SMEAR

In routine staining procedures the first step is the preparation of a **smear.** A loopful of liquid culture or fluid specimen is spread over an appropriate area of a clean glass slide to make a film, or a section of a colony taken with a needle from a plate or growth from a slant culture is emulsified in a loopful of clean water and spread over the required area. For best results the film should be homogeneous.

In either event the mixing and spreading should never be done vigorously; such treatment is potentially hazardous when working with pathologic material because it may create aerosols. In addition, harsh treatment will tend to destroy characteristic arrangements of cells, such as chains and clusters.

A smear ideally should always be permitted to dry in the air and then be heat fixed by passing gently through a Bunsen flame to promote adherence of the cells to the slide. Overheating causes distortion and should be avoided. The slide should then be allowed to cool.

When a large number of cultures are to be examined by a routine staining procedure, such as the Gram stain, several small smears can be made on the same slide in areas appropriately marked with a glass-marking pencil.

STAINING OF BACTERIA
Simple stain

1. After fixation of the smear, draw a vertical line on the glass slide about 1 inch from the left or right end, using a wax pencil. This will keep the stain away from the fingers when holding the slide.
2. Flood the smeared area with the stain to be used (crystal violet, fuchsin, methylene blue, or safranin).
3. Allow to react for the appropriate time, 30 seconds to 3 minutes, depending on the stain.
4. Wash off with a gentle stream of cool water. Avoid having the water fall directly on the smear with any force.
5. Blot dry between sheets of bibulous paper and examine under the oil-immersion lens.
6. It is good practice to sterilize blotting paper used in this simple procedure involving pathogenic microorganisms. The slides, if they are not to be retained for reference, should be placed in disinfectant.

Special staining procedures
Gram stain

The Gram stain ranks among the most important stains for bacteria. Devised by

Hans Christian Gram, a Dane, in 1884, it allows one to distinguish broadly between various bacteria that may exhibit a similar morphology. There are various modifications of the technique; the one described in Chapter 42 has been found reliable, even in the hands of a beginner.

On microscopic examination of a gram-stained smear containing a mixed bacterial flora, the differential features of the method become apparent. Many bacteria retain the violet-iodine combination and stain **purple** (gram positive), others stain **red** by the safranin (gram negative). Thus, in using this procedure not only are form, size, and other structural details made visible, but the microorganisms present can be grouped into gram-positive and gram-negative types by their reactions. This is an important diagnostic tool in subsequent identification procedures.

Probably the difference in the staining reaction between gram-positive and gram-negative bacteria can be attributed to their different chemical compositions. Gram-negative cell walls have a higher lipid content[1] than gram-positive cell walls, and although a crystal violet–iodine complex is formed in both kinds of cell, the alcohol removes the lipid from the gram-negative cells, thereby increasing the cell permeability and resulting in the loss of the dye complex. The complex is retained, however, in the gram-positive cells, in which dehydration by the alcohol causes a decrease in permeability.

Acid-fast stain

Another differential staining technique that depends on the chemical composition of the bacterial cell is the acid-fast stain, which is used in staining tubercle bacilli and other mycobacteria. Since these microorganisms are difficult to stain with the ordinary dyes, basic dyes are used in the presence of controlled amounts of acid and are generally applied with heat. Once stained, the tubercle bacillus is resistant to subsequent treatment with **acid** alcohol, whereas most other

bacteria are decolorized. For convenience in viewing, a counterstain of contrasting color is usually applied. The method most commonly employed is that of Ziehl and Neelsen, although the fluorochrome procedure also is recommended. These are described, as used at the Center for Disease Control in Atlanta,[2] in Chapter 42.

Acid-fast organisms are difficult to stain and once stained are not easily decolorized. Many theories have been introduced to account for this property of acid fastness. The most acceptable concept is that acid fastness is determined by the selective permeability of the cytoplasmic membrane. The brilliance of the red color is due to the retention of the dye, carbol-fuchsin, in solution within the cell. Should the cell be mechanically disrupted, its acid-fast property is lost. That this may not be a precise explanation of the reaction does not invalidate the usefulness of the technique for detecting fundamental differences in bacterial genera.

Capsule stain

The capsule of bacteria does not have the same affinity for dyes as do other cell components, which necessitates the use of special staining procedures. Some are designed to stain the cell and its background but not the capsule, so that the surrounding envelope is seen by contrast, as in the Anthony method. Other procedures produce a differential staining effect wherein the capsule will take a counterstain, as in the Muir method. Another procedure involves the negative stain principle in which the capsule shows up as a clear halo against a dark background, as in the India ink method.

The various procedures are given in Chapter 42.

Flagella stain

Flagella are fragile, heat-labile appendages and require careful handling and staining by special procedures to observe them. Success in staining depends on the freshness and effectiveness of a mordant. Since flagella are beyond the resolving power of the ordinary light microscope, their size must be increased to bring them into view. This may be accomplished by coating their surfaces with a precipitate from an unstable colloidal suspension (the mordant). This precipitate then serves as a layer of stainable material, and the threadlike structure of the flagellum shows up when the appropriate stain is applied.

Various methods may be used, but we have found the Gray method very satisfactory (see Fig. 20-2). The procedure may be found in Chapter 42.

Staining of metachromatic granules

Various methods for staining metachromatic granules have been devised. Some are preferred to others. We have found that two stains yield highly satisfactory results. One of these is the methylene blue stain and the other is Albert's stain (see Fig. 26-1). Both procedures are outlined in Chapter 42.

Spore stain

Spores are relatively resistant to physical and chemical agents and are not easily stained. In routine staining procedures, such as the Gram stain, they may appear as unstained refractile bodies. Heat is usually required during a special procedure for staining to promote penetration of the stain into the spores. Spores may be detected by darkfield examination.

The Dorner method or a modification thereof gives good results. The procedure is given in Chapter 42.

Relief staining

Relief staining does not stain microbial cells but produces an opaque or darkened background against which the cells stand out in sharp relief as white or light objects. It simulates darkfield microscopy and allows a rapid study of morphology.

Staining of spirochetes

On the whole, spirochetes have a poor affinity for the standard dyes and usually require special stains, such as the silver impregnation method. The Fontana stain usually gives good results. The procedure may be found in standard texts.

Staining of rickettsiae

Although rickettsiae are gram negative, they do not respond well to the Gram stain and are more successfully stained by the Giemsa or Castañeda methods. These procedures are found in Chapter 42.

Staining of yeasts and fungi

Fixed smears of yeast can be stained quite readily with crystal violet or methylene blue when these dyes are applied for 30 to 60 seconds or by Gram's method. Wet mounts of yeast can be stained effectively with methylene blue or Gram's iodine. Cells can be emulsified in a drop of either stain and covered with a coverglass. Alternatively, a loopful or small drop of the dye may be placed at the edge of the coverglass covering an unstained suspension (wet mount), and the dye will spread rapidly beneath it. Lactophenol cotton blue is excellent for staining fungi. The procedure is described in Chapter 42.

MOUNTING OF STAINED SMEARS

Microbiologists often want to develop or add to a slide collection for reference and teaching purposes as well as for demonstrations. For such purposes, permanent slides may be prepared by a relatively simple procedure.

If a slide has been examined under oil, blot it with bibulous paper (**do not wipe**) and place a drop of Canada balsam or Permount on a desirable area of the smear. Place a clean coverglass carefully over this to avoid trapping air bubbles. With the butt end of an inoculating loop or needle, press on the coverglass to force the mounting material to the sides of the coverglass to completely cover the area beneath.

Place the slides in a horizontal position away from dust to permit proper setting of the balsam. When the slides are completely dry, wipe away excess dye with cheesecloth dampened with xylene and cut away excess balsam with a razor blade. Some persons prefer to add a border of nail polish or other appropriate substance to the coverglass as a protec-

tive device. Most permanently mounted slides will last for years without deterioration.

REFERENCES

1. Salton, M. R. J.: The bacterial cell wall, Amsterdam, 1964, Elsevier Press, Inc.
2. Vestal, A. L.: Procedures for the isolation and identification of mycobacteria, DHEW Pub. No. 75-8230, Atlanta, 1975, Center for Disease Control.

CHAPTER 4

Methods of obtaining pure cultures

The importance of **pure cultures** cannot be overemphasized in diagnostic microbiology. They are essential to the accurate determination of colony characteristics, biochemical properties, morphology, staining reaction, immunologic reactions, and susceptibility to antimicrobial agents.

In general, pure cultures are best obtained by using solid media, either in streak plates or in pour plates. Both methods are useful, and each has certain advantages.

STREAK PLATE METHODS

The streak plate method, if properly performed, is probably the most practical and most useful for obtaining discrete colonies and pure cultures. Streaking methods may vary from one laboratory to another, but the following techniques may be used for inoculating any type of agar plate. These plates may be prepared in advance in any desired quantity and stored at room temperature or in a refrigerator until the time of inoculation. The agar surface should be free of beads of condensation before the streaking is done. The plate may be dried by inverting it and propping it on the lid (Fig. 4-1, *A*).

Method 1

Method 1 (Figs. 4-1, *B*, and 4-2), although designed for transfers from broth cultures, may also be used for cultures from agar plates or agar slants. Should either of the latter be used, a reduced amount of inoculum is recommended and the overlapping should be moderated.

1. Place a loopful of the inoculum near the periphery of the plate as indicated.

2. With the loop, spread the inoculum over the top quarter of the plate's surface.
3. Stab the loop into the agar several times and continue streaking, overlapping the previous streak as shown.
4. Stab the loop as before and continue streaking.
5. Flame the loop, allow it to cool, overlap the previous streak, and complete the streaking.
6. Lift the loop and streak the center of the plate with zigzag motions.

Discrete colonies should be found in the central portion of the plate, whereas additional information on hemolytic activity may be gained from the effect of a reduced oxygen tension on the organisms **stabbed** into the agar. For example, beta-hemolytic streptococci may appear to be alpha hemolytic on the surface.

Method 2

The second method of streaking (Fig. 4-1, *C*) may be used either for cultures growing on solid media (colonies or growth on slants) or for heavy broth cultures. The technique is primarily designed, however, for the former.

1. Select the desired colony from a crowded plate or pick up growth from a slant with a needle. Streak this carefully on a restricted area of the plate as indicated. Flame the needle and put it away.
2. With a sterile, cool loop make one light sweep through the needle-inoculated area and streak the top quarter of the plate's surface with close parallel strokes. Flame the loop and **allow it to cool.**
3. Turn the plate at right angles, make one light sweep with the loop

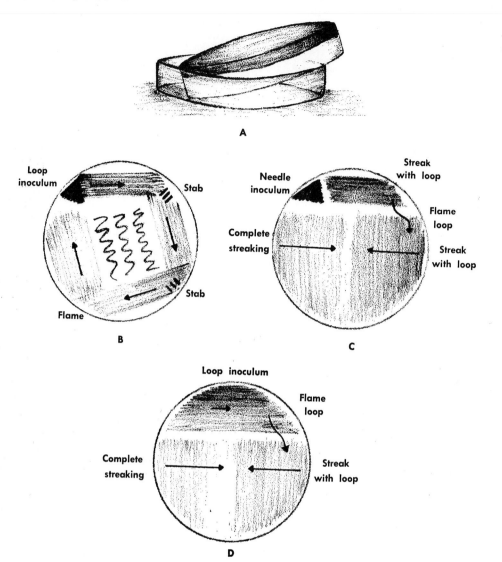

Fig. 4-1. Streak plate technique. **A,** Drying an agar plate in the laboratory. **B,** Method 1. **C,** Method 2. **D,** Method 3.

through the lower portion of the second streaked area, and streak, as before, approximately one half of the remaining portion. Do not overlap any of the previously streaked areas.

4. Turn the plate 180 degrees and streak the remainder of the plate with the same loop, avoiding any areas previously streaked.

5. Discrete colonies should appear in the lower portions of the plate.

Method 3

Method 3 (Fig. 4-1, *D*) is the simplest technique and is used largely for broth cultures, but it may also be used for cultures from solid media.

1. Place a loopful of the inoculum near the periphery of the plate and cover approximately one fourth of the plate with close parallel streaks. Flame the loop and allow it to cool.

2. Make one light sweep through the

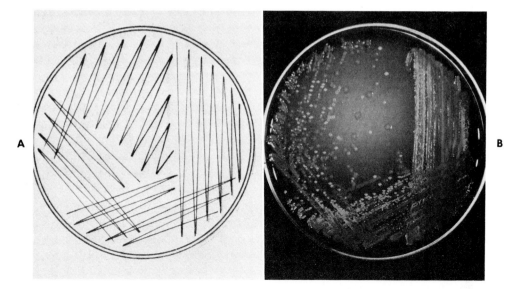

Fig. 4-2. A, Diagram of procedure for streaking to obtain well-isolated colonies. **B,** Mixed flora separated by technique illustrated in **A**.

lower portion of this streaked area, turn the plate at right angles, and streak approximately one half of the remaining portion without overlapping previous streaks.

3. Turn the plate 180 degrees and streak the remainder of the plate, avoiding previously streaked areas.
4. The appearance of this plate will resemble that obtained in method 2.

POUR PLATE METHODS

Pour plates are generally used in the laboratory as a means of determining the approximate number of viable organisms in a liquid such as water, milk, urine, or broth culture, and data are expressed as the number of **colony-forming units** per milliliter of the substance examined. Pour plates are also used to determine the hemolytic activity of deep colonies of bacteria, such as the streptococci, by using an agar medium containing blood. The pour plate also lends itself to pure culture study when the components of a mixed culture are to be separated. Differences are recognized by size, shape, and color of the colonies.

The pour plate method consists of the preparation of serial dilutions of the specimen in sterile water or other diluent, from which a prescribed volume may be pipetted into tubes of melted agar medium or directly into sterile Petri dishes. The former are then poured into Petri dishes and allowed to set, whereas melted agar medium is added to the latter. After a suitable incubation period the colonies (surface or subsurface) may be examined for differences in morphology or counted, depending on the exercise. The melted agar medium used should **not** exceed 50 C, and the dilutions should be adequately mixed with the medium. Additional tests on isolates may be carried out as desired.

The usefulness of the pour plate depends to a great extent on the number of colonies developing on the plate. This is particularly important when studying organisms for their hemolytic activity or where colonial characteristics aid in the separation of various types. The greatest accuracy in counting is obtained with plates that contain between 30 and 300 colonies. Above this number, counting is difficult, differential characteristics are not readily distinguishable, and colony selection without contamination is almost impossible.

A procedure that may be followed in counting overcrowded plates to estimate numbers present involves the use of a **counting plate.** The agar plate may be placed on the counting plate, which is ruled off in square centimeters. In this way, five or more representative square-centimeter areas may be counted, the average count calculated, and this figure then multiplied by 62.5 to give the estimated number of colonies per plate. The average Petri dish (inside diameter, 90 mm) has an area of 62.5 sq cm.

If a specimen is suspected of containing a large number of organisms, more than several organisms per microscopic field, an overcrowded pour plate may be avoided by using the following procedure, which is recommended for hemolytic streptococci:

1. Inoculate 8 to 10 ml of broth, sterile buffered saline, or water with one loopful of the original material and mix well.
2. Transfer one loopful of this dilution to a tube of melted agar at about 45 C.
3. Using a sterile pipet, add about 0.5 to 1 ml of sterile defibrinated blood to the inoculated tube and twirl between the palms of the hands to thoroughly mix the contents; then pour into a sterile Petri dish. Alternatively, the medium and inoculum may be combined and poured into the dish containing blood and the entire contents mixed by rotating the plate with the lid in place.
4. Allow the medium to gel, then incubate the plate in an inverted position (lid on bottom). This prevents collection of condensation on the agar surface. Unless the surface is dry, it will be difficult to obtain discrete surface colonies.

USE OF ENRICHMENT MEDIA

In the examination of fecal material for the presence of pathogenic bacteria, it is frequently necessary to use an **enrichment** medium (to be differentiated from an enriched medium, which contains a nutritive supplement). These media, because of their chemical composition, inhibit or kill off the normal intestinal flora, such as the coliforms (commensals), and permit salmonellae and some shigellae, which may be present only in small numbers, to grow almost unrestricted, thus enriching the population of such forms. After incubation, subcultures from these media must be made to solid plating media to obtain isolated colonies for further study. Tetrathionate broth and selenite broth are examples of enrichment media and are used primarily in enteric bacteriology.

USE OF DIFFERENTIAL AND SELECTIVE MEDIA

Differential media, by virtue of their chemical compositions, characterize certain bacterial genera by their distinctive colonial appearances in culture. Differential media, such as eosin–methylene blue agar (EMB) and MacConkey agar, contain lactose and a dye or indicator in the decolorized state. Bacteria that ferment the lactose with the production of acid or aldehyde produce red colonies or colonies with a metallic sheen, depending on the medium. This distinguishes lactose-fermenting from lactose-nonfermenting organisms.

Selective media are complex plating media that may also serve to differentiate among certain genera but in addition are highly selective in their action on other organisms. Such media as Salmonella-Shigella agar, deoxycholate-citrate agar, and bismuth sulfite agar will inhibit the growth of the majority of coliform bacilli, along with many strains of *Proteus* (those developing are prevented from swarming), and permit selective isolation of enteric pathogens. The coliform colonies that do develop are readily differentiated from lactose-nonfermenting organisms by color and opacity.

Phenylethyl alcohol blood agar is a selective medium for the isolation of gram-positive cocci in specimens or cultures contaminated with gram-negative organisms, particularly *Proteus* species. Although many gram-negative rods form visible colonies on this medium, spreading or swarming does not occur.* On this medium, colonies of gram-positive cocci are similar to those produced on ordinary media, whereas the gram-negative organisms that do grow will produce very small colonies. To avoid contamination when selecting colonies, only those that are well separated from the small forms should be used.

Infusion agar containing potassium tellurite and either serum or blood may be used as a selective plating medium for the isolation of members of the genus *Corynebacterium*, particularly in the laboratory diagnosis of diphtheria. On this medium, staphylococci and streptococci are usually inhibited.

Pathogenic staphylococci that do develop appear as black colonies on tellurite medium and as golden yellow colonies with yellow halos on mannitol salt agar, a medium containing a high salt concentration (see Chapter 16).

USE OF MEDIA CONTAINING ANTIBIOTICS

Media containing antibiotics are selectively inhibitory and may be effectively used for isolating pathogenic species from a mixed population. Sabouraud dextrose agar containing cycloheximide and chloramphenicol supports the growth of dermatophytes and most fungi causing systemic mycoses, although it greatly inhibits the majority of saprophytic fungi and bacteria found in clinical specimens.

An enriched chocolate agar medium containing vancomycin, colistin, and nystatin has proved effective for the selective isolation of *Neisseria* and *Haemophilus* species from clinical material.* The antibiotics suppress saprophytic neisseriae, *Acinetobacter* species, and *Pseudomonas* species.

OTHER METHODS

The isolation of species of the genus *Clostridium* from clinical material is frequently complicated by the presence of rapidly growing facultative organisms, such as coliforms, *Pseudomonas* species, and spreading *Proteus* species. Anaerobic blood agar plates may be overgrown by these organisms, and the isolation of clostridia becomes a difficult task. If the anaerobe produces spores in thioglycollate medium (used in the primary culturing of wound specimens), the heat resistance of these spores ("heat shock" method) may be utilized in the following manner:

1. Inoculate a fresh tube of thioglycollate medium[1] with 0.1 ml of the original culture or specimen containing the suspected species of *Clostridium*.
2. Heat this tube at approximately 80 C for 10 to 15 minutes. This will kill all vegetative forms but will not destroy spores, if present.
3. Incubate the heated culture and an unheated control tube for 24 to 48 hours.
4. Examine daily by making gram-stained smears: if gram-positive bacilli are found, proceed to step 5.
5. Subculture to two blood agar plates, incubating one aerobically and one anaerobically.

Clostridium perfringens rarely can be recovered successfully by this heating procedure, however, since it fails to sporulate in most media. A method that will

---

*Swarming may also be inhibited by increasing the agar content of the medium to 5%.

*Thayer-Martin selective medium, Difco Laboratories, Detroit; Baltimore Biological Laboratory, Cockeysville, Md.

aid in the successful isolation of this organism from mixed cultures is as follows:

1. Incubate the thioglycollate medium containing the original mixed culture for a minimum of 48 to 72 hours.
2. Streak several blood agar plates serially from this culture and incubate aerobically and anaerobically.
3. It will be found occasionally that the plates are overgrown with gram-negative organisms. In most cases, however, these will be fewer in number, and isolated colonies of suspected *Clostridium perfringens* will be readily obtained.
4. Transfer such colonies to tubes of thioglycollate medium for subsequent pure-culture study.

Other methods for the isolation of pure cultures, based on physical and chemical alterations, may be used. For example, variation of incubation temperatures, changes in the relative acidity or alkalinity of the medium, or variations in oxygen tension or gas concentration may be employed for the isolation of specific microorganisms. Indeed, in attempting to isolate offending food-poisoning organisms, it may take several of these methods to recover the responsible agent from a food product. Isolation of certain pathogenic bacteria from clinical material frequently may be accomplished more readily by animal inoculation. The techniques of these various methods are more fully described in subsequent chapters. For further information, however, the reader may refer to reference 2.

REFERENCES

1. Brewer, J. H.: Clear liquid mediums for the aerobic cultivation of anaerobes, J.A.M.A. **115:** 598-600, 1940.
2. Lennette, E. H., Spaulding, E. H., and Truant, J. P., editors: Manual of clinical microbiology, ed. 2, Washington, D. C., 1974, American Society for Microbiology.

PART II

Recommended procedures with clinical specimens

CHAPTER 5

Philosophy and general approach to clinical specimens

REJECTION OF SPECIMENS

It is important that criteria be set up for specimen rejection; such criteria are indicated in appropriate places in this book. However, it is an important rule to always talk to the requesting physicians **before** rejecting specimens, since they are primarily responsible for the patients' welfare. They need support and advice. They may know or think they know that an unorthodox specimen may provide useful information. If this is not the case, the microbiologist must explain why not and work with physicians to resolve the dilemma.

Frequently it is necessary and important to do the best possible job on a less than optimal specimen. It may be necessary to educate the clinician, and the best way to do this is to review individual specimens in a friendly and cooperative manner.

EXTENT OF IDENTIFICATION REQUIRED

We lean heavily toward definitive identification. For example, speciation of the *Bacteroides fragilis* group may be useful. *B. fragilis* and *B. thetaiotaomicron* are frequently found as pathogens, whereas recovery of *B. ovatus* may mean contamination of a specimen with normal bowel flora (or, in a blood culture, transient incidental bacteremia).

An outbreak of infantile diarrhea of serious proportion persisted for many months in a large general hospital in the Southwestern United States because stool cultures did not reveal any pathogens. Until the *Escherichia coli* isolates were tested for enterotoxin production and found to be positive, all therapeutic and control measures attempted proved futile.

Definitive identification of the unusual blood culture isolates from a number of outbreaks—and these were not really recognized initially as outbreaks—finally permitted determination that commercial intravenous fluid bottles were contaminated; this led to definitive control measures.[3] Many cases of bacteremia and many outbreaks across the country were attributable to this source.

Identification of a clostridial blood culture isolate as *Clostridium septicum* will alert the clinician to the possibility of malignancy in a patient because of the strong association between these two events.[1] In a patient without underlying serious disease, not receiving any antineoplastic or corticosteroid therapy, identification of an *Aspergillus* as *A. fumigatus* makes it more likely to be a clinically significant isolate than if it were another species. Identification of an organism suspected initially of being *B. fragilis* in a patient with bacteremia of unknown source as actually being a *B. splanchnicus* could indicate to the clinician that the source is probably in the gastrointestinal tract, whereas if it were *B. fragilis* it might also be in the genital tract or even another site.

Precise identification of organisms from a patient with two distinct episodes of infection at an interval of some months might permit establishing that the second infection represents a recurrence rather than a new infection, or it might permit excluding that possibility. Recurrent infection may suggest a foreign body or an abscess from prior surgery; in the absence of prior surgery, it suggests the possibility of malignancy or other underlying process. Definitive identification of an atypical mycobacterium may permit more accurate assessment as to whether the organism is truly involved in the disease, what the source might be, its drug susceptibility pattern, and the likelihood of its responding to the therapeutic regimen chosen.

In the case of certain organisms, the anaerobes especially, incomplete identification may lead to very serious errors in identification, since the anaerobes defy all the usual rules. Things that we learn in our first course of microbiology—that certain organisms produce spores, some are gram positive and some are gram negative, and so on—do not necessarily help with the anaerobes. It may be very difficult to demonstrate spores in sporulating anaerobes. Gram-positive anaerobes are very commonly decolorized and may appear gram negative, even early in the course of growth. Cocci may be mistaken for bacilli, or, more commonly, bacilli may be mistaken for cocci. It may even be difficult to define what an anaerobe actually is, considering that certain organisms, such as some clostridia (aerotolerant), are capable of relatively good growth under aerobic conditions. There are several outstanding examples in the literature of serious errors of this type in identification. For example, *B. melaninogenicus* has been incorrectly reported as being an anaerobic gram-positive streptococcus, and even such a great anaerobist as Prévot mistakenly classified as a *Fusobacterium (F. biacutum)* an organism subsequently shown to be gram positive and a *Clostridium.* Determination of the specific species of a group D streptococcus (e.g., *S. bovis* as opposed to *S. faecalis*)[2] may permit the use of less toxic and less expensive therapeutic remedies. Organisms as diverse as *Listeria*

monocytogenes and *Actinomyces viscosus* have been incorrectly identified as diphtheroids (even in reports in the literature) when shortcuts were taken.

At the same time, certain shortcuts and use of limited identification procedures in certain cases is necessary in most clinical laboratories. Careful application of knowledge of the significance of various organisms in these situations and thoughtful use of limited approaches will keep expenses in line and keep the laboratory's workload manageable, while providing for optimum patient care.

Careful bacteriologic studies permit us to define the role of various organisms in different infectious processes, the prognosis associated with these, and so forth. Finally, definitive identification is important in educating clinicians as to the role of various organisms in infectious processes and, perhaps most important, prevents deterioration of the skills and interest of the microbiologist.

QUANTITATION OF RESULTS

The concept of **quantitation** has not been adequately stressed in microbiology except in the case of urine cultures. Quantitation is useful in other situations as well. It may help to distinguish between an organism present as a "contaminant" and an organism actively involved in infection. It is often useful in determining the relative importance of different organisms recovered from mixed infections. Ordinarily, formal quantitation by dilution procedure or even by means of a quantitative loop is not necessary or desirable. Numbers of organisms present can be graded as "many" ("heavy growth"), "moderate numbers," or "few" ("light growth") on the basis of Gram-stain appearance and the amount of growth on a plate. For example, does a given colony type extend to the secondary streak, tertiary streak, and so on? The Gram stain is important because differences in amount of growth or rate of growth of different organisms may be related to the fastidiousness of the organism.

EXPEDITING RESULTS

The need for **speed** in identification and susceptibility testing is another crucial area that has been neglected. Few diseases are as dynamic and rapid acting as the infectious diseases. In certain serious infections, such as bacteremia, endocarditis, meningitis, and certain pneumonias, delay of a few hours (or even less in more critical cases) in providing proper therapy may lead to death. It is not always feasible or safe to try to cover all possible types of infecting organisms with antimicrobial therapy. Clearly, the clinician must use this type of empirical approach initially, but the microbiologist must be prepared to assist in choosing a rational approach through knowledge of the role of various organisms in different disease processes and by interpretation of direct smears. The microbiologist in such cases (and, of course, the clinician must communicate with the laboratory about problem cases) should also offer something more than routine daily inspection and subculture so that any necessary or desirable modification of therapy can be made as early as possible. Cultures can be examined at 6- to 12-hour intervals, and the processing can be speeded up considerably for selected cases.

Somewhat related to this point is the necessity for round-the-clock coverage. Patients do develop pneumonia on Saturday and Sunday and they do go into septic shock after 5 PM and even between midnight and 7 AM. In one manner or another, **reliable** (and we emphasize reliable, because many times emergency laboratory technicians know little or no bacteriology) bacteriology must be available at all times. This includes setting up cultures, inspecting cultures set up earlier, and interpreting direct smears. A qualified microbiologist might be on call for problems encountered by the emergency laboratory crew. If available, infectious disease house officers may be able to set up certain special cultures, interpret Gram stains, and so forth.

In conclusion, we acknowledge that the ideal approach to microbiology can and should be modified according to circumstances. A large teaching hospital or clinic will need to do more, and should do more, than is done in smaller hospitals. In a very small community hospital, the services of a qualified consulting microbiologist should be sought. More must be done for the seriously ill patient, and it must be done as quickly as possible. Every patient has the right to optimal medical care, and therefore every hospital either must have all the necessary capabilities or must have access to them and must know when and how to refer specimens or cultures.

REFERENCES

1. Alpern, R. J., and Dowell, V. R., Jr.: *Clostridium septicum* infections and malignancy. J.A.M.A. **209:**385-388, 1969.
2. Facklam, R. R.: Streptococci. In Lennette, E. H., Spaulding, E. H., and Truant, J. P., editors: Manual of clinical microbiology, ed. 2, Washington, D.C., 1974, American Society for Microbiology.
3. Maki, D. G., Rhame, F. S., Goldmann, D. A., and Mandell, G. L.: The infection hazard posed by contaminated intravenous infusion fluid. In Sonnenwirth, A. C., editor: Bacteremia: Laboratory and clinical aspects, Springfield, Ill., 1973, Charles C Thomas, Publisher.

CHAPTER 6

Collection and transport of specimens for microbiologic examination

COLLECTION PROCEDURES

Generally, a report from the bacteriologic laboratory can indicate only what has been found by microscopic and cultural examination. An etiologic diagnosis is thus confirmed or denied. Failure to isolate the causative organism, however, is not necessarily the fault of inadequate technical methods; it is frequently the result of faulty collecting or transport technique. In a busy hospital the collection of specimens is too often relegated to personnel who have neither the knowledge nor understanding of the requirements and consequences of such procedures. The microbiologist may also deserve criticism on occasion for negligence in providing adequate supplies or proper instructions, which may result in poorly collected samples. The following are **general considerations** regarding the collection of material for culture. Specific instructions for the handling of a variety of specimens are given in subsequent chapters.

Whenever possible, specimens should be obtained **before antimicrobial agents have been administered.** Often a purulent cerebrospinal fluid reveals no bacterial pathogens on smear or culture when an antibiotic has been given within the previous 24 hours. A patient with salmonellosis may have a negative stool culture if the specimen has been collected while he or she was receiving suppressive antibacterial therapy, only to reveal a positive culture several days after therapy has been terminated. If the culture has been taken after initiation of antibacterial therapy, the laboratory should be informed so that specific counteractive measures, such as adding penicillinase or merely diluting the specimen, may be carried out.

It is axiomatic that material should be collected where the suspected organism is **most likely to be found and with as little external contamination as possible.** The skin and all mucosal surfaces are populated with an indigenous flora and may often also acquire a transient flora or even become colonized for extended periods with potential pathogens from the hospital environment. The latter is particularly true of individuals who are quite ill, especially if they are receiving antimicrobial therapy (resistant organisms colonize as normal flora is suppressed). Accordingly, special procedures must be employed to help distinguish between organisms involved in an infectious process and those representing normal flora or "abnormal" colonizers that are not actually causing infection. Four major approaches are utilized to resolve this problem.[7]

1. Cleanse the area with a disinfectant, using enough friction for mechanical cleansing as well. Start centrally and go out in ever enlarging circles. Repeat several times, using a new swab each time. Alcohol (70%) is satisfactory for skin, but a full 2 minutes of wet contact time is needed. Iodine (2%) and povidone iodine work more quickly and are effective against spore-forming organisms.
2. Bypass areas of normal flora entirely (e.g., percutaneous transtracheal aspiration rather than coughed sputum).

3. Culture only for a specific pathogen (e.g., group A streptococci in the throat).
4. Quantitate culture results as a means of determining likelihood of organisms being involved in infection (e.g., quantitative urine culture). Less formal quantitation may also be satisfactory and should **routinely** be used; this may involve grading on a scale of 4+ to 1+, or simply heavy growth, moderate growth, light growth, four colonies, and so forth.

Another factor contributing to the successful isolation of the causative agent is the **stage of the disease** at which the specimen is collected for culture. Enteric pathogens are present in much greater numbers during the **acute,** or diarrheal, stage of intestinal infections and are more likely to be isolated at that time. Viruses responsible for causing meningoencephalitis are isolated from cerebrospinal fluid with greater frequency when the fluid is obtained soon after the **onset** of the disease rather than when the symptoms of acute illness have subsided.

There are occasions when patients must participate actively in the collection of a specimen, such as a sputum sample. They should be given full instructions, and cooperation should be encouraged by the ward attendant. Too often a container is placed on the patient's bedside table, with the only instructions being to "spit in this cup when you cough anything up."

Specimens should be of a **quantity sufficient** to permit complete examination and should be placed in sterile containers that preclude subsequent contamination of patient, nurse, or ward messenger. A serious danger to the laboratory worker, as well as to all others involved, is the soiled outer surface of a sputum container, a leaking stool sample, or possible contact with blood-containing exudates. The hazard of spread of disease by inadequately trained nonprofessional workers is frequently overlooked. Its control requires continued education and constant vigilance by those in responsible and supervisory positions.

Provision must be made for the **prompt delivery** of specimens to the laboratory if the subsequent results of analysis are to have validity. It is difficult, for example, to isolate *Shigella* from a fecal specimen that has remained on the hospital ward too long, permitting overgrowth by commensal organisms and an increasing death rate of the shigellae. In some instances it may be necessary to take culture media and other equipment to the patient's bedside to ensure prompt inoculation of the specimen. This is an unusual circumstance and requires prior arrangements with the laboratory.

Although not a function of specimen collection, it is an essential prerequisite that the **laboratory be given sufficient clinical information** to guide the microbiologist in selection of suitable media and appropriate techniques. Likewise, it is important for the clinician to appreciate the **limitations and potentials** of the bacteriology laboratory—to realize that a negative report does not exclude the correctness of diagnosis. It is essential that close cooperation and frequent consultation between the clinician, nurse, and microbiologist be the rule rather than the exception.

SPECIMEN CONTAINERS AND THEIR TRANSPORT

In the microbiologic examination of clinical specimens it is essential that the container bearing the specimen does not contribute its own microbial flora. Furthermore, the original flora should neither multiply nor decrease because of prolonged standing on the ward or prolonged refrigeration in the laboratory. In other words, **a sterile container should be used and the specimen should be plated as soon as possible.** Although these are not hard-and-fast rules, any deviation should be the responsibility of the microbiologist.

A variety of containers* have been devised for collecting bacteriologic specimens. Many of these can be used repeatedly, after proper sterilization and cleaning, whereas others must be incinerated after use. Apart from the sterile Pyrex Petri dish and its modern counterpart, the presterilized and disposable plastic dish, the most used (but not necessarily the most desirable) piece of collecting equipment is the cotton-, calcium alginate–, or polyester-tipped wooden applicator stick. These swabs are best prepared by autoclaving the wooden applicator sticks in Sorensen buffer, pH 7.2, for 5 minutes, drying them, and then tipping them with polyester batting,† calcium alginate,‡ or long-fibered medicinal cotton. One combination packaged as a sterile outfit consists of a Pyrex tube (20 by 150 mm), either cotton plugged or with a stainless-steel or disposable plastic cap,§ containing the applicator stick, and a small tube (10 by 75 mm) with several drops of a transport broth (not needed with a polyester tip). This may be used for the collection of material from the throat, nose, eye, ear, wound and operative sites, urogenital orifices, and the rectum, but is not recommended for optimal recovery of anaerobes (see below). The swab, inoculated with material from the patient, is placed in the inner broth tube to prevent drying out, and the whole outfit is properly labeled and promptly sent to the laboratory.

Another approach uses a sterile disposable culture unit (Culturette)‖ consisting of a plastic tube containing a sterile polyester-tipped swab and a small glass ampule of modified[1] Stuart's holding medium. The unit is removed from its sterile envelope, and the swab is used to collect the specimen. It is then returned to the tube, the ampule is crushed, and the swab is forced into the released holding medium. This will provide sufficient moisture for storage up to 72 hours at room temperature.

Some cotton used for applicators may contain fatty acids that may be detrimental to microbial growth.[8] An excellent substitute is calcium alginate wool, derived from alginic acid, a natural plant product.[5] This silky-fibered material dissolves in certain solutions that are compatible with bacterial preservation (such as dilute Ringer's solution with sodium hexametaphosphate) to form a soluble sodium alginate. Calcium alginate–tipped wooden applicators or flexible aluminum nasopharyngeal swabs, as well as the citrate or hexametaphosphate diluents, are available commercially.

A modification of the cotton applicator uses 28-gauge Nichrome or thin aluminum wire in place of the wooden stick. Because of its flexibility, the wire applicator is recommended for collecting specimens from the nasopharynx. To ensure that the small amount of cotton adheres to the wire during passage through the nasal tract, the end of the wire must be bent over and some collodion applied before the wrapping with cotton; otherwise, the outfit remains the same as that previously described. A commercially prepared sterile wire swab also is available.*

A variety of transport media have been devised for prolonging the survival of microorganisms when a significant delay occurs between collection and culturing. Stuart and others[10,11] advocated a medium that has proved effective in preserving the viability of pathogenic agents in clinical material. The medium consists of buffered semisolid agar devoid of nutrients and containing sodium thioglycollate as a reducing agent and is used in conjunction with cotton swabs. Stuart's medium†

---

*Falcon Plastics, Oxnard, Calif.
†Dacron polyester filling, half-pound bags, Sears, Roebuck and Co.
‡Colab Laboratories, Inc., Chicago Heights, Ill.
§Morton culture tube closure from Scientific Products, Division of American Hospital Supply Corp., Evanston, Ill.
‖Scientific Products, Division of American Hospital Supply Corp., Evanston, Ill.

---

*Falcon Plastics, Oxnard, Calif.
†Baltimore Biological Laboratory, Cockeysville, Md.

maintains a favorable pH and prevents both dehydration of secretions during transport and oxidation and enzymatic self-destruction of the pathogen present. However, the glycerophosphate present permits multiplication of certain organisms.

A report on a transport medium* by Cary and Blair[3] indicates that salmonellae and shigellae can be recovered from fecal specimens for as long as 49 days and *Vibrio cholerae* for 22 days; *Yersinia pestis* survives for at least 75 days.

The use of polyester-tipped swabs for delayed recovery of group A streptococci from throat cultures is discussed in Chapter 8. A report by Hosty and co-workers[6] indicates that the incorporation of a small amount of silica gel in the glass tube containing the polyester swab will further maintain the viability of group A streptococci in throat swabs for as long as 3 days before plating.

The collection of specimens for anaerobic culturing poses a special problem in that the conventional methods previously described will not lead to optimal recovery of these air-intolerant microorganisms. A crucial factor in the final success of anaerobic culturing is the transport of clinical specimens: the lethal effect of atmospheric oxygen must be nullified until the specimen has been processed anaerobically in the laboratory.[12] We recommend the use of a double-stoppered collection tube or vial, gassed out with oxygen-free CO_2 or N_2*†; the specimen (pus, body fluid, or other liquid material) is injected through the rubber stopper after first expelling all air from the syringe and needle. In the laboratory, the material is aspirated from the transport container by needle and syringe and inocu-

lated to media and incubated under anaerobic conditions. If only a swab specimen can be obtained, it may be collected on a swab that has been prepared in a "gassed out" tube,* and then transferred to a rubber-stoppered tube half filled with prereduced and anaerobically sterilized transport medium,* such as the Carey and Blair medium previously described.

An Anaerobic Culturette is now available.† One of us (S. M. F.) has made a preliminary evaluation of this item as well as the Marion Anaerobic Culture/Set.† To date, the Anaerobic Culturette has not provided optimal conditions for delicate anaerobes, but the principle is a good one and modifications are being evaluated. The Anaerobic Culture/Set, however, has been most impressive. In addition, it offers remarkable flexibility. It is not only suitable for swabs (for short periods; no specific means are incorporated to maintain a moist environment, nor is there a holding medium), but can be used to transport specimens in plastic syringes and smaller specimens aspirated (through an attached syringe) into small-gauge sterile plastic tubing.

It is important to note that the materials recommended for anaerobic transport actually constitute ideal **universal transport setups,** since **all** types of microorganisms should survive well in them. It should be stressed that, although the swab is the most widely used transport vehicle, it is preferable to submit a larger specimen whenever possible (e.g., a filled syringe or tube). When organisms may be scarce (as in some forms of tuberculosis) the larger the specimen the better. Further details on anaerobic culture techniques are given in Chapters 13 and 27.

Containers for special purposes, such as those used for the collection of 24-hour

---

*Available commercially from Scott Laboratories, Inc., Fiskeville, R. I.; Gibco Laboratories, Grand Island, N. Y.; and BioQuest, Cockeysville, Md.
†A convenient type of tube, stopper, and screw cap for laboratories wishing to make their own oxygen-free tubes is the Anaerobic Culture Tube, Hungate type, No. 2047-16125, available from Bellco Glass, Inc., Vineland, N.J.

---

*Available commercially from Scott Laboratories, Inc., Fiskeville, R. I.; Gibco Laboratories, Grand Island, N. Y.; and BioQuest, Cockeysville, Md.
†Marion Scientific Corp., Kansas City, Mo., Catalog No. 26-02-06.

urine samples, can usually be devised from equipment found around the hospital; or they may be purchased from a surgical supply house.

Although most pathogenic microorganisms are not greatly affected by small changes in temperature, they are generally susceptible to drying out, particularly when on cotton applicator sticks. However, some bacteria, such as the meningococcus in cerebrospinal fluid, are quite sensitive to low temperatures and require immediate culturing. On the other hand, clinical material likely to contain abundant microbial flora may in most instances be held at 5 C in a refrigerator for several hours before culturing if it cannot be processed right away. This is particularly true with such specimens as urine, feces, and sputum samples and material on swabs taken from a variety of sources, with the exception of wound cultures, which may contain oxygen-sensitive anaerobes. These should be inoculated promptly. Not only will refrigeration preserve the viability of most pathogens, it will minimize overgrowth of commensal organisms, increased numbers of which could make the isolation of a significant microbe more difficult. Objectively, however, the sooner an organism leaving the sheltered environment of its host is transferred to an appropriate artificial culture medium, the better are the chances of its survival and subsequent multiplication.

On occasion it is necessary to submit specimens to a **reference laboratory** in a distant city, thus requiring transportation by mail or express. If, for example, there is no viral diagnostic service available in the immediate area, it may be necessary to ship specimens at a low temperature to preserve their viability. This is especially true of virus-containing material, such as cerebrospinal fluid, throat and rectal swabs, stools, and tissue, which should be refrigerated immediately and shipped in a Styrofoam box with commercial refrigerant packs (e.g., 3M Cryogel). Whole blood is **not** shipped. Rather, the serum is separated and sent in a sterile tube.

If a culture of an isolated organism is to be sent to a reference laboratory (state or public health laboratory), it is not necessary to send it in the frozen state.

In most instances microbiologic specimens can be satisfactorily shipped through the mails, provided that **special precautions** are taken against breakage and subsequent contamination of the mailing container. According to the Code of Federal Regulations,* a viable organism or its toxin or a diagnostic specimen of a volume less than 50 ml shall be placed in a securely closed, watertight specimen container, which is enclosed in a second durable watertight container. The entire space between the containers should contain sufficient absorbent material to absorb the entire contents in case of breakage or leakage. The containers are then enclosed in an outer shipping container of corrugated fiberboard, cardboard, wood, or such. The outer container shall be labeled with the Etiologic Agents/Biomedical Material red-and-white label (available commercially) (see Chapter 33 and Fig. 33-1).

Double mailing containers must be used when the microbiologist considers the specimen to be of a hazardous nature or thinks it may constitute a definite danger to the person handling the container. This includes clinical specimens, cultures for identification, and so forth.

For the shipment of fecal specimens containing salmonellae or shigellae over long distances, the filter paper method[2] may also be employed. In collecting the specimen by this technique, fresh fecal material must be spread fairly thinly over a strip of filter paper or clean blotting paper and allowed to dry at room temperature. Using forceps, the smeared strip is then folded inward from the ends in such a way that the fecal material is covered. The folded specimen may then be inserted in a plastic envelope (polyethylene is recommended) and placed in a container to conform with postal regulations. It is

---

*Section 72.25 of Part 72, Title 42, amended.

possible to ship a large number of such specimens in one container through the mail. The pathogens are unaffected by this treatment, whereas the normal intestinal flora dies off. On receipt at the laboratory the paper specimen may be cut into three pieces. One piece is placed in physiologic saline for suspension and direct plating, and the remaining pieces are placed in selenite and tetrathionate broths, respectively, for enrichment and subsequent plating.

When it is necessary to ship fecal specimens or when such specimens must be held for some time before culturing, it is recomended that they be placed in a preservative solution. The buffered glycerol-saline solution of Sachs[9] has proved satisfactory. (Its final pH should be 7.4; it should be discarded if it becomes acid.) Approximately 1 g of feces is emulsified in not more than 10 ml of preservative, and the preserved specimen is shipped in a heavy glass container (universal type) with a screw cap and contained in the regular double mailing container just described.

HANDLING OF SPECIMENS IN THE LABORATORY

In previous sections the importance of a properly collected specimen was stressed and the responsibility of personnel in its collection was indicated. In the following section the subsequent handling of the specimen in the laboratory is considered. In addition, the responsibility of the laboratory worker is pointed out.

Since it is not always practical for many specimens to be inoculated as soon as they arrive in the laboratory, refrigeration at 4 to 6 C offers a safe and dependable method of storing many clinical samples until they can be conveniently handled. However, some may require immediate plating (such as specimens that might contain gonococci or *Bordetella pertussis*), whereas others must be immediately frozen (e.g., spinal fluid that is to be held for subsequent attempts at virus isolation).

The length of time of refrigeration varies with the type of specimen: swabs from wounds (except for anaerobic cultures), the urogenital tract, throat, and rectum and samples of feces or sputum can be refrigerated for 2 to 3 hours without appreciable loss of pathogens. Urine specimens for culture may be refrigerated for 24 to 48 hours without affecting the bacterial flora (except the tubercle bacillus, which may be adversely affected by the urine); on the other hand, cerebrospinal fluid from a patient with suspected meningitis should be examined **at once.**

Specimens submitted for the isolation of virus should be refrigerated immediately; if they must be held for over 24 hours, they should be frozen (at −70 C if possible). Specimens of clotted blood for virus serology may be refrigerated but never frozen.

Gastric washings and resected lung tissue submitted for culture of *Mycobacterium tuberculosis* should be processed soon after delivery, since tubercle bacilli may die rapidly in either type of specimen.

Pieces of hair or scrapings from the skin and nails submitted for the isolation of fungi may be kept at room temperature (protected from dust) for several days before inoculation. On the other hand, sputum, bronchial secretions, bone marrow, and purulent material from patients suspected of having systemic fungal infection should be inoculated to appropriate media as soon as possible, especially when diagnosis of histoplasmosis is considered.

REFERENCES

1. Amies, C. R.: A modified formula for the preparation of Stuart's transport medium, Canad. J. Public Health, **58:**296-300, 1967.
2. Bailey, W. R., and Bynoe, E. T.: The "filter paper" method for collecting and transporting stools to the laboratory for enteric bacteriological examination, Canad. J. Public Health **44:**468-475, 1953.
3. Cary, S. G., and Blair, E. B.: New transport medium for shipment of clinical specimens, J. Bacteriol. **88:**96-98, 1964.
4. Forney, J. E., editor: Collection, handling, and

shipment of microbiological specimens, Public Health Serv. Publ. No. 976, Washington D.C., Nov. 1968, U.S. Government Printing Office.

5. Higgins, M.: A comparison of the recovery rate of organisms from cotton-wool and calcium alginate wool swabs, Ministry Health Public Lab. Serv. Bull. 43, 1950.

6. Hosty, T. S., Johnson, M. B., Freear, M. A., Gaddy, R. E., and Hunter, F. R.: Evaluation of the efficiency of four different types of swabs in the recovery of group A streptococci, Health Lab. Sci. 1:163-169, 1964.

7. Matsen, J. M., and Ederer, G. M.: Specimen collection and transport, Human Pathol. 7:297-307, 1976.

8. Pollock, M. R.: Unsaturated fatty acids in cotton plugs, Nature 161:853, 1948.

9. Sachs, A.: Difficulties associated with bacteriological diagnosis of bacillary dysentery, J. Roy. Army Med. Corps 73:235-239, 1939.

10. Stuart, R. D.: The diagnosis and control of gonorrhea by bacteriological cultures, Glasgow Med. J. 27:131-142, 1946.

11. Stuart, R. D., Tosach, S. R., and Patsula, T. M.: The problem of transport of specimens for culture of gonococci, Canad. J. Public Health 45: 73-83, 1954.

12. Sutter, V. L., Vargo, V. L., and Finegold, S. M.: Wadsworth anaerobic bacteriology manual, ed. 2, Los Angeles, 1975, UCLA Extension Division.

PART III

Cultivation of pathogenic microorganisms from clinical material

CHAPTER 7

Microorganisms encountered in the blood

The organisms most likely to be found in blood cultures include the following:

Staphylococci (coagulase positive, coagulase negative), micrococci
Coliform bacilli and other enteric organisms
Alpha- and beta-hemolytic streptococci
Pneumococci
Enterococci
Haemophilus influenzae
Clostridium perfringens and related organisms
Pseudomonas species
Bacteroides fragilis, other gram-negative anaerobes
Anaerobic cocci
Opportunistic fungi, such as *Candida* and *Torulopsis*
Neisseria meningitidis
Salmonella species
Brucella species
Francisella tularensis
Listeria monocytogenes
Streptobacillus moniliformis
Leptospira species
Campylobacter fetus and related vibrios
Diphtheroid bacilli

BLOOD CULTURE—GENERAL CONSIDERATIONS

Blood for culture is probably the most important single specimen submitted to the microbiology laboratory for examination. It is unique in that it helps provide a clinical diagnosis as well as a specific etiologic diagnosis. The presence of living microorganisms in the patient's blood almost always reflects active and possibly spreading infection in the tissues. The prognosis of such a bacteremia or septicemia may well depend on its prompt recognition by bacteriologic means. Subsequent initiation of specific therapy based on laboratory findings may well prove to be lifesaving. Likewise, a negative culture would prove helpful in ruling out microbial etiology.

It is important to know which infections are likely to show a positive blood culture, and thus further reading on bacteremias is recommended.[6,21] A **transient bacteremia** frequently occurs during the course of many diseases, including pneumococcal pneumonia, bacterial meningitis, urinary tract infection, typhoid fever, and generalized salmonella infections. Wound infections caused by a beta-hemolytic streptococcus, by *Staphylococcus aureus*, and by *Bacteroides* species, infections of the gallbladder and biliary tract, osteomyelitis, peritonitis, and puerperal sepsis may be accompanied by bacteremia, which is also a frequent concomitant of operative manipulation in chronically infected areas, as in instrumentation of the urinary tract. It is generally unrewarding, however, to attempt to isolate the causative organism by culturing the blood in cases of tetanus, diphtheria, shigellosis, or tuberculosis.

Although a mild transitory bacteremia is a frequent finding in many infectious diseases, the persistent, continuous, or recurrent type of bacteremia is more serious. When the classic syndrome of a **septicemia** (chills, fever, prostration) due to the presence of actively multiplying bacteria and their toxins in the bloodstream is found, it is rarely difficult to isolate the causative organism. When localizing signs of infection are notably absent and a persistent positive blood culture is demonstrated, the possibility of a serious intravascular infection, such as bacterial endocarditis, must be considered.

In some diseases the probability of obtaining a positive blood culture depends on the **stage of the disease** at which the culture is made. For example, bacteria can be cultured from the blood during the first several days of illness in typhoid fever,* tularemia, plague, and anthrax, but the organisms disappear from the blood during the latter course of the disease. Thus, the early recovery of these bacteria from the blood is important, since it may be the only reliable means available to the clinician for making a diagnosis.

One of the more challenging systemic diseases from the clinical and bacteriologic points of view is **subacute bacterial endocarditis** (SBE). Confirmation of the diagnosis is almost entirely dependent on isolation of an organism from the blood. Frequently, the causative bacteria are present only in small numbers.

Furthermore, isolation of the causative organism may be complicated because the patient received prior antibiotic therapy. This may not have been at a drug level high enough to eradicate the microorganism but one that, when carried over in the blood sample, may prove sufficient to inhibit growth in the blood culture bottle.† For this reason it is recommended that a **minimal dilution** of 1 volume of blood in 10 volumes of broth be made as a safeguard against such an eventuality. This dilution also aids in counteracting the bactericidal effect of normal serum.

Since there is usually a lag period of 1 to 2 hours between the time of entrance of the bacteria into the circulation and the subsequent chill,[6] blood for cultures should be drawn, ideally, **shortly before** the expected temperature rise. The frequency with which blood samples are taken for culture is also important. In patients with suspected bacterial endocarditis who have not received any antibacterial agents, a total of three to four cultures taken during a 24- to 48-hour period should be adequate to establish the diagnosis in most cases, provided that techniques suitable for recovery of anaerobes and other fastidious organisms are included. In endocarditis, the bacteremia is **constant,** so blood cultures may be drawn at any time.

Austrian[4] believes that after appropriate antibacterial therapy it is incumbent on the attending physician to obtain negative blood cultures to document the absence of bacteremia. In patients with subacute bacterial endocarditis he recommends blood culture at the termination of therapy and again 2 weeks later, the period during which those patients who will most likely suffer a bacteriologic relapse are observed to do so. It is important to obtain follow-up blood cultures in bacteremic patients even if they become afebrile on antibacterial therapy.

In **brucellosis,** positive blood cultures are usually obtained during exacerbation of symptoms and elevation of temperature, although even during the acute phase isolation of the organism is not regularly successful.

Occasionally, culture of the **bone marrow** has yielded positive results when peripheral blood cultures have been sterile in cases of brucellosis, subacute bacterial endocarditis, histoplasmosis, and, after the first week, in typhoid fever. Concurrent bacterial and fungal cultures are recommended when bone marrow biopsies are performed for other diagnostic purposes.

A still more difficult problem for the bacteriologist is the generalized infection caused by a **rarely isolated** organism that requires special methods for its isolation. The causative agents of diseases such as leptospirosis, rat-bite fever, listeriosis, and infection by pasteurellae require complex culture media, and in some instances animal injection, for their isolation. Some of these special procedures are discussed in a later section.

---

*If negative, repeat weekly for several weeks if patient continues to be febrile.[18]

†Aberrant forms of bacteria may occur in blood cultures of patients receiving antibiotic therapy.[39]

Blood cultures should be made **immediately** in all cases of unexplained shock, particularly in those following genitourinary tract manipulation. A single negative blood culture should **never** be depended on to eliminate the possibility of a bacteremia, because a single specimen may be sterile even though the bloodstream as a whole is infected. A minimum of three cultures taken at half-hour intervals is recommended. In patients with **septic shock,** therapy must be begun immediately. Three separate blood cultures may be obtained, one right after another, by three separate venipunctures at different sites and with different syringes. Blood cultures should be made on all patients with **unexplained or persistent fever.** Chills and fever in patients with urinary tract infections, infected burns, soft-tissue infections, postoperative wound sepsis, or with indwelling venous catheters make them candidates for having blood cultures taken. Debilitated patients who develop fever while undergoing prolonged therapy with antibiotics, corticosteroids, immunosuppressives, antimetabolites, or parenteral hyperalimentation also should have blood drawn for culture.

The diagnosis of bacteremia can be made only by growing out pathogenic agents from culture media suitably inoculated with adequate amounts of the patient's blood. In the bacteriologic examination of the blood, several important principles must be considered:

1. An acceptable blood collection technique, which minimizes chances of contamination, is mandatory.
2. Enriched **aerobic** and **anaerobic** culture media should be utilized, and conditions of incubation must provide optimal conditions for bacterial growth.
3. Results of presumptively positive findings must be reported **by telephone without delay** to the attending physician.

Despite one negative report (Carlson and Andersen, J.A.M.A. **235:**1465-1466, 1976), Gram stain of the buffy coat (white blood cells) is a worthwhile procedure in critically ill patients, as it may immediately reveal organisms (usually gram-positive cocci) in the blood.

ROLE OF ANTICOAGULANTS AND OSMOTIC STABILIZERS IN BLOOD CULTURING

As Rosner[47] so aptly observed, the most important attribute of an effective blood culture system is its ability to support rapid growth of a wide variety of microorganisms so that their prompt recovery and identification can be accomplished.

One of these elements would be a means of preventing the blood sample from clotting in the culture bottle, since bacteria become entrapped in clotted blood.[67] Various anticoagulants have been recommended; the most effective one to date appears to be sodium polyanethol sulfonate (SPS) (Liquoid), a synthetic polyanionic agent. First reported by Von Haebler and Miles[58] in 1938 as a useful anticoagulant in blood culturing, it is now apparent that SPS* in a concentration of 0.03% will inhibit clotting, neutralize the bactericidal effect of human serum,[41] prevent phagocytosis,[48] and at least partially inactivate certain antibiotics, such as streptomycin, kanamycin, gentamicin, and polymyxin B.[55,56] Furthermore, SPS is stable at autoclave temperatures. It is not recommended, however, in agar pour plate cultures or membrane filter culture systems,[23] and may be inhibitory to some strains of *Peptostreptococcus anaerobius, Neisseria meningitidis,* and *N. gonorrhoeae.* Sodium amylosulfate (SAS) offers some theoretic advantages over SPS but was no better in a clinical blood culture trial and may be inhibitory to certain gram-negative bacilli (e.g., *K. pneumoniae*).[28]

The inhibitory effect of SPS may be

---

*Available in a 5% sterile aqueous solution as Grobax from Roche Diagnostics, Nutley, N.J.

neutralized by adding gelatin (1.2%) to blood culture media (Wilkins, T. D., and West, S. E. H.: J. Clin. Microbiol. 3:393-396, 1976; Eng, J., and Holten, E.: J. Clin. Microbiol. 6:1-3, 1977). A recent study (Traub, W. H.: J. Clin. Microbiol. 6:128-131, 1977) showed that SAS was less effective than SPS in neutralizing serum bactericidal activity.

Sodium citrate in a concentration of 0.5% to 1.0% is another anticoagulant. However, it may be inhibitory to gram-positive cocci[44] and is not recommended.

Another factor in the successful recovery of small numbers of circulating microorganisms in the blood is the use of an osmotic stabilizer, such as 10% to 30% sucrose.[48] This is especially true of gram-negative rod bacteremias in patients receiving antimicrobial therapy, where low levels of bacteria (10 colonies per milliliter or less) can be expected.[22] The sucrose may exert a protective effect on organisms having undergone cell wall damage (preventing osmotic pressure changes.)[48] However, hypertonic media also counteract normal bactericidal mechanisms in blood. Most microorganisms with intact cell walls also will grow well in enriched hypertonic media.[21] Although a number of studies found benefit in the use of hypertonic sucrose media, better results were obtained without sucrose in the case of fungi[45] and certain bacteria in studies from the Mayo Clinic.[63]

BLOOD CULTURE PROCEDURES

When blood for culture is drawn, scrupulous care must be exercised in preparing the site for venipuncture. A 2% solution of tincture of iodine is applied to the skin over the area of the selected vein by means of several saturated cotton-tipped applicator sticks or swabs. Start centrally over the planned site of venipuncture and, exerting moderate pressure, move out in concentric circles. The iodine is allowed to dry and then is removed with sponges saturated with 80% isopropyl alcohol. The entire process is repeated, a tourniquet is applied, and 20 ml of blood are withdrawn by a sterile 21-gauge needle on a 20-ml syringe. Many hospitals use 10 ml of blood (two aliquots of 5 ml, each placed into 50 ml of medium). Since bacteremia often involves very small numbers of organisms and cultures may be negative with smaller volumes of blood, the larger blood volume is recommended. In infants and children, volumes of 1 to 5 ml are satisfactory (see pp. 51-52). The blood is transferred **immediately** to two small sterile tubes or bottles fitted with rubber stoppers and containing 3.4 ml of 0.35% SPS in sterile physiologic saline.*[19] The blood may be drawn directly into the Vacutainer tubes (this is preferable) if the possibility of backflow into the patient is avoided by keeping the arm dependent. Sterilization of these evacuated blood collection tubes is advised, since some have been found contaminated with potential pathogens.[11,][62] An alternative method that is preferable is the direct inoculation of blood into culture media at the patient's bedside, employing a transfer set or a needle and syringe. It is then mixed well and promptly transported to the laboratory, where it is immediately incubated. If it is necessary to hold the specimen for a short interval, the blood may be refrigerated for not more than 1 to 2 hours.

For culture, 10 ml of the blood (the contents of one of the Vacutainer tubes) is transferred to a bottle containing 100 ml of brain-heart infusion broth and a brain-heart infusion agar slant (biphasic)[10,46,50] medium, and the remainder is distributed in a bottle containing 100 ml of trypticase soy broth or tryptic soy broth using aseptic technique throughout. These bottles, available commercially,†

---

*B-D Vacutainer tube No. S3208 XF308, Becton, Dickinson & Co., Rutherford, N.J.

†Among sources of supply are BBL, Cockeysville, Md.; Becton, Dickinson & Co., Rutherford, N.J.; Case Laboratories, Chicago; Difco Laboratories, Detroit; Gibco Laboratories, Grand Island, N.Y.; Pfizer Diagnostics, Clifton, N.J.; and others.

should contain 10% carbon dioxide and a vacuum and 0.03% SPS. One of the two bottles (the trypticase soy broth) may be made hypertonic with sucrose, if desired. After inoculation with the blood, the brain-heart infusion medium bottle is vented. The vent is removed after the vacuum has disappeared; this permits retention of the carbon dioxide. The other (anaerobic) bottle should **not** be vented. Agar pour plates may provide more rapid growth and earlier identification than is possible with broth cultures, but are not practical for routine use in the clinical laboratory. A commercially available prereduced medium* in a bottle with a gas-tight stopper proved to be no more effective for recovery of anaerobes than tryptic soy broth or thioglycollate medium.[64]

Since conventional methods for recovering microorganisms from the blood may require several days for growth and identification, a rapid method using membrane filters was introduced[65] (Fig. 7-1). This procedure proved cumbersome and time consuming and has recently been modified by lysing the blood rather than agglomerating the red cells prior to filtra-

*Brain-heart infusion broth supplemented (with or without Liquoid), prereduced and anaerobically sterilized, Scott Laboratories, Fiskeville, R.I.

tion. This technique requires less time (5 minutes) and a smaller blood sample (10 ml), making it practical for routine clinical usage. It proved very effective in a clinical blood culture trial,[53] yielding distinctly more positives and quicker growth and identification as compared with broth and pour plate techniques. Unfortunately, the disposable filtering apparatus used in these studies was never marketed. Nonetheless, the procedure can be carried out with commercially available reusable apparatus. More recently, another approach that also uses SPS to block antibacterial activity of blood and a technique for lysing erythrocytes and then density gradient centrifugation has been proposed.[16] Still another lysis-filtration system transfers the filter to broth and then uses impedance detection, which can be automated (Zierdt, C. M., et al., J. Clin. Microbiol. **5:**46-50, 1977; Kagen, R. L., et al., J. Clin. Microbiol. **5:**51-57, 1977).

If the patient is receiving penicillin therapy at the time the blood is drawn for culture, the concentration of the antibiotic in the blood may be high enough to inhibit the growth of susceptible bacteria in the sample in conventional broth blood cultures. To recover these organisms, it may be necessary to add penicillinase* to the blood culture, thereby nullifying the effect of the penicillin. Some workers, however, do not recommend the addition of penicillinase to the culture on the theory that it may increase the chance of contamination. Furthermore, if the blood is sufficiently diluted in the broth medium (1:10 to 1:30), it is thought that the inhibitory effect of the penicillin (as well as complement, natural antibody, and other antibacterial components of serum) is reduced to a level of no consequence.[3]

Hamburger and co-workers[29] reported that magnesium ion will antagonize the antistaphylococcal action of tetracycline

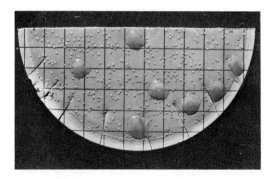

Fig. 7-1. Double bacteremia with *S. aureus* and *K. pneumoniae.* Membrane filter on EMB plate. Small colonies are *S. aureus;* large, mucoid colonies (lactose positive) are *K. pneumoniae.*

*Baltimore Biological Laboratory, Cockeysville, Md.; Difco Laboratories, Detroit.

and recommended the inclusion of magnesium sulfate in the blood cultures of patients receiving tetracycline antibiotics.

By the use of the aforementioned techniques and media, most of the common pathogens can be readily cultivated. The latter include streptococci, staphylococci, pneumococci, meningococci, coliform bacilli, anaerobic cocci, and members of the genera *Salmonella, Haemophilus, Bacteroides, Clostridium, Acinetobacter, Pseudomonas, Alcaligenes,* and *Candida.* **Special methods** or special media are necessary for the isolation of *Actinomyces, Brucella, Pasteurella, Leptospira, Listeria* and *Spirillum.* These are described in subsequent sections. See p. 45 for a special medium for gonococci.

Blood culture in patients with indwelling intravenous catheters

The problem of bacteremia secondary to indwelling intravenous catheters has been mentioned. In such patients, it is common practice on removing the offending catheter to aseptically cut off the indwelling end and drop it into broth for culture. In some institutions, such catheter-tip cultures are performed on removal of the catheter even if there has been no suggestion of infection. Maki and co-workers (N. Engl. J. Med. **296:**1305-1309, 1977) point out that it may be difficult to distinguish between true colonization of a catheter site, seeding of the catheter from a distant infected focus, and contamination during the culturing process and that semiquantitative culture (rolling the catheter tip across a blood agar plate and counting colonies subsequently) is helpful in this situation. In cases of catheter-related sepsis, there were always 15 colonies to confluent growth; in other cases, there was no growth or only one to seven colonies. The question also arises in suspected catheter-associated sepsis as to whether to draw blood cultures through the catheter or by separate venipuncture. In a

study of this question (Handsfield, J.A.M.A. **236:**2944, 1976; Tonnesen and Lockwood, J.A.M.A. **236:**2944, 1976; Tonnesen et al., J.A.M.A. **235:**1877, 1976), it was found that there was a high rate of isolation of *Staphylococcus epidermidis* from catheter-drawn cultures but not from venipuncture cultures. This high rate of false-positive cultures from catheter-drawn cultures indicates that it is preferable to perform venipunctures for blood cultures.

Other sources of false-positive blood cultures, in addition to contaminated evacuated blood collection tubes previously mentioned, include contaminated benzalkonium solution used for skin disinfection[36] (iodine and alcohol skin preparation avoids this problem) and contaminated commercial blood culture media.[42]

Despite the comments above, quantitative cultures on blood drawn through a catheter positioned at various sites proximal and distal to heart valves (sometimes in conjunction with dye dilution studies, angiography, and so forth) may permit distinction between bacteremia from a peripheral source and endocarditis and may permit determination of the infected site within the heart in the case of endocarditis.[43]

Incubation and examination of cultures

Blood cultures are incubated aerobically at 35 C (except where otherwise indicated) and are examined daily throughout the first week (more frequently when the laboratory is informed that the patient is seriously ill and a specific etiologic diagnosis has not yet been made) for evidence of growth. When a culture is **positive,** the **broth** may assume one of several appearances. In one type of growth, characteristic of gram-negative rods, the medium above the red cell layer becomes uniformly turbid, and gas bubbles caused by fermentation of glucose may be present. When pneumococci or meningococci are present, a similar although less distinct turbidity is produced, usually accompanied by a greenish tint in the medium. Streptococci grow as "cotton ball" colonies on top of the sedimented red cells, whereas the upper layer of broth may remain clear. This commonly occurs in blood cultures from patients with subacute bacterial endocarditis caused by alpha-hemolytic streptococci. If beta-hemolytic streptococci or other hemolytic organisms are present, marked hemolysis of the blood may be observed, along with turbidity. Pathogenic species of staphylococci will grow out readily and may produce a large jellylike coagulum throughout the broth due to coagulase production. Grossly visible discrete *Staphylococcus* colonies will also develop in the medium and impart turbidity. If the culture is contaminated with *Bacillus* species or a saprophytic fungus, the broth becomes hemolyzed and a fairly thick pellicle forms on its surface. *Clostridium* species may produce marked hemolysis, an unpleasant odor, and gas under pressure. *Bacteroides* species produce less gas, but a foul odor. Growth of *Haemophilus influenzae* does not generally produce any change in the medium and may be detected **only** by subculturing to chocolate agar.

In an **agar slant** (biphasic) bottle, bacterial growth generally appears either as large or small discrete colonies or as confluent growth on the slant, with cloudiness in the broth. Colonies of pneumococci are small, translucent, and difficult to recognize, as are colonies of group A streptococci. Staphylococcus colonies, on the other hand, are quite large, opaque, and easily seen, as are those of coliform bacilli, yeasts, and so forth. Growth of clostridia generally appears between the glass side wall and the agar slant and produces large bubbles of gas. *Proteus* and *Pseudomonas** colonies generally appear

as a translucent film on the agar slant. Otherwise, growth is similar to that previously described.

A **negative** blood culture usually remains clear, but it may develop cloudiness with continued incubation, sometimes caused by shaking the bottle before making smears, which in this instance will show no bacteria. After prolonged incubation (10 days or more) the bottle may develop a turbidity from fibrin or agar particles. This occurs usually above the red cell layer and may suggest bacterial growth. In either case turbidity alone is not a reliable guide; a gram-stained smear should be examined to determine the presence of microorganisms.

It is very important that **blind subcultures** be made from blood cultures showing no macroscopic growth. Failure to do this will result in significant delay in identifying bacteremia in many cases. Washington and his colleagues[30] recommend two blind subcultures (morning and afternoon) on the first day of incubation (the day of collection) of a blood culture, providing the culture is incubated at least 3 hours. Still another blind subculture is recommended 48 hours after the culture is first drawn. For the subculture, an aliquot (a few drops) of broth is aspirated by needle and syringe and inoculated to a chocolate agar plate for incubation under 3% to 10% carbon dioxide.

When growth appears on **any** medium, the culture should be inoculated to two blood agar plates for aerobic and anaerobic incubation and a chocolate agar plate under 3% to 10% carbon dioxide, and a gram-stained smear should be examined. If gram-negative rods are seen in the original bottles, an eosin–methylene blue (EMB) or a MacConkey agar plate should be streaked also. These will permit rapid preliminary identification if the organism is a coliform.

Failure to hold cultures long enough (in special instances, such as fungemia) or to examine them frequently enough may

---

*Pseudomonas and yeasts may require 3 to 4 days to grow out and are best recovered in an **air-vented** bottle.

result in a delay in recognizing the presence of pathogenic microorganisms or may even result in their loss of viability. Frequent openings of the culture bottles, on the other hand, may also result in contaminating the contents. The following schedule is offered as a desirable **routine** for the examination of regular blood cultures. Special methods for handling the less common pathogens are described subsequently.

Broth culture procedure

1. On the first day of incubation and after 48 to 72 hours incubation, shake the flask well and subculture (use sterile syringe and needle) a few drops of the blood broth to a chocolate agar plate. Reincubate the blood culture. Examine the blood culture bottle carefully, without agitating it, daily for 7 days.
2. Incubate plate in 3% to 10% CO_2 and examine for growth in 24 and 48 hours. Discard thereafter if negative.
3. After 7 days' incubation, repeat the subculture. Discard the broth flask if the plate shows no growth after 48 hours.
4. In special situations, as in suspected **bacterial endocarditis, fungemia, brucellosis,** or **anaerobic bacteremia**, it may be necessary to hold blood cultures for 2 to 4 weeks. However, holding routine cultures for more than 7 days merely permits recovery of skin contaminants, such as diphtheroids, which may lead to difficulties in interpretation.

Identification

In line with earlier comments about the seriousness of bacteremia and the importance of instituting specific therapy at the earliest possible moment, it is desirable to determine the identification and antimicrobial susceptibility of organisms from positive blood cultures at the earliest possible moment. The membrane filter system offers real advantages here, in addition to earlier appearance of growth. Growth is present in the form of colonies with typical morphology, pigmentation, hemolysis, and reactions on differential media (at Wadsworth Hospital Center, we typically place a portion of the membrane filter on EMB agar). Availability of colonies permits such rapid tests as the catalase and slide coagulase tests on gram-positive cocci and the oxidase test on gram-negative rods. These same advantages may also be present in pour plate cultures. With broth cultures, subculturing and expeditious use of processing schemes should be carried out. **Direct** or "emergency" inoculation to an API strip or other appropriate identification test systems and antimicrobial susceptibility testing should be done on macroscopically positive bottles, using the blood culture bottle contents directly as inoculum, assuming only one organism is present. This can subsequently be rechecked in the conventional manner.

The most frequently isolated skin contaminants include diphtheroid bacilli, which may be microaerophilic. These organisms are strongly catalase-positive, nonmotile, gram-positive rods, which may be confused morphologically with *Listeria monocytogenes*, a distinct pathogen from which they must be differentiated. The reader is referred to Chapter 26 for a discussion of the biochemical reactions of the latter organism.

Conclusion

If all cultures and subcultures are negative, the blood culture may be reported as **"No growth, aerobic or anaerobic, after 7 days' incubation."**

It is strongly recommended that a stock culture be made, on appropriate media,* of all significant organisms isolated from blood cultures. Such stocks should be

---

*CTA medium, Baltimore Biological Laboratory, Cockeysville, Md., is recommended. The culture, after overnight incubation, may be stored in the refrigerator.

maintained for at least several months. This is important for subsequent antimicrobial susceptibility testing by the tube dilution technique or for more extensive studies, such as serologic tests, animal inoculation, or referral to a diagnostic center, which may be indicated on occasion.

PNEUMOCOCCAL BACTEREMIA

Pneumococcal bacteremia can be demonstrated in about 25% of adult patients with pneumococcal pneumonia,[5] less frequently in infants and children.[8] In many cases, the blood culture may be the **only** specimen from which pneumococci are recovered; it therefore provides the most useful and readily available procedure for confirming a clinical impression of bacteremia or meningitis and should be obtained from **any** patient with a suspected pneumococcal infection.

Although the procedure previously outlined is satisfactory for isolating pneumococci from the blood, certain precautions should be observed in the handling of these cultures. After overnight incubation, gram-stained smears of the broth should be made and examined, regardless of the appearance of the flask. The pneumococci, if present, will have grown out in this period, and longer incubation may result in a loss of the organism through autolysis in a fluid medium. Pneumococci will appear as gram-positive cocci in chains resembling streptococci; some may show the typical lancet-shaped diplococcus with evidence of a capsule.

Whether or not organisms are seen in the smear of the blood broth, a subculture should be made. This demonstrates the presence of pneumococci in numbers too small to be seen in the smear. Methods for the identification of the pneumococcus may be found in Chapter 18.

ISOLATION OF NEISSERIA AND HAEMOPHILUS

The isolation of *Neisseria meningitidis* and *Haemophilus influenzae* from the blood culture of all patients with purulent meningitis should be attempted. Through use of the procedures previously described, both organisms may be recovered after 18 to 24 hours' incubation* by subculturing from the broth bottle to a **chocolate agar** plate and following with incubation for 18 to 24 hours in a candle jar. If this subculture shows no growth, the procedure is repeated after incubating the original broth flask for an additional 5 to 10 days. Gram-stained smears of *Neisseria* and *Haemophilus* may be difficult to interpret because of the presence of gram-negative debris and a heavy red background. Consequently, the practice of subculturing the broth flasks routinely may be very valuable. In the case of *N. gonorrhoeae*, CDC's Venereal Disease Control Division (*Criteria and Techniques for the Diagnosis of Gonorrhea,* 1976) recommends culturing blood in an enriched broth, such as trypticase soy broth supplemented with 1% IsoVitaleX, 10% horse serum, and 1% glucose.

Attention is called to a *Haemophilus* species, *H. aphrophilus,* which is an infrequent cause of bacterial endocarditis.[66] The organism is a gram-negative, nonmotile rod that does not require V factor but does require X factor and an increased carbon dioxide tension for its growth (see Chapter 23 for a description of this organism).

SEPTICEMIA AND BACTEREMIC SHOCK CAUSED BY GRAM-NEGATIVE RODS

The association between severe and frequently fatal septicemia caused by gram-negative enteric bacilli and confinement in the hospital is becoming increasingly apparent.[15,17] In the early antibiotic era the primary concern was for sepsis caused by *Staphylococcus aureus;* in the past decade, however, many institutions are reporting a great increase in infections caused by *E. coli* and members of

---

*N. gonorrhoeae may require 7 or more days.

the genera *Klebsiella, Enterobacter, Serratia, Proteus,* and *Pseudomonas.*[54]

Factors that have played a primary role in initiating these infections include:

1. Those compromising the resistance of the host, such as the use of adrenal corticosteroids, immunosuppressive agents, and antimetabolite drugs
2. The widespread utilization of broad-spectrum antimicrobial agents, which may upset the endogenous flora, with emergence of resistant strains
3. Procedures producing new artificial avenues of infection, such as the use of indwelling bladder and venous catheters, tracheostomies, and mechanical ventilators
4. More extensive surgical procedures
5. The increasing number of elderly, chronically debilitated patients admitted to the hospital
6. Nosocomial infections from contaminated IV fluids*
7. More prolonged survival of patients with serious, ultimately fatal diseases

Since mortality of 45% to 50% accompanies these gram-negative septicemias,[33] the importance of early diagnosis and the initiation of specific therapy cannot be overemphasized.

ANAEROBIC BACTEREMIA

An increasing interest on the part of both the microbiologist and the clinician in the recovery of anaerobes from clinical material has resulted in numerous reports of the isolation of these bacteria from the blood. In retrospective study of bacteremia at the Mayo Clinic involving over 3,000 positive blood cultures, Washington[63] found anaerobes in 11%, representing 20% of the patients with bacteremia. Sullivan and co-workers,[52] while studying bacteremia in 300 patients following uro-

logic procedures, found a 31% positive incidence in those having transurethral resections. Although enterococci and *Klebsiella* were isolated most frequently, there were also a relatively large number of anaerobic cocci, *Bacteroides,* and so forth recovered. The authors reported the highest isolation rate in osmotically stabilized broth containing SPS; a membrane filter system also was useful for detecting multiple organism bacteremia (26% in this series).

In papers from the Center for Disease Control, Felner and Dowell[20] and Alpern and Dowell[2] have reported the isolation of such anaerobes as *Bacteroides fragilis,* fusobacteria, clostridia (nonhistotoxic), and peptococci from the blood of patients with symptoms of endocarditis, gram-negative rod septicemia, or neonatal sepsis. Alpern and Dowell[1] also studied 27 patients with documented *Clostridium septicum* septicemia associated with malignant disease.

In suspected actinomycosis, freshly prepared brain-heart infusion medium without venting is preferred.

INFECTIONS CAUSED BY SPECIES OF BRUCELLA

The diagnosis of *Brucella* infections in humans is best made by the isolation of the causative organism. *Brucella* is most often found in the bloodstream, particularly in the febrile period, although it is not recovered as readily as the more commonly isolated pneumococcus, streptococcus, or staphylococcus. Two reasons have been advanced for this: (1) the relatively complex nutritional requirements of brucellae and (2) their presence in the blood only in small numbers because of their intracellular proclivity. The first difficulty has been overcome by the introduction of commercially available culture media, such as tryptose, trypticase soy, and Brucella broth and agar,* which will

---

*See special supplement to vol. 20, no. 9, Morbidity and Mortality Reports, Atlanta, 1971, Center for Disease Control.

*Difco Laboratories, Detroit; Baltimore Biological Laboratory, Cockeysville, Md.; Pfizer Diagnostics, Clifton, N.J.

readily support the growth of small inocula. The problem of the small numbers of brucellae can be resolved partly by culturing large quantities of blood (while maintaining a blood-to-broth ratio of at least 1:10) or by obtaining numerous cultures over a period of several days. Bone marrow cultures have also proved useful in some instances when blood cultures have been negative.

A simple method for cultivating brucellae and permitting examination of cultures by inoculation of agar has been described by Ruiz-Castañeda[10] and modified by Scott.[50] This method utilizes a bottle containing Brucella or trypticase soy broth and a Brucella or trypticase agar slant, which permits direct observation of colonies developing from the broth culture.* The bottle is under a partial vacuum and is closed with a thin rubber diaphragm in a screw cap. When blood is to be cultured for brucellae, the bottle is taken to the patient's bedside. Immediately after obtaining the specimen by venipuncture, 10 ml of blood are injected directly into a bottle containing 100 ml of broth through the previously sterilized (by iodine and alcohol) rubber diaphragm, using the same syringe and needle. The blood and broth are mixed well to prevent clotting, and the agar slant is inoculated by tilting the bottle. The bottle is then labeled, the rubber seal pierced with a sterile 18-gauge needle containing a wisp of cotton in the hub, and the bottle is incubated in a candle jar or carbon dioxide incubator (*B. abortus* requires increased carbon dioxide for initial growth) in an upright position at 35 C. Some laboratories prefer to incorporate carbon dioxide in the bottle directly.

The culture is examined at regular intervals by careful inspection of the agar slant for discrete colonies or confluent growth. The latter may not be readily apparent. If no growth is observed, the bottle is gently shaken and tilted (so that the blood broth flows over the agar slant

again) and is then replaced in the candle jar or carbon dioxide incubator. This procedure is repeated every 48 hours, and the culture is held for at least 21 days before it is discarded as negative. This method reduces the possibility of contaminating the culture or infecting the worker by repeated opening and subculturing of the bottle.

If growth occurs and consists of small coccoid or short bacillary gram-negative organisms, it may be identified as a species of *Brucella* by agglutination with type-specific antisera, hydrogen sulfide production, and susceptibility to certain dyes. These characteristics are described in Chapter 23.

INFECTIONS CAUSED BY SPECIES OF FRANCISELLA, PASTEURELLA, AND YERSINIA

The organisms comprising this group are nonmotile* gram-negative rods that usually exhibit polar staining. They are biochemically inactive and moderate to strictly fastidious in their growth requirements. They are primarily involved in diseases of lower animals, where they cause a highly fatal type of hemorrhagic septicemia; but closely related species can cause bubonic plague and tularemia in humans.

The classification of the genus *Pasteurella* has been altered in accordance with information on the characteristics of these bacteria. *P. tularensis* is now classified as *Francisella tularensis; P. pestis* and *P. pseudotuberculosis* are now in the genus *Yersinia*, which also includes *Y. enterocolitica*.

Tularemia, caused by *F. tularensis*, is a naturally occurring disease in animals such as wild rodents, rabbits, and deer. Human beings contract the disease accidently as a result of handling infected animals (chiefly rabbits) or from the bites of infected deerflies or ticks. The organism will not grow on ordinary media, differing in this respect from *Y. pestis*, the

---

*Baltimore Biological Laboratory, Cockeysville, Md.

Y. pseudotuberculosis is motile at 25 C.

causative agent of plague. It may be cultured on a coagulated egg medium, on cystine-glucose blood agar, or in heart infusion broth containing glucose, cystine, and hemoglobin. *F. tularensis* is difficult to cultivate from infected material, but successful isolation from the blood, pleural fluid, or pus from unopened skin lesions may be achieved by the following procedure:

1. Inoculate six cystine-glucose blood agar slants and one plain agar control slant with 0.5 to 1 ml of the patient's blood, freshly drawn by venipuncture. Distribute the blood over the slants before it clots and incubate the slants on their sides aerobically at 35 C. Pus or pleural fluid should be similarly spread over the slants. Precautions should be taken to restrict moisture formation, as this is deleterious to the organism.

2. Observe the slants daily for the presence of growth. Minute droplike colonies may appear in 3 to 5 days but can take as long as 10 days to develop.

3. Examine a stained smear of the colony for small gram-negative coccobacilli. If present, the culture may be identified by specific agglutination.*

4. Do not report the culture as negative before 3 weeks of incubation.

Note: Great caution is **mandatory** in handling infectious material from patients and cultures! Use syringes or hand-actuated pipets.

Guinea pigs can be infected without difficulty, and subcutaneous or intraperitoneal injection of a heavy saline suspension of blood or ground-up tissue is the preferred procedure when only small numbers of organisms may be present. Animals usually survive 5 to 7 days but may die in 2 days; they show enlarged caseous lymph nodes, enlarged spleens,

*The identification is best carried out by a reference laboratory, such as the Center for Disease Control, Atlanta.

and necrotic foci in the liver and spleen. Accidental laboratory infection is an ever-present hazard; animal injection, therefore, should not be attempted unless the laboratory worker has been previously immunized against tularemia and a safe animal facility (plastic cages with filter tops, vented hoods, and so forth) is available.

Y. pestis infection of humans is most commonly of the **bubonic plague** type, and the bacilli may be cultivated from material obtained from buboes, the spleen, or blood. In the pneumonic type the sputum is most likely to yield the organism. Blood cultures taken during the first 3 days of bubonic disease are positive in most cases. Blood specimens are first cultured in broth and then subcultured to blood agar plates daily. On this medium after 48 hours the colonies are small (1 mm in diameter), glistening, and transparent, with round granular centers and a notched margin. Cultures are identified by biochemical reactions and specific agglutination, although specific bacteriophage lysis is considered more reliable. Again, the laboratory worker must use extreme caution in handling specimens, cultures, and animals. Suspected and confirmed cases of plague must be reported to local health departments immediately.

Other members of this group that may infect humans are *P. multocida*, *Y. pseudotuberculosis*, and *Y. enterocolitica*. All are primarily animal pathogens, and humans are accidental hosts. Approximately one half of the cases of *P. multocida* infection in humans result from the bites or scratches of dogs or cats; the remainder are either pulmonary infections resulting from probable animal contact or from miscellaneous causes.[34] Bacteremia is apparently a rare occurrence, although a case of endocarditis caused by *Pasteurella*, nov. sp., following a cat bite was reported.[27] Both *Y. pseudotuberculosis* and *Y. enterocolitica* probably are involved in infection much more commonly than has been appreciated. They may cause septicemia, enteritis or enterocolitis, mesenteric lymphadenitis, and infection in various other parts of the body. Erythema nodosum and arthritis may be part of the clinical picture.

P. multocida grows readily on blood or chocolate agar plates, giving rise to large raised colonies, which show a tendency to become confluent. These organisms are very susceptible to penicillin. Further descriptive bacteriology is found in Chapter 23.

BACTEREMIA CAUSED BY LEPTOSPIRA SPECIES

Human leptospiral infection results principally from contact with an environment that has been contaminated by pathogenic leptospires from an animal urinary shedder or other reservoir hosts. The organisms enter the body through the abraded skin or the mucosal surfaces of the mouth, nasopharynx, conjunctivae or genitalia[57]; their principal sources are contaminated waters, soils, vegetation, foodstuffs, and so on. Occupational exposure is a primary factor in leptospirosis; the disease is most frequently associated with such people as farm workers, veterinarians, livestock handlers, abattoir attendants, and sewer workers. The disease is also associated with recreational pursuits such as swimming and camping.

The laboratory diagnosis of leptospirosis depends on isolating the organism from the blood or demonstrating a rise in antibody titer in the serum. Leptospirae may be recovered from the blood and occasionally the spinal fluid during the first week of the initial febrile illness, before circulating antibodies appear. Cultural methods or the inoculation of young guinea pigs or weanling hamsters may be employed. The organism is readily cultivated on relatively simple media enriched with rabbit serum, such as Fletcher's or Stuart's medium.* Blood cultures are made at the patient's bedside or

*Difco Laboratories, Detroit.

in the laboratory, using venous blood directly from the syringe or defibrinated blood in a sterile bottle. The semisolid medium is dispensed in screw-capped tubes or rubber-stoppered vaccine bottles, and a minimal inoculum of blood is introduced directly into the bottle by puncturing the rubber diaphragm at the patient's bedside.

RAT-BITE FEVER CAUSED BY SPIRILLUM MINOR AND STREPTOBACILLUS MONILIFORMIS

Two kinds of human disease may follow the bite of an infected rat: one, known as **sodoku** in Japan, is caused by a spiral organism presently classified as *Spirillum minor*; the other, more common in the United States, is caused by a highly pleomorphic organism, *Streptobacillus moniliformis*. These diseases present similar symptoms, including a primary ulcer, regional lymphadenopathy, fever, and malaise in addition to a characteristic skin rash. The latter infection may also be accompanied by an impressive polyarthritis.

A definitive diagnosis of rat-bite fever depends on the demonstration of the causative organism in the blood, lesion exudate, joint fluid, and so forth. In sodoku the presence of spirilla can be demonstrated by injecting 1 to 2 ml of the patient's whole blood intraperitoneally into several white mice or guinea pigs known to be free from naturally occurring spirilla. Mouse peritoneal exudate or the supernate of defibrinated guinea pig blood is obtained and examined weekly for a period of up to 4 weeks, either under darkfield microscopy or by special stains, for the presence of spirilla.

In rat-bite fever caused by *Streptobacillus*, the blood is cultured in the routine manner. After several days of incubation at 35 C, the organism grows out as "cotton balls" on the surface of the red cell layer in the blood culture bottle. Whether or not growth occurs, the blood culture should be subcultured to fresh tubes of the same medium, using a thin sterile capillary pipet for the transfer. When growth is observed, Gram stains and Wayson stains are made and examined for the presence of pleomorphic gram-negative rods. Coccobacillary, rod-shaped, branching cells and large swollen club-shaped cells are seen; special methods may also reveal the presence of L forms of *Streptobacillus*. These are fragile, filterable forms lacking a rigid cell wall; they require a complex medium for growth and are capable of reverting to the parent bacterial cell. Agglutinins are formed against the *Streptobacillus,* and a titer of 1:80 is regarded as of diagnostic significance.

INFECTIONS CAUSED BY LISTERIA MONOCYTOGENES

Listeriosis** is a bacterial infection of humans and animals that is being recognized with increasing frequency. The causative organism, *Listeria monocytogenes,* was first isolated from rabbits in 1926, and the first human infection was reported in 1929. Although the disease is common in domestic animals, the natural habitat of the organism or its epidemiology or epizoology has not been clearly defined.

Listeriosis is primarily an infection of the infant and the newborn baby, in whom it causes a purulent meningitis or meningoencephalitis or a granulomatous septicemia acquired in utero or during delivery. In the adult it may cause meningitis, abortion, or an acute septicemia.

L. monocytogenes is most frequently isolated from the blood and cerebrospinal fluid of human patients, although it may be recovered from meconium, vaginal and cervical secretions, discharge from the eye, or tissue collected at autopsy (particularly from the brain, liver, and

spleen).* A procedure for isolating *L. monocytogenes* from the blood (Seeliger and Cherry[51]) is as follows:

1. After careful disinfection of the skin, obtain 15 ml of blood by venipuncture, using a sterile syringe and needle.
2. Immediately inoculate 5 ml into each of two flasks containing 100 ml of a good fluid medium, such as trypticase or tryptic soy broth or brain-heart infusion broth; mix well to prevent clotting.
3. Incubate one flask in a candle jar or CO_2 incubator at 35 C. Inspect for growth and subculture to sheep blood agar at 2-day intervals for at least 1 week, incubating plates in CO_2 for 48 hours before discarding.
4. Refrigerate the second flask at 4 C, and if no growth occurs from the 35 C culture, subculture from the refrigerated flask to blood agar at weekly intervals for 4 weeks.
5. Transfer 5 ml of blood to a sterile tube, allow to clot, and separate serum; inactivate at 56 C for 30 minutes and store in freezer for subsequent serologic testing.

Agglutination tests on acute- and convalescent-phase sera may be carried out, although most children under the age of 10 years, as well as many adults, have normal antibody titers to *Listeria*. Titers below 1:160 should not be considered positive.

INFECTIONS CAUSED BY CAMPYLOBACTER FETUS

Campylobacter (Vibrio) fetus was first recognized as a causative agent of contagious abortion in cattle by McFaydean and Stockman in England in 1909. It was not until 1948, however, that the first hu-

---

**The reader is referred to an excellent monograph on the subject by Gray and Killinger,[26] and also the CDC report on human listeriosis in the United States, 1967-1969.[9]

**Directions for culturing these specimens may be found in Lennette, E. H., Spaulding, E. H., and Truant, J. P., editors: Manual of Clinical Microbiology, ed. 2, Washington, D.C., 1974, American Society for Microbiology, Chapter 13.

man infection was reported in the United States by Ward.[60] Up to 1970, *C. fetus* had been isolated from human sources in at least 74 instances,[7] in many cases from the blood. Although the organism has been found in the mouth and in the female genital tract, the epidemiology in humans remains unclear; however, a bovine origin is suspected. Human infection is probably more common than is generally believed.

Since there are no distinctive clinical features of human infection, except for an undulating fever of unknown origin, signs of sepsis, and dysentery in persons who may have had contact with infected domestic animals or in those who have had acute thrombophlebitis or recent dental extractions, a definitive diagnosis must depend on the isolation and identification of *C. fetus* from the blood or other body sites.

C. fetus is readily isolated from blood cultured by the same techniques as those used in cases of suspected brucellosis. The organism grows well in tryptose phosphate or trypticase soy broth. Since the bacterium is a microaerophile, primary incubation in 5% to 10% carbon dioxide is **essential**. Strains may tend to die out on repeated subculture; aliquots of the original blood cultures frozen at –70 C may prolong survival. See Chapter 24 for a description of the cultural characteristics of this microorganism.

POLYMICROBIAL BACTEREMIA

The finding of two or more different bacterial species in a blood culture is reportedly an uncommon occurrence and may cause concern as to the possibility of contamination. It has been the experience of a number of workers, however, that such findings are not unusual, particularly in patients with gastrointestinal tract, hepatic, or urinary tract abnormalities or with hematologic disorders.[20,31,32] The incidence of two or more isolates from a single culture is generally reported as 6% to 10%; the organisms implicated include the gram-negative enteric bacilli, bacteroides, clostridia, and the gram-positive cocci. In these instances, the key to diagnosis and definitive antibacterial therapy rests with the bacteriology laboratory; as Elizabeth King[37] so aptly pointed out, "If people really looked, we would find that mixed infections are not as rare as previously supposed. Many workers consider any organism isolated from blood or spinal fluid to be in pure culture."

ENDOCARDITIS CAUSED BY COAGULASE-NEGATIVE STAPHYLOCOCCI, MICROCOCCI, AND DIPHTHEROIDS

Although these organism are generally considered as contaminants when isolated from blood cultures, they can be responsible for serious illness, such as bacterial endocarditis. Until about 1955, the incidence of coagulase-negative staphylococci, for example, in this disease was reported to be around 3%, but a study by Geraci and co-workers[25] in 1968 indicated an incidence of 10% in bacterial endocarditis. The etiologic significance of coagulase-negative staphylococci or micrococci and of diphtheroids from two or more blood cultures, therefore, should not be overlooked. It has been recommended (Bartlett, R. C., et al., Cumitech 1, Am. Soc. Microbiol., 1974), however, that the two isolates be tested simultaneously by biochemical tests and antimicrobial susceptibilities; if results are clearly dissimilar, their probable clinical significance is thereby reduced.

BLOOD CULTURES ON NEWBORN INFANTS

Multiple blood cultures (three or more) are preferred for the diagnosis of bacteremias in adults; in newborn infants, however, this becomes impractical for obvious reasons. A retrospective study of single blood cultures taken on infants less than 28 days old was reported by Franciosi and Favara.[24] Approximately 1 to 3 ml of heparinized blood were collected in a bottle containing 5 ml of trypticase soy broth and agar slant; of 56 patients with positive cultures, 28 were considered to have septicemia, and a single culture was considered satisfactory for diagnosis. Nevertheless, two blood cultures should be obtained when feasible. Since clinical symptoms of neonatal septicemia frequently are minimal, with the possibility of rapid deterioration and death within hours after birth, blood culture can be lifesaving in the newborn.

INFECTIONS CAUSED BY L FORMS OF BACTERIA AND MYCOPLASMA

Cell-wall defective forms of microorganisms, or L forms, have been isolated from the blood of patients treated with antibiotic agents, such as penicillin or methicillin, which interfere with cell-wall synthesis.[39] These fragile forms either do not survive or fail to grow in conventional isotonic blood culture media, and the culture remains sterile. However, in an osmotically stabilized medium, such as the previously described hypertonic sucrose broth with SPS, these aberrant forms may revert to their parent organism and subsequently be recovered and identified by conventional means.

Rarely, *Mycoplasma hominis* may be recovered from the blood of women with postpartum fever.

INFECTIONS CAUSED BY FUNGI (FUNGEMIA)

The incidence of bloodstream invasion by fungi, particularly members of the genus *Candida*, has increased significantly. This infection, once considered a rarity, is now a frequent complication in patients requiring therapy with adrenal corticosteroid hormones, cytotoxic agents, irradiation, or multiple antibiotics. Fungemia has also been reported in patients after prolonged venous catheterization[49] or parenteral hyperalimentation[13] and as a postoperative complication

of renal transplantation or cardiac valve prosthesis insertion.[12] Candidal endocarditis in heroin addicts is also an increasing problem.[40]

The diagnosis of fungemia depends primarily on isolation of the organism on blood culture, the results of which are not always readily available. Species of *Candida*, particularly *C. albicans* and *C. tropicalis*, will grow in the brain-heart infusion broth agar slant bottle previously described (pp. 40-41) within 1 to 4 days, especially if the bottle is vented to the air. Other yeasts, however, such as *C. parapsilosis*, *C. guilliermondii*, and *Torulopsis glabrata* may require 1 to 2 weeks' incubation before growing out. Blood cultures should be held for 21 to 30 days at 30 C in cases of suspected fungemia.[46]

Sternal marrow or the buffy coat of venous blood may be examined for fungi by methods similar to those used in hematologic studies, using the Giemsa stain or other stain. Cultures should be made of the aspirated material on Sabouraud dextrose agar and brain-heart infusion agar without added antibiotics. For best results this is done by transferring 0.5 to 1.0 ml of the specimen directly to several slants of the media at the patient's bedside, using the syringe and needle with which the marrow aspirate was obtained.* These media are incubated at room temperature and at 35 C and examined for the presence of growth at frequent intervals for several weeks.

Appropriate methods for the identification of isolates may be found in Chapter 32.

NONCULTURAL TECHNIQUES FOR DETECTION OF BACTEREMIA

A special method for early detection of bacteremia using a radioisotope has been recommended.[14] The method is

---

*This procedure may also be used in culturing blood for fungi, especially for prompt isolation of *Candida* species and for *Histoplasma capsulatum* in disseminated disease.[38] Inoculate slants directly with 1 to 2 ml of blood.

based on the ability of an organism to release CO_2 from ^{14}C-labeled glucose or other labeled substrate, and the $^{14}CO_2$ formed is detected by sampling the atmosphere above the blood culture by radiometric means. Early detection of positive blood cultures is the major theoretic advantage of this procedure. Reports of its effectiveness are conflicting, although generally earlier detection of growth was noted in a portion of the cultures. However, not all organisms are recovered as often or as early by this system as by conventional blood cultures. The amount of blood used and the composition of the media utilized may be factors. This procedure, it should be noted, offers no clue as to the probable identity of the organism; consequently, conventional procedures must be used thereafter. Terminal subculture is necessary (Strauss, R. R., et al.: J. Clin. Microbiol. 5:145-148, 1977).

Another method makes use of gas chromatography to detect metabolic end products resulting from bacterial growth in blood cultures after a short incubation period. In a limited study, however, it was shown that certain end products, such as volatile fatty acids, were detected, but never before the time that the blood culture was positive by gross inspection.[21] The procedure requires further study of its applicability in the clinical laboratory.

CONTAMINATED BANKED BLOOD AND PLASMA

Of the many complications of blood transfusion, including incompatibility due to mismatched blood and serum hepatitis, those caused by bacterially contaminated blood are usually the most dangerous to the patient. Not only will severe reactions (chills, fever, and a fall in blood pressure) ensue immediately, but peripheral collapse may occur within an hour after the transfusion of massively contaminated blood, and the patient frequently dies. It is important, therefore, that the microbiology laboratory offer a procedure for the detection of contamination in preserved blood and its products **before** they are administered to a patient.

The organisms most frequently encountered in fatal transfusion reactions comprise a small group of **psychrophilic gram-negative rods** (often from the donor's skin) that have the ability to multiply in blood stored at refrigerator temperature.[59] These organisms and their optimal temperatures for growth are:

1. *Pseudomonas* species (other than *Pseudomonas aeruginosa*), which grow slowly at 4 C, rapidly at room temperature (22 to 25 C), and slowly or not at all at 35 C.
2. *Citrobacter* species, which grow best at room temperature and at 35 C.
3. Other late lactose-fermenting members of the family Enterobacteriaceae, which will multiply at all three temperatures.
4. *Achromobacter* species, which grow best at room temperature and at 35 C.

These gram-negative organisms are generally found in the air, soil, water, and dust and in the intestinal tract of humans and animals. They can be transferred readily to the skin of humans, where they may become a part of the normal transient flora of that tissue. These organisms will multiply in the refrigerated blood, autolyze, and release **endotoxins** that may produce a toxic effect when administered in a blood transfusion. The reaction that follows is the result of this phenomenon rather than the effect produced by the introduction of the bacteria into the circulatory system or their multiplication within the tissues of the patient.

Contamination of the blood may be caused in various ways: (1) by the use of improperly sterilized equipment and reagents for the collection, (2) by some flaw in aseptic technique during collection, and most frequently (3) by the inadequate cleansing and disinfection of the skin over the site of venipuncture. For procedures to obviate these difficulties, the reader is referred to a manual on technical methods and procedures issued by the American Association of Blood Banks.

Sterility test procedure

The following methods for carrying out sterility tests are recommended by the Food and Drug Administration (Code of Federal Regulations, Federal Register, No. 640.2, Washington, D.C., 1976, U.S. Government Printing Office):

a. Sterility tests shall be performed at regular intervals and not less than once monthly. Blood intended for transfusion shall not be tested by a method that entails entering the container. A record shall be kept of the results of sterility tests.

b. Technic of sterility testing: Each month at least one container of normal-appearing blood shall be tested within the 18th to 24th day after collection. The test should be performed by inoculating 10 ml of blood into ten times the volume of fluid thioglycollate medium in a manner that introduces blood throughout the medium without destroying the anaerobic conditions in the lower portions of the broth. The sample should be incubated for 7 to 9 days at 30 to 32 C, or separate samples should be incubated at 18 to 20 C and 35 to 37 C, respectively, and examined every work day for evidence of microbial growth. A subculture of 1 ml inoculated into 10 ml of thioglycollate medium should be made on the third, fourth, or fifth day. The subculture should be incubated for 7 to 9 days as described above and examined daily for evidence of microbial growth.

Culturing plastic packs

Only when blood or plasma is collected in an open system, that is, where the blood container is entered, is it necessary to carry out a periodic sterility check.* After a transfusion reaction, however, it is strongly recommended that the following procedure, described by Walter and associates,[59] be carried out in the culturing of plastic blood packs and plasma packs. To secure a representative sample, the bag should be turned gently end over end to mix the contents; the integral donor tube should also be stripped several times. A 5-inch segment is then sealed with a dielectric sealer and cut free in the center of

---
*Public Health Service, Section 73.3004, amended March 25, 1972, Federal Register vol. 37, no. 59.

the seal. Both ends of the segment are then dipped in 70% ethyl alcohol, flamed, and with a pair of scissors similarly flame sterilized, one seal is cut off. This cut end then is held over a tube of thioglycollate medium, the top seal nicked to admit air, and the sample (approximately 1 ml) allowed to run in.

Cultures should be made in duplicate when possible and held for at least 10 days at room temperature and at 35 to 37 C. At the end of the incubation period, a loopful of the culture is streaked on a blood agar plate or placed in another tube of thioglycollate medium and held at the preceding temperatures for 24 hours to confirm sterility. Stained smears of the original cultures also may be examined.

Culture of donor arm bleeding site

In addition to the periodic sterility checks of blood and the equipment and solutions used in its collection, the culture of the venipuncture site before and after the arm "prep" is also recommended as a quality control measure. The technique described by Joress[35] may be used:

1. A cotton applicator moistened in 1 ml of pH 7.5 sterile phosphate buffer (see Chapter 43 for preparation) is rubbed vigorously over a 1-inch-square area of the venipuncture site for 1 minute. This is done in duplicate, before and after the "prep."
2. The applicator is then twirled in a tube (15 to 20 ml) of melted and cooled (45 C) soybean-casein digest agar and squeezed out against the inside of the tube.
3. Approximately 1 ml of sterile defibrinated sheep blood is added; the contents are mixed well by inversion and are poured into a sterile Petri dish.
4. The plates are examined after 48 hours' incubation at 35 C, and colony counts are made, together with the identification of significant isolates, if indicated.

5. The number of bacteria per square inch of skin should be reduced almost to zero if an effective degerming has been achieved.

Signs of contamination

Contamination of a bottle of blood often may be suspected from gross examination of the sample. The following changes should be looked for:

1. Change in the color of blood to a bluish black
2. Zone of hemolysis in the lowest portion of the supernatant plasma at the cell-plasma interface
3. Appearance of a large quantity of free hemoglobin in plasma when shaken
4. Appearance of small fibrin clots in a bottle, which had formerly appeared satisfactory, due to the utilization of citrate by the organisms previously described
5. Turbidity of the supernatant plasma.

The transfusion of contaminated blood is followed (usually after the first 50 ml) by a sudden onset of chills, fever, hypotension and shock. This may progress to convulsions and coma. An immediate Gram stain of blood from the blood container may reveal the presence of organisms; three separate blood cultures from the patient also should be obtained. Residual blood from the container of blood causing the reaction should be cultured at 18 to 20 C and 35 to 37 C. Prompt diagnosis and initiation of specific therapy may be lifesaving.

REFERENCES

1. Alpern, R. J., and Dowell, V. R., Jr.: *Clostridium septicum* infections and malignancy, J.A.M.A. **209:**385-388, 1969.
2. Alpern, R. J., and Dowell, V. R., Jr.: Nonhistotoxic clostridial bacteremia, Am. J. Clin. Pathol. **55:**717-722, 1971.
3. Anderson, T. G.: Personal communication, 1960.
4. Austrian, R.: The role of the microbiological laboratory in the management of bacterial infections, Med. Clin. North Am. **50:**1419-1432, 1966.
5. Austrian, R.: Current status of bacterial pneumonia with special reference to pneumococcal infection, J. Clin. Pathol. **21** (Suppl. 2), 1968.
6. Bennett, I.: Bacteremia, Veterans Admin. Tech. Bull. TB 10/104, Washington, D.C., 1954, U.S. Government Printing Office.
7. Bokkenheuser, V.: *Vibrio fetus* infection in man, Am. J. Epidemiol. **91:**400-409, 1970.
8. Burke, J. P., Klein, J. O., Gezon, H. M., and Finland, M.: Pneumococcal bacteremia, Am. J. Dis. Child. **121:**353-359, 1971.
9. Busch, L. A.: Human listeriosis in the United States, 1967-1969, J. Infect. Dis. **123:**328-332, 1971.
10. Castaneda-Ruiz, M.: Practical method for routine blood cultures in brucellosis, Proc. Soc. Exp. Biol. Med. **64:**114-115, 1947.
11. Center for Disease Control: False-positive blood cultures related to the use of evacuated nonsterile blood-collection tubes, Morbid. Mortal. Rep. **24:**387-388, 1975.
12. Chandhuri, M. R.: Fungal endocarditis after valve replacements, J. Thorac. Cardiovasc. Surg. **60:**207-214, 1970.
13. Curry, C. R., and Quie, P. G.: Fungal septicemia in patients receiving parenteral hyperalimentation, N. Engl. J. Med. **285:**1221-1225, 1971.
14. DeBlanc, H. J., de Land, F., and Wagner, H. N., Jr.: Automated radiometric detection of bacteria in 2,967 blood cultures, Appl. Microbiol. **22:**846-849, 1971.
15. Dodson, W.H.: *Serratia marcescens* septicemia, Arch. Intern. Med. **121:**145-150, 1968.
16. Dorn, G. L., Haynes, J. R., and Burson, G. G.: Blood culture technique based on centrifugation. Developmental phase, J. Clin. Microbiol. **3:**251-257, 1976; Clinical evaluation, **3:**258-263, 1976.
17. Edmonson, E. B., and Sanford, J. P.: The Klebsiella-Enterobacter (Aerobacter)-Serratia group; a clinical and bacteriological evaluation, Medicine **46:**323-340, 1967.
18. Edwards, P. R., and Ewing, W. H.: Identification of Enterobacteriaceae, ed. 3, Minneapolis, 1972, Burgess Publishing Co.
19. Ellner, P. D.: System for inoculation of blood in the laboratory, Appl. Microbiol. **16:**1892-1894, 1968.
20. Felner, J. M., and Dowell, V. R., Jr.: Anaerobic bacterial endocarditis, N. Engl. J. Med. **283:**1188-1192, 1970.
21. Finegold, S. M.: Early detection of bacteremia. In Sonnenwirth, A. C., editor: Septicemia: laboratory and clinical aspects, Springfield, Ill., 1973, Charles C Thomas, Publisher.
22. Finegold, S. M., White, M. L., Ziment, I., and Winn, W. R.: Rapid diagnosis of bacteremia, Appl. Microbiol. **18:**458-463, 1969.
23. Finegold, S. M., Ziment, I., White, M. L., Winn, W. R., and Carter, W. T.: Evaluation of polyan-

ethol sulfonate (Liquoid) in blood cultures. In Hobby, G. L., editor: Proceedings of the Sixth Interscience Conference on Antimicrobial Agents and Chemotherapy, Philadelphia, 1966, Washington, D.C., 1967, American Society for Microbiology.

24. Franciosi, R. A., and Favara, B. E.: A single blood culture for confirmation of the diagnosis of neonatal septicemia, Am. J. Clin. Pathol. **57:** 215-219, 1972.

25. Geraci, J. E., **Hanson, K. C.,** and Giuliani, E. R.: Endocarditis caused by coagulase-negative staphylococci, Mayo Clin. Proc. **43:**420-434, 1968.

26. Gray, M. L., and Killinger, A. H.: *Listeria monocytogenes* and listeric infections, Bacteriol. Rev. **30:**309-382, 1966.

27. Gump, D. W., and Holden, R. A.: Endocarditis caused by a new species of *Pasteurella,* Ann. Intern. Med. **76:**275-278, 1972.

28. Hall, M. M., Warren, E., Ilstrup, D. M., and Washington, J. A. II: Comparison of sodium amylosulfate and sodium polyanetholsulfonate in blood culture media, J. Clin. Microbiol. **3:**212-213, 1976.

29. Hamburger, M., Carleton, J., and Harcourt, M.: Reversal of the antistaphylococcal activity of tetracycline by magnesium, Antbiot. Chemother. (N.Y.) **7:**274-278, 1957.

30. Harkness, J. L., Hall, M., Ilstrup, D. M., and Washington, J. A. II: Effects of atmosphere of incubation and of routine subcultures on detection of bacteremia in vacuum blood culture bottles, J. Clin. Microbiol. **2:**296-299, 1975.

31. Hermans, P. E., and Washington, J. A. II: Polymicrobial bacteremia, Ann. Intern. Med. **73:**387-392, 1970.

32. Hochstein, H. D., Kirkham, W. R., and Young, V. M.: Recovery of more than one organism in septicemias, N. Engl. J. Med. **273:**468-474, 1965.

33. Holloway, W. J., and Scott, E. G.: Gram-negative rod septicemia, Del. Med. J. **40:**181-185, 1968.

34. Holloway, W. J., Scott, E. G., and Adams, Y. B.: *Pasteurella multocida* infection in man, Am. J. Clin. Pathol. **51:**705-708, 1968.

35. Joress, S. M.: A study of disinfection of the skin: a comparison of povidone-iodine with other agents used for surgical scrubs, Ann. Surg. **155:** 296-304, 1962.

36. Kaslow, R. A., Mackel, D. C., and Mallison, G. F.: Nosocomial pseudobacteremia: positive blood cultures due to contaminated benzalkonium antiseptic, J.A.M.A. **236:**2407-2409, 1976.

37. King, E. O.: Personal communication, 1965.

38. Koch, M. L. Bacteriologic problems in blood banking, Bull. Am. Assoc. Blood Banks **12:**397-401, 1959.

39. Lorian, V., and Waluslka, A.: Blood cultures showing aberrant forms of bacteria, Am. J. Clin. Pathol. **57:**406-409, 1972.

40. Louria, D. B., Hensle, T., and Rose, J.: The major medical complications of heroin addiction, Ann. Intern. Med. **67:**1-22, 1967.

41. Lowrance, B. L., and Traub, W. H.: Inactivation of the bactericidal activity of human serum by Liquoid (sodium polyanetholsulfonate), Appl. Microbiol. **17:**839-842, 1969.

42. Noble, R. C., and Reeves, S. A.: *Bacillus* species pseudosepsis caused by contaminated commercial blood culture media, J.A.M.A. **230:** 1002-1004, 1974.

43. Pazin, G. J., Peterson, K. L., Griff, F. W., et al.: Determination of site of infection in endocarditis, Ann. Intern. Med. **82:**746-750, 1975.

44. Rammell, C. G.: Inhibition by citrate of the growth of coagulase-positive staphylococci, J. Bacteriol. **84:**1123-1125, 1962.

45. Roberts, G. D., Horstmeier, C. D., and Ilstrup, D. M.: Evaluation of a hypertonic sucrose medium for the detection of fungi in blood cultures, J. Clin. Microbiol. **4:**110-111, 1976.

46. Roberts, G. D., and Washington, J. A. II: Detection of fungi in blood cultures, J. Clin. Microbiol. **1:**309-310, 1975.

47. Rosner, R.: Effect of various anticoagulants and no anticoagulant on ability to isolate bacteria directly from parallel clinical blood specimens, Am. J. Clin. Pathol. **49:**216-219, 1968.

48. Rosner, R.: A quantitative evaluation of three blood culture systems, Am. J. Clin. Pathol. **57:** 220-227, 1972.

49. Salter, W., and Zinneman, H. H.: Bacteremia and candida septicemia, Minn. Med. **50:**1489-1499, 1967.

50. Scott, E. G.: A practical blood culture procedure, J. Lab. Clin. Med. **21:**290-294, 1951.

51. Seeliger, H. R. P., and Cherry, W. B.: Human listeriosis, Washington, D.C., 1957, U.S. Government Printing Office.

52. Sullivan, N. M., Sutter, V. L., Attebery, H. R., and Finegold, S. M.: Bacteremia after genital tract manipulation: bacteriological aspects and evaluation of various blood culture systems, Appl. Microbiol. **23:**1101-1106, 1972.

53. Sullivan, N. M., Sutter, V. L., and Finegold, S. M.: Practical aerobic membrane filtration blood culture technique. Development of procedure, J. Clin. Microbiol. **1:**30-36, 1975; Clinical blood culture trial, **1:**37-43, 1975.

54. Thoburn, R., Fekety, F. R., Cluff, L. E., and Melvin, V. B.: Infections acquired by hospitalized patients, Arch. Intern. Med. **121:**1-10, 1968.

55. Traub, W. H.: Antagonism of polymyxin B and kanamycin sulfate by Liquoid (sodium polyanetholsulfonate) *in vitro,* Experientia **25:**206-207, 1969.

56. Traub, W. H., and Lowrence, B. L.: Media-dependent antagonism of gentamicin sulfate by

Liquoid (sodium polyanetholsulfonate), Experientia **25:**1184-1185, 1969.

57. Turner, L. H.: Leptospirosis, Br. Med. J. **1:**231-235, 1969.

58. Von Haebler, T., and Miles, A. A.: The action of sodium polyanethol sulfonate (Liquoid) on blood cultures, J. Pathol. Bacteriol. **46:**245-252, 1938.

59. Walter, C. W., Kundsin, R. B., and Button, L. N.: New technic for detection of bacterial contamination in a blood bank using plastic equipment, N. Engl. J. Med. **259:**364-369,1957.

60. Ward, B. R.: The apparent involvement of *Vibrio fetus* in an infection in man, J. Bacteriol. **55:**113-114, 1948.

61. Washington, J. A. II: Comparison of two commercially available media for detection of bacteremia, Appl. Microbiol. **22:**604-607, 1971.

62. Washington, J. A. II: The microbiology of evacuated blood collection tubes, Ann. Intern. Med. **86:**186-188, 1975.

63. Washington, J. A. II, Hall, M. M., and Warren, E.: Evaluation of blood culture media supplemented with sucrose or with cysteine, J. Clin. Microbiol. **1:**79-81, 1975.

64. Washington, J. A. II, and Martin, W. J.: Comparison of three blood culture media for recovery of anaerobic bacteria, Appl. Microbiol. **25:**70-71, 1973.

65. Winn, W. R., White, M. L., Carter, W. T., Miller, A. B., and Finegold, S. M.: Rapid diagnosis of bacteremia with quantitative differential-membrane filtration culture, J.A.M.A. **197:**539-548, 1966.

66. Witorsch, P. and Gordon, P.: *Hemophilus aphrophilus* endocarditis, Ann. Intern. Med. **60:**957-961, 1964.

67. Wright, H. D.: The bacteriology of subacute infective endocarditis, J. Pathol. Bacteriol. **28:**541-578, 1925.

CHAPTER 8

Microorganisms encountered in respiratory tract infections

The pathogens most likely to be found in the upper and lower respiratory tract include the following:

Group A beta-hemolytic streptococci
Corynebacterium diphtheriae
Streptococcus pneumoniae
Haemophilus influenzae
Neisseria meningitidis
Mycobacterium tuberculosis, other mycobacteria
Fungi including *Candida* species, *Histoplasma capsulatum, Coccidioides immitis*, and others
Klebsiella pneumoniae, other coliform bacilli
Pseudomonas aeruginosa, other pseudomonads
Bacteroides melaninogenicus, Fusobacterium nucleatum
Anaerobic and microaerophilic cocci
Bordetella pertussis

THROAT AND NASOPHARYNGEAL CULTURES

Throat cultures and nasopharyngeal cultures are important as an aid in the diagnosis of certain infections, such as streptococcal sore throat, diphtheria, or candidal infections of the mouth (thrush); in establishing the focus of infection in diseases such as scarlet fever, rheumatic fever, and acute glomerulonephritis; and also in the detection of the carrier state of organisms such as group A beta-hemolytic streptococci, meningococci, *Staphylococcus aureus*, and the diphtheria bacillus.

To obtain the best results it is important that material for a throat culture be obtained **before** antimicrobial therapy and taken in a proper manner. A satisfactory method is as follows:

1. Use a sterile throat culture outfit such as a commercially available disposable unit* that contains a polyester-tipped applicator and an ampule of modified Stuart transport medium.
2. With the patient's tongue depressed and the throat well exposed and illuminated, rub the swab firmly over the back of the throat (the posterior pharynx), both tonsils or tonsillar fossae, and any areas of inflammation, exudation, or ulceration. Care should be taken to avoid touching the tongue, cheeks, or lips, with the swab.
3. Replace the swab in the inner tube, crush the ampule, force the swab into the released holding medium, and send it to the laboratory immediately, where it should be plated as soon as possible. If it is necessary to hold the culture for more than an hour, it should be refrigerated until plates can be streaked.

If diphtheria is suspected, **two** swabs should be taken; one should be inoculated to Loeffler medium immediately and the other handled as described in step 3 above.

Nasopharyngeal cultures are recommended when attempting the isolation of meningococci, because these organisms are found more commonly in the nasopharynx than in the nose or throat. Nasopharyngeal swabs also are essential for the recovery of *Neisseria meningitidis* from suspected carriers or *Bordetella pertussis* from suspected cases of whooping

---

*Culturette, Scientific Products, Evanston, Ill.

cough. Such cultures are best taken on cotton-tipped nichrome or stainless steel wire applicators (B. & S. 28 gauge)* in a sterile tube containing a few drops of broth. With the patient's head firmly held, a nasal speculum is inserted; the wire swab is **gently** inserted (without force) through the nose to the posterior nasopharynx, where it is rotated gently and then allowed to remain for 20 to 30 seconds, and then deftly withdrawn. The wire swab may also be bent at a right angle near the cotton tip (bending against the inside of the sterile tube container) and then inserted through the mouth and behind the uvula and soft palate into the nasopharynx. Care must be taken to avoid mouth and throat contamination of the swab. In either case the inoculated swab is returned to the inner tube containing the broth or to the transport medium of a Culturette and thus transported promptly to the laboratory. A soft rubber catheter inserted in the nasopharynx is an alternative to a swab for diagnosis of pertussis.

Anterior nares cultures are occasionally required, particularly for the study of staphylococcal or streptococcal nasal carriers. They are taken by introducing a cotton swab moistened with broth about 1 inch into the nares, gently rotating the swab against the nasal mucosa on both sides, and returning it to the broth tube previously described.

ROUTINE PROCEDURES, INCLUDING ISOLATION METHODS FOR HEMOLYTIC STREPTOCOCCI

The throat or nasopharyngeal swab is used for the direct inoculation of blood agar plates, and several methods are available. Since the routine throat culture is taken chiefly to detect the presence of

group A **beta-hemolytic streptococci,** the latter's significance will be stressed in the following procedures:

1. If polyester-tipped* swabs are used (as is preferred), direct inoculation of a **moist sheep blood agar** plate (without added glucose, which inhibits characteristic hemolysis) is recommended. Roll the swab firmly over a small portion of the agar surface and streak by using a sterile wire loop. Spread the inoculum over the remainder of the plate to obtain well-isolated colonies. There are many techniques for spreading the inoculum; one is shown in Fig. 4-1. Make a few **stabs** into the agar with the loop for observation of **subsurface hemolysis** by strains producing beta-type hemolysis under reduced oxygen tension. This procedure may be used on either a freshly made culture or on one in which the polyester swab has remained in its original container for several days (must be refrigerated).

2. If an applicator immersed in broth has been used, twirl the swab vigorously against the side of the broth tube to obtain as much material in suspension as possible. It has been shown[8] that the flora of such a throat swab can be determined better by inoculating a broth suspension than by directly streaking the swab on a blood agar plate. Using a sterile loop, transfer a loopful of the suspension to a blood agar plate and streak as described previously.

3. If a **pour plate** is desired, the cotton-tipped applicator is placed in 2 ml of trypticase soy broth and the suspension prepared as previously described.

   (a) Transfer a loopful of this suspen-

---

*These are cut in 8-inch lengths, and a loop is made at one end to act as a handle. The tip is tightly bent over itself, dipped in collodion, and carefully covered with a tuft of fine eye cotton. An excellent substitute is the calcium alginate swab on a soft flexible aluminum wire, Falcon No. 2050, Oxnard, Calif.

---

*This fiber appears to prolong the survival of beta-hemolytic streptococci from throat swabs and such swabs may be mailed in an envelope to the laboratory without noticeable loss of streptococci.[18]

sion to a 20- by 150-mm screw-capped tube containing approximately 15 ml of melted and **cooled** (45 to 50 C) trypticase soy agar.

(b) Pipet 0.8 ml of sterile defibrinated sheep blood into the agar and mix the entire contents of the tube by inversion. The sheep blood may be conveniently stored in 1-ml amounts in small sterile tubes in a refrigerator, where it will remain stable for 7 to 10 days.

(c) Flame the lip of the tube and pour the agar into a sterile Petri dish; allow to cool and solidify.

Pour plates are useful in revealing certain strains of group A streptococci that may appear to be nonhemolytic when growing on the surface of blood agar plates because these strains produce streptolysin O only, which is inactivated by atmospheric oxygen. Most strains of group A streptococci, however, show beta-hemolysis around surface colonies due to their production of streptolysin S, which is oxygen stable.

All throat cultures are incubated at 35 C **aerobically**[21] and nasopharyngeal culture plates (for meningococci) are incubated in a candle jar (or 10% carbon dioxide incubator) for 18 to 24 hours. Some workers recommend incubation under anaerobic conditions (see Facklam, R. R.: CRC Crit. Rev. Clin. Lab. Sci. **6**:287-317, 1976).

Incorporation of sulfamethoxazole and trimethoprim in sheep blood agar enhanced recovery of group A and B streptococci from throat cultures (Gunn, B. A., et al.: J. Clin. Microbiol. **5**:650-655, 1977).

The results obtained from throat cultures on blood agar depend to an extent on the **type of blood** used in preparing the plates. For reasons described in Chapter 17, sheep blood agar plates are recommended for these cultures (horse blood is also acceptable); human blood, either fresh or outdated bank blood, should be used only when sheep blood is not available.

READING OF PLATES

Blood agar plates inoculated with throat cultures are sometimes confusing to the inexperienced bacteriologist. A knowledge of the **normal flora,** of the organisms that may be significant, and of the colony characteristics of both groups will aid in solving the problem. The organisms that may be encountered in the **normal** throat, in order of their predominance, are as follows:

Alpha-hemolytic streptococci
Normal throat species of *Neisseria*, including pigmented forms
Coagulase-negative staphylococci; occasionally *Staphylococcus aureus*
Haemophilus haemolyticus (may be confused with beta-hemolytic streptococci); *H. influenzae*
Pneumococci
Nonhemolytic (gamma) streptococci
Diphtheroid bacilli
Coliform bacilli (particularly after antimicrobial therapy)
Yeasts, including *Candida albicans*
Beta-hemolytic streptococci other than group A

Microorganisms that may be encountered in an **infected** throat are the following:

Beta-hemolytic streptococci of group A; occasionally groups B, C, and G
Corynebacterium diphtheriae
Meningococcus

Most throat cultures are taken for the detection of group A streptococci. Therefore, either the blood agar streak or pour plate should be carefully examined for the presence of colonies of beta-hemolytic streptococci after overnight incubation. A description of these colonies is detailed in Chapter 17.

Any colony suggestive of **beta-hemolytic streptococci** should be subcultured by stabbing with a straight needle and streaked on a sector of a sheep blood agar plate or inoculated to a tube of blood broth. This will usually yield a pure culture; if it does not, the process should be repeated. Further identification of these colonies may be carried out by use of the bacitracin disk sensitivity test and confirmed, preferably by the fluorescent-antibody (FA) procedure. These procedures are also described in Chapter 17.

BACTERIOLOGIC DIAGNOSIS OF PERTUSSIS

The diagnosis of **whooping cough** is confirmed by the isolation of *Bordetella pertussis* from the respiratory secretions. Nasopharyngeal swabs are used to obtain material for culture, and a special medium, Bordet-Gengou agar containing 15% to 20% sheep blood, is used for cultivating the organism.

Nasopharyngeal swabs are obtained in the following manner. While the child's head is immobilized, the special wire applicator (see discussion of throat and nasopharyngeal cultures elsewhere in this chapter) is carefully passed through the nasal aperture along the nasal floor until it touches the posterior pharyngeal wall. It is left there during several induced coughs before withdrawal, whereupon it is immediately streaked on the surface of the Bordet-Gengou medium, and the inoculum is spread over the plate with a sterile loop or spreader. Penicillin (0.5 unit per milliliter) may be incorporated in the medium at the time of preparation; this inhibits the growth of penicillin-sensitive organisms normally present in the nasopharynx, usually without affecting the growth of pertussis organisms. The plates are incubated aerobically at 35 C for 4 or 5 days and are examined with a hand lens for the presence of typical colonies resembling mercury droplets surrounded by a hemolytic zone. *B. pertussis* is identified by staining reaction, FA, and agglutination with specific antiserum, as indicated in Chapter 23.

The yield of positive isolations from clinical cases of pertussis varies from 20% to 98%. This variation is apparently related to the procedures used, the stage of the disease, and the prior administration of antibiotics. Using the FA technique, Whitaker and associates[27] were able to confirm the diagnosis in 94% of pertussis cases during the first week of illness. Hol-

werda and Eldering[15] employed the FA technique in direct staining of early growth from Bordet-Gengou plates* and were able thereby to identify *B. pertussis* 1 day earlier. The FA technique is an aid in diagnosis; however, it should always be used in conjunction with conventional culture procedures.

BACTERIOLOGIC DIAGNOSIS OF DIPHTHERIA

The diagnosis of **diphtheria** is a clinical problem, and the responsibility should rest primarily on the attending physician. It is unreasonable for a clinician to place the responsibility on the bacteriologist and expect a definitive diagnosis within 15 minutes based on the examination of a direct smear of the lesion. Although the appearance of diphtheria bacilli in stained smears from lesions and membranes is highly characteristic, these organisms cannot be identified by morphology alone, since other nonpathogenic diphtherialike bacilli may be indistinguishable from *Corynebacterium diphtheriae*. The chance of error in either direction is too great when diagnosis is based exclusively on microscopic examination; such a reading should be accepted only as **presumptive** evidence of infection. The identity and type of the organism in question is obtained **only** after isolating toxigenic *C. diphtheriae* from the patient. Determination of the presence of these organisms requires a careful cultural and microscopic study of several days' duration followed by the performance of a virulence test in suitable animals or by an in vitro test.

In primary isolation, *C. diphtheriae* is best cultivated on appropriate media, such as Loeffler medium (three parts animal serum and one part dextrose broth, coagulated as slants) or modified Pai medium, where growth occurs after 18 to 24 hours' incubation at 35 C. Differential

---

*This appears to be the chief value of the FA technique in diagnosing pertussis.

and selective media, such as cystine or chocolate-tellurite agar, also give a high proportion of recoveries and should be included in primary inoculation; however, the microscopic morphology of the diphtheria bacillus from chocolate-tellurite agar is not as distinctive as that from Loeffler or Pai medium. These media are described in Chapter 41.

Material received in the laboratory for diphtheria culture is usually sent on swabs taken from the **throat** and **nasopharynx.** These are promptly inoculated to the surface of a Loeffler or Pai slant and streaked on cystine-tellurite agar plates or modified Tinsdale's medium.[13] Since severe throat infections caused by beta-hemolytic streptococci may produce lesions simulating diphtheria, it is recommended that sheep blood agar plates be inoculated also and examined for group A streptococci, along with the Loeffler slant and the tellurite plate. After the media are inoculated, a smear can be prepared by rolling the swab on a slide, fixing it, and staining it with **methylene blue stain.** It is preferable to have a separate swab for the smear so the other swab may be left on the Loeffler slant after inoculation. The slide should be examined for the presence of irregularly stained gram-positive pleomorphic organisms suggestive of diphtheria bacilli (see below) and for the presence of the gram-negative spiral and fusiform organisms found in Vincent's infection, which also may resemble the diphtheritic lesion in the throat.

The inoculated media are incubated at 35 C. On a Loeffler slant that has been incubated for a short period (2 to 8 hours), *C. diphtheriae* will frequently outgrow other organisms present and will exhibit a characteristic morphology when stained with **methylene blue.** The bacilli are deeply stained and appear septated, with wedge-shaped ends and tapered points. The cells are often at angles to each other, as in the letters V and Y or Chinese letters, and do not palisade as is seen with some of the diphtheroid bacilli.

If organisms resembling diphtheria bacilli are found in smears from the Loeffler slant but not from the tellurite plate, the growth from the slant is suspended in 1 to 2 ml of sterile broth, and a loopful of this is streaked on a fresh tellurite plate to isolate the organism. All plates should be held and examined for 48 hours before being discarded as negative.

FA procedures for the detection of *C. diphtheriae* in known or suspected cases of clinical diphtheria have been used as a rapid presumptive diagnostic procedure. The conjugate used cross-reacts with toxigenic strains of *C. ulcerans* (rarely found in the United States) and does not differentiate between toxigenic and nontoxigenic strains of *C. diphtheriae.* It is recommended that the procedure be used in conjunction with conventional culturing procedures to achieve the greatest degree of sensitivity and specificity.

LABORATORY DIAGNOSIS OF VINCENT'S INFECTION (FUSOSPIROCHETAL DISEASE)

Vincent's gingivitis or stomatitis is a pseudomembranous or ulcerative infection of the gums, mouth, or pharynx. It had been thought to be caused by a gram-negative spirochete, *Borrelia vincentii,* and a straight or slightly curved, anaerobic gram-negative rod with sharply pointed ends, *Fusobacterium nucleatum,* but is undoubtedly a complex mixed infection involving also *F. necrophorum, B. melaninogenicus,* and anaerobic cocci. Experimental work indicates that *B. melaninogenicus* is probably of special importance. A reduction in local tissue resistance is generally a precursor of the infection.

The pseudomembrane present on tonsillar lesions in the anginal form of this infection has been mistaken for a similar lesion in diphtheria, and it is important that they be differentiated. Fusospirochetal organisms can be readily demonstrated by staining smears with **crystal violet** (that used in Gram stain) for 1 minute and looking for large numbers of the characteristic forms previously described, along with numerous pus cells. Since these organisms are often present in the normal mouth and gums, their presence in smears must be correlated with the clinical findings to be of significance. Cultures are of no diagnostic value and should not be done.

NASOPHARYNGEAL CULTURES FOR MENINGOCOCCI

Nasopharyngeal cultures are important in demonstrating the presence of *Neisseria meningitidis* in cases of suspected meningococcemia and meningococcal meningitis and in the detection of meningococcal carriers. The specimen may be collected through the mouth by passing a wire swab under and beyond the uvula to the posterior wall of the nasopharynx or through the nostril by passing the wire swab along the floor of the nasal passage to the posterior wall. In either case it is desirable to secure a bit of the mucus by gently twirling the swab while in place against the nasopharyngeal wall. It should be noted that *N. meningitidis* can be recovered from healthy people; carrier rates among population groups experiencing a meningitis outbreak may rise as high as 80%. Under these circumstances it is doubtful that a positive culture is significant. A **negative** culture from a patient convalescing from meningococcal meningitis who may be in contact with younger susceptible persons on returning home is perhaps more meaningful. *N. meningitidis* is extremely sensitive to cold, dehydration, unfavorable pH, or the inhibitory activity of other microflora; therefore, culture material should be inoculated as soon as possible.

Meningococci may be isolated on moist blood agar used in routine throat and nasopharyngeal cultures or modified Thayer-Martin medium (preferred), where they appear after 18 to 24 hours' incubation in a candle jar as small, gray, mucoid, nonhemolytic colonies of gram-negative diplococci.

Like the gonococcus, the meningococcus is **oxidase positive**; this provides a helpful test that can be used for detecting the presence of these colonies in a mixed culture. It must be noted, however, that the nonpathogenic neisseriae of the oropharynx also will give a positive oxidase test. To identify *N. meningitidis*, therefore, it is necessary to carry out a slide agglutination test with a polyvalent antiserum,* FA, or a capsular swelling (quellung) reaction test. Fermentation reactions with several carbohydrates are also essential for identification. Refer to Chapter 19 for further information.

THROAT CULTURES FOR SPECIES OF CANDIDA

The yeastlike organism *Candida albicans* may cause an intraoral infection called **thrush** in infants and neonates, diabetics, and individuals receiving antimicrobial or corticosteroid therapy. Material for Gram stain and culture is taken from the whitish, loosely adherent membrane attached to the inner cheek, palate, or other portions of the oral mucosa, using the technique described for throat cultures. On receipt in the laboratory, the swab is inoculated to slants of Sabouraud dextrose agar, with and without added antibiotics (see Chapter 34), and incubated both at room temperature and at 35 C. Primary growth of *Candida* occurs in 2 to 3 days as a yeastlike colony with a characteristic fermentative odor. Identification of *C. albicans* depends on the demonstration of germ tubes or typical chlamydospores on a special medium and other tests. The procedure is described in Chapter 34.

EPIGLOTTITIS

Epiglottitis is an uncommon but severe life-threatening infection of the epiglottis. The swelling of the epiglottis result-

---

*Difco Laboratories, Detroit.

ing from the infection may close off the airway, presenting an emergency that requires immediate tracheotomy or endotracheal intubation. It occurs primarily in infants or young children, but is seen in adults occasionally. The most common pathogen by far is *Haemophilus influenzae,* with group A streptococci, pneumococci, and staphylococci occasionally responsible. There are reports of isolated cases due to *H. parainfluenzae*[7] and to a beta-lactamase–producing ampicillin-resistant *H. paraphrophilus.*[16]

Blood cultures are frequently positive in epiglottitis, and cultures of the epiglottis will yield the offending organism.

SPUTUM CULTURES

Bacterial pneumonia, pulmonary tuberculosis, and chronic bronchitis constitute a most important group of human diseases. Since specific treatment frequently depends on a bacteriologic diagnosis, the prompt and accurate examination of a properly collected sputum specimen by smear, culture, and antimicrobial susceptibility testing is imperative. This is particularly true in pneumonia caused by *Klebsiella pneumoniae* and *Staphylococcus aureus,* which may be more rapidly fatal than that caused by *Streptococcus pneumoniae.*[1]

Furthermore, other bacterial species, including the gram-negative enteric bacilli, *Pseudomonas aeruginosa,** and *Haemophilus influenzae,* are being reported in hospital-associated pneumonia in the debilitated patient.[10] These also may cause pneumonia, complicating conditions such as hematologic disorders, virus infections, alcoholism, and pregnancy,[24,25] and only proper bacteriologic examination of the sputum will reveal their presence. Many other organisms of varying types (e.g., *N. meningitidis, Nocardia, Yersinia, Mycoplasma,* fungi, various

mycobacteria, various anaerobes, Legionnaires' disease agent, and viruses) may cause pneumonia, empyema, lung abscess, or other pulmonary infections.

Recovery of an etiologic agent from sputum or other appropriate specimen depends not only on the laboratory methods used but also on the **care taken** in securing the specimen. Too often, the culturing of unsuitable material results in misleading information for the clinician, because the true infecting agent is missed entirely or an incidental pathogen is identified. It should require no special courage on the part of the microbiologist to discard a specimen that is obviously saliva as being **unsatisfactory** for bacteriologic examination.

The collection of sputum for culture also requires the cooperation of the patient and should include instructions to obtain material from a deep cough (tracheobronchial sputum), which is expectorated directly into a sterile Petri dish or other satisfactory container.* The volume of the specimen need not be large; 1 to 3 ml of purulent or mucopurulent material will be sufficient for most examinations except mycobacteriologic culturing. On receipt in the laboratory the specimen should be examined without delay, especially if histoplasmosis is suspected. However, refrigeration of the specimen for not more than 1 to 3 hours is satisfactory for the recovery of most pathogens if immediate culturing is not feasible.

The organisms most frequently associated with **acute** bacterial infections of the lower respiratory tract are:

Pneumococcus *(Streptococcus pneumoniae)*
Klebsiella pneumoniae
Haemophilus influenzae
Staphylococcus aureus
Coliform bacilli, *Pseudomonas, Proteus* species
Mycoplasma pneumoniae

---

*Some of these organisms are most frequently involved in patients requiring mechanical ventilation, including reservoir nebulizers, tracheostomy, or endotracheal intubation.

*The use of superheated hypertonic saline aerosols for sputum induction is recommended when the cough is not productive. Proper decontamination of the equipment must be carried out when using this procedure.[11]

With the exception of *Mycobacterium tuberculosis*, it is generally possible to recover the important pulmonary pathogens from the upper respiratory tract of apparently healthy people. The recovery of these organisms, therefore, is not always sufficient evidence of their etiologic role in a particular infection. In acute bacterial pneumonia, however, the pathogen is generally present in large numbers in optimum specimens. Sputum culture is **never** satisfactory for diagnosis of anaerobic pulmonary infection; it should never be cultured anaerobically.

EXAMINATION OF GRAM-STAINED SMEARS

The first step in examination of a sputum Gram stain is to determine whether the specimen is likely to be reliable. The slide should be scanned at 100× magnification and the number of squamous epithelial cells and polymorphonuclear leukocytes noted. Murray and Washington[20] noted that the presence of more than 10 epithelial cells per 100× field correlated with significant oral contamination (on the basis of culture results compared with cultures of transtracheal aspirates); they felt such specimens should be rejected. Van Scoy,[26] also of the Mayo Clinic, reviewed their data and decided that cultures yielding viridans streptococci, S. *epidermidis*, *Neisseria* species, *Haemophilus* species, yeast, and *Corynebacterium* (as unlikely causes of pneumonia in adults) could be ignored and that then a more reliable criterion of a valid specimen was the presence of more than 25 leukocytes per high-power field. While *Haemophilus* and *Neisseria* are uncommon causes of pneumonia in adults and most of the other organisms rarely, if ever, cause pneumonia, viridans streptococci are not at all uncommon etiologic agents in aspiration pneumonia (often as part of a mixed flora also involving anaerobes). Additional studies, with careful clinical evaluation and anaerobic cultures (of **transtracheal aspirates**) are undoubtedly

needed. In the meantime, it may be reasonable to **reject specimens with both more than 10 epithelial cells and less than 25 leukocytes per high-power field.** In effect, this will usually be selecting on the basis of Van Scoy's recommendations.

It must be remembered that severely leukopenic patients (peripheral white blood cell count of 500/mm³ or less) may not be able to muster sufficient polymorphonuclear leukocytes to meet the above criteria for sputum. Exceptions must be made in such cases. In general, a microbiologist should never discard a specimen outright, except obvious saliva. Rather, problem specimens should be discussed with the clinician responsible for the patient.

An early presumptive diagnosis of bacterial pneumonia can often be made by examining a direct smear of sputum properly stained by the Gram method. The presence of many encapsulated, lancet-shaped cocci occurring singly, in pairs, or in short chains, along with pus cells, revealed by a capsule stain, may suggest a pneumococcal infection in typical cases. Likewise, the use of Omniserum for the quellung reaction (swelling of capsules on exposure to antiserum) increases the likelihood of accurate identification of pneumococci on direct observation considerably[9] (see Chapter 18). A stained preparation showing great numbers of gram-positive cocci in clusters is suggestive of staphylococcal pneumonia. Detection of numerous short, fat, encapsulated gram-negative rods suggests *Klebsiella* infection.

However, the **presumptive** aspect of these findings should be emphasized— only by the isolation of pneumococci, staphylococci, pseudomonads, or klebsiellae from cultures of sputum, blood, or pleural fluid can such an impression be confirmed.

CULTURE FOR PNEUMOCOCCUS

Since *Streptococcus pneumoniae* continues to be the major cause of bacterial pneumonia, the examination of a fresh sputum specimen is the first step in the diagnostic procedure.

Blood-tinged (rusty) mucopurulent sputum freshly obtained from patients with lobar pneumonia may be handled in the following manner:

1. Select purulent or bloody flecks and prepare direct smears as follows. Transfer a loopful to a clean slide, press another slide over it, squeeze the slides together, and then pull them apart. Flame both slides as soon as they are dry. Stain one of them by the Gram method and one by the acid-fast technique.

2. Streak a sheep blood agar plate* with a small loopful of a **selected** portion of the specimen (not saliva!). Place the blood plate in a candle jar†; incubate both the jar and the thioglycollate culture at 35 C for 18 to 24 hours.

3. If the sputum is unusually tenacious, transfer a mucopurulent mass to a Petri dish containing sterile physiologic saline and swish it around to remove the excess saliva. Transfer to a clean Petri dish, add a small volume of saline, and repeatedly aspirate and expel the specimen in a sterile syringe to emulsify it. **Exercise caution** to avoid aerosols. Use a loopful of this suspension to inoculate the media previously indicated.

4. If after incubation of the blood plate, small, shiny, greenish, transparent, flat colonies with depressed centers and raised edges are seen, they are most likely colonies of pneumococci. These may be cultured and further identified by determining their solubility in bile salts or their susceptibility to Optochin. These tests are described in Chapter 18.

5. Although pneumococci usually predominate in sputum cultures from patients with early lobar pneumonia, the blood plate may also show a large number of colonies of the usual throat flora, consisting of alpha-hemolytic streptococci, nonpathogenic neisseriae, and so forth. Before discarding such a plate as showing only normal throat flora, it is recommended that a search for pneumococcus colonies be made with a **hand magnifying lens** (3× to 8×). If these colonies are present, an attempt should be made to recover them in pure culture by fishing to sectors of a blood agar plate.

QUANTITATIVE SPUTUM CULTURE

Quantitative sputum culturing, by means of liquefying the specimen with a mucolytic agent, homogenizing it, and plating tenfold serial dilutions of it on appropriate solid culture media,[19] has had numerous proponents. The technique, however, is cumbersome and time consuming, requires a considerable number of plates, and appears to offer few advantages over conventional methods. N-acetyl cysteine should **not** be used as a liquefying agent, as it has antibacterial activity (Parry & Neu: J. Clin. Microbiol. **5:**58-61, 1977). A technique employing preliminary washing gives much more reliable information[4] and is worth considering in acutely ill patients in whom transtracheal aspiration is not possible.

TRANSTRACHEAL ASPIRATION IN PNEUMONIA

Since severe and sometimes fatal necrotizing pneumonia caused by gram-negative enteric bacilli and other pathogens

---

*The worker is well advised to streak serially two or more blood agar plates and a MacConkey or eosin–methylene blue (EMB) plate (if *Klebsiella* is suspected) to obtain isolated colonies when the bacterial population of the sample is unusually high. Dilworth et al.[9] found that use of sheep blood agar containing 5 μg per milliliter of gentamicin was distinctly better than either conventional blood agar or mouse inoculation.

†Austrian[2] has shown that a small percentage of pneumococci require 3% to 10% CO_2 for recovery on agar plates at the time of primary inoculation.

(including anaerobes) is occurring more frequently,[5,23] the reliability of a routine sputum culture (contaminated by mouth flora) has been questioned. In 1959 Pecora[22] proposed transtracheal aspiration, in which a small-gauge catheter is threaded into the trachea through a needle introduced at the cricothyroid membrane, as an alternative procedure for obtaining material from the lower respiratory tract, uncontaminated by oropharyngeal microorganisms. The technique has proved safe and the results bacteriologically reliable[6,14,17] and is recommended in patients with pneumonia who are unable to raise a satisfactory sputum specimen[12] and in suspected anaerobic pulmonary infection (lung abscess, aspiration pneumonia, bronchiectasis).[6] Results of gram-stained smears of tracheal aspirates correlate much better with their corresponding cultures than do smears of expectorated sputum from the same patient. The technique is **not** suitable for individuals with bleeding tendency or significant ventilation problems.

CULTURE FOR OTHER PATHOGENS

Sputum from acute or chronic pulmonary infections caused by agents other than pneumococci can vary from a thick tenacious specimen that reveals species of *Klebsiella* on culture to a foul-smelling purulent aspirate from a patient with a lung abscess, from which species of *Bacteroides* and other anaerobes may be recovered. As a rule, the isolation and identification of klebsiellae pose no problem, although serologic identification by means of capsular swelling test may require the aid of a reference laboratory.* The anaerobes are more difficult to recover and identify; they should be looked for especially in the foul-smelling† material from patients with aspiration pneumonia, lung abscess, bronchiectasis, and empyema. Transtracheal aspirates from these patients should be streaked on two **fresh** blood agar plates and inoculated to two tubes of enriched thioglycollate medium. One set of media is incubated in an **anaerobic jar** subsequently handled as described in Chapter 13; the other set is handled as described for pneumococcus isolation.

The examination of both stained and unstained wet smears prepared directly from the sputum may be informative, particularly when fungus elements can be demonstrated. Darkfield illumination may also be helpful in revealing the presence of spirilla and spirochetes. See Chapter 13 for a description of anaerobic culture procedures.

It is frequently difficult to evaluate the bacterial flora of **bronchial aspirates** because these specimens are usually contaminated with upper respiratory tract secretions. Even with special collection techniques, cultures of lower respiratory tract secretions obtained at bronchoscopy, including fiberoptic bronchoscopy, are apt to contain normal or colonizing flora from the mouth or upper respiratory tract.[3] Another problem is the use of large amounts of topical anesthetic, which exerts antibacterial effects.[3] Perhaps quantitative culture would be helpful. Cultures for pathogenic fungi and tubercle bacilli should be set up on all bronchoscopic samples.

The agent of Legionnaires' disease, a serious type of pneumonia, is difficult to recover. Most isolations have been by egg yolk sac or guinea pig inoculation. However, Mueller-Hinton agar supplemented with 0.01% soluble ferric pyrophosphate* and 0.05% L-cysteine hydrochloride at a pH of 6.9 to 7 is a satisfactory growth medium. Pleural fluid or transtracheal aspirates would be the best specimens to utilize. If this agent is suspected, a biologic safety hood should be used. Incuba-

---

*Some antisera are available from Difco Laboratories, Detroit.
†The foul odor is an important clue when present, but is noted in only about 50% of anaerobic pleuropulmonary infections; thus, absence of such odor does not exclude anaerobic infection.[5]

---

*Mallinckrodt Inc., St. Louis.

tion should be under 5% CO_2 for at least 1 week (growth may be noted in 3 to 4 days). The CDC should be notified of any suspect isolates.

The etiologic significance of the **pathogenic fungi** in pulmonary disease has been recognized increasingly in recent years. In the light of present knowledge, it becomes a necessity for the modern diagnostic microbiology laboratory to be able to isolate and at least recognize the agents of candidiasis, histoplasmosis, cryptococcosis, sporotrichosis, and the opportunistic fungi, such as *Aspergillus* and *Mucor* species. Because of the complexities of the techniques involved, a special section of medical mycology has been included in Chapter 34.

Since the introduction and widespread use of the broad-spectrum antibiotics, the laboratory rarely receives tonsillar or adenoidal tissue for culture. **Tissue specimens** must be minced with sterile scissors and ground with sterile alundum, using a tissue grinder made for this purpose (see Chapter 34). The resulting suspension is streaked on a sheep blood agar plate, and a loopful is inoculated to a tube of enriched thioglycollate medium. A blood agar pour plate, prepared from a broth dilution of the original suspension, is also recommended. The media are incubated and examined as described for throat cultures.

Special methods for anaerobes, fungi, and mycobacteria are described in their respective chapters.

REFERENCES

1. Austrian, R.: The role of the microbiology laboratory in the management of bacterial infections, Med. Clin. N. Am. **50**:1419-1432, 1966.
2. Austrian, R., and Collins, P.: Importance of carbon dioxide in the isolation of pneumococci, J. Bacteriol. **92**:1281-1284, 1966.
3. Bartlett, J. G., Alexander, J., Mayhew, J., Sullivan-Sigler, N., and Gorbach, S. L.: Should fiberoptic bronchoscopy aspirates be cultured? Am. Rev. Respir. Dis. **114**:73-78, 1976.
4. Bartlett, J. G., and Finegold, S. M.: Improved technique to process coughed sputum for culture, Clin. Research **19**:183, 1971.
5. Bartlett, J. G., and Finegold, S. M.: Anaerobic infections of the lung and pleural space, Am. Rev. Respir. Dis. **110**:56-77, 1974.
6. Bartlett, J. G., Rosenblatt, J. E., and Finegold, S. M.: Percutaneous transtracheal aspiration in the diagnosis of anaerobic pulmonary infection, Ann. Intern. Med. **79**:535-540, 1973.
7. Chow, A. W., Bushkell, L. L., Yoshikawa, T. T., and Guze, L. B.: Haemophilus parainfluenzae epiglottitis with meningitis and bacteremia in an adult, Am. J. Med. Sci. **267**:365-368, 1974.
8. Committee on Acute Respiratory Diseases: Problems in determining the bacterial flora of the pharynx, Proc. Soc. Exp. Biol. Med. **69**:45-52, 1948.
9. Dilworth, J. A., Stewart, P., Gwaltney, J. M., Jr., Hendley, J. O., and Sande, M. A.: Methods to improve detection of pneumococci in respiratory secretions, J. Clin. Microbiol. **2**:453-455, 1975.
10. Feingold, D. S.: Hospital-acquired infections, New Eng. J. Med. **283**:1384-1391, 1970.
11. French, M. L. V., Dunlop, S. G., and Wentzler, T. F.: Contamination of sputum induction equipment during patient usage, Appl. Microbiol. **21**:899-902, 1971.
12. Hahn, H. H., and Beaty, H. N.: Transtracheal aspiration in the evaluation of patients with pneumonia, Ann. Intern. Med. **72**:183-187, 1970.
13. Hermann, G. J., and Bickham, S. T.: *Corynebacterium*. In Lennette, E. H., Spaulding, E. H., and Truant, J. P. (editors): Manual of clinical microbiology, ed. 2, Washington, D.C., 1974, American Society for Microbiology.
14. Hoeprich, P. D.: Etiologic diagnosis of lower respiratory tract infections, Calif. Med. **112**:1-8, 1970.
15. Holwerda, J., and Eldering, G.: Culture and fluorescent-antibody methods in diagnosis of whooping cough, J. Bact. **86**:449-451, 1963.
16. Jones, R. N., Slepack, J., and Bigelow, J.: Ampicillin-resistant *Haemophilus paraphrophilus* laryngo-epiglottitis, J. Clin. Microbiol. **4**:405-407, 1976.
17. Kalinske, R. W., Parker, R. H., Brandt, D., and Hoeprich, P. D.: Diagnostic usefulness and safety of transtracheal aspiration, New Eng. J. Med. **276**:604-608, 1967.
18. Lattimer, A. D., Siegel, A. C., and De Celles, J.: Evaluation of the recovery of beta hemolytic streptococci from two mail-in methods, Am. J. Public Health **53**:1594-1602, 1963.
19. Louria, D. B.: Uses of quantitative analyses of bacterial populations in sputum, J.A.M.A. **182**:1082-1086, 1962.
20. Murray, P. R., and Washington, J. A. II: Microscopic and bacteriologic analysis of expectorated sputum, Mayo Clin. Proc. **50**:339-344, 1975.

21. Murray, P. R., Wold, A. D., Schreck, C. A., and Washington, J. A. II: Effects of selective media and atmosphere of incubation on the isolation of group A streptococci, J. Clin. Microbiol. **4:**54-56, 1976.

22. Pecora, D. V.: A method of securing uncontaminated tracheal secretions for bacterial examination, J. Thorac. Surg. **37:**653, 1959.

23. Pierce, A. K., Edmonson, B., McGee, G., Ketchersid, J., Loudon, R. G., and Sanford, J. P.: An analysis of factors predisposing to gram-negative bacillary necrotizing pneumonia, Amer. Rev. Resp. Dis. **94:**309-315, 1966.

24. Tillotson, J. R., and Lerner, A. M.: Pneumonia caused by *Escherichia coli* and *Pseudomonas,* Antimicrob. Agents Chemother. **6:**198-201, 1966.

25. Turck, M.: Current therapy of bacterial pneumonias, Med. Clin. North Am. **51:**541-548, 1967.

26. Van Scoy, R. E.: Bacterial sputum cultures: a clinician's viewpoint, Mayo Clin. Proc. **52:**39-41, 1977.

27. Whitaker, J. A., Donaldson, P., and Nelson, J. D.: Diagnosis of pertussis by the fluorescent-antibody method, New Eng. J. Med. **263:**850-851, 1960.

CHAPTER 9

Microorganisms encountered in the gastrointestinal tract

Members of the family Enterobacteriaceae make up a large part of the facultatively aerobic microflora of the human intestinal tract. These include the intestinal commensals (the coliforms and species of *Proteus*), as well as the enteric pathogens of the *Salmonella* and *Shigella* genera and related species, intestinal streptococci (enterococci and others), species of *Bacteroides* (the predominant microorganisms in the normal stool), anaerobic cocci, non-spore-forming anaerobic gram-positive bacilli, clostridia, various yeasts (including *Candida albicans*), and occasionally pathogenic staphylococci. The vibrio of cholera may be isolated from cases of this disease, and rarely it is necessary to attempt to isolate *Mycobacterium tuberculosis* from fecal material.

At least 65 species of virus have been recovered from the intestinal tract of humans, and the number is constantly increasing. These viruses consist of three main groups: the polioviruses, the Coxsackie viruses, and the echoviruses* (enteric cytopathogenic human orphan). All require special techniques for recovery.

COLLECTION AND TRANSPORT

1. A plastic or waxed container is sufficient if the specimen can be transported promptly to the laboratory.
2. If a delay of over 2 to 3 hours is expected, use 0.033 M phosphate buffer mixed with equal parts of glycerol or Amies, Cary-Blair, or Stuart transport medium.

---

*Although not considered enteroviruses, the viruses of infectious hepatitis are present in the feces.

3. If a delay of many hours is expected or if the specimen is to be sent by mail, use 0.033 M phosphate buffer with an equal part of glycerol.
4. When stool specimens are not readily obtainable in outbreaks of enteric disease, **rectal swabs** afford the most practical method of securing material for culture, although they should not be relied on for maximum recovery. Rectal swabs are of less value than fecal specimens in the examination of convalescent patients or in carrier surveys.

ISOLATION OF SALMONELLA AND SHIGELLA

Salmonella and *Shigella* are present in the stool in appreciable numbers only during the **acute stage** (first 3 days) of a diarrheal disease; therefore, fecal specimens should be obtained within this period whenever possible. Bits of bloody mucus or epithelium should be selected from the freshly voided specimen for inoculation to appropriate media. In **chronic** dysentery, however, the immediate culturing of material obtained during **proctoscopic** examination offers the best means of obtaining positive *Shigella* isolations. In the aforementioned situations it is well to remember that rapid proliferation of normal bacterial commensals occurs and that certain forms, particularly *Shigella*, may die off rather rapidly after the specimen is collected. These samples, therefore, should be **inoculated to appropriate culture media soon after collection.** Whenever possible, **multiple** stool specimens should be examined; numerous investigations have demonstrated the value of this procedure.[7]

A great variety of culture media have been devised for the isolation of salmonellae and shigellae from fecal specimens. These may be divided into several general groups (discussed in more detail in Chapters 20 and 41) without any sharp division among groups.

1. **Differential mildly selective media,** such as eosin–methylene blue (EMB) agar, MacConkey agar, or Leifson desoxycholate agar, contain certain carbohydrates, indicators, and chemicals that are inhibitory to many gram-positive bacteria. Differentiation of enteric bacteria is achieved through the incorporation of lactose (and sucrose in some brands of EMB agar), since the organisms that attack lactose will form colored colonies, whereas those that do not ferment lactose will appear as colorless colonies. Among the latter are found the salmonellae and shigellae, as discussed in Chapter 20.

2. **Differentially moderately selective media,** such as Salmonella-Shigella (SS) agar, xylose lysine desoxycholate (XLD) agar, Hektoen enteric (HE) agar, and desoxycholate citrate agar, and highly selective media, such as Kauffmann brilliant green agar and Wilson-Blair bismuth sulfite agar, are complex combinations of nutrients and chemicals that serve to inhibit many coliforms, especially strains of *Escherichia* and *Proteus,* but permit the growth of most *Salmonella* and many *Shigella* organisms isolated in the United States. However, not all *Shigella* will do well on SS agar, and these organisms are completely inhibited on brilliant green agar and bismuth sulfite agar; therefore, if one anticipates *Shigella,* less inhibitory media, such as MacConkey agar, should also be used.[9] These pathogens, along with slow or lactose-nonfermenting organisms, generally form colorless colonies, although some may appear as black or greenish colonies on certain media (see Chapter 20). The lac-

tose-fermenting organisms that are not inhibited grow as pink, orange, or red colonies on media other than bismuth sulfite agar. Coliforms, if they develop, may appear as small black, brown, or greenish colonies on bismuth sulfite agar.

3. **Selective enrichment media,** such as Leifson selenite broth, Mueller tetrathionate broth, or Hajna GN broth, incorporate nutrients and such chemicals as selenium salts, tetrathionate (by oxidation of thiosulfate through addition of iodine just prior to use), or desoxycholate and citrate. Such agents inhibit the growth of gram-positive organisms and temporarily (12 to 18 hours) limit the growth of coliforms and *Proteus,* while encouraging the multiplication of salmonellae and shigellae.

4. **Highly selective media.** It is recognized that selective plating media used for isolating enteric pathogens, by virtue of their inhibitory action on coliforms, may also inhibit less hardy pathogens. Thus, whereas media containing brilliant green, bismuth sulfite or bile salts and citrate may be excellent for isolation of salmonellae, their value may be limited in the recovery of some *Shigella* species.

XLD agar,* as modified by Taylor[11] is recommended for general application, but with particular reference to *Shigella* and *Providencia.* The differentiation of the various species is based on xylose fermentation, lysine decarboxylation, and hydrogen sulfide production. In a comparison of the recovery efficiency of various enrichment broths and plating media in a series of more than 1,000 fecal specimens, Taylor and Schelhart[12] concluded that the most successful combination utilized GN or selenite-F enrichment medium, followed by plating on XLD agar. A study by Rollender and associates[10] indicated a definite superiority of XLD agar over MacConkey agar in the recovery of salmonellae and shigellae by

---

*Difco Laboratories, Detroit; Baltimore Biological Laboratory, Cockeysville, Md.

direct inoculation of stools and urine specimens. On the other hand, Dunn and Martin (Appl. Microbiol. **22**:17-22, 1971) found, in their comparison of media for isolation of salmonellae and shigellae from fecal specimens, that although the shigellae were best isolated by direct inoculation, the salmonellae were isolated in greater numbers after tetrathionate (without brilliant green) enrichment with subsequent culturing to plating medium. These authors recommend the use of a variety of plating media to enhance recovery of a larger number of enteric pathogens. All of the enrichment broths should be subcultured to differential and selective plating media to demonstrate colonies of enteric pathogens.

From the foregoing discussion it should be most apparent that **no single medium can be used for all purposes;** a variety of plating and enrichment media results in a higher number of positive isolations than will be obtained when only one medium is used. The following procedures are recommended for culturing stool and rectal specimens for the presence of *Salmonella* and *Shigella*[1]:

1. Using a cotton-tipped swab, **heavily** inoculate one or more of the following media: SS agar, Hektoen enteric agar (HE), brilliant green agar, XLD agar, and desoxycholate citrate agar with fresh material. Streak with the swab or a wire spreader to provide good distribution of isolated colonies over a major portion of each plate (see Chapter 4). A suspension of fecal material in enrichment medium also may be used to inoculate these plates (see step 4).
2. At the same time **lightly** inoculate one or more of the following differential media: EMB, MacConkey agar, or desoxycholate agar plate and streak as in step 1.
3. Since bismuth sulfite agar is the best medium at present for the isolation of typhoid bacilli, its use is recommended whenever typhoid fever or the carrier state is suspected. **Heavily** streak a bismuth sulfite agar plate as described here. In addition, make two pour plates, using about 0.1 ml and 5 ml of a heavy fecal suspension in broth mixed with the melted and cooled (45 C) agar medium.
4. Also inoculate **heavily** a tube of selenite enrichment or GN broth with the fecal specimen (usually one part to ten parts). After overnight incubation, and depending on the enrichment broth(s) used, transfer two or three loopfuls from these broth media as follows: selenite to EMB or MacConkey, and GN to XLD plate.
5. Incubate all media at 35 C for 18 to 24 hours and for 48 hours if indicated (bismuth sulfite plates may require a longer period). Examine the plates in a good light for suspicious colonies.

Salmonellae and shigellae produce typical **colorless** colonies on these media, except on XLD agar, where they appear as **red** colonies, sometimes with black centers. On HE agar, however, coliforms appear as salmon to orange in color, whereas salmonellae and shigellae appear bluish green. *Salmonella typhi* forms **black,** opaque colonies on bismuth sulfite agar. It should be kept in mind that atypical colonies may appear, especially on selective media. On such plates great care must be exercised in picking colonies, since microscopic growth of other inhibited bacteria may be present on or near the ones selected, resulting in mixed cultures. Refer to Chapter 20 and discussion of the Enterobacteriaceae for further identification procedures.

S. typhi may be recovered from the duodenum during acute typhoid fever, using a string capsule device.[2]

ENTEROPATHOGENIC AND ENTEROTOXIGENIC ESCHERICHIA COLI

It is now widely accepted that certain toxin-producing or invasive *Escherichia coli* are responsible for outbreaks and sporadic cases of newborn or infantile diarrhea and diarrhea, including "traveler's diarrhea," in adults. Although the value of serotyping of enteropathogenic *E. coli* has been questioned, the fact that other pathogenic mechanisms are also involved and the disagreement between experts suggest that serotyping should not be totally abandoned at this time.

Specimens should be collected early in the course of the illness and before any antibiotics have been administered. Cultures may be obtained from freshly soiled diapers or by means of a rectal swab and must be cultured without delay. Since enteropathogenic or enterotoxigenic *E. coli* organisms are inhibited by the selective media used for the isolation of salmonellae and shigellae (SS agar, desoxycholate citrate agar, and so forth), it is necessary to utilize the less inhibitory media, such as EMB or MacConkey agar, to recover them. Blood agar plates also should be inoculated because pure cultures are easier to obtain on this medium. It may be necessary to inoculate other media, however, so that the presence of salmonellae and shigellae is not overlooked. Ten colonies, including all different morphotypes, are picked for serotyping and for toxin studies.

ISOLATION OF PATHOGENIC STAPHYLOCOCCI

It has been well demonstrated that the administration of broad-spectrum antibiotics to patients may result in a disturbance of the normal microflora of the lower intestinal tract. This may result in the development of large numbers of coagulase-positive staphylococci in this area; if these happen to be enterotoxin producers, they may induce a syndrome characterized by fever, abdominal pain, and a fulminating diarrhea (staphylococcal pseudomembranous enterocolitis).

Through the use of a selective medium containing a high concentration of sodium chloride, such as staphylococcus medium 110,* mannitol salt agar,* or other inhibitory agents, such as polymyxin staphylococcus medium, pathogenic staphylococci can be isolated from the stool. By using this technique the carrier rate of coagulase-positive staphylococci found in normal adults may be shown to be as high as 15% to 20%.[4] This figure may be higher in infants or in patients hospitalized for more than several days or in those who have been given antibiotics. A diagnosis of staphylococcal enteritis, based only on the demonstration of coagulase-positive staphylococci in the stool, therefore, is questionable in light of these findings.

On the other hand, our own experience parallels that of Hinton and associates.[4] When coagulase-positive staphylococci are isolated in **large numbers** from a stool specimen by employing a medium such as phenylethanol blood agar, and a **Gram stain** of a direct smear of the fecal specimen reveals **large numbers** of gram-positive cocci in clusters, this is likely to be of considerable clinical significance.

ISOLATION OF VIBRIO CHOLERAE AND V. PARAHAEMOLYTICUS

Asiatic cholera is confined chiefly to India, Pakistan, Burma, Nepal, and China, but there was considerable global spread between 1961 and 1975. Infection persisting in small numbers of human cases (apparently there is no chronic carrier state) between epidemics keep the disease smoldering in these areas. This, then, spreads geographically (generally westward), producing yearly epidemics. Only infrequently is a laboratory in the continental United States called on to attempt isolation and culture of this microorganism.

Cholera vibrios are present in enormous numbers in the characteristic **rice water stool** produced by the profuse diarrhea in the early stages of the disease. At

---

*Difco Laboratories, Detroit; Baltimore Biological Laboratory, Cockeysville, Md.

times the vibrios may be demonstrated in direct smears prepared from flecks of mucus from the stool, but prompt culture of the specimen on the nonselective taurocholate gelatin agar or the selective thiosulfate citrate bile sucrose (TCBS) agar yields good results. After overnight incubation, colonies of *Vibrio cholerae* appear small, smooth, and translucent, distinct from the rough coliform colonies. SS and EMB media inhibit the growth of the organism. *Vibrio parahaemolyticus*, recovered from several outbreaks of enteritis in the United States, is also recovered effectively on TCBS.[8] *V. cholerae* and *V. parahaemolyticus* are discussed in more detail in Chapter 21. Sucrose teepol tellurite (STT) agar has recently been recommended as more effective than TCBS (Chatterjee, B. D., et al.: J. Infect. Dis. **135**:654-657, 1977).

ISOLATION OF YERSINIA ENTEROCOLITICA

Yersinia enterocolitica causes gastroenteritis, among other types of illness. It is difficult to isolate from feces because it grows slowly at 35 C; it may be mistaken for other Enterobacteriaceae. Cold-temperature enrichment may be very useful in recovery of this organism.[3]

ISOLATION OF MYCOBACTERIUM TUBERCULOSIS FROM THE STOOL

The examination of fecal specimens for *Mycobacterium tuberculosis* is not routine and is of questionable value when carried out. Tubercle bacilli isolated from intestinal contents probably reflect the presence of pulmonary tuberculosis, because sputum is often swallowed. The method recommended is described by Kubica and Dye[6] and is outlined in Chapter 31.

CANDIDA ALBICANS IN THE STOOL

Although the yeastlike fungus *Candida albicans* (and other species of *Candida*) can be found normally on the skin, in the mouth, and in the intestinal tract, intensive use of antibiotics may lead to overgrowth by *Candida* in the colon. It is rare, however, that this is of clinical significance.

REFERENCES

1. Edwards, P. R., and Ewing, W. H.: Identification of enterobacteriaceae, ed. 3, Minneapolis, 1972, Burgess Publishing Co.
2. Gilman, R. H., and Hornick, R. B.: Duodenal isolation of *Salmonella typhi* by string capsule in acute typhoid fever, J. Clin. Microbiol. **3**:456-457, 1976.
3. Greenwood, J. R., Flanigan, S. M., Pickett, M. J., and Martin, W. J.: Clinical isolation of *Yersinia enterocolitica*: cold temperature enrichment, J. Clin. Microbiol. **2**:559-560, 1975.
4. Hinton, N. A., Taggart, J. G., and Orr, J. H.: The significance of the isolation of coagulase-positive staphylococci from stool, Amer. J. Clin. Path. **33**:505-510, 1960.
5. Kozinn, P. J., and Taschdjian, C. L.: Enteric candidiasis, Pediatrics **30**:71-85, 1962.
6. Kubica, G. P., and Dye, W. E.: Laboratory methods for clinical and public health mycobacteriology, Public Health Service Pub. No. 1547, Washington, D.C., 1967, U.S. Government Printing Office.
7. McCall, C. E., Martin, W. T., and Boring, J. R.: Efficiency of cultures of rectal swabs and fecal specimens in detecting *Salmonella* carriers; correlation with numbers of salmonellas isolated, J. Hyg. **64**:261-269, 1966.
8. McCormack, W. M., DeWitt, W. E., Bailey, P. E., Morris, G. K., Soeharjono, P., and Gangarosa, E. J.: Evaluation of thiosulfate–citrate–bile salts–sucrose agar, a selective medium for the isolation of *Vibrio cholerae* and other pathogenic vibrios, J. Infect. Dis. **129**:497-500, 1974.
9. Rahaman, M. M., Huq, I., and Dey, C. R.: Superiority of MacConkey's agar over Salmonella-Shigella agar for isolation of *Shigella dysenteriae* Type 1, J. Infect. Dis. **131**:700-703, 1975.
10. Rollender, W., Beckford, O., Kostroff, B., and Belsky, R.: Comparison of xylose lysine deoxycholate agar and MacConkey agar for the isolation of salmonella and shigella from clinical specimens, paper M 238, Bacteriol. Proc., p. 106, 1968.
11. Taylor, W. I.: Isolation of shigellae; xylose lysine agars; new media for isolation of enteric pathogens, Am. J. Clin. Pathol. **44**:471-475, 1965.
12. Taylor, W. I., and Schelhart, D.: Isolation of *Shigella*. VI. Performance of media with stool specimens, paper M 242, Bacteriol. Proc., p. 107, 1968.

CHAPTER 10

Microorganisms encountered in the urinary tract

Bacterial infections of the urinary tract affect patients of all age groups and both sexes, and they vary in severity from an unsuspected infection to a condition of severe systemic disease. The clinical diagnosis of pyelonephritis is frequently overlooked because of the absence of urinary tract symptoms and pyuria.[8] The correlation of bacteriuria with unsuspected active pyelonephritis at autopsy has been demonstrated.[14] The role of the indwelling catheter in the development of bacteriuria, frequently accompanied by a gram-negative rod bacteremia, has also been revealed.[6] The demonstration of bacteria by appropriate cultural methods is the only reliable means of making a specific diagnosis. Readers interested in reviews of clinical and other aspects of urinary tract infection are referred to the excellent monographs by Kunin[11] and Stamey[23]; Barry and co-workers present a very good summary of laboratory diagnosis of urinary tract infection.[1]

Urine is an excellent culture medium for the common pathogens of the urinary tract, and when bacteria are deposited in the urine, they multiply readily, often exceeding 1 million per milliliter. Since specimens of urine, either clean voided or catheterized, are frequently contaminated on collection, the recovery of organisms, even known pathogens, does not necessarily establish the diagnosis of a urinary tract infection. Studies by Kass, Sanford, MacDonald, Beeson, and others, summarized in an extensive bibliography and review of this topic,[19] indicate that bacterial counts on fresh, voided urine from **infected** patients ordinarily show more than 100,000 (10[5]) organisms per milliliter,* whereas specimens from noninfected or normal persons may be sterile or contain up to 1,000 organisms per milliliter. It should be pointed out, however, that bacterial counts of less than 10[5] per milliliter may occur in patients who are receiving specific antibacterial therapy or in patients who are excessively hydrated, with a consequent dilution of urine.[2] Other factors accounting for low counts despite true infection include fastidious organisms with a slow growth rate (rare) and infection above an obstruction (such as a ureteral stone). To evaluate the clinical significance of a **positive** urine culture, therefore, it is strongly recommended that some means of estimating the number of organisms present in a specimen be part of a routine urine culture.

As will be noted from the following, the bacterial flora of **normal** voided urine differs from that of infected specimens. The organisms listed in the first group are most frequently encountered in normal urine. These, of course, represent contaminants from the urethra, vagina, and so forth, as bladder urine is normally sterile.

Staphylococci, coagulase negative
Diphtheroid bacilli
Coliform bacilli
Enterococci
Proteus species
Lactobacilli
Alpha-hemolytic and beta-hemolytic streptococci
Saprophytic yeasts
Bacillus species

---

*This figure is based on first voided urine in the morning; specimens collected later may give lower counts, due to hydration and to insufficient "incubation time" in the bladder. First morning specimens will have had overnight incubation in the bladder.

The flora of **infected** urine, as a rule, will include one or more of the following:

Escherichia coli, Klebsiella, Enterobacter, Serratia

Proteus mirabilis and other Proteus species; Providencia species

Pseudomonas aeruginosa and other Pseudomonas species

Enterococci (Streptococcus faecalis)

Staphylococci, coagulase positive and coagulase negative

Alcaligenes species

Acinetobacter species

Candida albicans, Torulopsis glabrata, other yeasts

Haemophilus species (probably Haemophilus vaginalis)

Mycobacterium tuberculosis, other mycobacteria

Beta-hemolytic streptococci, usually groups B and D

Neisseria gonorrhoeae

Salmonella and Shigella species

COLLECTION OF SPECIMENS

The distal portion of the human urethra and the perineum are normally colonized with bacteria, particularly in the female. These organisms will readily contaminate a normally sterile urine on voiding, but contamination can be prevented by using the "clean catch" technique. In this, the periurethral area (tip of penis, labial folds, vulva) is carefully cleansed with two separate washes with plain soap and water or a mild detergent and **well rinsed** with warmed sterile water to remove the detergent, with the glans penis or labial folds retracted. The urethra is then "flushed" by passage of the first portion of the voiding, which is discarded. The subsequent midstream urine, voided directly into a sterile container, is used for culturing and colony counting.

It is recommended that two successive clean-voided midstream specimens be collected to approach a 95% confidence level when using a bacterial count of 10^5 per milliliter as an index of significant bacteriuria,[10] although this is not routinely done in clinical practice unless there is some question about the diagnosis.

To increase the accuracy of localizing urinary tract infections, Stamey and col-leagues[24] recommend the use of suprapubic needle aspiration of the female bladder and a technique of dividing voided urine from the male into urethral, midstream (bladder), and postprostatic massage cultures. In the former procedure, both urethral and vaginal contamination is eliminated; in the latter, localization of a lower urinary tract infection to the urethra or prostate gland is facilitated. Another method pertinent to the microbiologist for localizing urinary tract infection is the use of the direct immunofluorescent method to detect bacteria in urine that are coated with antibody. Early studies indicated a very significant correlation between such antibody coating and infection involving the upper urinary tract. However, a recent study indicates that the association may be less specific than originally thought.[21]

Since urine generally supports the growth of most of the urinary pathogens as well as broth media does, it is absolutely essential for culture purposes that urine be processed within an hour of collection or **stored in a refrigerator** at 4 C until it can be cultured. Studies have indicated that such specimens may be kept in the refrigerator for extended periods without significantly reducing their bacterial content,[18] and counts remain stable for at least 24 hours at the refrigerator temperature. Specimens received at various times during a laboratory work day may be placed in a refrigerator as received, then set up in a batch at some designated hour in the afternoon.

CULTURE PROCEDURES

Although anaerobes are involved in urinary tract infection on occasion, voided urine should **never** be cultured anaerobically because of the presence of anaerobes in indigenous flora of the urethra and surrounding areas. In suspected anaerobic urinary tract infection, urine must be obtained by percutaneous suprapubic bladder aspiration or from a nephrostomy tube if one is in place.

A common practice in the past in a number of laboratories has been to centrifuge urine specimens and inoculate either liquid or solid media with the sediment or to inoculate the urine directly into a tube of broth. This should never be done. Centrifugation will concentrate contaminants as well as pathogens; the introduction of only a few contaminating bacteria by directly inoculating broth will invariably result in a positive culture. For these reasons it is strongly recommended that one of the following **quantitative** methods be used for culturing urine specimens.

Calibrated loop-direct streak method

The calibrated loop–direct streak method, first reported by Hoeprich[7] in 1960, uses a calibrated bacteriologic loop commonly employed in dairy bacteriology to inoculate and streak plates of standard or differential culture media. After proper incubation the number of colonies present is estimated and reported as a measure of the degree of bacteriuria. Used properly, this technique is reliable for clinical purposes and is simple and rapid. The delivery volume of calibrated loops may change with use; this can be checked by a procedure outlined by Barry and associates.[1]

1. Using a flame-sterilized and cooled platinum loop* calibrated to deliver 0.001 ml, hold the loop **vertically** and immerse just below the surface of the well-mixed urine specimen. Deliver one loopful to (a) a blood agar plate and (b) an EMB or a MacConkey agar plate.
2. Streak both plates by making a straight line down the center of the plate and then streaking for isolation with a regular loop by a series of very close passes at a 90° angle through the original line (Mayo Clinic technique).

---

*Platinum inoculating loop, Cat. No. 7011-J20, Arthur H. Thomas Co., Philadelphia, or Cat. No. N2075-2, Scientific Products, Evanston, Ill.

3. Incubate both plates overnight at 35 C and read the following morning. Examine and count the colonies on both plates; estimate the total count from the blood plate and gram-negative bacterial count from the EMB or MacConkey plate. In each case the colony count is multiplied by 1,000 (0.001 ml used) to give an estimate of the number per milliliter of urine. One hundred colonies would represent 10^5 cells present, for example. If a 0.01-ml loop is used, multiply colony count by 100.

After determining the count, proceed with identification of the organisms present* and determine their susceptibility to antimicrobial agents, as described in subsequent sections.

4. If no growth occurs in 24 hours, hold the plates for another day and, if still negative, report as **"No growth after 48 hours."**

Pour plate method

Although the pour plate method has been criticized as not entirely reflecting the true bacterial count of a urine specimen (clumps of organisms can give rise to single colonies),[20] it remains the most accurate procedure for measuring the degree of bacteriuria. The procedure is as follows:

1. Prepare three tenfold dilutions of well-mixed urine in sterile distilled water in sterile screw-capped tubes as follows:

   1:10 (10^{-1}) = 1.0 ml urine + 9.0 ml water
   1:100 (10^{-2}) = 1.0 ml of 1:10 dilution + 9.0 ml water
   1:1000 (10^{-3}) = 1.0 ml of 1:100 dilution + 9.0 ml water

---

*Some laboratories consider that a urine culture yielding three or more isolates should be considered a contaminated specimen and therefore do not identify the isolates. This is not necessarily true. Such polymicrobic bacteriuria is not uncommon in individuals with **indwelling catheters;** indeed, such bacteriuria may be of special importance in that it may show a significant association with bacteremia.[5]

Use separate 1-ml serologic pipets for each dilution, with adequate shaking between dilutions.

2. Pipet 1 ml of each of the dilutions into appropriately labeled, sterile Petri dishes.

3. Overlay each dilution with 15 to 20 ml of melted and cooled (50 C) nutrient or infusion agar; mix well by carefully swirling the dishes.

4. When solidified, invert the dishes and incubate overnight at 35 C.

5. Using a Quebec colony counter, enumerate the number of colonies in the plate yielding 30 to 300 colonies; multiply by the dilution to obtain the total number of bacteria per milliliter of urine.

Other methods

Attention has been given to chemical tests for the rapid detection of bacteriuria. These include the reduction of triphenyltetrazolium chloride (TTC) by metabolizing bacteria[22]; the Griess nitrite test, based on the rapid reduction of nitrate by members of the Enterobacteriaceae and staphylococci; the glucose oxidase test, dependent on metabolism of the small amount of glucose present in normal urine by bacteria[12]; and the catalase test,[17] in which rapid gas production from urine reacting with hydrogen peroxide indicates bacteriuria. These tests, however, have been considered unsatisfactory due to substantial numbers of false-negative results and sometimes lack of specificity as well. Commercial test kits based on these principles likewise have shown correspondingly equivocal results.

A rapid (2 to 4 hr) impedance-based screening test for bacteriuria has been described by Zafari and Martin (J. Clin. Microbiol. **5:**545-547, 1977).

A number of **screening culture** tests for bacteriuria have been proposed. In one such method (dip-slide,[3,4] tube,[16] spoon,[15] paddle, pipet, or cylinder[12]), an agarcoated vehicle is dipped into a freshly voided midstream urine specimen; after overnight incubation, the colonial growth is counted or compared with density photographs of known colony counts and the significance of the culture is determined.* A similar procedure utilizes a filter paper strip[13] that is dipped into the urine specimen, then transferred to the surface of either a miniature or a conventional agar plate. After incubation, the number of colonies growing in the inoculum area are counted and their significance is evaluated. Multiple specimens can be cultured on a single conventional agar plate.

These methods—chemical or dip-culture—are not appropriate for a clinical microbiology laboratory and have their principal value as a screening procedure for bacteriuria in a doctor's office or for field surveys. A duplicate urine specimen held in the refrigerator can be submitted for conventional culture and sensitivity testing if the screening test proves to be positive.

Interpretation of colony counts

Generally, counts of less than 10^3 are suggestive of contamination, whereas urine containing between 10^3 and 10^5 organisms suggests possible infection (and may be an indication for repeat culture), and counts of 10^5 and greater are indicative of infection, although no single concentration of bacteria distinguishes infected from contaminated specimens. Counts in infected patients may be low when the rate of urine flow is rapid, or the patient is receiving suppressive therapy, or occasionally when the urine pH is less than 5 and the specific gravity of the urine is less than 1.003.[20]

---

*Commercial kits utilizing these techniques may be obtained from Smith, Kline & French Laboratories, Philadelphia; Flow Laboratories, Rockville, Md.; Abbott Laboratories, South Pasadena, Calif.; Ayerst Laboratories, New York; BBL, Cockeysville, Md.; and others.

Examination of direct smears

The examination of a direct smear of fresh **uncentrifuged** urine has long been used as a screening device to distinguish between true bacteriuria and contamination.[9] It is done by making a smear of the well-mixed urine, allowing the smear to air dry, fixing it with heat, and performing a Gram stain. A positive smear is one showing **one or more bacteria** in the majority of oil-immersion fields.[10]

If 10^5 organisms represent a significant colony count, the correlation of the direct smear has been variously reported as 75% to 95%.[7,9,18,20] If lower counts are compared with the direct smear (counts between 10^4 and 10^5), the accuracy falls to around 50%.[18]

REFERENCES

1. Barry, A. L., Smith, P. B., and Turck, M. In Gavan, T. L., editor: Laboratory diagnosis of urinary tract infections, Cumitech 2, Washington, D.C., 1975, American Society for Microbiology.
2. Clapp, M. P., and Grossman, A.: The quantitative evaluation of bacteriuria and pyuria, Amer. J. Med. Sci. **248:**158-163, 1964.
3. Cohen, S. N., and Kass, E. H.: A simple method for quantitative urine culture, N. Engl. J. Med. **277:**176-180, 1969.
4. Grettman, D., and Naylor, G. R. E.: Dip-slide aid to quantitative urine culture in general practice, Br. Med. J. **3:**343-345, 1967.
5. Gross, P. A., Flower, M., and Barden, G.: Polymicrobic bacteriuria: significant association with bacteremia, J. Clin. Microbiol. **3:**246-250, 1976.
6. Hodgin, U. G., and Sanford, J. P.: Gram-negative rod bacteremia, Am. J. Med. **39:**952-960, 1965.
7. Hoeprich, P. D.: Culture of the urine, J. Lab. Clin. Med. **56:**899-907, 1960.
8. Kaitz, A. L., and Williams, E. J.: Bacteriuria and urinary-tract infection in hospitalized patients, N. Engl. J. Med. **262:**425-430, 1960.
9. Kass, E. H.: Asymptomatic infections of the urinary tract, Trans. Assoc. Am. Physicians **69:** 56-64, 1956.
10. Kass, E. H.: Pyelonephritis and bacteriuria: a major problem in preventive medicine, Ann. Intern. Med. **56:**46-53, 1962.
11. Kunin, C. M.: Detection, prevention and management of urinary tract infections, ed. 2, Philadelphia, 1974, Lea & Febiger.
12. Kunin, C. M.: New methods in detecting urinary tract infections, Urol. Clin. North Am. **2:** 423-432, 1975.
13. Leigh, D. A., and Williams, J. D.: Method for the detection of significant bacteriuria in large groups of patients, J. Clin. Pathol. **17:**498-503, 1964.
14. MacDonald, R. A., Levitin, H., Mallory, G. K., and Kass, E. H.: Relation between pyelonephritis and bacterial counts in urine; autopsy study, N. Engl. J. Med. **256:**915-922, 1957.
15. Mackay, J. P., and Sandys, G. H.: Laboratory diagnosis of infections of urinary tract in general practice by means of dip-inoculation transport medium, Br. Med. J. **2:**1286-1288, 1965.
16. Mackay-Scollay, E. M.: A simple quantitative and qualitative microbiological screening test for bacteriuria, J. Clin. Pathol. **22:**651-653, 1969.
17. Montgomerie, J. Z., Kalmanson, G. M., and Guze, L. B.: Use of catalase test to detect significant bacteriuria, Am. J. Med. Sci. **251:**184-187, 1966.
18. Mou, T. W., and Feldman, H. A.: The enumeration and preservation of bacteria in urine, Am. J. Clin. Pathol. **35:**572-575, 1961.
19. Prother, G. C., and Sears, B. R.: In defense of the urethral catheter, J. Urol. **83:**337-344, 1960.
20. Pryles, C. V.: The diagnosis of urinary tract infection, Pediatrics **26:**441-451, 1960.
21. Rumans, L. W., and Vosti, K. L.: Clinical syndrome vs. urinary antibody-coated bacteria in unselected populations, Clin. Research **25:** 157A, 1977.
22. Simmons, N. A., and Williams, J. D.: A simple test for significant bacteriuria, Lancet **1:**1377-1378, 1962.
23. Stamey, T. A.: Urinary infections, Baltimore, 1972, The Williams & Wilkins Co.
24. Stamey, T. A., Gavan, D. E., and Palmer, J. M.: The localization and treatment of urinary tract infections; the role of bactericidal urine levels as opposed to serum levels, Medicine **44:**1-36, 1965.

CHAPTER 11

Microorganisms encountered in the genital tract

The microbial flora of the normal human genital tract consists chiefly of non-pigmented staphylococci or micrococci and gram-positive rods. In normal females the vaginal flora varies considerably with the pH of the secretions and the amount of glycogen present in the epithelium; these factors in turn depend on ovarian function. In most instances, however, microaerophilic lactobacilli (the Doederlein bacillus groups[21]) predominate, together with diphtheroids, *Bifidobacterium*, gram-negative enteric bacilli, species of *Bacteroides*, microaerophilic and anaerobic streptococci and cocci, clostridia, enterococci, species of *Corynebacterium*, and coagulase-negative staphylococci.

The normal cervix also has a relatively profuse flora. The organisms present are identical with those found in the upper vagina.

The bacterial flora of the vulva is a mixture of organisms present on the skin of this area, including the acid-fast saprophyte *Mycobacterium smegmatis*, and other bacteria descending from the vagina.

The organisms most frequently encountered in the genital tract are:

Coliform bacilli, enterococci, *Bacteroides* species, *Clostridium*, *Peptostreptococcus*, and other organisms indigenous to the genital tract
Lactobacilli
Corynebacterium (Haemophilus) vaginale
Trichomonas vaginalis
Candida albicans, other *Candida* species, and saprophytic yeasts
Beta-hemolytic streptococci of groups A, B, and D
Mycobacterium tuberculosis
Nonpathogenic mycobacteria

Neisseria gonorrhoeae
Treponema pallidum
Haemophilus ducreyi
Mycoplasma species

ISOLATION OF THE GONOCOCCUS (NEISSERIA GONORRHOEAE)

The laboratory diagnosis of **gonorrhea** depends on the demonstration of intracellular diplococci in smears and on the isolation and identification of *Neisseria gonorrhoeae* by culture procedures. As a rule, gonococci may be found readily in smears of pus from **acute** infections, particularly in the male. In **chronic** infections, especially in the female, the value of the smear findings decreases; cultural methods generally yield a higher percentage of positive results.

Collection of specimens and primary inoculation of media

In the **female,** the best site to obtain a culture is the **cervix,** and this should be collected with care by an experienced professional. A sterile bivalve speculum is moistened with warm water (the usual lubricants contain antibacterial substances that may be lethal to gonococci) and inserted, and the cervical mucus plug is removed with a cotton ball and forceps. Gentle compression of the cervix between the speculum blades may produce endocervical exudate. If not, a sterile alginate or cotton-tipped applicator is then inserted into the endocervical canal and a rotating motion used to force exudate from endocervical glands. Only swabs made of cotton specified as being of low toxicity should be used. The swab is immediately inoculated to a modified Thayer-Martin chocolate agar plate (directions for preparation in Chapter 41) or a Transgrow bottle* (this must be inoculated in an upright position to prevent escape of contained carbon dioxide gas). The modified Thayer-Martin plate is streaked in a Z pattern with the swab and thereafter cross-streaked closely with a sterile bacteriologic loop, preferably in the clinic; the Transgrow bottle is inoculated over the entire agar surface, working from the bottom up, after moistening the swab with excess moisture contained in the bottle. The modified Thayer-Martin plate should be placed in a candle jar (use only white candles) (approximately 3% CO_2) within 1 hour and incubated at 35 C for 20 hours before examining. The Transgrow bottle, used primarily for convenience in a physician's office, VD clinic, or small laboratory, should be incubated at 35 C overnight before shipping; this provides sufficient growth for survival during prolonged transport. It is then sent to the diagnostic facility by mail or other convenient means, avoiding any marked temperature changes.

Anal canal infection is common in the female, so anal cultures are also recommended, especially when the cervical culture is negative, and as a follow-up of treatment efficacy. This specimen is easily obtained without using an anoscope by carefully inserting an alginate or cotton-tipped applicator approximately 1 inch into the anal canal and moving it from side to side to obtain material from the crypts. Obtaining material under direct visualization at anoscopy is preferred when feasible. If the swab shows the presence of fecal material, it is discarded and another swab is used. The swab is inoculated to modified Thayer-Martin medium as previously described.

In special situations where a cervical specimen is not indicated, for example, in children or hysterectomized patients, a urethral or vaginal culture may be substituted. Urethral cultures in the female are also recommended in the case of acute symptomatic disease, Bartholin abscess, pelvic inflammatory disease, pharyngitis, or disseminated infection.[22]

In the **male** with a purulent urethral exudate, the examination of a gram-stained direct smear (made by **rolling** the swab over the slide to preserve cell morphology) is usually sufficient to confirm a clinical diagnosis of gonorrhea. However, since an appreciable number of males may be asymptomatic and the stained smear probably negative, a **urethral culture** is recommended. This is readily obtained by gently inserting a thin alginate urethrogenital swab* (which may be moistened with sterile water) 2 cm into the urethra, gently rotating it, and then immediately inoculating a Thayer-Martin plate or Transgrow bottle as previously described. In homosexuals, anal and pharyngeal specimens should also be obtained.

Occasionally, material from other sites may be submitted for gonococcal culture; these include throat swabs, freshly voided urine, joint fluid, swabs from the eye, and so forth. The swabs and sediments from centrifuged fluids should be immediately inoculated to modified Thayer-Martin medium and to supplemented chocolate agar and handled as indicated above. A recent study (Feng et al., J.A.M.A. **237**:896-897, 1977) showed that culture of uncentrifuged first-voided urine (10-20 ml) for *N. gonorrhoeae* in males compared favorably with urethral swab culture and was less expensive and better accepted by patients.

Direct smears for gonococci

To obtain the highest percentage of positive findings in cases of suspected gonorrhea, **both** smears and cultures should be made. In the case of purulent

---

*Available from Baltimore Biological Laboratory, Cockeysville, Md.; Difco Laboratories, Detroit; shelf-life at room temperature in excess of 3 months.

*Falcon No. 2050 sterile alginate nasopharyngeal applicator.

material, slides should be prepared by the examining physician at the time the cultures are taken. Such smears should be carefully prepared by **rolling** the swab over the slide rather than by rubbing it on. All aspects of the swab head should come in contact with the slide. This distributes the pus cells into layers, thereby permitting accurate observation of intracellular organisms. The practice of submitting moist material stuck between two slides is condemned for obvious reasons. Smears of urine and other fluid specimens should be prepared from the centrifuged sediment of these materials. Smears should cover at least 1 square centimeter of the slide, if possible. They should be air dried and heat fixed.

Slides must be carefully stained by the Gram method and examined for the presence of characteristic **gram-negative diplococci** with their adjacent slides flattened (coffee-bean shaped). These organisms are generally found inside pus cells in acute gonorrhea, but in very early infections or in chronic gonorrhea they may be found extracellularly only, and frequently as single cocci. It should also be noted that recent administration of a specific antimicrobial agent such as penicillin may either eradicate or alter the morphology and staining reaction of *N. gonorrhoeae*. Care should be taken to distinguish gonococci from bipolar-staining, gram-negative coccobacilli, *Moraxella osloensis*, which have been mistaken for neisseriae.[41] Biochemical tests must be carried out to differentiate them with certainty. *N. meningitidis* may also be found rarely in the genitourinary tract and in the anus. Positive smears should be reported as **"Intracellular (extracellular) gram-negative diplococci resembling gonococci found; many (few) pus cells present."** Presence of epithelial cells should also be noted. All elements (bacteria, cells) should be semiquantitated.

Culture media

Many different and complex media have been introduced for the isolation of the gonococcus, but excellent results may be obtained by using the medium introduced by Thayer and Martin[42] in 1964. The original formula, an enriched chocolate agar medium containing the antibiotics ristocetin and polymyxin B, was recommended for the isolation of *Neisseria gonorrhoeae* and *N. meningitidis*. The authors reported that the medium showed an inhibitory action against other neisseriae and most species of Mimeae (*Moraxella*), and it suppressed *Pseudomonas* and *Proteus* species. The removal of one of the antibiotics from the market necessitated a suitable substitute, and the above authors reported the successful use of vancomycin, sodium colistimethate (colistin), and nystatin[43] (V-C-N inhibitor*). Comparison of the new medium with the original formula showed comparable growth of *N. gonorrhoeae* from both male and female patients, along with a greater inhibition of staphylococci and saprophytic neisseriae.

A further improvement in Thayer-Martin chocolate agar was the incorporation of a chemically defined supplement* that resulted in as good or better recovery of gonococci than was achieved with previous media.[30] The use of Imferon, an iron-dextran complex, as a replacement for ferric nitrate enhanced the growth of both the gonococcus and the meningococcus (Payne, S. M., and Finkelstein, R. A.: J. Clin. Microbiol. **6:**293-297, 1977).

Recently a modification of the Thayer-Martin medium—Transgrow—was introduced, prepared according to the formulation of Martin and Lester.[31] Recommended for transport and growth of pathogenic neisseriae (while maintaining their viability for at least 48 hours at room temperature), the medium supresses contaminating organisms in a manner similar to the Thayer-Martin formula. The agar content of Transgrow was increased to 2% and the glucose content to 0.25%; the medium also contains trimethoprim

---

*Available from Baltimore Biological Laboratory, Cockeysville, Md.; Difco Laboratories, Detroit.

lactate, which is inhibitory to *Proteus* species. It is available commercially,* packaged in a tightly closed flat bottle under a partial CO_2 atmosphere, and has a shelf-life of 3 to 4 months. Transgrow medium appears to be superior to other transport media for maintaining the viability of pathogenic neisseriae and can be mailed to the reference laboratory after overnight incubation without appreciable loss of gonococci. The procedure for inoculation of Transgrow bottles has been described previously. Comparative studies with Thayer-Martin medium by CDC's Venereal Disease Research Laboratory have indicated good correlation, both in the laboratory and in field trials.[35]

Another medium that is very good for the isolation of pathogenic *Neisseria* was described in 1973 by Faur and associates.[9,10] Known as New York City (NYC) medium, it consists of a proteose peptone-cornstarch-agar–buffered base containing a supplement of horse plasma and hemoglobin solution, glucose, yeast dialysate, and four antimicrobial agents (vancomycin, colistin, nystatin or amphotericin, and trimethoprim lactate). It is a clear medium that provides 24-hour luxuriant growth of pathogenic *Neisseria* as well as being highly selective.

A modification of NYC medium, which gave good performance as a culture and transport medium without the addition of ambient CO_2, was also reported by Faur and co-workers.[11] The addition to NYC medium of filter-sterilized yeast dialysate provided a carbon source for CO_2 development and release. They found that when it was packaged in an airtight container, it permitted growth of *N. gonorrhoeae* from clinical specimens without provision of ambient CO_2.

Although Transgrow medium has proved to be of value, it appears to have some important deficiencies. The commercial preparations of Transgrow have been found to contain variable CO_2 concentra-

tions.[4] Different batches of the medium vary in ability to support the growth of gonococci. Moisture accumulating inside the bottles may cause poor visibility and may contribute to the spreading of contaminants and interfere with development of isolated colonies. Furthermore, the narrow bottle neck is inconvenient for inoculation and subculturing of colonies.

A new test system, called Gono-Pak, for growth of *N. gonorrhoeae* has been described by Martin and associates.[29] Consisting of a Petri plate containing modified Thayer-Martin medium, the specimen is inoculated, sealed, and placed in a plastic bag along with a CO_2-generating tablet. Employing modified NYC transport medium,[11] Symington[39] found the Gono-Pak to be suitable for both transportation and incubation of *N. gonorrhoeae*.

A more recent development has been that of a rectangular plastic plate that contains a small well to accommodate a CO_2-generating tablet. Known as the Jembec plate,* this plate and a CO_2-generating tablet sealed in a plastic "zip-lock" pouch provides a CO_2 environment enclosure. Jembec plates that contain prepoured modified Thayer-Martin medium are known as Neigon plates.†

In evaluating improved transport systems for *N. gonorrhoeae* in clinical specimens, Symington[40] studied the efficiency of the following: (1) Amies charcoal transport medium, (2) Jembec chambers containing Neigon plates of modified Thayer-Martin medium, and (3) Jembec chambers containing plates of modified NYC transport medium. She found that for up to 2 days in transit, the three systems were not significantly different. However, after 3 days in transit, modified NYC-Jembec chambers were better than the other two (highly significant statistically). The modified NYC-

---

*Baltimore Biological Laboratory, Cockeysville, Md.; Difco Laboratories, Detroit.

*Available from Ames Co., Division of Miles Laboratories, Elkart, Ind.
†Available from Flow Laboratories, Inc., Rockville, Md.

Jembec chambers withstood 241 miles of postal transit during winter months, with 80% of the *N. gonorrhoeae* present in clinical specimens remaining viable from 2 to 5 days under these conditions. The CO_2 generated by the tablet in the Jembec chamber was sufficient to support the growth of *N. gonorrhoeae* if the chambers were incubated at 36 C immediately after inoculation. If delayed in transit, however, the chambers had to be incubated in 5% to 10% CO_2 to promote the growth of *N. gonorrhoeae*.

Handling of specimens in the laboratory

Ideally, specimens for the isolation of gonococci are submitted on Thayer-Martin or NYC medium plates (or in Transgrow bottles) that have been properly inoculated previously and incubated in candle jars overnight. In a hospital situation, however, the specimen for culture is usually received on a swab in holding medium such as the Culturette,* having been obtained from the patient a short time before. Occasionally, one may also receive purulent material from an aspirated joint, a freshly voided urine sample, or other specimens. In any case, the specimen (or sediment following centrifugation) should be inoculated **immediately** to a freshly prepared modified Thayer-Martin or supplemented chocolate agar plate (brought to room temperature), using the Z and cross-streaking procedure previously described. Swab specimens should not be left at room temperature unless in Transgrow or other transport medium, and **under no circumstances** in a 35 C incubator; gonococci either die off or are rapidly overgrown by commensal bacteria. However, the swab may be refrigerated at 4 to 6 C for not longer than 3 hours before inoculation without a noticeable reduction in recovery of gonococci (not true of meningococci, which are notably cold sensitive).

All cultures, including previously in-

oculated Thayer-Martin plates (or Transgrow bottles received with loosened caps), are placed in a CO_2 incubator (5% to 10% CO_2) or in a candle jar with a tight-fitting lid* containing a short, thick, white smokeless candle (affixed to a glass slide) that is lighted inside the jar before closing with the lid. This will generate approximately 3% CO_2 before extinguishing itself.

The plates or bottles are incubated at **35 C**† overnight and examined for growth; cultures showing no growth are returned to the incubator for a total period of not less than 48 hours.‡ Typical growth of *N. gonorrhoeae* appears as translucent, mucoid, raised colonies of varying size; in Transgrow bottles, colonies may or may not appear typical. The **oxidase test** is used to verify the presence of gonococcus colonies and, along with the Gram stain of those colonies reacting positively, serves to presumptively identify *N. gonorrhoeae* from a urogenital site. These procedures and further identification tests by carbohydrate fermentation and immunofluorescent (FA) technique are discussed in Chapter 19.

Isolation of gonococci from synovial fluid

CDC's Venereal Disease Control Division (*Criteria and Techniques for the Diagnosis of Gonorrhea,* 1976) recommends that synovial fluid be cultured for gonococci in an enriched broth, such as trypticase soy broth supplemented with 1% IsoVitaleX, 10% horse serum, and 1% glucose.

An interesting paper from Holmes and colleagues[19] in Seattle reports on the re-

---

*Available from Scientific Products, Evanston, Ill.

*Available from Erno Products Co., Philadelphia; EC jar (clear) 120-mm mouth opening #C-3118, with metal screw-cap lid. This will hold approximately twelve 90-mm plates and a candle.

†The incubator should be adjusted to this temperature because many strains of *N. gonorrhoeae* will not grow well at 37 C.

‡Nearly one half of positive cultures require this incubation period.

covery of *N. gonorrhoeae* from "sterile" synovial fluid in gonococcal arthritis by means of a medium developed for the propagation of L forms of those organisms in vitro. The medium was a modification of that introduced by Bohnhoff and Page[1] for *N. meningitidis* L forms and consisted of trypticase soy broth supplemented with 10% sucrose and 1.25% agar, to which was added 20% inactivated horse serum, after sterilization. After inoculation, the surface of the medium was overlaid with 0.5 ml of the same medium containing one half the agar concentration, and following 72 hours' incubation, oxidase-positive, reverting L form–type colonies developed. On subculture to chocolate agar, tiny colonies that were identified as *N. gonorrhoeae* by FA staining and by carbohydrate fermentation grew out. No growth occurred on regular chocolate agar or Thayer-Martin agar inoculated at the same time, after 96 hours' incubation.

The authors hypothesized that the gonococci may have existed in the synovial fluid in a cell wall–deficient, osmotically fragile state. A further trial of this medium in patients with suspected gonococcal arthritis is certainly warranted.

MISCELLANEOUS VENEREAL DISEASES
Syphilis

Although full discussion is not within the scope of this text, mention should be made of the serologic diagnosis of **syphilis** and the methods most frequently used in making this diagnosis. These tests are basically of two types: the **nontreponemal** antibody tests, such as the VDRL and rapid plasma reagin (RPR) flocculation tests, and the Kolmer complement fixation test; or the **treponemal** antibody tests, such as the fluorescent-antibody-absorption (FTA-ABS) test. Further discussion of these may be found in Chapter 38.

Mention should be made of **darkfield** microscopy as an aid to the diagnosis of early syphilis. An ample drop of tissue fluid expressed from a primary lesion is placed on a coverglass, which is inverted over a glass slide and pressed to make a thin film. This preparation is then examined microscopically, using the oil immersion objective and a properly adjusted darkfield illumination. The presence of motile treponemes of characteristic morphology in a typical early lesion establishes the diagnosis. Difficulty may be experienced in darkfield examination of oral or rectal lesions, since these sites may contain motile treponemes from the indigenous flora.

A fluorescein-labeled *Treponema pallidum* conjugate* is available for the direct identification of *T. pallidum* by the immunofluorescent technique. Material from the lesion is applied to a slide and air dried; the conjugate is then added, rinsed off, and dried; a coverslip is applied, and the preparation is examined by fluorescent microscopy. Preparations may be made, dried, and examined later, eliminating the necessity of immediate darkfield examination.

Chancroid

Chancroid is a soft chancre of the genitalia of venereal origin, caused by *Haemophilus ducreyi* (Ducrey bacillus), a very small gram-negative rod occurring singly or in small clumps.

H. ducreyi can be demonstrated by the following procedure.[2] Ten ml of the patient's blood is distributed in 5-ml amounts in sterile screw-capped tubes. After clotting has occurred, the serum is removed, transferred aseptically to another sterile tube, and inactivated in a 56 C water bath for 30 minutes. After cooling, the serum is inoculated with material obtained from the undermined border of the genital lesion (previously cleansed with saline-moistened gauze) and incubated at 35 C for 48 hours. A gram-stained smear of the growth is then examined for the presence

---

*Available from Baltimore Biological Laboratory, Cockeysville, Md., Catalog No. 40806.

of coccobacillary gram-negative organisms in tangled chains or in long parallel strands ("schools of fish"), which are diagnostic of *H. ducreyi*.

Granuloma inguinale

Granuloma inguinale is an infection of the genital region characterized by a slowly progressive ulceration and caused by *Calymmatobacterium granulomatis*. This organism appears within the cytoplasm of large mononuclear cells as a small (1 to 2 μm), plump, heavily encapsulated coccobacillus and is best observed in impression smears of cleansed tissue obtained by punch biopsy at the edge of the lesion, stained with Wright's stain. A diligent search of smears prepared on several occasions may be necessary to find the organisms.

Lymphogranuloma venereum (LGV)

Lymphogranuloma venereum is caused by one of a large group of gram-negative intracellular parasites responsible for some important human diseases, including psittacosis and trachoma. Once considered to be large viruses, they are now classified as **chlamydiae**.[18] The diagnosis is made clinically and by demonstrating delayed skin hypersensitivity to an antigen prepared from infected yolk sacs and a fourfold rise in complement fixation titer on paired (acute and convalescent) sera, using a similar antigen. A serologic test for syphilis (STS) should also be done on these patients to rule out syphilis.

TRICHOMONAS INFECTION

The most practical method of confirming a diagnosis of *Trichomonas* infection is by the demonstration of **actively motile** flagellates in a saline suspension of vaginal or urethral discharges. The preparation must be examined shortly after collecting, or the characteristic motility and morphology may not be apparent. Occasionally, cultures of vaginal specimens, semen, urine, and other materials will reveal trichomonads when wet smears are

negative. The Kupferberg[28] medium is recommended for this purpose, with the addition of 5% human serum and antibiotic (chloramphenicol, 100 μg per milliliter of medium). The base medium may be obtained in dehydrated form.* Cultures are examined after 2 to 4 days' incubation.

CANDIDA INFECTION

The yeastlike fungus *Candida albicans* is also a frequent cause of vaginitis, and it can readily be identified by the examination of gram-stained smears and growth on Sabouraud dextrose agar slants. Methods are described in Chapter 34.

ISOLATION OF HAEMOPHILUS (CORYNEBACTERIUM) VAGINALIS

The bacteriologic examination of cases of nonspecific vaginitis and urethritis frequently fails to provide the clinician with any specific agent against which to direct therapy. Gardner and Dukes[12] reported the isolation of a small, pleomorphic, gram-negative bacillus from many of their cases of nonspecific vaginitis, and they named it *Haemophilus vaginalis*. Smears of vaginal discharge in such cases showed epithelial cells covered with these bacilli—the so-called "clue cells." The organism grew out on sheep blood agar plates that had been incubated for 24 to 48 hours in a candle jar as colorless, transparent, pinpoint colonies, best seen by oblique illumination, and frequently surrounded by a clear zone of hemolysis. Characteristic "puff ball" colonies appeared in thioglycollate medium enriched with ascitic fluid. Some strains are obligately anaerobic.

Zinnemann and Turner[47] have reported the successful cultivation of this organism on a medium containing neither hemin (X factor), nicotinamide adenine dinucleotide (V factor), nor other coenzymes required by members of the genus *Hae-*

---

*Baltimore Biological Laboratory, Cockeysville, Md.; Difco Laboratories, Detroit.

mophilus and proposed reclassification as *Corynebacterium vaginale*. Their work has been confirmed by Dunkelberg and co-workers[7] at the Venereal Disease Research Laboratory, CDC, who also suggest that the bacterium be reclassified as *C. vaginale*. However, the organism's cell wall does not contain DAP and contains 6-deoxytalose in place of arabinose.[3]

Haemophilus vaginalis is thought to be associated with vaginitis and has also been implicated in puerperal fever, nongonococcal urethritis in the male, and other genitourinary tract infections. However, the etiologic role of this organism is not universally accepted.

NONGONOCOCCAL (NONSPECIFIC) URETHRITIS AND CERVICITIS

Nongonococcal urethritis is occasionally caused by *Trichomonas*. Nonspecific urethritis is the term used to designate the approximately 90% of nongonococcal cases in which no pathogen is noted by either direct examination or by cultural methods.

Nonspecific urethritis is widespread. For example, it is the most frequently recorded sexually transmitted disease in England. Although not a reportable disease in the United States, data obtained from two venereal disease clinics associated with the Center for Disease Control showed that nonspecific urethritis accounted for 30% of urethritis in black men and up to 70% among white men.[44] The incidence of nonspecific urethritis in women is uncertain because they generally do not have clinically recognizable symptoms at presentation.

The cause of nonspecific urethritis has been generally regarded as unknown. In the past few years a great deal of research effort has focused on the *Chlamydia* and *Mycoplasma*.

As obligate intracellular organisms, *Chlamydia* are cultured like viruses and are known to be responsible for trachoma, inclusion conjunctivitis, and lymphogran-

uloma venereum. The technique of culture of *Chlamydia* in irradiated McCoy cells[5,15] provides a more convenient method for testing large numbers of specimens than does the yolk sac inoculation procedure. Based on a count of inclusion bodies, the McCoy cell method is more rapid and allows quantitative measurement of the level of infection.[14] It was also found to be three to four times more sensitive than the detection of inclusions in smears or their isolation in the yolk sac of embryonated eggs.[6,14] Gordon and associates[13] reported that the infectivity of *Chlamydia* was enhanced by centrifugation of the inoculum onto the irradiated cells. Wentworth and Alexander[45] found that 5-iodo-2-deoxyuridine enhanced the susceptibility of McCoy cell cultures to chlamydial infection; elimination of the need for irradiation simplifies the procedure considerably.

Other studies have shown that *Chlamydia* infection can be increased 45 times in DEAE-treated McCoy cells, compared with untreated controls.[36] Moreover, comparative studies with human cervical specimens have shown HeLa 229 cells to be equal to McCoy cells in susceptibility to infection.[27] Wentworth and her associates[46] reported that DEAE-dextran increases the level of infectivity of *Chlamydia* 100 times in HeLa cells but by less than twice in McCoy cells.

Dunlop and co-workers[8] reported the isolation of *Chlamydia* from approximately 40% of men with nonspecific urethritis by inserting a swab into the urethra for a distance of 2 to 5 cm. That these organisms have a significant etiologic role in nongonococcal urethritis in a midwestern community has been recently reported by Smith and his associates.[38] From October 1973 through August 1974, these investigators inoculated 335 genitourinary tract specimens from patients with urethritis into McCoy cell cultures. They obtained 45 *Chlamydia* isolates, of which 42 were recovered when glass vials, rather than plastic microtiter plates, were used as cell culture vessels. Herpes simplex virus was isolated 15 times. *Chlamydia* have also been observed in 30% of women seen in a venereal disease clinic.[34] Furthermore, colonization with this organism was observed in 66% of women who were regular sexual partners of men with *Chlamydia*-associated urethritis, whereas in control groups of men and women isolation rates of less than 5% were obtained. *Chlamydia* may be responsible for clinically manifest cervicitis.

The microorganisms of the *Mycoplasma* group, often called PPLO, are the smallest free-living organisms that can be cultivated on artificial media. They lack a rigid cell wall, require protein-enriched media for growth, and exhibit a characteristic "fried egg" colony when growing on solid media.

Currently, seven mycoplasmas of human origin have been described.[24] Of these, *M. hominis, M. fermentans,* and T strains *(Ureaplasma urealyticum)* are found in the human genitourinary tract. T-strain mycoplasmas have been associated with from 60% to over 90% of cases of nongonococcal urethritis in males, whereas their natural occurrence in a similar group of normal controls ranged from 21% to 26%, as reported by Shepard.[37] Nevertheless, it is not established that T strains of mycoplasma play an important etiologic role in this genital infection.

Other studies show no difference in incidence of T strains between patients and controls.[32] Handsfield and his colleagues[17] found *Chlamydia* to be the agent most frequently involved in nonspecific urethritis. Recently, Jacobs and co-workers (Ann. Intern. Med. **86:**313-314, 1977) noted that *Chlamydia* was found in 30% of patients with nonspecific urethritis, as compared with 12% of control patients; the difference was statistically significant.

MICROBIAL AGENTS IN HUMAN ABORTION

Listeria monocytogenes is most frequently associated in humans with septic perinatal infections in the female (some with bacteremia) and purulent meningitis, meningoencephalitis, or miliary granulomatosis in the neonate. The mother may exhibit flulike symptoms with a low-grade fever in the last trimester of pregnancy and subsequently may abort or deliver a stillborn baby.[23] Postpartum cultures from the vagina, cervix, or urine may be positive for *L. monocytogenes* from a few days to several weeks, although the patient may appear asymptomatic.[16] In uncontaminated material (blood, cerebrospinal fluid), *L. monocytogenes* can be readily isolated by conventional techniques, but contaminated specimens or tissue may require selective media and **cold enrichment** procedures. Refer to Chapter 26 for a further description of these methods.

Although *Campylobacter (Vibrio) fetus* clearly causes abortion in cattle, its role in human abortion is not conclusive. In describing 17 human infections caused by *C. fetus,* Elizabeth King[25] noted that three of these patients (from the French literature) had accompanying problems of pregnancy. Hood and Todd[20] of Charity Hospital in New Orleans reported what is apparently the first human case of *C. fetus* infection involving pregnancy in the Western Hemisphere. The bacterium was isolated from the placenta of the mother and the brain of the aborted fetus.

A suggested role for *Mycoplasma* in human reproductive failure[26] has not been established, but it appears that *M. hominis* is of etiologic importance in some women with fever following abortion.[33]

The role of *Toxoplasma* infection in abortion must also be considered uncertain.

It is apparent that the clinical microbiologist should be aware of all these agents and, through close communication with the clinician, offer microbiologic aid in their early detection.

REFERENCES

1. Bohnhoff, M., and Page, M. I.: Experimental infection with parent and L-phase variants of *Neisseria meningitidis*, J. Bacteriol. **95:**2070-2077, 1968.
2. Borchardt, K. A., and Hoke, A. W.: Simplified laboratory technique for diagnosis of chancroid, Arch. Dermatol. **102:**188-192, 1970.
3. Buchanan, R. E., and Gibbons, N. E.: Bergey's Manual of determinative bacteriology, ed. 8, Baltimore, 1974, The Williams & Wilkins Co.
4. Chapel, T., Smeltzen, M., Printz, D., Dassel, R., and Lewis, J.: An evaluation of commercially supplied Transgrow and Amies media for the detection of *Neisseria gonorrhoeae,* Health Lab. Sci. **11:**28-33, 1973.
5. Darougar, S., Kinnison, J. R., and Jones, B. R.: Simplified irradiated McCoy cell culture for isolation of chlamydiae. Excerpta Medica International Congress Series No. 223, 1970, pp. 63-70.
6. Darougar, S., Treharne, J. D., Dwyer, R. St. C., Kinnison, J. R., and Jones, B. R.: Isolation of TRIC agent *(Chlamydia)* in irradiated McCoy cell culture from endemic trachoma in field studies in Iran: comparison with other laboratory tests for detection of *Chlamydia,* Br. J. Ophthalmol. **55:**591-599, 1971.
7. Dunkelberg, W. E., Skaggs, R., and Kellogg, Jr., D. S.: A study and new description of *Corynebacterium vaginale (Haemophilus vaginalis),* Am. J. Clin. Pathol. **53:**370-377, 1970.
8. Dunlop, E. M. C., Vaughan-Jackson, J. D., Darougar, S., and Jones, B. R.: Chlamydial infection. Incidence in "nonspecific" urethritis, Br. J. Vener. Dis. **48:**425-428, 1972.
9. Faur, Y. C., Weisburd, M. H., Wilson, M. E., and May, P. S.: A new medium for the isolation of pathogenic *Neisseria* (NYC medium) I. Formulation and comparisons with standard media, Health Lab. Sci. **10:**44-54, 1973.
10. Faur, Y. C., Weisburd, M. H., and Wilson, M. E.: A new medium for the isolation of pathogenic *Neisseria* (NYC medium) II. Effect of amphotericin B and trimethoprim lactate on selectivity, Health Lab. Sci. **10:**55-60, 1973.
11. Faur, Y. C., Weisburd, M. H., and Wilson, M. E.: A new medium for the isolation of pathogenic *Neisseria* (NYC medium) III. Performance as a culture and transport medium without addition of ambient carbon dioxide, Health Lab. Sci. **10:**61-74, 1973.
12. Gardner, H. L., and Dukes, C. D.: *Haemophilus vaginalis* vaginitis, Am. J. Obstet. Gynecol. **69:**962-976, 1955.
13. Gordon, F. B., Dressler, H. R., and Quan, A. L.: Relative sensitivity of cell culture and yolk sac

14. Gordon, F. B., Harper, I. A., Quan, A. L., Treharne, J. D., Dwyer, R. St. C., and Garland, J. A.: Detection of *Chlamydia (Bedsonia)* in certain infections of man. I. Laboratory procedures: comparison of yolk sac and cell culture for detection and isolation, J. Infect. Dis. **120:**451-462, 1969.
15. Gordon, F. B., and Quan, A. L.: Isolation of the trachoma agent in cell culture, Proc. Soc. Exp. Biol. Med. **118:**354-359, 1965.
16. Gray, M. L., and Killinger, A. H.: *Listeria monocytogenes* and listeric infections, Bacteriol. Rev. **30:**309-382, 1966.
17. Handsfield, H. H., Holmes, K. K., Wentworth, B. B., Pedersen, A. H. B., Turck, M., and Alexander, E. R.: Etiology and treatment of nongonococcal urethritis. Abstracts of the Twelfth Interscience Conference on Antimicrobial Agents and Chemotherapy [Abstr. 122], Atlantic City, N.J., September 26-29, 1972.
18. Hanna, L., Schacter, J., and Jawetz, E.: Chlamydiae (Psittacosis-lymphogranuloma venereum-trachoma group). In Lennette, E. H., Spaulding, E. H., and Truant, J. P., editors: Manual of clinical microbiology, ed. 2, Washington, D.C., 1974, American Society for Microbiology.
19. Holmes, K. K., Gutman, L. T., Belding, M. E., and Turck, M.: Recovery of *Neisseria gonorrhoeae* from "sterile" synovial fluid in gonococcal arthritis, N. Engl. J. Med. **284:**318-320, 1971.
20. Hood, M., and Todd, J. M.: *Vibrio fetus;* a cause of human abortion, Am. J. Obstet. Gynecol. **80:**506-511, 1960.
21. Hunter, C. A., Jr., Long, K. R., and Schumacher, R. R.: A study of Doderlein's vaginal bacillus, Ann. N.Y. Acad. Sci. **83:**217-226, 1959.
22. Kellogg, D. S., Jr., Holmes, K. K., and Hill, G. A.: Laboratory diagnosis of gonorrhea, Cumitech 4, Washington, D.C., 1976, American Society for Microbiology.
23. Kelly, C. S., and Gibson, J. L.: Listeriosis as a cause of fetal wastage, Obstet. Gynecol. **40:**91-97, 1972.
24. Kenny, G. E.: Mycoplasma. In Lennette, E. H., Spaulding, E. H., and Truant, J. P., editors: Manual of clinical microbiology, ed. 2, Washington, D.C., 1974, American Society for Microbiology.
25. King, E. O.: Human infection with *Vibrio fetus* and a closely related vibrio, J. Infect. Dis. **101:**119-128, 1957.
26. Kundsin, R. B., and Driscoll, S. G.: Mycoplasmas and human reproductive failure, Surg. Gynecol. Obstet. **131:**89-92, 1970.
27. Kuo, C.-C., Wang, S.-P., Wentworth, B. B., and Grayston, J. T.: Primary isolation of TRIC orga-

nisms in HeLa 229 cells treated with DEAE-dextran, J. Infect. Dis. **125:**665-668, 1972.

28. Kupferberg, A. B., Johnson, G., and Sprince, H.: Nutritional requirements of *Trichomonas vaginalis,* Proc. Soc. Exp. Biol. Med. **67:**304-308, 1948.

29. Martin, J. E., Jr., Armstrong, J. H., and Smith, P. B.: A new system for cultivation of *N. gonorrhoeae,* Appl. Microbiol. **27:**802-805, 1974.

30. Martin, J. E., Jr., Billings, T. E., Hackney, J. F., and Thayer, J. D.: Primary isolation of *N. gonorrhoeae* with a new commercial medium, Public Health Rep. **82:**361-363, 1967.

31. Martin, J. E., and Lester, A.: Transgrow, a medium for transport and growth of *Neisseria gonorrhoeae* and *Neisseria meningitidis,* HSMHA Health Reports **86:**30-33, 1971.

32. McCormack, W. M., Braun, P., Lee, Y. H., Klein, J. D., and Kass, E. H.: The genital mycoplasmas, N. Engl. J. Med. **288:**78-89, 1973.

33. McCormack, W. M., and Lee, Y.-H.: Genital mycoplasmas. In Charles, D., and Finland, M., editors: Obstetric and perinatal infections, Philadelphia, 1973, Lea & Febiger.

34. Oriel, J. D., Reeve, P., Powis, P., Miller, A., and Nicol, C. S.: Chlamydial infection. Isolation of *Chlamydia* from patients with non-specific genital infection, Br. J. Vener. Dis. **48:**429-436, 1972.

35. Printz, D. W., VD Branch, Center for Disease Control, Atlanta. Personal communication, 1972.

36. Rota, T. R., and Nichols, R. L.: Infection of cell cultures by trachoma agent: enhancement by DEAE dextran, J. Infect. Dis. **124:**419-421, 1971.

37. Shepard, M. C.: Nongonococcal urethritis associated with human strains of "T" mycoplasmas, J.A.M.A. **211:**1335-1340, 1970.

38. Smith, T. F., Weed, L. A., Segura, J. W., Pettersen, G. R., and Washington, J. A. II: Isolation of *Chlamydia* from patients with urethritis, Mayo Clin. Proc. **50:**105-110, 1975.

39. Symington, D. A.: An evaluation of New York City transport medium for detection of *N. gonorrhoeae* in clinical specimens, Health Lab. Sci. **12:**69-75, 1975.

40. Symington, D. A.: Improved transport system for *Neisseria gonorrhoeae* in clinical specimens, J. Clin. Microbiol. **2:**498-503, 1975.

41. Svihus, R. H., Lucero, E. M., Mikolajczyk, R. J., and Carter, E. E.: Gonorrhea-like syndrome caused by penicillin-resistant Mimeae, J.A.M.A. **177:**121-124, 1961.

42. Thayer, J. D., and Martin, J. E., Jr.: A selective medium for the cultivation of *N. gonorrhoeae* and *N. meningitidis,* Public Health Rep. **79:**49-57, 1964.

43. Thayer, J. D., and Martin, J. E., Jr.: Improved medium selective for cultivation of *N. gonorrhoeae* and *N. meningitidis,* Public Health Rep. **81:**559-562, 1966.

44. Volk, J., and Kraus, S. J.: Nongonococcal urethritis: a venereal disease as prevalent as epidemic gonorrhea, Arch. Intern. Med. **134:**511-514, 1974.

45. Wentworth, B. B., and Alexander, E. R.: Isolation of *Chlamydia trachomatis* by use of 5-iodo-2-deoxyuridine-treated cells, Appl. Microbiol. **27:**912-916, 1974.

46. Wentworth, B. B., Bonin, P., Holmes, K. K., Gutman, L., Weisner, P., and Alexander, E. R.: Isolation of viruses, bacteria, and other organisms from venereal disease clinic patients: methodology and problems associated with multiple isolations, Health Lab. Sci. **10:**75-81, 1973.

47. Zinnemann, K., and Turner, G. C.: The taxonomic position of *"Haemophilus vaginalis"* (*Corynebacterium vaginale*), J. Pathol. Bacteriol. **85:**213-219, 1963.

CHAPTER 12

Microorganisms encountered in cerebrospinal fluid

Bacteriologic examination of the spinal fluid is an essential step in the diagnosis of any case of suspected meningitis. The specimen must be collected under sterile conditions and transported to the laboratory **without delay.**

Acute bacterial meningitis is an infection of the meninges—the membranes covering the brain and spinal cord—and is caused by a variety of gram-positive and gram-negative microorganisms, predominately *Haemophilus influenzae, Neisseria meningitidis,* and *Streptococcus pneumoniae.* Bacterial meningitis also may be secondary to infections in other parts of the body, and rarely, *Salmonella* species, coliform bacilli, staphylococci, or mycobacteria may be recovered; in neonatal meningitis, *Escherichia coli* is the most frequent organism isolated, along with group-B beta-hemolytic streptococci, an increasingly important pathogen, and *Listeria monocytogenes* (which may also affect adults).

In bacterial meningitis the cerebrospinal fluid is usually **purulent,** with an increased white cell count (generally greater than 1,000 per cubic millimeter), a predominance of polymorphonuclear cells, and a reduced concentration of spinal fluid glucose. Amebic meningoencephalitis may closely simulate bacterial meningitis (see Chapter 33). On the other hand, in meningitis caused by the tubercle bacillus or by nonbacterial agents, such as viruses or fungi, the fluid is usually nonpurulent, with a low cell count of the mononuclear type and a normal or moderately reduced glucose content. However, in early or partially treated bacterial meningitis the cell count may be low and without a polymorphonuclear response, whereas in early tuberculosis or viral meningitis such cells may predominate.

Since it may be difficult or impossible for the clinician to differentiate the meningitides clinically, rapid and accurate means of identifying the etiologic agents must be available in the laboratory. It is therefore strongly recommended that a complete microbiologic workup, including smear and culture, be carried out on all cerebrospinal fluid specimens from cases of suspected meningitis, whether the fluid is clear or cloudy.

Generally, in a case of suspected meningitis the fluid is submitted for chemical and cytologic examination as well as for culturing. Since the amount of fluid provided is usually small, it is suggested that the specimen be centrifuged for 15 minutes (except where *Cryptococcus neoformans* is suspected) at high speed as soon as it is received. The supernatant fluid is then removed with a sterile Pasteur pipet to another tube for chemical or serologic studies, leaving the sediment and a few drops of fluid for culture procedures. In this way the entire specimen may be concentrated by centrifugation, obviating a need for its division into several different tubes for the other examinations.

Purulent (cloudy) fluids should be examined **immediately** by a gram-stained smear (using a **sterile** slide), followed by appropriate culturing procedures. In these acute forms of meningitis the etiologic agent frequently can be demonstrated in stained films of the specimen. In some cases, as in *Haemophilus influ-*

enzae and pneumococcus infections, **the organism can be identified immediately by a capsular swelling (quellung) test with specific antiserum.** Refer to Chapters 18 and 37 for procedures.

The following organisms (in approximate order of frequency) are isolated from cerebrospinal fluid:

Haemophilus influenzae, type b (infants and children)
Neisseria meningitidis (meningococcus), (most common in Great Britain)
Streptococcus pneumoniae (pneumococcus)
Mycobacterium tuberculosis
Staphylococci, streptococci (including enterococci)
Listeria monocytogenes
Cryptococcus neoformans
Viruses, other fungi, and other agents
Coliform bacilli, *Pseudomonas* and *Proteus* species
Leptospira species
Edwardsiella tarda
Anaerobic bacteria
Acinetobacter calcoaceticus

ROUTINE CULTURE FOR COMMONLY ISOLATED PATHOGENS

For the cultivation of *Haemophilus influenzae*, meningococci, pneumococci, streptococci and other gram-positive cocci, and also the aerobic and anaerobic gram-negative bacilli from cerebrospinal fluid, the following procedures are recommended (keep specimens at **35** C while awaiting examination):

1. Using a loopful of the fluid or sediment obtained by centrifugation, inoculate the following media: (a) two blood agar plates, (b) a chocolate agar plate, (c) a tube of enriched thioglycollate medium and other anaerobic media, if indicated.

2. Make a thin smear for Gram staining and, if appropriate, an India ink preparation (described in Chapter 42) for

encapsulated *Cryptococcus neoformans.*

3. To the sediment remaining in the tube, add approximately 5 ml of dextrose ascitic fluid semisolid agar (see Chapter 41). This is done by flaming the mouths of both tubes, cooling, and pouring the semisolid medium over the remaining sediment. In the experience of one of us (EGS), the diagnosis of a bacterial meningitis has been established frequently by demonstrating growth in this semisolid medium when streak plates and smears have been negative.

4. Incubate one blood agar plate in an anaerobic jar and the other plates in a candle jar and the fluid media under ordinary conditions at 35 C. Examine all but the anaerobic jar after overnight incubation (and daily thereafter) for the presence of growth. The semisolid medium will become turbid and turn yellow (due to the production of acidity that is indicated by the phenol red present) when growth has occurred. Make a gram-stained smear of this medium and subculture to appropriate solid media after reading the smear. It is important to include antimicrobial sensitivity disks on these plates for obvious reasons. The anaerobic jar is opened after 48 hours.

5. Hold all cultures for 72 to 96 hours (5 to 7 days when antibiotics have been given before culture was obtained) before discarding as negative. Anaerobic cultures should be held at least 7 to 10 days.

6. Identify all organisms isolated by the methods described in subsequent sections on the various genera.

Hypertonic culture media have led to recovery of organisms from spinal fluid that could not otherwise be recovered.[18,21]

---

*Available from Difco Laboratories, Detroit, as pools and some single types. Antisera prepared by Erna Lund, Statens Serum-institut, Copenhagen, Denmark, are also recommended. Particularly useful is **Omni-serum**, a polyvalent pneumococcus antiserum that gives capsular swelling reactions with 84 types.

---

*Artifacts resembling yeasts cells have been observed in the spinal fluid of patients who have had recent myelograms.[1]

This was not necessarily related to prior therapy with cell wall–active antibiotics. One of us (SMF) has had very rapid recovery of conventional bacteria from spinal fluid, as compared with routine culture procedures, using a membrane filter procedure.

In all cases of purulent meningitis an attempt should be made to detect the **primary focus** of infection. For this purpose, three blood cultures should be taken and cultures also should be made from underlying foci of infection, such as draining ears or purulent sinusitis, **before** antibacterial therapy is initiated. Petechial rash is common in meningococcal meningitis; the organism can sometimes be demonstrated by smear and culture of the petechiae.

HAEMOPHILUS INFLUENZAE MENINGITIS

Haemophilus influenzae type **b,** is the most frequent cause of acute purulent meningitis in children between the neonatal period and 6 years of age and only rarely the cause in adults (probably due to acquisition of *H. influenzae* antibodies). The organism enters by way of the respiratory tract, where it produces a nasopharyngitis, sinusitis, or middle-ear infection. It may reach the bloodstream from these sites and be carried to the meninges, where it produces the characteristic symptoms of acute meningitis.

When morphologically typical, short, slender gram-negative rods appear in sufficient numbers in the cerebrospinal fluid (either uncentrifuged or in the sediment), a direct capsular swelling procedure may be carried out using type **b** rabbit antiserum.*

The limulus lysate test for endotoxin is reliable in spinal fluid and almost always is positive in *H. influenzae* and other gram-negative meningitis.[22] Reagents are available commercially.*

The initial culture procedures outlined

in the previous section are satisfactory for the isolation of *H. influenzae;* further cultural details will be found in Chapter 23.

MENINGOCOCCAL MENINGITIS

Neisseria meningitidis is the cause of epidemic bacterial meningitis and is involved in sporadic cases as well. All ages may be involved, but cases are seen more frequently among infants and young children. Epidemics may be a special problem in the military. The three stages of meningococcal infection are the local or nasopharyngeal (of no clinical importance to the individual), the invasive or bacteremic stage, and the meningeal phase.

On spinal fluid Gram stain, the organisms appear primarily as intracellular gram-negative coffee bean–shaped diplococci. Capsular swelling tests* may be done successfully when sufficient organisms are present. A slide agglutination test and the FA procedure are also useful. See Chapters 18 and 38 for further details on these procedures and for additional information on cultivation and identification procedures.

PNEUMOCOCCAL MENINGITIS

Streptococcus pneumoniae is the third most common cause of bacterial meningitis. It is seen most often in infants and elderly persons. In infancy, pneumococcal meningitis is usually secondary to otitis media, mastoiditis, and pneumonia. In older persons, the disease is seen most frequently in alcoholics and may accompany extensive pneumococcal pneumonia. Sickle cell disease and previous skull fractures may also predispose to pneumococcal meningitis.

The quellung reaction, counterimmunoelectrophoresis and other direct tests, and cultural and identification procedures are discussed in Chapters 18 and 38.

---

*Difco Laboratories, Detroit.

*A variety of antisera are available from Difco Laboratories, Detroit.

MYCOBACTERIUM TUBERCULOSIS AS A CAUSE OF MENINGITIS

Cerebrospinal fluid from patients with tuberculous meningitis contains, at best, only a few recoverable tubercle bacilli. Acid-fast stains of the sediment, therefore, rarely reveal their presence. The chance of recovering the organisms increases in proportion to the volume of spinal fluid submitted for examination. About 10 ml are required for an adequate culture.*

Although cultural procedures are preferred for the recovery of tubercle bacilli from cerebrospinal fluid, the clinical significance of finding any acid-fast bacilli by smear is obvious. Because of the urgency to discover an etiologic agent, two methods for demonstrating mycobacteria in stained smears are included:

1. Preparations may be made from the pellicle or coagulum ("fibrin web") that forms in some specimens. This formation takes place when the spinal fluid is allowed to stand undisturbed at room temperature or in a refrigerator overnight. Carefully remove the formed coagulum with a **new** loop (a rough, old wire loop is inadequate, since it balls up the material and makes removal difficult). Crush the coagulum by pressure between an albumin-coated slide and a coverglass. Air dry, flame, and then stain the resulting films by the acid-fast or fluorochrome method.

2. Acid-fast bacilli occasionally can be demonstrated in the sediment obtained by centrifuging the cerebrospinal fluid at 3,000 rpm or higher for 15 to 30 minutes. Since this method requires the use of the entire sediment and is not always reliable, it is recommended that the sediment be used instead for inoculation of culture media or guinea pig injection. Techniques for these procedures will be found in Chapter 31 on *Mycobacterium tuberculosis.*

Two recent papers from the CDC (Brooks, J. B., et al.: J. Clin. Microbiol. **5:**625-628, 1977; Craven, R. B., et al.: J. Clin. Microbiol. **6:**27-32, 1977) indicate that gas chromatography may be useful in diagnosing tuberculous meningitis.

LISTERIA MONOCYTOGENES AS A CAUSE OF MENINGITIS

In recent years attention has been focused on the role of *Listeria monocytogenes* in acute meningitis and meningoencephalitis in human beings. In a study of more than 420 cases of human listeriosis in the United States, Gray and Killinger[12] found that over 80% occurred as meningitis or meningoencephalitis in neonates or in adults over 40 years of age. Less than 7% were perinatal infections, with only an influenzalike illness of the mother, but frequently premature delivery of a stillborn or acutely ill infant. The organism is also occasionally recovered from the blood or spinal fluid of debilitated patients or those undergoing chemotherapy for tumor or leukemia.[17] It has also emerged as a significant opportunist pathogen in the immunosuppressed host. The organism is probably too often discarded in the laboratory as a contaminating diphtheroid when isolated from cerebrospinal or subdural fluid, blood, or other clinical specimens; a high index of suspicion on the part of the microbiologist is the best aid to its recovery. The reader is referred to Chapter 26 on *Listeria,* where identification procedures may be found.

CRYPTOCOCCUS NEOFORMANS AS A CAUSE OF MENINGITIS

Cryptococcus neoformans, a yeastlike fungus, has been responsible for a number of meningeal infections in humans, some of which have been diagnosed erroneously as brain tumors or as tuberculous meningitis. Both immunosuppressed and debilitated and normal hosts may be af-

---

*Use of the membrane filter culture procedure increases the likelihood of recovery of tubercle bacilli.

fected. Budding yeast cells with capsules occur in masses in tissue; the cerebrospinal fluid, although generally not grossly purulent, may contain numbers of yeast cells as well as variable numbers of lymphocytes. The organisms may be mistaken for lymphocytes in spinal fluid cell counts, since cryptococci resemble them in size and shape; the correct diagnosis may be overlooked unless cultures are made. The organism is heavily encapsulated, however, and can be identified readily by mixing a loopful of sediment from the centrifuged spinal fluid with a loopful of India ink on a slide, covering with a thin coverglass, scanning under the high dry objective, and confirming under oil immersion.* Typically, large, clear hyaline **capsules** are seen surrounding the yeast forms, some of which show single buds. Since the number of cryptococci in cerebrospinal fluid may be of the magnitude of only one cell per 15 to 20 ml of fluid, Utz[26] has recommended that the **entire specimen** be inoculated directly to a series of culture slants. He has frequently made an etiologic diagnosis only after inoculating 40 to 50 ml of fluid collected during encephalography. As in tuberculous meningitis, use of the membrane filter procedure, with as much spinal fluid as possible, increases the likelihood of recovery of fungi. The test for cryptococcal antigen is an especially important and useful diagnostic tool. Gas chromatography may also be useful.[23]† Further identification procedures are found in Chapters 32 and 36.

LEPTOSPIRA AS A CAUSE OF MENINGITIS

Human **leptospirosis** is a disease of protean manifestations, varying from a mild

---

*Some authors believe that centrifugation may injure the fragile cryptococcal cells and recommend inoculating the uncentrifuged specimen directly to appropriate culture slants.
†A recent paper by Craven and co-workers (J. Clin. Microbiol. **6:**27-32, 1977) also discusses the use of gas chromatography in the diagnosis of cryptococcal meningitis.

and inapparent infection to a severe and sometimes fatal illness with deep jaundice and profound prostration. It is transmitted by contact, either directly or indirectly, with the urine of animal carriers. The central nervous system may be involved in many cases of severe infection, and leptospirae may be recovered from the spinal fluid. In a report by Heath and co-workers[13] of 483 cases of human leptospirosis in the United States, the central nervous system was the organ system most frequently involved (235 cases, or 68%). Because of the diversity of clinical expressions, however, the authors concluded that many human infections, particularly of a mild type, remain undiagnosed in the United States.

Recovery of the organism (see description of techniques, Chapters 7 and 24) from the cerebrospinal fluid or demonstration of a rise in specific antibody may lead to a definitive diagnosis. All suspected cultures should be confirmed by a leptospirosis reference laboratory, such as the Center for Disease Control (CDC).

MENINGITIS CAUSED BY GRAM-NEGATIVE BACTERIA

Meningeal infections caused by coliform bacilli, species of *Pseudomonas* or *Proteus*, and other gram-negative enteric organisms are not common. In a series of 294 cases of bacterial meningitis studied at the Mayo Clinic[8,11] only 23 (7.7%) were caused by these facultative gram-negative bacilli. In nine of the 23 cases, *Escherichia coli* was the responsible pathogen, causing death in three of four infected newborn infants, who appear to have a great susceptibility to this organism; other bacterial agents included *Klebsiella, Pseudomonas, Proteus,* and *Alcaligenes.* In 21 of the patients an underlying disease process was present; the infection was controlled in only eight of the 23 patients with meningitis. In adults, gram-negative bacillary meningitis may be seen secondary to gram-negative bacteremia, often arising from the urinary tract.

Acinetobacter calcoaceticus

A disease simulating meningococcal meningitis, both clinically and bacteriologically, may be produced by a gram-negative coccobacillus previously known as *Mima polymorpha*,[20] first described by de Bord in 1948[6] and presently classified as *Acinetobacter calcoaceticus.* This organism appears coccoid on solid media and shows both bacillary and coccal forms in liquid media. It stains gram negative, frequently shows bipolar staining, is nonmotile, and does not reduce nitrates.

Moraxella osloensis, formerly classified as a member of de Bord's Mimeae (*M. polymorpha* var. *oxidans*) is similar in morphology and gram reaction to *Neisseria meningitidis* and is also oxidase positive. It has been incriminated in meningitis[27] and is best characterized by biochemical reactions, as described in Chapter 22.

Flavobacterium species

Flavobacterium species are gram-negative bacilli whose natural habitat is the soil and water. They have also been isolated from a variety of other sources, including hospital sink traps, nursery equipment, and water supplies.

A report by Brody and others[2] recorded two outbreaks of meningitis among newborn infants in hospital nurseries in which a gram-negative bacillus was isolated from the spinal fluid of 17 of the 19 infants; there were 15 deaths. The organism, first described in 1944, was studied by King, who found the organisms to be thin, nonmotile, gram-negative rods, proteolytic and nitrite negative, producing a small amount of indole (tested with Ehrlich reagent) and fermenting carbohydrates in nutrient broth after some delay. After overnight incubation at 35 C, a lavender-green discoloration is seen on blood agar. Another nursery outbreak was described by Cabrera and Davis,[4] who re-

corded 14 clinical cases of neonatal meningitis, with 10 deaths. The source of this nosocomial epidemic appeared to be a faulty sink trap in the premature nursery, and the organism isolated was similar to that described by King.

Because the organisms satisfy the requirements for inclusion in the genus *Flavobacterium*, King suggested the name *Flavobacterium meningosepticum* for them. For a more complete biochemical and serologic characterization, the reader is referred to King's paper[16] and to Chapter 22.

Edwardsiella

A report by Sonnenwirth and Kallus[24] details the first case of meningitis due to *Edwardsiella tarda*, a member of the family Enterobacteriaceae. Variously named "Asakusa group," "Bartholomew group," and "biotype 1483-59," the genus *Edwardsiella* (with a single species, *E. tarda*) was proposed by Ewing and colleagues[9] in 1965.

The organism is characterized by its marked production of hydrogen sulfide, a negative urease and beta-galactosidase (ONPG) reaction, and the fermentation of glucose and maltose but not lactose and mannitol. The reader is referred to Chapter 20 and to the references cited for a further description of *E. tarda*.[7]

Pasteurella multocida

The pleomorphic gram-negative coccobacillus *Pasteurella multocida* has been implicated in central nervous system infections, including meningitis.[15] Because of its appearance in stained films of spinal fluid, it has led to incorrect diagnoses of meningococcal, *Haemophilus influenzae*, or "*M. polymorpha*" types of meningitis. Unlike most of the gram-negative rods, *P. multocida* is **very susceptible to penicillin** in vitro (either by the disk technique or by tube dilution). If such an organism is isolated, if it grows poorly or not at all on EMB agar or MacConkey agar, and if it produces no change in the butt of a TSI or

KIA agar slant within 24 hours but does produce an acid reaction throughout the medium in 48 to 72 hours, the presence of *P. multocida* should be considered. This may be confirmed biochemically and serologically by submitting the culture to a reference laboratory. See Chapter 23 for a further description of the organism.

MENINGITIS DUE TO ANAEROBIC BACTERIA

Although meningitis involving anaerobes is not common, it has undoubtedly been overlooked many times. Nonetheless, there are well over 200 cases reported in the literature.[10] Anaerobic meningitis is not uncommonly part of a more extensive intracranial infection; brain abscess or subdural or extradural empyema are frequently associated. By far the most common underlying process is otitis media or mastoiditis, usually chronic. *Fusobacterium* and *Bacteroides* species, including *B. fragilis,* and anaerobic and microaerophilic cocci are the most commonly encountered anaerobes, but clostridia (chiefly *C. perfringens*) and *Actinomyces* are recovered as well.

Since a number of other agents that cause meningitis may grow at least as well, and sometimes better, under anaerobic conditions as aerobically, it is clearly desirable that all spinal fluid be cultured anaerobically as well as by other techniques. Further details on cultivation and identification of anaerobes are given in Chapters 13 and 27-30.

MENINGITIS CAUSED BY MORE THAN ONE BACTERIUM

In the series previously cited,[8] cultures of the cerebrospinal fluid of 40 patients (13.6%) revealed two or more organisms isolated simultaneously or subsequently during the course of the illness. These included *Streptococcus faecalis,* coagulase-negative staphylococci, *Staphylococcus aureus,* gram-negative enteric bacilli, and so forth in various combinations. In some instances the clinical signifi-cance of the additional organism was not readily determined.

In a similar study of 534 infants and children conducted at St. Louis Children's Hospital,[14] 20 patients (3.7%) were found to show two different bacterial species in initial cerebrospinal fluid cultures. *Haemophilus influenzae,* combined variously with *N. meningitidis, S. pneumoniae, E. coli, S. aureus, E. aerogenes,* and other *Enterobacter* species, occurred most frequently; the meningococcus or pneumococcus was also accompanied by other bacterial species. The authors pointed out that those having the responsibility for reading cerebrospinal fluid cultures should be aware of the possibility of the occurrence of simultaneous mixed bacterial meningeal infections.

ASEPTIC MENINGITIS

Aseptic or nonbacterial meningitis is a **clinical syndrome** rather than a disease of specific etiology. It is generally associated with a viral agent, although approximately one fourth of the cases may remain undiagnosed.

The disease is characterized by rapid onset with fever and symptoms referable to the central nervous system. In the majority of patients with viral meningitides, the spinal fluid cell count is increased, with lymphocytes predominating (polymorphonuclear cells may appear at the onset). The protein content is generally elevated, while the spinal fluid glucose level is usually normal, an important diagnostic differential from bacterial meningitis, in which the glucose level is depressed.

Aseptic meningitis may be caused by a variety of viral agents, including Coxsackie group B (occasionally A) viruses, poliovirus, lymphocytic choriomeningitis and arthropod-borne encephalitis viruses, enteroviruses, mumps virus, and herpesvirus. Leptospirosis may also be present as a benign aseptic meningitis. The multiplicity of etiologic agents requires a variety of laboratory examinations available only at a diagnostic virology laboratory. These include virus isolation by animal injection or by inoculation of embryonated eggs or tissue cell cultures from specimens of stool, throat washings, and cerebrospinal fluid, as well as complement fixation and neutralization tests on acute and convalescent sera. Since the only contribution of the bacteriology laboratory in such cases may be the responsibility for the proper collection and transportation of the necessary clinical materials, these techniques are considered more fully in Chapter 33.

Aseptic meningitis may also occur after measles, mumps, vaccinia, chickenpox, and smallpox. It may be caused by a parameningeal focus of infection, including brain abscess, and by toxins, neoplasms, and allergens.[25]

EFFECT OF PRIOR ANTIBACTERIAL THERAPY ON RECOVERY OF BACTERIAL PATHOGENS

In a series of 310 cases of bacterial meningitis, Dalton and Allison[5] reported that approximately one half of the patients had received antibacterial drugs before admission to the hospital. This partial treatment reduced the recovery rate of the etiologic agent by approximately 30%, particularly in the isolation of *Neisseria meningitidis*, *Streptococcus pneumoniae*, and a miscellaneous group. Elevated spinal fluid lactic acid levels may be very useful in establishing the diagnosis of bacterial meningitis in this situation, but this does not pinpoint the specific infecting agent.[3]

There was also a reduction in the number of positive spinal fluid smears in the treated group, especially in meningococcal meningitis. Gram-positive organisms also tended to appear gram negative. The authors recommended caution in interpreting smears from these treated cases. False-positive Gram stains of spinal fluid may also be a problem, due to organisms in tubes used to collect or centrifuge the fluid and in stains.[19]

REFERENCES

1. Bartlett, R. C.: In Summary report, 1968, Chicago, Commission on Continuing Education (American Society of Clinical Pathologists).
2. Brody, B. H., Moore, H., and King, E. O.: Meningitis caused by an unclassified gram-negative bacterium in newborn infants, J. Dis. Child. **96:** 1-5, 1958.
3. Brook, I., Bricknell, K. S., Overturf, G., and Finegold, S. M.: Abnormalities in spinal fluid detected by gas liquid chromatography in meningitis patients, Annual Meeting, 1976, American Society for Microbiology, Abstract C107, p. 43.
4. Cabrera, H. A., and Davis, G. H.: Epidemic meningitis of the newborn caused by flavobacteria, J. Dis. Child. **101:**289-295, 1961.
5. Dalton, H. P., and Allison, M. J.: Modification of laboratory results by partial treatment of bacterial meningitis, Am. J. Clin. Pathol. **49:**410-413, 1968.
6. De Bord, G. G.: *Mima polymorpha* in meningitis, J. Bacteriol. **55:**764-765, 1948.
7. Edwards, P. R., and Ewing, W. H.: Identification of Enterobacteriaceae, ed. 3, Minneapolis, 1972, Burgess Publishing Co.
8. Eigler, J. O. C., Wellman, W. E., Rooke, E. D., Keith, H. M., and Svien, H. J.: Bacterial meningitis. I. General review (294 cases), Proc. Staff Meet. Mayo Clin. **36:**357-365, 1961.
9. Ewing, W. H., McWhorter, A. C., Escobar, M. R., and Lubin, A. M.: *Edwardsiella*, a new genus of Enterobacteriaceae based on a new species, *E. tarda*, Int. Bull. Bacteriol. Nomenclat. Taxon. **149:**33-38, 1965.
10. Finegold, S. M.: Anaerobic bacteria in human disease, 1977, New York, Academic Press, Inc.
11. Gorman, C. A., Wellman, W. E., and Eigler, J. O. C.: Bacterial meningitis. II. Infections caused by certain gram-negative enteric organisms, Proc. Staff Meet. Mayo Clin. **37:**703-712, 1962.
12. Gray, M. L., and Killinger, A. H.: *Listeria monocytogenes* and listeric infections, Bacteriol. Rev. **30:**309-382, 1966.
13. Heath, C. W., Jr., Alexander, J. D., and Galton, M. M.: Leptospirosis in the United States, N. Engl. J. Med. **273:**912-922, 1965.
14. Herweg, J. C., Middlekamp, J. N., and Hartmann, A. F., Sr.: Simultaneous mixed bacterial meningitis in children, J. Pediatr. **63:**76-83, 1963.
15. Hubbert, W. T., and Rosen, M. N.: *Pasteurella multocida* infections. II. *Pasteurella multocida* infection in man unrelated to animal bite, Am. J. Public Health **60:**1109-1116, 1970.

16. King, E. O.: Studies on a group of previously unclassified bacteria associated with meningitis in infants, Am. J. Clin. Pathol. **31**:241-247, 1959.

17. Louria, D. B., Blevins, A., and Armstrong, D.: *Listeria* infections, Ann. N.Y. Acad. Sci. **174**:545-551, 1970.

18. Louria, D. B., Kaminski, T., Kapila, R., Tecson, F., and Smith, L.: Study on the usefulness of hypertonic culture media, J. Clin. Microbiol. **4**:208-213, 1976.

19. Musher, D. M., and Schell, R. F.: False-positive gram stains of cerebrospinal fluid, Ann. Intern. Med. **79**:603-604, 1973.

20. Olaffson, M., Lee, Y. C., and Abernathy, T. J.: *Mima polymorpha* meningitis; report of a case and review of the literature, N. Engl. J. Med. **258**:465-470, 1958.

21. Rosner, R.: Comparison of isotonic and radio-metric-hypertonic cultures for the recovery of organisms from cerebrospinal, pleural and synovial fluids, Am. J. Clin. Pathol. **63**:149-152, 1975.

22. Ross, S., et al.: Limulus lysate test for gram-negative bacterial meningitis, J.A.M.A. **233**: 1366-1369, 1975.

23. Schlossberg, D., Brooks, J. B., and Shulman, J. A.: Possibility of diagnosing meningitis by gas chromatography: cryptococcal meningitis, J. Clin. Microbiol. **3**:239-245, 1976.

24. Sonnenwirth, A. C., and Kallus, B. A.: Meningitis due to *Edwardsiella tarda;* first report of meningitis caused by *E. tarda,* Am. J. Clin. Pathol. **49**:92-95, 1968.

25. Swartz, M. N., and Dodge, P. R.: Bacterial meningitis; a review of selected aspects, N. Engl. J. Med. **272**:898-902, 1965.

26. Utz, J. P.: Recognition and current management of the systemic mycoses, Med. Clin. North Am. **51**:519-527, 1967.

27. Waite, C. L. and Kline, A. H.: *Mima polymorpha* meningitis; report of case and review of the literature, J. Dis. Child. **98**:379-384, 1959.

CHAPTER 13

Microorganisms encountered in wounds; anaerobic procedures

Although the microbial flora of infected wounds frequently is varied, **organisms most frequently isolated from wounds** are:

> *Staphylococcus aureus*
> *Streptococcus pyogenes*
> Coliform bacilli (from the lower half of body)
> *Bacteroides* species and other anaerobic nonsporing gram-negative and gram-positive rods
> *Proteus* species
> *Pseudomonas* species
> *Clostridium* species
> Anaerobic cocci *(Peptococcus, Peptostreptococcus)*
> Enterococci

Another group, which might be labeled **organisms rarely isolated from wounds,** includes:

> *Clostridium tetani, C. botulinum*
> *Francisella tularensis, Pasteurella multocida*
> *Mycobacterium tuberculosis, M. marinum,* and other mycobacteria
> *Corynebacterium diphtheriae*
> *Bacillus anthracis*
> Systemic fungi *(Sporothrix* and so forth)
> *Erysipelothrix insidiosa*
> *Actinomyces, Nocardia* species

Since **anaerobic** microorganisms are the predominant microflora of humans and are constantly present in the intestinal, upper respiratory, and genitourinary tracts, it is not unexpected to find them invading both usual and unusual anatomical sites, giving rise to severe and often fatal infections.[10] Although anaerobes are commonly involved in wounds and this chapter is a convenient place to discuss anaerobic procedures, it should be kept in mind that anaerobes may participate in all varieties of infections and may involve any organ or tissue of the body. Therefore, it seems that anaerobes deserve more attention than they have been given and it behooves the microbiologist and technologist to familiarize themselves with the techniques for the isolation and identification of these indigenous bacteria.

Although anaerobic procedures are not more difficult to carry out than those used in aerobic bacteriology, only a strict adherence to basic principles and a degree of patience will ensure successful recovery of these pathogens. These principles include the following:*

1. **Proper selection of specimens.** As suggested by Sutter and co-workers,[23] certain bacteriologic clues should prompt the microbiologist to carry out anaerobic culturing of selected clinical material. Some of these specimens include:

a. Pus from any deep wound or soft-tissue abscess, especially if associated with a **foul or fetid** odor, or containing "sulfur granules"

b. Necrotic tissue or debrided material from suspected gas gangrene or less serious gas-forming or necrotizing infections

c. Material from infections close to a mucous membrane

d. Material from abscesses of the brain, lung, liver, or other organ or from intraabdominal, perirectal, subphrenic, or other sites

e. Aspirated fluids from infections of normally sterile sites, including blood and peritoneal, pleural, synovial, or amniotic fluids

---

*The reader is referred to several excellent anaerobic laboratory manuals,[8,12,23] for a more complete discussion of the subject.

f. Material from an infected human or animal bite
g. Exudates with black discoloration or red fluorescence under ultraviolet light
h. Material which, on Gram stain, shows organisms with unique morphology suggestive of anaerobes
i. Food suspected of causing botulism or food poisoning

2. **Proper specimen collection.** Since anaerobic organisms are part of the normal flora of body sites such as the skin, oropharynx, intestinal tract, or genitalia, the following are **not** cultured anaerobically:

a. Swabs from the throat, nose, urethra, vagina, cervix, or rectum
b. Expectorated sputum, bronchoscopic specimens, voided or catheterized urine, feces, and gastric contents

Furthermore, all other specimens should be collected to preclude or minimize contamination by normal anaerobic microflora, and **collection by needle aspiration rather than by swab is recommended.** This is particularly appropriate in the following clinical situations:

a. Pus from a closed abscess
b. Pleural fluid (by thoracentesis)
c. Urine (by suprapubic bladder aspiration)
d. Pulmonary secretions (by transtracheal aspiration)
e. Sinus tract material (by insertion of a small-gauge pediatric intravenous type of plastic catheter through a decontaminated area and aspiration with a syringe)

All air and gas should be expelled from the syringe and needle and their contents injected directly into a "gassed out" sterile tube, described later. If a swab **must** be used, a two-tube system has been recommended,[23] one tube containing the swab in oxygen-free CO_2, the other containing a prereduced and anaerobically sterilized (PRAS) transport medium, such as Cary and Blair semisolid medium.* After collection, the swab should be inoculated to appropriate culture media as soon as possible. As noted in Chapter 6, the Anaerobic Culture/Set† is excellent for maintaining an anaerobic environment during transport of swabs, small amounts of fluid in plastic catheters, larger specimens in syringes, and so forth.

3. **Proper specimen transport.** As indicated previously, once a clinical specimen has left a body site where the Eh (oxidation-reduction potential) may be as low as or lower than –250 mV,[12] any anaerobes present must be protected from the toxic effect of atmospheric oxygen until the specimen is properly set up anaerobically. While larger volumes of frankly purulent material require little or no protection, it is wise to **routinely** use optimal transport procedures. For this reason the use of a "gassed out" collection tube is strongly recommended, since it will provide an anaerobic environment during transport to the laboratory. The outfit consists of a tube with a recessed butyl rubber stopper and screw cap; it is prepared with prereduced nonnutritive medium containing cysteine and resazurin indicator, then flushed out with oxygen-free CO_2 and sterilized by autoclaving. The tube can be prepared by hand with a modified Hungate method[2] or in an anaerobic chamber or glove box; although expensive, it is available commercially.‡ See Chapter 6 regarding the Bellco anaerobic tube, which is very convenient to use. The specimen is injected through the rubber stopper by syringe and needle and subsequently in the laboratory is removed in the same way. As noted above and in Chapter 6, specimens can be transported effectively in plastic syringes in the Anaerobic Culture/Set. For short periods of transport (up to 20

---

*Available commercially from Scott Laboratories, Fiskeville, R.I.; Gibco, Grand Island, N.Y.
†Available commercially from Marion Scientific Corp., Kansas City, Mo., Catalog No. 26-02-06.
‡Anaport, Scott Laboratories, Fiskeville, R.I.

minutes), the syringe alone is satisfactory. Tissue can be transported in the Anaerobic Culture/Set or in the anaerobic minijar.[23]

4. **Use of fresh culture media.** For the primary inoculation of specimens for anaerobic culturing, all media should be fresh. There is no evidence that storage of plating media under anaerobic conditions is necessary, but it would likely extend the effective life of the medium.[11,18] If such storage is to be used for over 72 hours, the atmosphere should **not** contain CO_2.[11] Freshly made media may be stored in folded-over Mylar bags (the plastic bags that Petri dishes come in) in the refrigerator for up to 2 weeks.[18] In suspected actinomycosis, media prepared the same day should be used.

5. **Provision of a proper anaerobic environment.** Numerous methods have been introduced for providing an anaerobic environment, including the PRAS roll-tube method of Hungate,[13] the anaerobic chamber[20] or glove box,[1] and the anaerobic jar.[4,7] It appears that the use of the anaerobic jar with a catalyst and hydrogen, nitrogen, and 5% to 10% CO_2 offers the most satisfactory and practical method for achieving an anaerobic environment in the clinical laboratory; this is discussed further in the next section. Other methods, including techniques that depend on displacement of oxygen by inert gases alone, by chemical means, or by cultivation in a vacuum are not recommended in routine clinical practice.

METHODS OF OBTAINING ANAEROBIOSIS
Anaerobic jar method

The simplest method for the cultivation of anaerobes is the use of the **anaerobic jar,** a tightly sealed container in which the oxygen is completely eliminated by various means, including hydrogen and a catalyst. This principle was first applied by Laidlaw[15] in 1915 and later adapted by McIntosh and Fildes.[16] Many modifications of the McIntosh and Fildes jar have

been introduced, including the Brewer,* GasPak,* and Torbal jars.*

Brewer and Torbal jars

The Brewer[7] modification of the Brown jar is no longer being sold, but many are still in use. In it the oxygen is removed by means of an electrically heated platinized catalyst with the electrical connection outside the jar (eliminating the danger of explosion). The reaction chamber within the lid is shielded by a heavy wire mesh screen. An H_2-CO_2-N_2 gas mixture is recommended, and is employed as follows.[8] The jar is loaded with the media to be incubated, along with an indicator for anaerobiosis (see following section), and an evenly rolled length of Plasticine with ends joined is placed on the jar rim. The jar lid is pressed down on the Plasticine to form a good seal and is clamped in place, and the whole assembly is placed in a metal safety shield.

After evacuating the jar to approximately 60 to 70 cm of mercury (by water or vacuum pump), the jar is slowly filled with a gas mixture containing 10% hydrogen, 10% carbon dioxide, and 80% nitrogen.† This evacuation and filling procedure is repeated two more times, the outlet tubing is clamped, the jar is disconnected from the vacuum-gas assembly, and the electrical element is connected and allowed to heat for 30 minutes. The jar is then disconnected and placed in the incubator. Nitrogen can be used alone for all but the final fill. Further details may be found elsewhere.[9] The Torbal jar is similar but uses a rubber O-ring rather than Plasticine and a catalyst active at room temperature (thus no electrical heating is required).

---

*Baltimore Biological Laboratory, Cockeysville, Md.; Torsion Balance Co., Clifton, N.J.; Baird & Tatlock, Ltd., London, England.

†These gases are available commercially in a range of cylinder sizes from Matheson Co., Inc., Joliet, Ill., East Rutherford, N.J., or Newark, Calif., along with the necessary regulators, gauges, valves, and so forth. Consultation with the hospital's oxygen therapy staff is suggested in setting up such an assembly.

GasPak jar

The introduction by Brewer and All-geier[3-5] of the GasPak anaerobic jar* for both 100- and 150-mm plates, a disposable hydrogen and carbon dioxide generator envelope,* and a disposable anaerobic indicator* makes possible the simplest and most practical system for the cultivation of anaerobes. The polycarbonate plastic anaerobic jar, used with the disposable hydrogen generator, has no external connections, thereby eliminating the need for vacuum pumps, gas tanks, manometers, and so forth. It uses a room-temperature catalyst (palladium-coated alumina pellets), which obviates the need for an electrical connection to heat the catalyst.

To use this system the inoculated media are placed in the jar, along with one hydrogen generator envelope with a top corner cut off and a methylene blue anaerobic indicator.[6] After 10 ml of water are introduced with a pipet into the envelope, the jar cover (containing the catalyst in a screened reaction chamber in the lid) is promptly placed in position and the clamps are applied and **screwed only hand tight** (rubber attachments on the clamp provide release of excessive pressure).

Hydrogen is generated in the jar by the following reaction:

$$Mg + ZnCl_2 + 2H_2O \xrightarrow{NaCl} MgCl_2 + Zn(OH)_2 + H_2 \uparrow$$

The hydrogen reacts with the oxygen in the presence of the catalyst:

$$2H_2 + O_2 \rightarrow 2H_2O$$

An anaerobic atmosphere is thereby produced.

As this environment is achieved, condensed water will appear as a visible mist or fog on the inner wall of the jar and the lid over the catalyst chamber will become warm. If this does not occur within 25 minutes, either the catalyst needs replac-

ing* or the lid was not secured properly. After overnight incubation the methylene blue indicator should appear colorless and the jar will be under a slight positive pressure.

It should be pointed out that the disposable hydrogen generator may also be used in the Torbal jar or Brewer jar (see the manufacturer's directions for such use).

Since hydrogen is an **explosive** gas, every precaution must be taken to prevent a laboratory accident when using it:

1. Any open flame in the vicinity must be extinguished.
2. All jars must be inspected for cracks and discarded if faulty.
3. The metal screen inside a Brewer jar lid must be intact.
4. A wooden or metal safety shield should be used whenever the catalyst is activated electrically.

Other methods

Other anaerobic methods mentioned previously include the use of PRAS culture media and roll tubes, introduced by Hungate,[13] with modifications by Holdeman and Moore,[12] and the anaerobic glove box[1] or chamber.[20]

In the Hungate technique, a closed tube of PRAS medium is inoculated through the rubber stopper by needle and syringe; in the VPI technique, an open tube of PRAS medium is inoculated in the presence of a continuous stream of oxygen-free gas introduced by a flame-sterilized cannula inserted in the tube.

The anaerobic chamber techniques utilize a plastic glove box or rigid chamber

---

*This should be done **after each use,** since excess moisture and hydrogen sulfide (from H_2S-producing organisms) will inactivate the catalyst. Workers at the VPI Anaerobe Laboratory[12] have suggested that the catalyst may be reactivated by being heated in a 160 C drying oven for 2 hours and subsequently stored in an airtight container with desiccant, such as a discarded antibiotic disc cartridge container. It is convenient to have extra baskets of catalyst for this purpose. These catalysts may be regenerated repeatedly over extended periods of time.

---

*Baltimore Biological Laboratory, Cockeysville, Md.

with attached and sealed gloves, which are used to manipulate the material inside this work area. The chamber is kept continuously anaerobic by a catalyst and hydrogen gas, and material is passed in and out through an interchange, a rigid appurtenance attached by a gastight seal to the chamber.

Several authors[14,18] have compared the three anaerobic systems—GasPak, glove box, and roll tube—for their effectiveness in the isolation of anaerobic organisms from clinical material and have concluded that recovery of anaerobes is comparable in all three systems and that the GasPak method was as effective as the other more complex methods. The interested reader is referred to the papers cited for further details.

INDICATORS OF ANAEROBIOSIS

The use of an **indicator** of oxidation-reduction potential is essential in anaerobic culturing. Among those available, the original Fildes and McIntosh methylene blue indicator is recommended (see Chapter 43 for preparation). The indicator is freshly prepared each time by mixing equal parts of solutions of methylene blue, glucose, and sodium hydroxide in a test tube and boiling until colorless (the methylene blue is reduced to its leuco base). This tube is immediately placed in the previously loaded jar, and the jar is then sealed and charged by the methods described. If anaerobic conditions are secured and maintained throughout the incubation period, the indicator solution will be **colorless**. Should the indicator turn blue, anaerobiosis is not achieved or maintained. A disposable indicator also may be used; it is available commercially* in a sealed envelope, which is opened and prepared at the time of use.

Smith[19] recommends the inclusion of a culture of a strict and fastidious anaerobe, such as *Clostridium haemolyticum* or *C. novyi* type b, as a monitor for the adequacy of anaerobic media and methods; if isolated colonies of these organisms are obtained by the procedures used, the techniques are satisfactory for all known anaerobic pathogens.

INOCULATION OF CULTURES

A variety of liquid and solid culture media, including various selective media, are available for primary inoculation of clinical specimens, and many are available commercially.*

The following media are inoculated **as soon as possible** after collection (methods previously described):

1. Swabs (or aspirated material from a syringe or a gassed-out tube) are used to inoculate the following **fresh** solid media†:

   1 trypticase soy blood agar plate (BA)
   1 brucella blood agar plate with added vitamin K_1 and hemin (BRBA)
   1 kanamycin-vancomycin laked blood agar plate (KVLBA) with added vitamin K_1 and hemin[25]
   1 phenylethanol blood agar (PEA) (optional)

   A direct smear for Gram stain is also prepared (see following section).

   The plates are streaked, to secure isolated colonies, with a platinum-iridium loop (**not** nichrome, which oxidizes the inoculum).

2. The swab is then placed directly into two tubes of enriched thioglycollate medium‡ (THIO) and gently rotated (avoiding agitation). Optional: freshly boiled and cooled chopped meat–glucose medium (CMG).†

3. If fluid material is submitted, it is inoculated with a capillary pipet; one or more drops are deposited on each

---

*GasPak disposable anaerobic indicator, Baltimore Biological Laboratory, Cockeysville, Md.

---

*Scott Laboratories, Fiskeville, R.I.; Gibco, Grand Island, N.Y.
†Preparation of these media is described in Chapter 41.
‡Freshly prepared, or boiled (10 minutes) and cooled, after which add vitamin K_1 (0.1 μg/ml), sodium bicarbonate (1 mg/ml), and hemin (5 μg/ml). Rabbit or horse serum (10%) or Fildes enrichment (5%) may also be added.[23]

of the plates, and several drops are introduced to the bottom of a tube of enriched thioglycollate medium, with as little agitation as possible.

4. If tissue is received, it is promptly transferred to a sterile tissue grinder and ground with sterile alundum or sand and thioglycollate, avoiding aeration (ideally in an anaerobic chamber or under flowing oxygen-free gas). The resulting homogenate is inoculated as a fluid specimen (step 3).

5. If clostridia are suspected, an egg yolk agar plate (EYA)* is also inoculated. Optional: Nagler egg yolk agar antitoxin plate.* If *Actinomyces* is anticipated, use a freshly made brain-heart infusion agar plate also.

In the event of a seriously ill patient, one may set up a **duplicate** KVLBA and Nagler plate and incubate for 12 to 24 hours (or until growth is visible through the plastic bag) in an Anaerobic Culture/Set. The GasPak jar should not be opened before 48 hours.

EXAMINATION OF DIRECT FILMS

It is important to prepare and examine a gram-stained film of the original specimen **after** inoculating the media (**before** if slide has been flame sterilized); usually sufficient residual material remains on the swab. When enough material is present, or if two swabs are submitted, it is desirable to prepare a Gram stain **before** inoculating the media. Examination of this smear may suggest the desirability of including certain selective or other media that would not be used routinely.

The type and approximate number of organisms present are noted, as well as the presence, shape, and location of spores (? clostridia), branching gram-positive elements (?*Actinomyces)*, gram-positive cocci in pairs or chains (? staphylococci, streptococci), and gram-negative

rods with round or pointed ends (? coliforms, fusobacteria). Pleomorphism and irregularity of staining are also seen with anaerobes.

A **preliminary report** of these findings should be submitted to the attending physician without delay, since this may aid in the selection of appropriate antimicrobial agents. It also serves as an important quality control feature for the laboratory.

INCUBATION OF CULTURES

Incubate all inoculated media at 35 C in the following manner (based on one culture):

1. The BA plate in a candle jar (approximately 3% CO_2), and the THIO* in air.† These are to be used for routine "aerobic" culture. EMB or MacConkey plates should also be used.

2. The KVLBA, BRBA, PEA (and EYA or Nagler, if used) plates in a GasPak jar with H_2-CO_2 generator, anaerobic indicator, and a paper towel in the bottom of the jar to absorb excess moisture formed during incubation.‡

3. The CO_2 and "aerobic" BA and the EMB or MacConkey plates are examined after overnight incubation, then subcultured; isolates are identified, and antibiotic susceptibility tests are set up, as required by aerobic culture methods delineated in subsequent chapters.

4. The THIO (or CMG) tube is examined after 48 hours. If Gram stain fails to show organisms that seem different from those recovered on solid media, the THIO is discarded. Otherwise, it is subcultured to a BRBA (and any other media that seem indicated) for anaerobic incubation.

---

*Preparation of these media is described in Chapter 41.

*This is useful as a "backup" tube and may be examined, when turbid, only when there is no growth on the primary plate (step 2) or there has been failure of anaerobiosis after incubation.
†Some people prefer to incubate the THIO, with loosened cap, in an anaerobic jar.
‡Note that a fresh charge of reactivated catalyst should be used each time a jar is loaded (p. 103).

5. All of the anaerobic plates are incubated for a **minimum of 48 hours** (slow-growing organisms may require 3 to 5 days or longer); if the jar is opened sooner, some fastidious strains may cease to grow, even if reincubated anaerobically.

EXAMINATION OF CULTURES

After appropriate incubation, remove the KVLBA, BRBA, and PEA (and EYA, if used) plates from the GasPak jar, emptying only one anaerobic jar at a time to avoid undue exposure to air. Examine the growth on each plate, using a hand lens. Describe and record on a work sheet the colony types observed; all colony types present on the anaerobic plates are handled as described below:

1. Each different colony type is picked and inoculated to the following:

    A one-fourth or larger sector of BRBA, to be incubated anaerobically

    A one-fourth sector of a chocolate agar plate, to be incubated under CO_2

    A one-fourth sector of BRBA incubated aerobically

    A tube of enriched thioglycollate medium (see footnote, p. 104)

    The anaerobic BRBA plates and THIO are incubated for 48 hours, the aerobic and CO_2 plates for 18 to 24 hours. While the above subculture procedures are being completed, a number of these inoculated plates can be safely held at room temperature in a glass jar under a continuous stream of CO_2 until they are set up anaerobically. The loose-fitting jar top is fabricated of ½-inch-thick Plexiglas and contains a small drill hole to permit escape of the CO_2, which is introduced to the bottom of the jar at a flow rate producing a steady stream of bubbles through a water bottle.* A double-vented jar is also available commercially.†

2. The primary KVLBA and BRBA plates are also examined under ultraviolet light* for the presence of colonies of *Bacteroides melaninogenicus,* which characteristically exhibit a **brick-red fluorescence.** Growth on KVLBA is presumptive evidence that the organism is a member of the genus *Bacteroides.* Biochemical and other tests for definitive identification are described in Chapter 27.

3. After overnight incubation, the CO_2 jar plate and aerobic plate are examined; if growth occurs, the subculture was probably not an anaerobe.† If the organism has not previously been picked up on aerobic plates, it is processed for identification.

4. After 48 hours' incubation, the anaerobic subcultures are examined for growth, hemolysis, pigmentation, pitting of agar, and colonial morphology. If a **pure culture** is present, a Gram stain (and a repeat THIO subculture if insufficient growth has occurred in THIO [step 1]) is made, examined, and recorded on the work sheet. Motility can be checked in a sealed hanging drop slide from a 4- to 6-hour culture in THIO.

5. The growth in pure culture of the THIO is also gram stained, examined, and the results noted, with particular attention paid to Gram reaction, cellular morphology, and arrangement of cells.

6. The EYA plate is examined for the presence of lecithinase activity, which is indicated by the formation of an **opaque zone** in the medium around the growth and for lipase activity, which is indicated by an

---

*Procedure adopted from Martin.[17]
†Baltimore Biological Laboratory, Cockeysville, Md.

---

*"Blak-Ray" ultraviolet lamp and viewbox, model UVL 56, from Ultra-Violet Products, San Gabriel, Calif.
†Some species of *Clostridium* and occasional other anaerobes are aerotolerant.

Table 13-1. Group identification from preliminary tests

Group	Gram reaction	Spores	Antibiotic disk identification C* 10 µg	E 60 µg	K 1,000 µg	2 U*	R 15 µg	Va 5 µg	Lecithinase	Lipase	Nagler	Indole	Catalase	Motility	Bile	SPS disk
Gram-positive cocci	+*	−	R*			S		S	−	−	−	V	V	−		I of *Peptostreptococcus anaerobius*
Gram-negative cocci	−	−	S			S		R	−	−	−	−	V	−		
B. fragilis group	−	−	R	S	R	R^S	S	R	−	−	−	V	V	−	No inhibition or stimulation	
B. melaninogenicus-oralis-ochraceus group	−	−	V	S	R	S^R	S	V	−	V	−	V	−	−	I	
B. corrodens	−	−	S	S	S	S	S	R	−	−	−	−	−	−	I	
F. mortiferum-varium group	−	−	S	R	S	S^R	R	R	−	−	−	V	−	−	No inhibition or stimulation	
Certain other *Fusobacterium* sp.	−	−	S	S^R	S	S	S	R	−	V	−	V	−	−	V	
Certain GPNSB*	+	−	R			S^R		S^R	−	−	−	V	V	−		
Clostridium sp.-α-toxin producers	+	+	R			S^R		V	+	−	+	V	−	V		
Certain other *Clostridium* sp.	+	+	R			S^R		S^R	V	V	−	V	−	V		

Modified from Sutter, V. L., Vargo, V. L., and Finegold, S. M.: Wadsworth anaerobic bacteriology manual, ed. 2, Los Angeles, 1975, Dept. Contin. Educ. Health Sci., Univ. Extens., School Med., UCLA.

*Gram-positive, nonsporeforming bacilli; +, positive or present; −, negative or absent; C, colistin; E, erythromycin; I, inhibition; K, kanamycin; P, penicillin; R, rifampin; Va, vancomycin; R, resistant, zones < 10 mm; R^S, usually resistant, sometimes sensitive; S, sensitive, zones ≥ 10 mm; S^R, usually sensitive, sometimes resistant; U, units; V, variable.

"oil-on-water" **sheen** on the surface of colonies. This plate can also be used for the detection of catalase by exposing it to air for 30 minutes and then dropping 3% hydrogen peroxide on the colony.[23] If subcultures are required, they should be made before exposing the plate to air. The Nagler reaction is discussed in Chapter 29.

7. Perform the spot indole reaction[21] by smearing a loopful of the growth from a trypticase soy blood agar plate on a filter paper saturated with 1% para-dimethylaminocinnamaldehyde in 10% hydrochloric acid.* A blue color indicates a **positive** reaction; no change or a yellow color indicates a **negative** reaction (check with tube test).

8. Prepare a culture in thioglycollate medium (BBL-135 C) enriched with either 5% Fildes extract plus 1 mg/ml sodium bicarbonate or 25% ascitic fluid; incubate 4 to 6 hours or until barely visible growth ensues (half of the density of McFarland Standard No. 1).

9. Inoculate 0.5 ml of the above on each of two blood agar plates with a capillary pipet and spread evenly over the surface with a cotton swab.†

10. Using three disks per plate, place disks of colistin 10 μg, erythromycin 60 μg, kanamycin 1,000 μg, penicillin 2 units, rifampin 15 μg, and vancomycin 5 μg, on the surface of the seeded plates. Only the colistin and penicillin disks are available commercially; the others may be prepared from stock antibiotics‡ (see Sutter and Finegold's paper[22]). The key disks

needed, if one wishes to simplify the procedure, are **vancomycin, kanamycin,** and **colistin.** All three can be placed on a single plate along with an SPS disk,* which should be placed close to the colistin disk.

11. Incubate the blood agar plates anaerobically in a GasPak jar for 48 hours and measure the diameters of the inhibition zones in millimeters (zones $\geq$ 10 mm, sensitive; $<$ 10 mm, resistant).

12. Determine effect of bile (thioglycollate BBL-135C + 2% dehydrated oxgall + 0.1% sodium deoxycholate) by inoculating the above plus a tube of thioglycollate **without** bile and deoxycholate and by observing inhibition or stimulation of the growth.

From the results of the above observations, the group to which the isolate belongs is determined (Table 13-1) and further tests are then carried out for definitive identification. These procedures are described in the following sections under specific headings, such as anaerobic cocci, Bacteroidaceae, anaerobic sporeformers, and so forth.

---

*SPS disk—5% sodium polyanethol sulfonate, 20 μl per disk. Inhibition zones $\geq$ 18 mm with this disk presumptively identify an anaerobic gram-positive coccus as *Peptostreptococcus anaerobius.*[24]

REFERENCES

1. Aranki, A., Syed, A., Kenney, E. B., and Freter, R.: Isolation of anaerobic bacteria from human gingiva and mouse cecum by means of a simplified glove box procedure, Appl. Microbiol. **17:** 568-576, 1969.
2. Attebery, H. R., and Finegold, S. M.: Combined screw-cap and rubber-stopper closure for Hungate tubes (pre-reduced, anaerobically sterilized roll tubes and liquid media), Appl. Microbiol. **18:**558-561, 1969.
3. Brewer, J. H., and Allgeier, D. L.: Disposable hydrogen generator, Science **147:**1033-1034, 1965.
4. Brewer, J. H., and Allgeier, D. L.: Safe self-contained carbon dioxide–hydrogen anaerobic system, Appl. Microbiol. **14:**985-988, 1966.
5. Brewer, J. H., and Allgeier, D. L.: A disposable

---

*Stable up to 9 months when stored in a brown bottle at 4 C.
†Acme Cotton Company.
‡These antibiotic disks may be obtained on written request to Baltimore Biological Laboratories, Cockeysville, Md.

anaerobic system designed for field and laboratory use, Appl. Microbiol. **16:**848-850, 1968.

6. Brewer, J. H., Allgeier, D. L., and McLaughlin, C. B.: Improved anaerobic indicator, Appl. Microbiol. **14:**135-136, 1966.

7. Brewer, J. H., and Brown, J. H.: A method for utilizing illuminating gas in the Brown, Fildes and McIntosh or other anaerobe jars of the Laidlaw principle, J. Lab. Clin. Med. **23:**870-874, 1938.

8. Dowell, V. R., Jr., and Hawkins, T. M.: Laboratory methods in anaerobic bacteriology, CDC Laboratory manual, DHEW Pub. No. (CDC) 74-8272, 1974, Washington, D. C., U. S. Government Printing Office.

9. Finegold, S. M.: Gram-negative anaerobic rods —Bacteroidaceae. In Frankel, S., Reitman, S., and Sonnenwirth, A. C., editors, Gradwohl's clinical laboratory methods and diagnosis, ed. 7, St. Louis, 1970, The C. V. Mosby Co.

10. Finegold, S. M.: Anaerobic bacteria in human disease, New York, 1977, Academic Press, Inc.

11. Hanson, C. W., and Martin, W. J.: Evaluation of enrichment, storage, and age of blood agar medium in relation to its ability to support growth of anaerobic bacteria, J. Clin. Microbiol. **4:**394-399, 1976.

12. Holdeman, L. V., and Moore, W. E. C., editors: Anaerobe laboratory manual, ed. 3, Blacksburg, Va., 1975, Virginia Polytechnic Institute and State University.

13. Hungate, R. E.: A roll tube method for cultivation of strict anaerobes. In: Methods in microbiology, vol. 3B, New York, 1969, Academic Press, Inc.

14. Killgore, G. E., Starr, S. E., Del Bene, V. E., Whaley, D. N., and Dowell, V. R., Jr.: Comparison of three anaerobic systems for the isolation of anaerobic bacteria from clinical specimens, Am. J. Clin. Pathol. **59:**552-559, 1973.

15. Laidlaw, P. P.: Some simple anaerobic methods, Br. Med. J. **1:**497, 1915.

16. McIntosh, J., and Fildes, P.: Cultivation of anaerobes, Lancet **1:**768, 1916.

17. Martin, W. J.: Practical method for isolation of anaerobic bacteria in the clinical laboratory, Appl. Microbiol. **22:**1168-1171, 1971.

18. Rosenblatt, J. E., Fallon, A. M., and Finegold, S. M.: Recovery of anaerobes from clinical specimens, Appl. Microbiol. **25:**77-85, 1973.

19. Smith, L. DS. The pathogenic anaerobic bacteria, ed. 2, Springfield, Ill., 1975, Charles C Thomas, Publisher.

20. Socransky, S., MacDonald, J. B., and Sawyer, S.: The cultivation of *Treponema microdentium* as surface colonies, Arch. Oral Biol. **1:**171-172, 1959.

21. Sutter, V. L., and Carter, W. T.: Evaluation of media and reagents for indole-spot tests in anaerobic bacteriology, Am. J. Clin. Pathol. **58:**335-338, 1972.

22. Sutter, V. L., and Finegold, S. M.: Antibiotic disc susceptibility tests for rapid presumptive identification of gram-negative anaerobic bacilli, Appl. Microbiol. **21:**13-20, 1971.

23. Sutter, V. L., Vargo, V. L., and Finegold, S. M.: Wadsworth anaerobic bacteriology manual, ed. 2, Los Angeles, 1975, Dept. Contin. Educ. Health Sci., Univ. Extens., School Med., UCLA.

24. Wideman, P. A., Vargo, V. L., Citronbaum, D., and Finegold, S. M.: Evaluation of the sodium polyanethol sulfonate disc test for the identification of *Peptostreptococcus anaerobius*, J. Clin. Microbiol. **4:**330-333, 1976.

25. Wilkins, T. D., Chalgrin, S. L., Jiminez-Ulate, F., Drake, C. R., Jr., and Johnson, J. L.: Inhibition of *Bacteroides fragilis* on blood agar plates and reversal of inhibition by added hemin, J. Clin. Microbiol. **3:**359-363, 1976.

CHAPTER 14

Microorganisms encountered in the eye, ear, mastoid process, paranasal sinuses, teeth, and effusions

EYE CULTURES

Because of the constant washing activity of tears and their antibacterial constituents, the number of organisms recovered from cultures of many eye infections may be relatively low. Unless the clinical specimen is obviously purulent, it is recommended that a relatively **large inoculum** and a variety of media be used to ensure recovery of an etiologic agent.

The following organisms are **most frequently isolated** from infections of the eye:

> **Pathogens or potential pathogens**
> *Staphylococcus aureus*
> *Haemophilus* species
> *Streptococcus pneumoniae*
> *Neisseria gonorrhoeae*
> Alpha- and beta-hemolytic streptococci
> *Moraxella lacunata*
> *Acinetobacter calcoaceticus*
> Coliform bacilli, other enteric bacilli
> *Pseudomonas aeruginosa*
> *Corynebacterium diphtheriae*
> Viruses, chlamydiae
> Fungi
> Anaerobes
> *Mycobacterium*
> **Nonpathogens, or opportunists**
> *Corynebacterium xerosis*, other diphtheroid
> bacilli, *Propionibacterium*
> Coagulase-negative staphylococci
> Micrococci
> Saprophytic fungi

If a diagnosis of **purulent conjunctivitis** has been made, the purulent material is collected on a sterile cotton swab or surgical instrument (**before** the local application of antibiotics, irrigating solutions, or other medications) from the surface of the lower conjunctival sac and inner canthus of the eye. This purulent material should be inoculated immediately to blood agar and chocolate agar plates, which should be incubated in a candle jar, and to a tube of enriched thioglycollate broth. All media should be held for at least 48 hours, and any resulting growth should be identified by the appropriate methods.* One should be aware of the possibility of gonococcal infection in the eye of the newborn infant, and the oxidase test may be carried out on the chocolate agar plate to detect colonies of neisseriae. Sabouraud dextrose and brain-heart infusion blood agar slants should be inoculated if a mycotic infection is suspected. Whenever possible, a **gram-stained smear** of the purulent material present also should be examined. Frequently, the results may give sufficient information to the physician to confirm a clinical diagnosis and serve as a guide to proper therapy.

Since potentially pathogenic organisms may be present in an eye without causing disease, it may be very helpful to the clinician, when only one eye is infected, to culture organisms from both eyes. Differences in the bacteriology in the two eyes may be significant.

Scrapings of the conjunctivae for the presence of eosinophils or of ulcerative lesions of the cornea for demonstrating viral inclusion bodies must be taken by an ophthalmologist. These scrapings are

---

*Pseudomonas aeruginosa, Streptococcus pneumoniae, and other organisms may cause a severe and damaging corneal infection. All results should be reported without delay to the physician.

transferred to a glass slide, dried, and stained by the Wright-Giemsa method.

Moraxella lacunata (Morax-Axenfeld bacillus) is a short, thick, gram-negative diplobacillus that causes a subacute or chronic catarrhal conjunctivitis, which is particularly severe in the outer angle of the eye. The organism is best cultivated on Loeffler medium, where it causes characteristic pitting and proteolysis of the medium, although nonproteolytic strains are being isolated with some frequency. Identification methods are described in Chapter 22.

Haemophilus aegyptius (Koch-Weeks bacillus) is a small gram-negative bacillus, closely resembling *H. influenzae*, and is the causative agent of an acute epidemic conjunctivitis commonly called **pink eye.** The organism grows well on blood agar or chocolate agar plates. (See Chapter 23 for cultural characteristics.)

Corynebacterium diphtheriae may cause a pseudomembranous conjunctivitis; the organism must be demonstrated by culture and proved toxigenic before a diagnosis of diphtheria can be made. The organism may be readily cultivated on Loeffler or Pai medium. Their identifying characteristics and tests for virulence are described in Chapter 26.

EAR CULTURES

The following organisms are encountered **most frequently** from cultures of the ear:

Pathogens or potential pathogens
  Pseudomonas aeruginosa
  Staphylococcus aureus
  Proteus species
  Alpha- and beta-hemolytic streptococci
  Streptococcus pneumoniae
  Haemophilus influenzae
  Coliform and other enteric bacilli
  Aspergillus fumigatus, Candida albicans, and other fungi
  Bacteroides, Fusobacterium, anaerobic cocci
Nonpathogens, or opportunists
  Coagulase-negative staphylococci and micrococci
  Diphtheroids
  Bacillus species
  Saprophytic fungi

The following organisms are **rare or uncommon** pathogens in such cultures:

Corynebacterium diphtheriae
Actinomyces species
Mycobacterium tuberculosis, other mycobacteria
Mycoplasma pneumoniae

Material from the ear, especially that obtained after perforation of the eardrum, is best collected by an otolaryngologist, using sterile equipment and a sterile cotton or polyester swab. Discharges from the ear in **chronic** otitis media usually reveal the presence of pseudomonads and *Proteus* species, but often the major pathogens in chronic otitis media are anaerobes and enteric bacilli.[2] **Acute** or **subacute** otitis usually yields pyogenic cocci. In external otitis the external ear should be cleansed with a 1:1,000 aqueous solution of benzalkonium chloride or other detergent to free the skin of contaminating bacterial flora before a culture is taken, if the results are to be of clinical significance. Otherwise, a variety of nonpathogenic bacteria and saprophytic fungi will be recovered. *P. aeruginosa* and *S. pyogenes* are commonly isolated. Specimens are cultured as described in the preceding section, with the addition of phenylethanol blood agar, to recover other organisms in the presence of spreading *Proteus* species.

MASTOID PROCESS AND PARANASAL SINUS CULTURES

The widespread use of antimicrobial agents in the treatment of acute infections of the middle ear (otitis media) has resulted in a significant decrease in the incidence of acute mastoiditis, an infection of the mastoid process and surrounding structure. The offending organisms, usually originating from a suppurative otitis, generally are pyogenic cocci. Chronic mastoiditis, like chronic otitis media, commonly involves anaerobic bacteria.[2]

Cultures from the mastoid region are generally taken on a cotton or polyester swab (before antibiotic therapy) and are handled in the laboratory as any other wound culture would be.

Acute suppurative sinusitis may follow a common cold or occur after water is forced into the nose while swimming or diving. The most frequent isolates in this infection include the streptococci, staphylococci, and pneumococci; *Klebsiella, Bacteroides,* and *Haemophilus influenzae* are occasionally isolated and may give rise to serious complications. In chronic purulent sinusitis, anaerobes again play a more prominent role.[2] The aerobic and anaerobic methods previously described for wound cultures (Chapter 13) are generally satisfactory.

CULTURES OF TEETH

The bacterial flora of the normal mouth is made up of a wide variety of microorganisms, including streptococci, filamentous gram-negative and gram-positive anaerobic rods, neisseriae, anaerobic cocci, and spirochetes, in addition to lactobacilli.[2] The role of anaerobes, including spirochetes and fusiform bacilli, in Vincent's infection is discussed on p. 62.

Effective methods and culture media are available for the bacteriologic examination of root canals, tooth sockets, periapical abscesses, and other dentoalveolar infections.[3,4] Various anaerobes, as well as streptococci, are the predominant pathogens in these infections.[1,2] Specimens are obtained by the dental surgeon, using a rigidly aseptic technique and sterile equipment. The apex of an extracted tooth is cut off with a pair of cutting forceps, and the apical fragment is transferred directly to a tube of enriched medium, such as brucella broth or enriched thioglycollate medium* (penicillinase is added if penicillin has been used). The medium is held at 35 C for several days and observed for growth. Any indication of growth (increase in turbidity) is confirmed by examining a gram-

stained smear, which also serves to guide in the selection of appropriate aerobic and anaerobic media to be used in subculturing. Cultures from sockets or abscesses, obtained either with a sterile curet or cotton applicator, are handled as wound cultures would be.

In culturing root canals, the following method[4] is recommended, using strict asepsis throughout. After a sterile field is established, the seal and previous dressing are removed and discarded. The canal is cleansed of any residual medicament by flushing with about 1 ml of sterile water and is dried by inserting a fresh absorbent point with a wiping motion, then removing and discarding it. Another fresh sterile point is then inserted into the apex and allowed to remain in place for about 1 to 2 minutes, then removed. If the tip appears to be moist with exudate or blood, it is dropped into a tube of culture medium as previously indicated. If the point appears to be dry on removal, it is discarded and a fresh point, aseptically moistened with the culture medium, is inserted into the root canal, left for several minutes, and cultured as before. After 48 hours' incubation at 35 C, the culture tube is examined for the presence of growth, which is evidenced by any increase in turbidity, especially around the tip or surface of the absorbent point. Generally, negative cultures are kept for 1 week, and two consecutive negative cultures are obtained before filling the root canal.

CULTURES OF EFFUSIONS

All fluids suspected of being exudates should be examined bacteriologically. Clear or slightly cloudy specimens should be centrifuged at 2,500 rpm for 30 minutes, the supernatant fluid removed aseptically, and the sediment examined by means of smears and cultures. Specimens that are grossly purulent should be examined **directly** by Gram stains of thin films. All specimens should also be inoculated to routine media, including blood

---

*Difco Laboratories, Detroit; Baltimore Biological Laboratory, Cockeysville, Md.

and chocolate agar plates, plates for anaerobic incubation, enriched thioglycollate broth, and isolation media for *Mycobacterium tuberculosis* and systemic fungi if indicated. In certain cases, the use of special media may be required. It should be noted that if an anticoagulant is required to prevent clotting of the specimen, a heparin preparation **free of preservatives** or SPS should be used.

Synovial, or joint, fluid should also be examined for the presence of gonococci by inoculating an enriched broth, such as trypticase soy broth supplemented with 1% IsoVitaleX, 10% horse serum, and 1% glucose, as described in Chapter 11. **Pleural and pericardial effusions** should be centrifuged and the sediment used for inoculation of culture media for mycobacteria and fungi, as well as being set up in routine aerobic and anaerobic culture. In **empyema** the fluid is purulent or seropurulent; it may yield pneumococci, streptococci, coagulase-positive staphylococci, *Haemophilus influenzae*, or various anaerobes (especially *B. melaninogenicus*, *F. nucleatum*, and anaerobic and microaerophilic cocci) on culture. Effusions from patients with **peritonitis** vary in character from the thin cloudy fluid found in tuberculous peritonitis to the foul-smelling purulent specimen from mixed anaerobic-aerobic peritonitis. In general, the methods for handling these specimens are those described for routine culturing of specimens from wounds (Chapter 13) and should include inoculation of both **aerobic** and **anaerobic media**.

REFERENCES

1. Burnett, G. W., Scherp, H. W., and Schuster, G. W.: Oral microbiology and infectious disease, ed. 4, Baltimore, 1976, The Williams & Wilkins Co.
2. Finegold, S. M.: Anaerobic bacteria in human disease, New York, 1977, Academic Press. Inc.
3. Grossman, L. I.: Endodontic practice, ed. 8, Philadelphia, 1974, Lea & Febiger.
4. Nolte, W. A., editor: Oral microbiology, ed. 3, St. Louis, 1977, The C. V. Mosby Co.

CHAPTER 15

Microorganisms encountered in material removed at operation and necropsy

EXAMINATION OF MATERIAL OBTAINED AT OPERATION

Although largely a neglected function of the microbiology laboratory in the past, the current availability of specific antimicrobial agents has made the microbiologic examination of surgical tissue an essential adjunct to the histopathologic diagnosis.

Likewise, the increasing use of more refined techniques in postmortem microbiology has contributed to determination of etiology of infectious processes.

Collection of specimen

To carry out a proper microbiologic examination of excised tissue, a thorough search for aerobic and anaerobic microorganisms, acid-fast bacilli, fungi, viruses, and other pathogenic agents must be made. An **adequate** specimen, therefore, is a prerequisite; the surgeon or pathologist **must** assume the responsibility for obtaining sufficient material at the time of operation.

It is also necessary that a specimen container of **sufficient size** be available to the operator. A sterile, wide-mouthed, screw-capped bottle with a neoprene liner in the lid or a suitable sterile plastic container is recommended. This receptacle should be conveniently placed on the instrument table at the time of surgery so that the operator may deposit material directly into it. This prevents the possibility of accidentally fixing the tissue with formaldehyde or other germicidal agent, and it also avoids any possible contamination by the surgical pathologist.

In the collection of material from chronic draining sinuses and ulcers, appropriate sterile equipment should be made available to the examining physician. This may include curets, scissors, syringes, needles, medications, dressings, containers, and so forth. With the aid of these instruments, **deep curettage** of the sinus tract may be carried out; it should include the procurement of a portion of the wall of the tract. For collecting material from sinus tracts, an intravenous plastic catheter is introduced into the tract as deeply as possible after appropriate decontamination of the skin site, and material is aspirated with a syringe. This is immediately transferred to a sterile plastic bag, placed in a plastic anaerobic bag transport setup (see Chapter 6), and submitted promptly to the laboratory.

Material obtained from an ulcer should contain tissue from the base as well as the edge of the lesion. **Closed** abscesses should be aspirated, using a 15-gauge needle when feasible (to secure necrotic tissue debris that may be present) and a large-caliber syringe, and the material should be transferred to an anaerobic transport tube.[7] When possible, a portion of the abscess wall should be sent for microbiologic study. Some organisms, such as *Nocardia asteroides*, are usually found in the abscess wall; others, like *Actinomyces israelii*, are more likely to be in the pus itself.[2] Inspection for the presence of **granules** in the aspirated pus and a Gram stain should be a part of every examination, since this may be the first clue to an infection caused by *Actinomyces*, *Arachnia*, or *Nocardia*.

In some instances, **contaminated** material may be submitted for microbiologic examination. Such specimens as tonsils, autopsy tissue, or similar material may be surface cauterized with an electric soldering iron or heated spatula or blanched by immersing in boiling water for 5 to 10 seconds to reduce surface contamination. The specimen may then be dissected with sterile instruments to permit culturing of the **center** of the specimen, which will not be affected by the heating.

All surgical specimens intended for microbiologic examination should be divided by the operator, using sterile instruments; one half is submitted for histologic examination, and the other half is sent to the microbiology laboratory. A detailed clinical history should accompany the histologic specimen to guide microbiologic studies. Since viable organisms may be few in number in tissue, especially in old chronic lesions, and since they may be irregularly distributed throughout the tissue, it is desirable to secure multiple specimens when the lesion is large enough.[10]

It is well to keep in mind that surgical specimens differ from other clinical material in that they are frequently obtained at **considerable risk and expense** to the patient. Furthermore, a specimen may represent the entire pathologic process. It is obvious, therefore, that supplementary specimens cannot be obtained with the ease with which similar specimens of blood, urine, or feces can be secured. It is strongly recommended that a portion of the tissue be kept moist in sterile nutrient broth and **refrigerated or frozen** for subsequent studies should preliminary or routine examination prove unproductive.

The routine culturing of all biopsy specimens for fungi has been recommended by Utz.[8] He has obtained positive cultures from the brain, spleen, liver, kidney, prostate gland, epididymis, testis, muscle, skin, synovium, and other tissues, as well as from ulcers of the nose, mouth, epiglottis, and larynx.

Preparation of tissues for culture

In the important step of preparing tissues for culture, it is not enough to merely scrape material from the surface of the specimen with a bacteriologic loop. One must thoroughly **grind** the tissue into a fine suspension, since only a few organisms may be present in the whole sample.*

To prepare this suspension the tissue should first be finely minced with sterile scissors and transferred to a sterile tissue grinder, of which several types are available. A small one is illustrated in Weed's paper[10] on the isolation of fungi from tissue; the use of the ten Broeck tissue grinder† also has proved satisfactory. The tissue is ground to a pasty consistency (10% to 20% suspension), using sterile sand and sterile broth. After settling, the supernatant fluid is transferred to another small tube by use of a sterile capillary or larger pipet. It is then inoculated to culture media, as discussed later, or, when indicated, injected into animals in the same manner as any other biologic fluid. Intraperitoneal or intramuscular injection of guinea pigs and intraperitoneal injection of several mice is recommended. These are examined daily and are cultured at the time of death.

Histologic examination

Although a discussion of histologic techniques is beyond the intended scope of this text, it should be pointed out that a thorough histologic examination of fixed tissue is an essential part of any pathologic diagnosis. The microscopic study of carefully selected and sectioned tissue, stained by both routine and special methods (such as the Gram, acid-fast, Gomori, Gridley, periodic acid–Schiff methods),

---

*Yeastlike fungi, such as *Cryptococcus neoformans,* may be macerated and rendered nonviable by grinding. It is therefore recommended that **minced** tissue be inoculated directly to media for fungi.
†ten Broeck tissue grinder, small size, heavy-walled Pyrex glass, Bellco Glass, Inc., Vineland, N.J.

serves two useful purposes. First, it demonstrates the histopathology of the lesion —whether it is of a neoplastic, inflammatory or other nature. If the lesion proves to be of noninflammatory origin, a microbiologic study may not be necessary. Second, if the appearance suggests infection, it may serve as a guide to the selection of appropriate culture media and isolation techniques. Granulomatous lesions whether showing caseation or not, should be cultured for mycobacteria, fungi, and *Brucella*[2] as well as for the usual aerobic and anaerobic pathogens. The use of frozen sections of surgical specimens provides early histopathologic information.

Selection of culture media

In the selection of culture media, one must decide not only whether the tissue specimen contains a mixed or contaminated flora, indicating the need for selective isolation media, but also whether there are **fastidious pathogens** present, which may require special media.

No general rules can be made regarding the particular kinds of media to use; the choice will depend largely on the organisms sought or suspected. Enriched media, such as blood agar, chocolate agar, heart infusion broth, anaerobic media, and media for the primary isolation of mycobacteria and fungi, must be considered. Incubation under both 3% to 10% CO_2 and anaerobic conditions should be carried out. This is discussed in earlier chapters.

Isolation of some fastidious pathogens

Although the proper techniques for isolation of fastidious pathogens are presented in other parts of the text, a number of items are relevant to the recovery of these microorganisms and are pertinent to this section. In the isolation of **pathogenic fungi** and *Actinomyces* and *Nocardia,* for example, the following points are important:

1. Tissue for mycologic examination should be minced, rather than ground, and transferred with the knife blade directly to mycologic media (see footnote, p. 115).
2. In severe, disseminated histoplasmosis, cultures of lymph nodes and biopsies of the liver, bone marrow, and upper respiratory tract mucosal ulcers are likely to yield *Histoplasma capsulatum.*
3. Most of the fungi, with the exception of cryptococci, can be isolated on media containing antibiotics. Some require enriched media and may grow at both room and incubator temperatures.
4. *Actinomyces* species are microaerophilic or obligately anaerobic, require enriched media, are inhibited by many antibiotics, and grow best at 35 C in an anaerobic environment.
5. *Nocardia* species, on the other hand, are aerobic, will grow on simple media (including Sabouraud's agar); they are also inhibited by many antimicrobial agents.

It is apparent that no one medium, no specified temperature of incubation, and no single standard technique in handling will be suitable for all microorganisms. Massive inoculation of the tissue should be made to multiple sets of both simple and enriched media and incubated at both room and incubator temperatures for at least 4 weeks before discarding. Specific recommendations are found in Chapter 26.

Acid-fast stains of surgical specimens may be positive in less than one half of those cases subsequently proved to contain pathogenic mycobacteria.[11] A negative stain is thus of little value in ruling out mycobacteria, and a positive smear does **not** always denote tuberculosis; therefore, adequate bacteriologic studies must be made. These methods are considered further in Chapter 31.

The isolation of brucellae from contaminated surgical specimens can be accomplished only by the use of enriched media containing appropriate antibiotics, since routine cultures will generally be overgrown by other microorganisms. A fresh-meat-extract agar with added glucose* and 5% animal blood has proved satisfactory for the isolation of *Brucella* species from surgical specimens.[9] Moreover, the addition of antibiotics increases the usefulness of this medium for isolating brucellae from clinical material containing a mixed flora.

Lung biopsy and autopsy specimens (and pleural fluid) may be cultured for the agent of Legionnaires' disease on Mueller-Hinton agar supplemented with 1% hemoglobin and 1% to 2% IsoVitaleX. Incubation should be in an atmosphere of 5% CO_2. Growth may not be visible before 3 to 5 days. Work with this agent must be done in a biologic safety hood.

Cultures of most of these fastidious microorganisms should be incubated for 3 to 6 weeks before discarding. Many of the organisms isolated from chronic infections—including *Brucella, Histoplasma, Coccidioides,* and mycobacteria—require long incubation periods before growth becomes apparent. During the first day or so after primary inoculation, the medium should be examined for overgrowth by *Proteus,* pseudomonads, or other presumed contaminants. It is recommended that a portion of the original tissue suspension be stored in a refrigerator until the cultures are obviously free of any contamination. Such material can also be used to set up additional cultures for unusual pathogens if the original conventional cultures are negative. *Listeria monocytogenes* sometimes can be recovered only after cold incubation at 4 C.

Final identification methods should be carried out on all isolates whenever possible. Special procedures involving virulence tests by animal injection, serologic analysis, and definitive identification may not be feasible for the small laboratory. In such cases pure cultures of these isolates should be sent to a reference laboratory for further study. Means for handling and rapid transport of specimens are described in Chapter 6.

EXAMINATION OF MATERIAL OBTAINED AT AUTOPSY

Although postmortem invasion of the bloodstream and organs by commensal organisms can occur, it is now recognized that this does not happen as rapidly as was formerly believed. Earlier studies interpreted the high incidence of positive autopsy cultures as evidence of either antemortem infection or agonal invasion of tissue. On the other hand, O'Toole and co-workers[4] indicate that contamination of tissue by the environment or personnel at the autopsy may be more significant.

The microscopic examination of direct smears stained by the Gram, Ziehl-Neelsen, or other methods, along with a thorough evaluation by culturing blood, tissue, or purulent material from infected areas, may reveal significant data related to the cause of death.

An evaluation of the results of autopsy blood cultures by Wood and associates[12] indicates a good correlation with the results of antemortem cultures or with anatomical data derived at autopsy. A higher degree of correlation is obtained when the blood for culture is obtained close to the time of death. Prior antimicrobial therapy does not necessarily reduce the recovery of pathogens from this source appreciably.

Blood for culture is collected with a sterile syringe and a 14- or 15-gauge needle after searing the heart or one of the great vessels with a soldering iron.* From 10 to 20 ml of blood are inoculated to broths in the manner described in Chap-

---

*Brucella agar (Pfizer Laboratories, Clifton, N.J.) and trypticase soy agar (Baltimore Biological Laboratory, Cockeysville, Md.) can also be used.

*Silver and Sonnenwirth[6] indicate that heart blood obtained through the closed chest and before bowel manipulation gives more significant results than do the usual methods.

ter 7. Small fragments of blood clots may appear, but these do not interfere with the culturing if large-bore needles are used. Some workers believe that culture of aspirated splenic pulp is of greater value in confirming a diagnosis of septicemia than is blood culture.

Cultures of other tissues may be obtained by searing the surface with a soldering iron or hot spatula and passing a needle and syringe, swab, or fine-pointed pipet through the seared area to an unheated and uncontaminated region.

Another method of obtaining autopsy culture material that compares favorably with results obtained by a sterile autopsy technique has been described by de Jongh and associates.[3] In this procedure the tissue surface is seared to dryness with a heated steel spatula, and 1 cubic centimeter of tissue is excised from the center of the seared area using sterile forceps and scissors. Tissue blocks thus obtained are minced and ground as described previously and inoculated to appropriate aerobic and anaerobic media, as indicated by a Gram stain of the homogenate.

If **bacterial endocarditis** is suspected, a small portion of the friable vegetation is removed with sterile scissors and forceps and submitted to the laboratory in a Petri dish or preferably in an anaerobic mini-jar[7] or a plastic bag anaerobic transport setup. Here it is gently washed in at least three changes of sterile saline solution and then ground with sterile sand and broth in a tissue grinder, as described previously. A smear is prepared of this suspension, gram stained, and examined; a portion also is inoculated to blood agar and chocolate agar plates and enriched thioglycollate broth medium. The plates are incubated in a candle jar and also anaerobically at 35 C for 48 hours and examined for the presence of growth. If a delay has occurred in obtaining the specimen (more than 4 hours postmortem), with probable overgrowth of contaminants, it is well to streak the suspension on a phenylethyl alcohol blood agar plate and other media. This is done to better isolate alpha-hemolytic streptococci or other gram-positive cocci that may be overgrown by *Proteus* and other gram-negative bacteria that may be present on the other plates. If indicated, anaerobic subcultures of the thioglycollate broth medium may be carried out.

It should be emphasized that reliable microbiologic information can often be obtained even when a body has been embalmed.[2] In addition to fungi and mycobacteria, streptococci and various gram-negative bacilli have been recovered from embalmed tissues.

Collection of autopsy specimens for virus isolation*

An effort should be made to obtain postmortem specimens in all fatal cases of central nervous system disease of suspected viral etiology, especially if antemortem studies were not carried out. Using sterile precautions at the time of autopsy, the whole brain is removed and refrigerated but **not** frozen, a 3-inch segment of the descending colon is tied off at both ends, and aliquots of various other tissues are obtained and placed in sealed sterile containers and refrigerated. A blood specimen also should be collected by cardiac puncture; the serum should be separated and then refrigerated. In a fatal case of respiratory disease, lung tissue and a tracheobronchial swab should be collected and refrigerated until shipment; a blood specimen also should be obtained. All specimens thus obtained for virus isolation should be transported **without delay** to the nearest virus reference laboratory; prior arrangements for shipment and handling of such specimens at the laboratory should be made.

---

*Abstracted from the instruction pamphlet by the New Jersey State Department of Health, Dr. Martin Goldfield, Assistant Commissioner.

Procedures for isolation of Listeria monocytogenes from tissue

The tissues of all fetuses, premature infants, and young babies coming to necropsy as a result of an infectious process should be cultured for listeriae for reasons given previously (see discussion of blood cultures). Specimens of the brain, liver, and spleen are most likely to contain the organism. The isolation procedure is given in detail by Seeliger and Cherry.[5]

1. Five to 10 ml of ground tissue are added to each of two flasks of infusion broth.
2. Incubate one flask at 35 C for 24 hours and inoculate a drop of this to a blood agar plate and a tellurite blood agar plate. Incubate these in a candle jar for at least 48 hours, along with the original broth flask.
3. Store the second flask in a refrigerator at 4 C. If the 35 C subcultures (step 2) are unsuccessful, subculture material from the refrigerated flask at weekly intervals for at least 1 month.
4. *Listeria monocytogenes* is identified by the procedures described in Chapter 26. If the material is likely to be contaminated, inoculate a plate of modified McBride medium with an 18- to 24-hour tryptose broth culture incubated at 35 C (Chapter 26) and incubate at 35 C for 24 hours in a candle jar, along with the original broth tube. Examine and identify as indicated in Chapter 26. If no listeriae are obtained from the 35 C cultures, repeat the process using the 4 C cultures. Subculture these at weekly intervals for 3 months

from the original broth culture held in the refrigerator.

REFERENCES

1. Bearns, R. E., and Girard, K. F.: On the isolation of *Listeria monocytogenes* from biological specimens, Am. J. Med. Techn. **25:**120-126, 1959.
2. Brewer, N. S., and Weed, L. A.: Diagnostic tissue microbiology methods, Human Pathol. **7:** 141-149, 1976.
3. De Jongh, D. S., Loftis, J. W., Green, G. S., Shively, J. A., and Minckler, T. M.: Postmortem bacteriology—a practical method for routine use, Am. J. Clin. Pathol. **49:**424-428, 1968.
4. O'Toole, W. F., Saxena, H. M. K., Golden, A., and Ritts, R. E.: Studies of postmortem microbiology using sterile autopsy technique, Arch. Pathol. (Chicago) **80:**540-547, 1965.
5. Seeliger, H. P. R., and Cherry, W. B.: Human listeriosis; its nature and diagnosis, Department of Health, Education, and Welfare, Washington, D.C., 1957, U.S. Government Printing Office.
6. Silver, H., and Sonnenwirth, A. C.: A practical and efficacious method for obtaining significant postmortem blood cultures, Am. J. Clin. Pathol. **52:**433-437, 1969.
7. Sutter, V. L., Vargo, V. L., and Finegold, S. M.: Wadsworth anaerobic bacteriology manual, ed. 2, Los Angeles, 1975, Dept. Contin. Educ. Health Sci., Univ. Extens., School Med., UCLA.
8. Utz, J. P.: Recognition and current management of the systemic mycoses, Med. Clin. North Am. **51:**519-527, 1967.
9. Weed, L. A.: Use of a selective medium for isolation of *Brucella* from contaminated surgical specimens, Am. J. Clin. Pathol. **27:**482-485, 1957.
10. Weed, L. A.: Technics for the isolation of fungi from tissues obtained at operation and necropsy, Am. J. Clin. Pathol. **29:**496-502, 1958.
11. Weed, L. A., McDonald, J. R., and Needham, G. M.: The isolation of "saprophytic" acid-fast bacilli from lesions of caseous granulomas, Proc. Staff Meet. Mayo Clin. **31:**246-259, 1956.
12. Wood, W. H., Oldstone, M., and Schultz, R. B.: A re-evaluation of blood culture as an autopsy procedure, Am. J. Clin. Pathol. **43:**241-247, 1965.

PART IV

Methods for identification of pathogenic microorganisms

CHAPTER 16

Facultative staphylococci and micrococci

Micrococci are ubiquitous and exist as free-living saprophytes, parasites, and pathogenic forms. The great majority of pathogenic micrococci fall within the genus *Staphylococcus*, but nonstaphylococci are being isolated from clinical sites with increasing frequency. Collectively, the organisms of the group are **spherical, gram-positive cocci** occurring singly, in pairs, tetrads, packets, and irregular clusters. Most are strongly **catalase positive;** they are nonmotile and occasionally pigmented. Nitrates are usually reduced to nitrites. With the exception of members of the genus *Peptococcus*, which are obligate anaerobes, the micrococci are **aerobic or facultative.**

Despite confusion in the classification of these organisms, clinical laboratories should have few problems in identifying the major pathogens. *Bergey's Manual*[4] places the tetrad-forming cytochrome-negative strains in the *Streptococcaceae* and recognizes five genera: *Streptococcus, Pediococcus, Aerococcus, Gemella,* and *Leuconostoc*. The International Subcommittee on Nomenclature of *Micrococcaceae*, however, considers the aerococci as part of the *Micrococcaceae* and includes *Pediococcus, Gaffkya,* and, presumably, *Gemella* in the genus *Aerococcus*. *Sarcina* is now limited to anaerobic forms. The scheme proposed by Facklam and Smith[7] (see outline below) appeals to us as practical for the clinical microbiologist. *Planococcus* and *Leuconostoc* are not considered, since they do not occur in human infection. Thus, we will consider the genera *Staphylococcus, Micrococcus,* and *Aerococcus* in this chapter and *Streptococcus* in Chapters 17 and 18. Anaerobic cocci of all types are covered in Chapter 28.

Although the catalase test is very simple and is widely used, there are occasional pitfalls in using it to differentiate the various groups discussed in the outline below. Some micrococci do not decompose hydrogen peroxide, and aerococci release oxygen from hydrogen peroxide without catalase mediation. The benzidine* test for cytochrome enzymes is entirely reliable. To carry it out, flood a 24- to 48-hour culture plate with benzidine reagent (see Chapter 43). After all microbial growth has been contacted, add an equal volume of 5% hydrogen peroxide. A **positive** test yields a blue-green or deep green color.

CLASSIFICATION OF GRAM-POSITIVE COCCI

I. Cytochrome enzymes present (benzidine positive); that is, catalase test yields strong, vigorous bubbling of hydrogen peroxide by culture—*Micrococcaceae*
   A. Acid from glucose anaerobically (Subcommittee media), susceptible to lysis by lysostaphin—*Staphylococcus*
   B. No acid from glucose anaerobically, nonsusceptible to lysis by lysostaphin—*Micrococcus*
II. No cytochromes present; that is, catalase reaction usually negative, but an occasional weak bubbling of hydrogen peroxide by culture will occur—*Streptococcaceae*
   A. Cellular arrangement primarily chains and pairs, acid from glucose in anaerobic culture in most

---

*Benzidine is a potential carcinogen; accordingly, caution should be used when working with this compound.

cases—*Streptococcus* (see Chapter 17)

B. Cellular arrangement primarily in tetrads and clumps, acid not formed from glucose in anaerobic culture—*Aerococcus*

Differentiation between *Staphylococcus* and *Micrococcus* can be made in two ways: *Staphylococcus* produces acid from glucose anaerobically and is susceptible to lysis by lysostaphin,* and *Micrococcus* is negative in both of these tests. Ordinary sugar fermentation bases **cannot** be used for the first of these tests (see Chapter 41 for fermentation medium for differentiating *Staphylococcus* and *Micrococcus*). *Staphylococcus* is lysed by 1 unit per milliliter of lysostaphin endopeptidase. Microorganisms formerly classified as *Micrococcus*, subgroups 1 through 4, have now been reclassified on the basis of their deoxyribonucleic acid and guanine plus cytosine content as members of the species *Staphylococcus saprophyticus*.[1] This organism is taxonomically intermediate between *Staphylococcus* and *Micrococcus*. It does not actively produce acid from glucose anaerobically, but may slowly produce small amounts of acid. Schliefer and Kloos[16] note that virtually all staphylococci and no micrococci produce acid aerobically from glycerol in the presence of 0.4 μg/ml of erythromycin. Smith's selective medium for corynebacteria, which contains 50 μg/ml of furoxone,† was found to support growth of *Micrococcus* and to prevent growth of *Staphylococcus*.[5]

AEROBIC AND FACULTATIVE MICROCOCCI
Genus Staphylococcus

Staphylococci are frequently found on the skin, on nasal and other mucous membranes of humans, and in various

---

*Schwarz/Mann Research Laboratory, Orangeburg, N.Y.
†Eaton Laboratories, Norwich, N.Y.

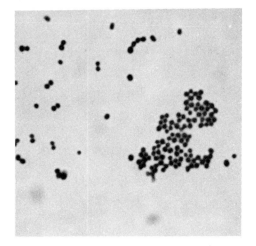

Fig. 16-1. *Staphylococcus aureus,* showing typical grapelike clusters. (1000×.)

food products. Three species are recognized: *S. aureus, S. epidermidis,* and *S. saprophyticus.*

Staphylococcus aureus

Staphylococcus aureus is a gram-positive, nonmotile coccus, occurring singly, in pairs, in short chains, or in irregular clusters. The last arrangement is probably the most characteristic (Fig. 16-1). The Greek word "staphyle," meaning a bunch of grapes, is used descriptively as the stem of the generic term. In older cultures the cells tend to lose their ability to retain the crystal violet and may appear gram variable or even gram negative. Typically, on initial isolation the organism produces a golden yellow pigment, which is soluble in alcohol and ether and is classed as a **lipochrome.** This characteristic, however, is variable; white or pale colonies may arise after laboratory cultivation and are not infrequently isolated from clinical sources.

Colonies are usually opaque, circular, smooth, and entire, with a butyrous consistency. The organism grows well on trypticase soy agar or nutrient agar but develops larger colonies on blood agar. The hemolytic activity is variable. Both surface and subsurface colonies of some strains are hemolytic. The surface colonies of many virulent strains are hemolytic, but hemolysis also occurs around subsurface colonies. The hemolytic property may be lost in stock cultures or after a number of transfers.

Most strains of *S. aureus* **ferment mannitol,** tolerate relatively high concentrations of salt (7.5% to 10%), and are relatively resistant to polymyxin. These characteristics help promote their isolation from material such as feces, in which a large and varied bacterial flora exists. Polymyxin staphylococcus medium or mannitol salt agar are recommended for the isolation of *S. aureus* when many other organisms are present. Colonies of salt-tolerant staphylococci appear on mannitol salt agar, surrounded by a yellow halo after 24 to 48 hours, which indicates mannitol fermentation. Mannitol fermentation, among other tests, helps to distinguish *S. aureus* from *S. epidermidis*. *S. saprophyticus* resembles *S. epidermidis* in most properties but may be differentiated from it on the basis of susceptibility to novobiocin; *S. saprophyticus* is typically resistant to novobiocin.[11] *S. aureus* and *S. epidermidis* are phosphatase positive; *S. saprophyticus* is negative.[2]

In the differentiation of *S. aureus* from *S. epidermidis*, hemolysis and pigmentation are **not** considered reliable or valid criteria. Deoxyribonuclease (DNase) activity, on the other hand, agrees well with coagulase activity, but the former test is more difficult to perform and therefore the coagulase test is preferred in most diagnostic laboratories. Production of acid from mannitol and from trehalose aerobically is characteristic of *S. aureus* and *S. saprophyticus,* whereas *S. epidermidis* is negative in this regard.[2]

The best single test for differentiating the staphylococci is the **coagulase tube test,** which demonstrates "free" coagulase. Citrated rabbit plasma,* 0.5 ml of a

1:4 dilution, in a small (12 × 100 mm) tube is inoculated heavily with one or two drops of an overnight culture of the organism and incubated at 35 C in a water bath. Complete or partial coagulation in 1 to 4 hours is interpreted as **positive.** A variation of this test may be performed by using a single colony in a 1:5 dilution of plasma; the test is observed for a period up to 24 hours.[15] Typical strains of *S. epidermidis* and *S. saprophyticus* are coagulase negative, whereas pathogenic strains of *S. aureus* are coagulase positive.

A word of caution should be offered to those who have adopted the practice of using plasma from outdated human bank blood for this test. Inhibitory factors are found in some human blood, and control tests with known strongly and weakly positive and negative strains should always be included.* This should be part of a daily quality control procedure in **all** coagulase testing.

The **coagulase slide test,** which demonstrates "bound" coagulase, or clumping factor, is used by many laboratories and gives results comparable to the tube test. It is performed by emulsifying growth from a typical colony in a drop of water on a slide and adding a loopful of fresh **human** plasma. This is mixed thoroughly for 5 seconds. If the reaction is **positive,** easily visible white clumps appear immediately; if no such clumping appears, the reaction is **negative** and **should be checked by the tube test,** since some strains of *S. aureus* may be negative by slide test. This is an excellent **screening** procedure, however, as most strains of *S. aureus* are positive by the slide test and false-positives do not occur, assuming that proper controls are used.

Some investigators report that a good

---

*Dehydrated sterile rabbit plasma is available commercially.

*Citrated plasma may also be coagulated by organisms other than *S. aureus,* such as some citrate-utilizing enterococci[3] or gram-negative bacilli, particularly when isolated colonies of staphylococci or overnight incubation are used. The use of EDTA (ethylenediamine tetra-acetate) in place of citrate prevents this problem.

correlation exists between **phosphatase** production and pathogenicity; yet some coagulase-negative strains are more strongly phosphatase positive than are some coagulase-positive strains. The amount or degree of phosphatase activity is also known to vary with the phage type of the staphylococci.

Strains of *S. aureus* isolated from human sources can elaborate a variety of metabolites. Some of these are toxic and of pathologic significance; others are either nontoxic or of low toxicity but have some diagnostic significance. There are recorded differences in the exotoxins produced by human and animal pathogenic strains. According to the report of Elek,[6] human pathogens normally produce alpha and delta lysins, whereas the animal pathogens produce alpha, beta, and delta lysins. Other exotoxins produced by the pathogenic staphylococci are **leukocidin,** which is probably the same as the delta lysin, a **dermonecrotizing toxin,** a **lethal toxin,** and an **enterotoxin.** Only strains of *S. aureus* that produce enterotoxin cause staphylococcal food poisoning or staphylococccal pseudomembranous enterocolitis. There are several types of enterotoxin, all of which are heat stable and resistant to the enzymes pepsin and trypsin.

Alpha hemolysin does not lyse human erythrocytes but does lyse rabbit and sheep red cells. Beta lysin lyses sheep red cells, and delta lysin lyses horse erythrocytes. Alpha hemolysin serves as a convenient index of virulence because the more virulent the strain, the more alpha hemolysin it produces. Filtrates of virulent broth cultures usually yield a high titer.

Tests for the dermonecrotizing toxin may be carried out by intradermal injection in a rabbit and tests for lethal toxin by intravenous injection. Tests for enterotoxin are best done by serologic procedures.

Among the nontoxic metabolites produced by the staphylococci are phosphatase and coagulase, to which reference was made previously, hyaluronidase, deoxyribonuclease, staphylokinase, or fibrinolysin, and in addition lipase, gelatinase, and protease. Most, if not all, of these are antigenic.

The correlation between deoxyribonuclease activity and coagulase activity commented on previously has been demonstrated by Weckman and Catlin[17] and confirmed by other investigators. Deoxyribonuclease activity may be tested for by streaking or spotting the surface of commercially available deoxyribonuclease test medium* with the test strain. Enzyme activity may be checked by flooding the surface with 1 N **hydrochloric acid** or 0.1% toluidine blue. The appearance of a **clear zone** around the streak (or spot) constitutes a **positive** test. Plates flooded with toluidine blue will show a bright **rose pink** zone around colonies of staphylococci that produce the enzyme deoxyribonuclease. The formula for this medium is given in Chapter 41.

S. aureus can develop resistance to a number of the widely used antibiotics, as well as to new ones being brought into use, with surprising facility. At least 90% of hospital strains of staphylococci are resistant to penicillin. Indeed, there seems to be a correlation between virulence and penicillinase production; the more resistant such strains are to penicillin, the more virulent they appear to be. Laboratory personnel are thus advised to consider each isolate or strain individually and to test it for its susceptibility to a number of antibiotics by one of the accepted methods of assay. It should be noted that resistance to two or more antibiotics is relatively rare in phage group II.† The reader is referred to Chapter 36 for information on methods of antimicrobial susceptibility testing of microorganisms.

---

*Baltimore Biological Laboratory, Cockeysville, Md.; Difco Laboratories, Detroit.
†Parker, M. T.: Personal communication (W.R.B.).

Staphylococcus epidermidis

Staphylococcus epidermidis resembles *S. aureus* in microscopic morphology. The colonies are circular, smooth, and usually a pale translucent white.

The organism is very salt tolerant, as is *S. aureus,* but differs from that species in being **coagulase negative, DNase negative,** and **mannitol negative.** In other biochemical reactions it resembles the pathogenic species. The organism is basically parasitic rather than pathogenic. However, it is unquestionably involved in certain clinical syndromes. It normally resides on the skin and mucous membranes of humans and other animals. Whereas the penicillinase-resistant penicillins and cephalosporins are very effective against *S. aureus, S. epidermidis* is often more resistant. In one study (Males et al.: J. Clin. Microbiol. **1:**256-261, 1975), it was noted that 13% of strains were resistant to methicillin and 1.8% to cephalothin. Resistance to other agents was also seen (erythromycin, 25%; kanamycin, 17%; and chloramphenicol, 3.9%). Vancomycin is generally active against all gram-positive cocci.

Staphylococcus saprophyticus

Staphylococcus saprophyticus, as noted above, resembles *S. epidermidis* closely but is said to be distinctive by virtue of its resistance to novobiocin.[11] Until recent years, urinary isolates of coagulase-negative staphylococci were regarded as "contaminants" or, occasionally, as opportunistic pathogens. Several reports have implicated these organisms as the etiologic agent in from 7% to 26% of urinary infections.[9,10] *S. saprophyticus* subgroup 3, formerly *Micrococcus* subgroup 3, has been noted to have a predilection for the urinary tract. Women aged 16 to 25 years have been most affected.[10] A critical appraisal of the significance of urinary isolates of coagulase-negative *Micrococcaceae* was given recently by Williams, Lund, and Blazevic.[18] Of 16,347 urine cultures submitted to their hospital laboratory, only 68 specimens (0.4%) from 50 patients yielded over 10^4 coagulase-negative staphylococci per milliliter in pure culture. A total of 62 of the 63 organisms available for their study could be classified as follows: 45 *S. epidermidis* (predominantly subgroup 1), 15 *S. saprophyticus* (subgroup 3), and two *S. aureus.* "Probable" urinary infections were noted in only 21 patients; eight patients had two or more positive urine cultures. All isolates from the same patient were identical by morphology, antibiotic susceptibility, and hemolytic pattern. Nine (75%) of their 12 isolates of *S. saprophyticus,* which were novobiocin resistant and nonhemolytic on a synergistic hemolysis test, were from patients with probable urinary infection. Eight were young women with acute symptoms and pyuria. Unfortunately, novobiocin resistance could not be relied on to differentiate their isolates of *S. saprophyticus* from *S. epidermidis* (two of 15 strains of *S. saprophyticus* were sensitive).

Phage typing of staphylococci

Since the recognition of prevalent hospital phage types is of interest in out-

Table 16-1. Lytic groups of the international staphylococcal typing phages*

Group	Phage types
I	29, 52, 52A, 79, 80
II	3A, 3B, 3C, 55, 71
III	6, 7, 42E, 47, 53, 54, 75, 77, 83A
IV	42D
Not allotted	81, 187

*Compiled from data provided by M. T. Parker, Cross Infection Laboratory, Central Public Health Laboratory, London, N.W. 9, England.

Phages 81 and 187 cannot be allotted to any of the lytic groups. Phage 81 lyses many strains that otherwise have group I patterns, but it also forms part of some group III patterns. Phage 187, however, lyses strains that are sensitive to this phage only.[12] Strains that are lysed only by phage 81 are placed in phage group I.

Additional information may be gained from the report in Int. J. System. Bacteriol. **21:**165, 167, 171, 1971.

breaks, some laboratories are involved in phage typing of staphylococci. The value of this procedure in epidemiologic studies cannot be questioned, but because of the rather involved and time-consuming procedures that are necessary, it is recommended that typing not be done as a routine.

Phage typing is carried out by the Laboratory Centre for Disease Control, Ottawa, by the Cross Infection Laboratory, C. P. H. Laboratory, London, by various state public health laboratories, and at the Center for Disease Control in Atlanta. Table 16-1 lists the lytic groups of the international set of staphylococcal basic typing phages.

Diseases caused by staphylococci

The **most common** disorders caused by the pathogenic staphylococci are cellulitis, pustules, boils, carbuncles, impetigo, secondary infection of acne, and postoperative wound infections. One of the most common types of food poisoning is staphylococcal food poisoning, caused by the enterotoxin elaborated in food. It has been reported that production of this toxic metabolite is restricted mainly to phage groups III and IV.[19]

Among the **most serious** staphylococcal infections are septicemia, endocarditis, meningitis, puerperal sepsis, pneumonia, and osteomyelitis. Staphylococcal pseudomembranous enterocolitis may develop with great rapidity after oral antibiotic therapy if resistant strains develop. There is suggestive evidence that certain syndromes may be associated with specific phage types of staphylococcus. For example, in phage group II, strains reacting only with phage 71 appear to be specifically associated with vesicular skin lesions, such as impetigo and pemphigus of the newborn.[13]

S. epidermidis is of particular importance clinically in infections such as bacterial endocarditis, particularly after cardiac surgery with valve replacement. Additionally, it is often the cause of a persistent bacteremia following ventriculoatrial shunts for the control of hydrocephalus.[14]

As noted earlier, *S. saprophyticus* may be involved in urinary tract infections in young women.

Blood cultures are important in confirming the diagnosis of staphylococcal endocarditis. If clinical observations suggest this syndrome, at least **three** blood cultures should be drawn before treatment. In patients needing immediate treatment, cultures should be set up from two separate venipunctures made within a period of 5 to 10 minutes.

Genus Micrococcus

As previously noted, *Micrococcus* is distinguished from *Staphylococcus* primarily by its failure to ferment glucose anaerobically in a special medium and its lack of susceptibility to lysis by lysostaphin. *Micrococcus* does have cytochrome enzymes (is benzidine positive). It is usually strongly catalase positive.

Members of this genus may cause endocarditis and other infections of the type caused by *S. epidermidis,* but are probably much less commonly involved.

Genus Aerococcus

Aerococcus grows primarily in **tetrads** and **clumps,** but occasional single or paired cells are noted. The genus is distinct from *Staphylococcus* and *Micrococcus* by virtue of a **negative** benzidine test. Aerococci may release oxygen from hydrogen peroxide, but this is not catalase mediated. Acid is **not** produced from glucose in anaerobic culture, whereas it usually is by *Streptococcus* (also benzidine negative).

Aerococcus may resemble enterococcal group D streptococci in that most strains tolerate 6.5% sodium chloride and 40% bile and some strains blacken bile esculin medium. However, *Aerococcus* does not possess the group D antigen, does not grow at 10 or 45 C, and does not hydrolyze arginine.

Aerococcus viridans is alpha hemolytic, but the greening may be delayed until 48 hours of incubation. Hippurate is split by most strains, often slowly.

Aerococcus-like organisms have been isolated from blood cultures of patients with subacute bacterial endocarditis and bacteremia, from urine cultures of patients with urinary tract infection, and, rarely, from empyema and wound infections (Colman, G.: J. Clin. Pathol. **20:** 294-297, 1967; and Parker, M. T., and Ball, L. C.: J. Med. Microbiol. **9:**275-302, 1976). The group is apparently much more susceptible to antibiotics than the enterococci are.

REFERENCES

1. Baird-Parker, A. C.: The basis for the present classification of staphylococci and micrococci, Ann. N.Y. Acad. Sci. **236:**7-14, 1974.
2. Baird-Parker, A. C., Hill, L. R., Kloos, W. E., and Kocur, M.: Appendix 1. Identification of staphylococci, Int. J. System. Bacteriol. **26:**333-334, 1976.
3. Bayliss, B. .G., and Hall, E. R. Plasma coagulation by organisms other than *Staphylococcus aureus,* J. Bacteriol. **89:**101-105, 1965.
4. Buchanan, R. E., and Gibbons, N. E.: Bergey's manual of determinative bacteriology, ed. 8, Baltimore, 1974, The Williams & Wilkins Co.
5. Curry, J. C., and Borovian, G. E.: Selective medium for distinguishing micrococci from staphylococci in the clinical laboratory, J. Clin. Microbiol. **4:**455-457, 1976.
6. Elek, S. D.: *Staphylococcus pyogenes*, London, 1959, E. & S. Livingstone, Ltd.
7. Facklam, R. R., and Smith, P. B.: The grampositive cocci, Human Pathol. **7:**187-194, 1976.
8. Finegold, S. M., and Sweeney, E. E.: New selective and differential medium for coagulase-positive staphylococci allowing rapid growth and strain differentiation, J. Bacteriol. **81:**636-641, 1961.
9. Mabeck, C. E.: Significançe of coagulase-negative staphylococcal bacteriuria, Lancet **2:**1150-1152, 1969.
10. Maskell, R.: Importance of coagulase-negative staphylococci as pathogens in the urinary tract, Lancet **1:**1155-1158, 1974.
11. Mitchell, R. G., and Baird-Parker, A. C.: Novobiocin resistance and the classification of staphylococci and micrococci, J. Appl. Bacteriol. **30:**251-254, 1967.
12. Parker, M. T.: Phage typing and the epidemiology of *Staphylococcus aureus* infection, J. Appl. Bacteriol. **25:**3, 1962.
13. Parker, M. T., and Williams, R. E. O.: Further observations on the bacteriology of impetigo and pemphigus neonatorum, Acta Paediatr. Upps. **50:**101-112, 1961.
14. Quinn, E. L., Cox, F., and Fisher, M.: The problem of associating coagulase negative staphylococci with disease, Ann. N.Y. Acad. Sci. **128:**428-442, 1965.
15. Recommendations, Subcommittee on Taxonomy of Staphylococci and Micrococci, Int. Bull. Bact. Nomenclat. Taxon. **15:**109-110, 1965.
16. Schliefer, K. H., and Kloos, W. E.: A simple test system for the separation of staphylococci from micrococci, J. Clin. Microbiol. **1:**337-338, 1975.
17. Weckman, B. G., and Catlin, B. W.: Desoxyribonuclease activity of micrococci from clinical sources, J. Bacteriol. **72:**747, 1957.
18. Williams, D. N., Lund, M. E., and Blazevic, D. J.: Significance of urinary isolates of coagulase-negative *Micrococcaceae*, J. Clin. Microbiol. **3:** 556-559, 1976.
19. Williams, R. E. O., Rippon, J. E., and Dowsett, L. M.: Bacteriophage typing of strains of *Staphylococcus aureus* from various sources, Lancet **1:**510-514, 1953.

CHAPTER 17

Facultative streptococci

Members of the genus *Streptococcus* are widely distributed in nature and may be found in milk and dairy products, water, dust, vegetation, and the normal respiratory tract and intestinal tract of various animals, including humans. The majority are probably saprophytic and nonpathogenic, but a number of species are pathogens for humans and animals.

The single streptococcal cell is characteristically spherical but may appear elliptical on occasion. The cells normally occur in **chains** of varying lengths. Cell size varies from 0.5 to 1 μm in diameter, depending on growth conditions and age of culture. Large cells are seen occasionally with normal-sized cells in a chain in an aged culture, whereas undersized cells may appear with normal cells in anaerobic culture.

Liquid cultures normally yield longer chains (Fig. 17-1) than cultures grown on agar. Beta-hemolytic streptococci may exhibit long chains in human infections and in milk from diseased cows.

Streptococci are characteristically **gram positive** but may become gram negative as the cells age. Although a few motile forms have been reported, the streptococci are primarily **nonmotile.** Virulent forms are usually encapsulated and contain an abundance of hyaluronic acid in the capsule. Streptococcal colonies are small, translucent to slightly opaque, circular, generally less than 1 mm in diameter, convex, and appear as minute beads of moisture on a moist agar surface. On drier surfaces, colonies are less moist and almost opaque. By comparison, pneumococcal colonies are flatter and translucent.

Colony variation is quite common, with **mucoid, smooth** or **glossy,** and **matte** or

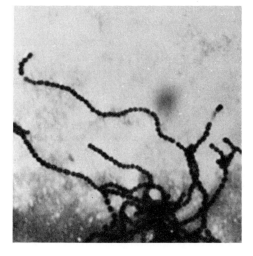

Fig. 17-1. *Streptococcus pyogenes*, showing chain arrangement. (1000×.)

rough forms. The mucoid and matte forms contain relatively large amounts of M protein and are virulent, whereas the smooth or glossy forms contain very little of this substance and are usually avirulent.

The organisms grow well on most enriched media and are generally facultative in relation to their oxygen requirements, although some strains are obligately anaerobic. They produce quite large amounts of lactic acid, without gas, from fermentable carbohydrates. **Inulin** is usually not fermented, and the organisms are **not soluble** in bile salts. The latter criterion, especially, distinguishes streptococci from pneumococci. They differ also from the staphylococci in being **catalase negative** (described on p. 123).

CLASSIFICATION OF STREPTOCOCCI

Various classifications for the streptococci have been used over the years. For example, *Bergey's Manual of Deter-*

minative Bacteriology (ed. 7)[2] lists the following groups: **pyogenic, viridans, enterococcus,** and **lactic.** Brown's[3] classification, based on the reactions obtained in **blood agar** and to which reference is made later, is another example. The Lancefield[9] system, based on the antigenic characteristics of **group-specific C substance,** is considered the most reliable. C substance is a cell-wall polysaccharide, which permits the arrangement of the streptococci into a number of antigenic groups identified as Lancefield groups A, B, C, D, and so forth.

HEMOLYTIC REACTIONS ON BLOOD AGAR

The most useful criterion for the preliminary differentiation of human streptococcal strains is their hemolytic activity on blood agar. As originally postulated by Brown, several types, including a recently described subtype, can be observed, particularly in subsurface colonies in pour or stabbed plates[8]:

Alpha(α)—an indistinct zone of partially lysed red cells surrounds the colony, frequently accompanied by a greenish or brownish discoloration, which is best seen around subsurface colonies.

Beta(β)—a clear, colorless zone surrounds the colony, indicating complete lysis of the red blood cells. This is best seen in deep colonies in a pour plate. Surface colonies, on the other hand, may appear as alpha or nonhemolytic, due to inactivation of one of the hemolysins, **streptolysin O,** which is **oxygen labile; streptolysin S,** an **oxygen-stable** hemolysin, may be present in only small amounts in these strains showing poor surface hemolysis.

Gamma(γ)—colonies show no apparent hemolysis or discoloration in either surface or subsurface colonies.

Alpha-prime(α'), or wide-zone alpha— a small zone of alpha hemolysis surrounds the colony, with a zone of complete or beta hemolysis extending beyond this zone into the medium. This can be confused with a beta-hemolytic colony when examined only macroscopically.

Hemolysis of mammalian erythrocytes by the streptococci involves a complex system of many variables, including the influence of the basal medium, the production of streptolysins O and S by a given strain, the effect of various types of blood (sheep, horse, rabbit, human), aerobic or anaerobic incubation, and so forth. To begin with, a good basal medium, such as soybean-casein digest agar* **without** dextrose (the acid produced by carbohydrate fermentation inactivates streptolysin S) at pH 7.3 to 7.4, to which 5% defibrinated sheep blood has been added, is recommended. This medium will support the growth of fastidious strains and permit good differentiation of the various types of hemolysis.

Although much has been written about the species of red blood cells and their effect on hemolytic patterns, it appears that this is restricted to the enterococci, of which over 90% show alpha hemolysis on sheep blood agar but are beta hemolytic on other mammalian blood agar media.[15] Sheep blood is recommended especially for its inhibitory action on the growth of *Haemophilus hemolyticus,* a normal throat commensal whose beta-hemolytic colonies may be confused with those of beta-hemolytic streptococci. When in doubt, the examination of a Gram stain of the suspected colony will rapidly differentiate gram-positive streptococci from gram-negative bacillary forms of *H. hemolyticus.* It should be stressed that outdated human blood bank blood is **not** recommended for preparation of blood agar, because it may contain inhibitory factors, such as antibacterial substances, antibiotics, or an excess of citrate ion.

Although incubation under increased CO_2 tension (candle jar or CO_2 incubator)

---

*BBL trypticase soy agar, Difco tryptic soy agar.

for 18 to 24 hours has been recommended for the isolation and recognition of beta-hemolytic streptococci on streaked and stabbed sheep blood agar plates, it should be noted that anaerobic incubation or prolonged aerobic incubation and subsequent overnight refrigeration may increase hemolytic activity of these organisms. However, Murray and associates[13] noted recently that significantly more non–group A beta-hemolytic streptococci grow in an anaerobic or CO_2 atmosphere. Accordingly, they recommended **aerobic** incubation (with stabbing of the medium) for throat swabs where **only** group A streptococci are of concern. Aerobic incubation is satisfactory for pour plates.

PREPARATION OF POUR PLATES

The ideal method of identification of hemolytic activity is by the **microscopic examination** of subsurface colonies in a blood agar **pour plate.** Although not always practical, it is not a difficult procedure, and may be carried out as follows:

1. Melt a tube of sterile soy-casein digest agar (15 to 20 ml) and cool to approximately 45 C.
2. Aseptically add 0.6 to 0.8 ml of sterile defibrinated sheep blood.
3. Inoculate this with one **small** loopful (drained against inner wall of tube) of a suspension prepared from the original throat swab suspended in 1 ml of sterile Todd-Hewitt broth.
4. Mix the inoculated medium by rotating the tube between the palms of the hands and pour into a sterile Petri plate. Allow to harden; incubate overnight at 35 C. If desired, the surface of the plate also may be streaked with a loopful of the suspension.

When colonies are examined microscopically, the illumination is reduced, and the low-power objective (100× total magnification) is focused on both the colony and the layer of blood cells in approximately the same plane. A colony near the bottom of the plate should be selected for examination to minimize the effect of the intervening agar in focusing on the inverted plate. Surface colonies are examined with the lid of the plate removed.

BETA-HEMOLYTIC STREPTOCOCCI

As previously indicated, beta-hemolytic streptococci produce surface or subsurface colonies surrounded by a clear, cell-free zone of hemolysis. Although seen best in pour plates, beta hemolysis may also be demonstrated by making several stabs with the inoculating loop into the agar at the time of plate streaking. This permits subsurface growth and participation by both streptolysins O and S, if present.

Although beta-hemolytic streptococci of groups A, B, C, D, E, F, G, H, K, L, M, O, and others are found in Lancefield's classification,* **group A strains** (*S. pyogenes*) are most often associated with communicable disease in humans and are the etiologic agents in streptococcal pharyngitis, scarlet fever (*Staphylococcus aureus* may also cause this), epidemic wound infections, and so forth. They are subdivided serologically into 44 M types and 26 T agglutination types, depending on type-specific surface antigens.[7] According to Moody and co-workers,[12] more than 90% of group A strains can be typed by a combination of M precipitin and T agglutination procedures.

Human pathogens may also occur among other serologic groups, including groups B and D. Group B strains (*S. agalactiae*) are normally present in the vagina and may be associated with maternal septicemia and neonatal meningitis,[1] as well as other infections in humans[14]; animal strains are involved in bovine mastitis. Group D strains, including enterococci, are discussed in a subsequent section.

---

*Antisera are available for these and other groups from Difco Laboratories, Detroit, and Burroughs Wellcome Co., Research Triangle Park, N.C.

Table 17-1. Identification of beta-hemolytic streptococci*

Characteristics	Group A	Group B	Group C human	Group D (enterococci)†
On blood agar				
Surface colonies	White to gray, opaque, hard, dry; 2-mm zones of hemolysis	Gray, translucent, soft; narrow hemolytic zone; a few RBC may be observed microscopically under the colonies	Similar to group A	Gray, translucent, soft; zone of hemolysis wider than the colony
Subsurface colonies	2- to 2.5-mm zones of hemolysis, with sharply defined edges; spindle shaped	0.5-mm zone of hemolysis after 24 hours; 1-mm zone of hemolysis after 48 hours; refrigeration produces double zones of hemolysis	Similar to group A	3- to 4-mm zones of hemolysis
Pigment production	None	Yellow-orange‡ (stab inoculation)	None	None
CAMP** test	Negative	Positive	Negative	Negative
Bacitracin susceptibility	Susceptible	Resistant§	Resistant	Resistant
Sodium hippurate	Not hydrolyzed	Hydrolyzed‖	Not hydrolyzed	Not hydrolyzed§
Growth at 10 C	−	−	−	+
Growth at 45 C	−	−	−	+
Bile esculin medium¶	−	−	−	+
6.5% NaCl broth	No growth	No growth	No growth	Growth
Source	Throat, blood, wounds, rarely spinal fluid	Urine, peritoneum, blood, throat, genital tract of females, cerebrospinal fluid	Throat, nose, vagina, intestinal tract	Urine, peritoneum, feces; milk and milk products
Pathogenicity	Septicemia, tonsillitis, scarlet fever, puerperal sepsis, pneumonia, cellulitis, erysipelas; sequelae—rheumatic fever, glomerulonephritis	Female genital tract infections; urinary tract infections, septicemia, endocarditis, meningitis, respiratory infections	Erysipelas, puerperal sepsis, throat infections; opportunist pathogen	Subacute bacterial endocarditis; urinary tract infections; bacteremia; female genital tract infections; intra-abdominal infections

*Note: Groups A, B, and C will not grow at pH 9.6 or in 0.1% methylene blue milk—two criteria that distinguish them from group D (enterococci).
†See Table 17-3, p. 137. May also be alpha or gamma hemolytic.
‡Merritt, K., and Jacobs, N. J.: J. Clin. Microbiol. **4:**379-380, 1976.
§95% are positive with the rapid test of Lee and Ederer (J. Clin. Microbiol. **5:**290-292, 1977).
‖A 2-hour test has been described by Hwang and Ederer (J. Clin. Microbiol. **1:**114, 1975), and a rapid test requiring no incubation was described by Edberg and Samuels (J. Clin. Microbiol. **3:**49-50, 1976).
¶Baltimore Biological Laboratory, Cockeysville, Md.; Difco Laboratories, Detroit.
A CAMP–disk test has been described by Wilkinson (J. Clin. Microbiol. **6:42-45, 1977).

Occasionally, strains of group C, G, H, K, and other streptococci have been reported in human infections, including bacteremia, and respiratory and genitourinary tract infection.[4] These strains generally produce either alpha or beta hemolysis on blood agar.

Table 17-1 shows some useful criteria for differentiation of some of the more important groups of beta-hemolytic streptococci.

Benzyl penicillin (penicillin G) and other penicillins, including ampicillin, are the drugs of choice in the treatment of infections due to streptococci of groups A, B, C, G, and others. Sulfonamides or erythromycin are sometimes used prophylactically against group A streptococcal infections in patients with rheumatic heart disease. Erythromycin is a good therapeutic alternative to penicillin in an individual with hypersensitivity to that drug.

THE BACITRACIN DISK TEST FOR GROUP A STREPTOCOCCI

A useful **presumptive** test for differentiating group A from other groups of beta-hemolytic streptococci is the bacitracin disk test, first introduced by Maxted[11] and modified by Levinson and Frank.[10] The test depends on the selective inhibition of group A streptococci on a blood agar plate by a paper disk containing 0.04 units of bacitracin.* (**Caution:** Use the differential disk containing 0.04 units, **not** the sensitivity disk containing 10 units.) A high degree of correlation (95%)† between bacitracin and serologic tests with group A streptococci exists if the following conditions are fulfilled[6,8]:

1. A **pure culture** of a beta-hemolytic colony is used—do not use on a primary plate of a mixed culture.

---

*Available from Baltimore Biological Laboratory (Taxo A); Difco Laboratories (Bacto Differentiation Disc Bacitracin).

†Approximately 0.5% of group A streptococci will be missed by this test, and about 4% of non-group A strains will be incorrectly identified as group A.

2. A fresh, **moist** blood agar plate must be used—an old, dried-out plate will reduce diffusion of the bacitracin, giving a false-negative reading.
3. The inoculum size must be such as to ensure **confluent growth**—a light growth of other streptococci may show inhibition zones.
4. The bacitracin disks must be stored in the refrigerator with a desiccant and should be checked biweekly for performance with known group A and non–group A strains. Each new lot of disks obtained also should be checked on receipt.
5. The test should be used only to differentiate **beta**-hemolytic gram-positive streptococci—some alpha strains may show moderate zones of inhibition (8 to 10 mm).
6. **Any zone of inhibition,** regardless of size, is positive. In reporting a positive test, the results should be worded "Beta-hemolytic streptococcus (probably group A)" or "Beta-hemolytic streptococcus, presumptively group A by bacitracin test."

Further identification procedures include the use of immunofluorescence technics and the Lancefield serologic procedures, described in Chapters 37 and 39.

ALPHA- AND GAMMA-HEMOLYTIC STREPTOCOCCI

Streptococci that generally do not possess group antigens and produce alpha or no hemolysis on blood agar are known collectively as the **viridans streptococcus group.** They are constantly present in the human oropharynx and include such species as *S. salivarius, S. mitis, S. mutans,* and *S. sanguis.* These organisms are the most frequent cause of subacute bacterial endocarditis, an insidious and fatal infection (if untreated) that usually follows dental or surgical procedures or instrumentation in patients with a previously damaged heart valve or other lesions of the endocardium. These organisms may also play a role in certain other serious

Characteristics	Streptococcus faecalis*	Streptococcus salivarius	Streptococcus MG†	Streptococcus sanguis	Streptococcus mitis	Streptococcus mutans‡
On blood agar Surface colonies	Relatively large, gray, shiny, translucent; no greening after 24 hours, slight greening at 48 to 72 hours	Small, raised, convex, opaque; narrow zone of hemolysis with or without greening, depending on blood used	Nonhemolytic after 24 hours but may give alpha appearance after 48 hours; variations due to type of blood used	Alpha hemolysis in 24 to 48 hours	Alpha hemolysis in 24 to 48 hours	Nonhemolytic (usually); often requires increased CO_2 for growth; when incubated at 35 C for 48 hours, colonies may pit and adhere to agar
Subsurface colonies	Large, non-hemolytic after 24 hours; definite greening after 48 to 72 hours; hemolysis evident after 24 hours of refrigeration	Typical alpha appearance—fixation of cells, greening, and hemolysis after 24 to 48 hours	Nonhemolytic after 24 hours, changing to alpha at 48 hours; refrigeration for 24 hours produces typical appearance	Alpha hemolysis in 24 to 48 hours	Alpha hemolysis in 24 to 48 hours	Nonhemolytic (usually)
Colonies on 5% sucrose agar	Small, compact	"Gum drop" (large, raised, mucoid); extracellular glucan (levan)	Small, compact; fluorescent in ultraviolet light	Small, compact; forms extracellular glucan (dextran)	Small, compact	Forms extracellular glucan (dextran)
Growth in 5% sucrose broth	No change	No change	No change	Forms glucan (jelling of medium)	No change	Forms glucan
Bile esculin medium	Growth	No growth	No growth	No growth	No growth	No growth§
6.5% NaCl broth	Growth	No growth	No growth	No growth	No growth	No growth
0.1% Methylene blue milk	Reduction	No reduction	No reduction	No reduction	No reduction	No reduction
Mannitol	+	–	–	–	–	+
Inulin	–	+	–	+	–	+
Source	Intestine and genitourinary tract, blood	Respiratory tract, blood	Respiratory tract	Blood	Respiratory tract, blood	Blood, bite wounds
Pathogenicity	Urinary tract infections, subacute bacterial endocarditis; opportunist pathogen	Respiratory infections, subacute bacterial endocarditis	Respiratory tract infections	Subacute bacterial endocarditis	Respiratory infections, subacute bacterial endocarditis	Subacute bacterial endocarditis

Streptococcus faecalis var. *liquefaciens* shows the same cultural characteristics as *S. faecalis*, but the former liquefies gelatin and the latter does not.
†S. MG can be identified specifically by the quellung reaction, using S. MG antiserum.
‡Facklam, R. R. Appendix (Review of *Streptococcus mutans*). In: Proficiency testing; summary analysis bacteriology, vol. 3, Atlanta, 1976, Center for Disease Control, pp. 11-18.
§Some strains may give a positive reaction on bile esculin medium.

infections, such as brain abscess, necrotizing pneumonia, or liver abscess. *Streptococcus MG* has been used in an agglutination test for the serologic diagnosis of *Mycoplasma pneumoniae* infections (primary atypical pneumonia).

Colonies of alpha-hemolytic streptococci must be distinguished from those of pneumococci or enterococci, both of which also may produce alpha hemolysis on blood agar. Table 17-2 shows some tests that are useful in identifying the alpha-hemolytic (or nonhemolytic) streptococci; other differential tests are indicated in subsequent sections.

Penicillin G is the antibiotic of choice in the treatment of streptococcal (nonenterococcal) subacute bacterial endocarditis. It is essential that the organism be isolated promptly from blood cultures (see Chapter 7) and its antimicrobial susceptibility (with bactericidal endpoint) determined.

This antimicrobial agent is also recommended as prophylactic therapy for patients with rheumatic or congenital heart disease who undergo dental or surgical procedures. In penicillin-allergic patients, erythromycin, a cephalosporin, or vancomycin are satisfactory substitutes.

GROUP D STREPTOCOCCI, INCLUDING ENTEROCOCCI

The streptococci that react serologically with Lancefield group D antisera comprise two different categories. The first contains the **enterococci**, *S. faecalis* and *S. faecium*, of human intestinal origin and important agents in human infections. The second category includes the **nonenterococci** of group D, *S. bovis* and *S. equinus*. Although most human group D clinical isolates are *S. faecalis*, *S. bovis* is being recovered increasingly from patients with subacute bacterial endocarditis and may be found in other infections as well. The differentiation is important, since *S. bovis* is penicillin susceptible and *S. faecalis* is not. The incidence of other members of the group has not been accurately determined.

Generally, group D streptococci appear as alpha or nonhemolytic colonies on sheep blood agar; occasional varieties of *S. faecalis* (var. *zymogenes*) will produce wide zones of beta hemolysis. Some group D strains have a distinct buttery odor on the medium. Traditionally, group D strains have been differentiated from other streptococci by their ability to grow at 45 C and by being thermostable at 60 C for 30 minutes; however, streptococci of other groups may demonstrate these characteristics, especially when standardized test conditions have not been met. The most accurate presumptive test for recognizing group D streptococci is the use of **bile esculin medium** (BEM).[5] Incorporating bile (oxgall) into an agar or broth medium containing esculin* makes the medium selective for the growth of enterococci (also *Listeria monocytogenes* and Enterobacteriaceae) that are capable of hydrolyzing esculin to 6,7-dehydroxycoumarin. This reacts with an iron salt in the medium to form a **dark brown or black** compound. The agar slant or broth is inoculated with a pure culture and examined after 48 hours' incubation at 35 C for the production of a brownish black color. Plus-minus reactions are read as negative.

Another useful biochemical test for the identification of enterococci is growth in heart infusion broth containing **6.5% sodium chloride** in 18 to 24 hours (see Table 17-3).

The results of these tests are interpreted as follows[5]:

1. BEM positive, salt tolerance positive—enterococcus
2. BEM positive, salt tolerance negative—group D streptococcus, not enterococcus
3. BEM negative, salt tolerance nega-

---

*Several bile esculin media are available: Difco BE agar contains 4% oxgall, Pfizer PSE agar and BBL enterococcus agar contain 1% oxgall plus sodium azide. Some viridans strains may grow and hydrolyze esculin in the latter medium. Accordingly, Facklam[5] recommends the BEM with 4% oxgall. Broth media also are available.

Table 17-3. Differentiation of group D streptococci*

Characteristics	Strepto-coccus faecalis	Strepto-coccus faecalis var. liquefaciens	Strepto-coccus faecalis var. zymogenes	Strepto-coccus faecium	Strepto-coccus durans	Strepto-coccus bovis
Hemolysis on sheep blood agar	Alpha to gamma	Alpha to gamma	Beta	Gamma	Alpha or gamma	Gamma
Bile esculin medium	Growth	Growth	Growth	Growth	Growth	Growth
6.5% NaCl broth	Growth	Growth	Growth	Growth	Growth	No growth
Growth at 10 C	+	+	+	+	+	−
Growth at 45 C	+	+	+	+	+	+(−)
Ammonia from arginine	+	+	+	+	+	
Gelatin hydrolysis	−(+)	+	+(−)	−	−	−
Acid in litmus milk	+	+	+	+	+	+
Acid from						
Glycerol	V	+	+	−	−	−
Mannitol	+	+	+	+	−	+(−)
Sorbitol	+	+	+	−(+)	−	−
Sucrose	+(−)	+	+	+	V	+

*Enterococci will survive a temperature of 60 C for 30 minutes and grow in pH 9.6 broth and 0.1% methylene blue milk. Other streptococcal groups may also react positively with these tests.[5] +, positive reaction; ( ), occasional; V, variable.

tive—non–group D streptococcus (serologic grouping suggested)

The therapy of choice in the treatment of serious enterococcal infections is penicillin (or ampicillin) plus streptomycin (or another aminoglycoside); if penicillin allergy exists, vancomycin may be used. Penicillin is the drug of choice for *S. bovis* infections.

REFERENCES

1. Braunstein, H., Tucker, E. B., and Gibson, B. C.: Identification and significance of *Streptococcus agalactiae*, Am. J. Clin. Pathol. **51**:207-213, 1969.
2. Breed, R. S., Murray, E. G. D., and Smith, N. R.: Bergey's manual of determinative bacteriology, ed. 7, Baltimore, 1957, The Williams & Wilkins Co.
3. Brown, J. H.: Monograph no. 9, New York, 1919, Rockefeller Institute for Medical Research.
4. Duma, R. J., Weinbert, R. T., Medrek, T. F., and Kunz, L. J.: Streptococcal infections, Medicine **48**:87-127, 1969.
5. Facklam, R. R.: Comparison of several laboratory media for presumptive identification of enterococci and group D streptococci, Appl. Microbiol. **26**:138-145, 1973.
6. Facklam, R. R.: Streptococci. In Lennette, E. H., Spaulding, E. H., and Truant, J. P., editors: Manual of clinical microbiology, ed. 2, Washington, D.C., 1974, American Society for Microbiology.
7. Griffith, F.: The serological classification of *Streptococcus pyogenes*, J. Hygiene **34**:542, 1934.
8. Hall, C. T., In: Summary analysis of results for the proficiency testing survey in bacteriology (Jan. 9, 1970), National Communicable Disease Center, May 7, 1970.
9. Lancefield, R. C.: A serological differentiation of human and other groups of hemolytic streptococci, J. Exp. Med. **57**:571-595, 1933.
10. Levinson, M. L., and Frank, P. F.: Differentiation of group A from other beta hemolytic streptococci with bacitracin, J. Bacteriol. **69**:284-287, 1955.
11. Maxted, W. R.: The use of bacitracin for identifying group A hemolytic streptococci, J. Clin. Pathol. **6**:224-226, 1953.
12. Moody, M. D., Padula, J., Lizana, D., and Hall, C. T.: Epidemiologic characterization of group A streptococci by T agglutination and M precipitation tests in the public health laboratory, Health Lab. Sci. **2**:149-162, 1965.
13. Murray, P. R., Wold, A. D., Schreck, C. A., and Washington, J. A. II: Effects of selective media and atmosphere of incubation on the isolation of group A streptococci, J. Clin. Microbiol. **4**:54-56, 1976.
14. Patterson, M. J., and Hafeez, E. B.: Group B streptococci in human disease, Bacteriol. Rev. **40**:774-792, 1976.
15. Updyke, E. L.: Laboratory problems in the diagnosis of streptococcal infections, Publ. Health Lab. **15**:78-80, 1957.

CHAPTER 18

Pneumococci

STREPTOCOCCUS PNEUMONIAE

The **pneumococcus** is a single species, *Streptococcus (Diplococcus) pneumoniae*. These organisms consist of **gram-positive lanceolate cocci**, characteristically appearing as diplococci but occasionally as short, tight chains or single cocci. In fresh specimens of sputum, spinal fluid, or other exudates, they are frequently surrounded by a capsule. Based on a specific capsular polysaccharide, 84 capsule types have been recognized.

The normal habitat of the pneumococcus is the upper respiratory tract of humans; from there it may invade the lungs and the systemic circulation. Consequently, the pneumococcus is the most frequent cause of lobar pneumonia (80% to 90%) in adults. Bacteremia occurs in about one fourth of these patients, usually early in the course of the disease; the organisms also may extend to the pleural cavity or disseminate to the endocardium and pericardium, the meninges, joints, and so forth, with ensuing complications. Pneumococci have also been implicated in infections of the middle ear, mastoid, or eye; they are occasionally isolated from peritoneal fluid, urine, vaginal secretions, wound exudates, and other clinical specimens. Carrier rates in the respiratory tract of healthy adults may vary from 30% to 70%, depending on the season of the year.

Pneumococci require an **enriched** medium for their primary isolation; trypticase or brain-heart infusion agar enriched with 5% defibrinated sheep, horse, or rabbit blood is recommended. Since 5% to 10% of strains require incubation under increased CO_2 tension to grow on primary plate culture, it is necessary to incubate such media in a candle jar or CO_2 incubator. Rare strains of pneumococcus are obligately anaerobic (Yatabe, J. A. H., Baldwin, K. L., and Martin, W. J.: J. Clin. Microbiol. **6:**181-182, 1977).

CLINICAL SPECIMENS

The specimens usually submitted for culture include sputum (from the lower respiratory tract—**not** saliva), transtracheal aspirates, blood, cerebrospinal fluid, nasopharyngeal swabs (particularly in the pediatric patient), purulent exudates, serous effusions, and so forth. Sputum, if tenacious, may be homogenized by repeated mixing with a small volume of broth in a sterile syringe without a needle attached. Otherwise, a carefully selected portion of purulent or bloody material is teased out with the split halves of a sterile wooden applicator stick and one portion is inoculated to blood agar. The other portion may be used for preparing a direct smear for gram staining. If quantitative studies of the bacterial flora of sputum are desired, a mucolytic agent, such as pancreatin or dithiothreitol,* may be used.

CULTURAL CHARACTERISTICS

After overnight incubation, typical pneumococcal colonies on blood agar are round and glistening with entire edges, **transparent,** mucoid, and about 1 mm in diameter. They are surrounded by an approximate 2 mm-zone of alpha hemolysis (beta hemolysis when incubated anaerobically[5]), due to the activity of hyaluronidase. Young colonies are usually dome

---

*Available commercially as Sputolysin, California Biochem., La Jolla, Calif.

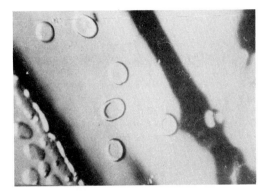

Fig. 18-1. Typical colony morphology of *S. pneumoniae.* (20×.)

shaped but on aging become flattened with a raised margin and depressed central portion (Fig. 18-1) or give the appearance of a **checker** or nail head. This is best seen with a dissecting microscope and oblique surface illumination. Alpha-hemolytic streptococci, by contrast, produce small, raised, **opaque** colonies. Colonies of type 3 pneumococcus (occasionally, other types) are larger, more mucoid and confluent than those already described, and resemble droplets of oil on the agar surface. These eventually flatten on drying and exhibit characteristics similar to those of other types.

IDENTIFICATION TESTS FOR PNEUMOCOCCI

In addition to the colony characteristics by which pneumococci may be tentatively identified, certain tests provide a means of differentiating these organisms from the alpha streptococci or other cocci. Among them are the tests for (1) **bile solubility,** (2) **inulin fermentation,** (3) **optochin susceptibility,** (4) **mouse virulence,** and (5) the **Neufeld (quellung) reaction.**

Bile solubility test. Surface-active agents, such as bile, bile salts (sodium deoxycholate or taurocholate), or sodium dodecyl sulfate ("Dreft"), act on the cell wall of pneumococci and bring about lysis of the cell. In the bile solubility test used at the Mayo Clinic, [8] a few drops of

10% sodium deoxycholate are placed directly onto a 24-hour-old colony on a blood agar plate. The colony typically dissolves if it is a pneumococcus. (Be sure that the colony has not just been floated away by the reagent.)

It should be noted that some strains of pneumococci are insoluble in bile, whereas some strains of alpha-hemolytic streptococci are soluble; the test, therefore, is not absolute.

Inulin fermentation test. Most strains of pneumococci will ferment this carbohydrate. However, because some strains of *Streptococcus sanguis* and *S. salivarius* will also ferment it, the test should not be used alone but in conjunction with other confirmatory tests, such as bile solubility or optochin sensitivity, as an identification procedure.

Optochin growth-inhibition test. This is currently the most widely used test for differentiating pneumococci from other alpha-hemolytic streptococci. It was first described by Bowers and Jefferies[3] in 1954 and modified by Bowen and co-workers[2] in 1957. The test is carried out by placing a 6-mm absorbent paper disk containing 5 µg of ethylhydrocupreine HCl (Optochin)* on a blood agar plate heavily inoculated with a pure culture of the suspected strain. After overnight incubation **aerobically** at 35 C, pneumococci will usually exhibit a zone of inhibition **greater than 18 mm** in diameter.

Experts consider the test over 90% reliable in differentiating pneumococci from other streptococci, with occasional strains not inhibited. The Center for Disease Control (CDC) has received cultures of "penicillin-resistant pneumococci" that proved to be alpha streptococci, exhibiting a moderate inhibition zone (10 to 12 mm) to Optochin, especially when light inocula were used.[4]

Note: It is important that each new lot

---

*Taxo P disks, Baltimore Biological Laboratory, Cockeysville, Md.; Optochin disks, Difco Laboratories, Detroit; and others.

of Optochin disks be checked with known strains of pneumococci and alpha-hemolytic streptococci before use.

Mouse virulence test. The white mouse is particularly susceptible to infection by small inocula of pneumococci and may be used to advantage in certain unusual situations, especially with mixed cultures. For example, if 4 to 6 hours after intraperitoneal injection of 1 ml of emulsified sputum the mouse's abdomen is entered with a sharp-tipped capillary pipet, the peritoneal fluid will contain pneumococci in pure culture, which may be used for capsular swelling tests, culture, and other procedures. It should be noted that some pneumococcus types, including type 14, may be avirulent for mice.

Avery's "artificial mouse" also may be used with mixed cultures. This culture medium (1% glucose and 5% defibrinated rabbit blood in meat infusion broth at pH 7.8) promotes rapid growth and encapsulation of pneumococci after overnight incubation at 35 C and provides a good source for capsular swelling tests.

Neufeld (quellung) reaction. The **quellung** reaction is the most accurate, reliable, and specific test for the identification of pneumococci and pneumococcal types. However, it is infrequently used because the treatment of pneumococcal infections is no longer dependent on identification of the specific capsular type. The procedure, nevertheless, has proved of value in epidemiologic investigations and can be of considerable value in providing rapid, definitive identification of pneumococci in clinical specimens. The consensus is that the **capsular swelling reaction,** as it is popularly referred to, is not an actual swelling. The phenomenon is probably caused by a change of the refractive properties brought about by the union of the specific antiserum with the capsule (a precipitin reaction), which makes the outline of the capsule more readily visible. Fig. 18-2 illustrates capsules seen in the quellung reaction. Capsular antisera for the pneumococci are available com-

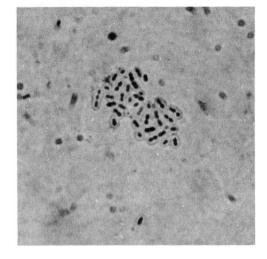

Fig. 18-2. *Streptococcus pneumoniae*, showing capsular swelling. (1000×.)

mercially*; an omnivalent serum, which reacts with all pneumococcal types, has been prepared by Lund and Rasmussen.[6] This "omniserum" has proved very useful as a reliable reagent for the rapid identification of pneumococci in clinical material, such as sputum, cerebrospinal, pleural, or synovial fluid, or in positive blood culture bottles. Merrill and associates[7] found the direct quellung (omni) test on sputum smears considerably more accurate than Gram stains when correlated with culture results.

Directions for performance of the capsular swelling test are given in Chapter 37.

SIGNIFICANCE OF RECOVERY OF PNEUMOCOCCI

Because of the relatively high seasonal carrier rate of pneumococci in the respiratory tract, their recovery from this site may be difficult to interpret. However, isolation, particularly of the recognized virulent strains,[1] such as types 1, 3, 4, 7, 8, or 12 from sputum of a patient with lobar

---

*Six pools of 33 types are available from Difco Laboratories, Detroit; Omniserum, 9 pools and 46 monotypic sera are available from the State Serum Institute, Copenhagen.

pneumonia is likely to be of clinical significance. Further, isolation of pneumococci from blood or spinal fluid not only affords definite evidence of a pneumococcal infection but isolation from spinal fluid suggests a guarded prognosis and has therapeutic implications.

Pneumococci are typically highly susceptible to penicillin, which remains the drug of choice in treating infections by this organism; however, recent isolates from South Africa and one from the United States have been resistant. Erythromycin is the second-choice agent in the penicillin-allergic patient.

Since the advent of antibiotics, especially penicillin, the overall mortality of adults with pneumococcal pneumonia has been reduced from 84% to 17%[1] (type 3 infections cause a higher mortality); results in the pediatric age group appear to be even better. Pneumococcus types 14, 3, 6, 18, 19, and 23 are isolated most frequently from infections of infants and children.

It should be stressed that pneumococci are usually not recoverable from sputum culture soon after adequate penicillin therapy has been begun.[1] It is therefore imperative that the clinician secure material for culture **before** initiation of any antimicrobial therapy.

REFERENCES

1. Austrian, R.: The current status of pneumococcal pneumonia and prospects for prophylaxis, Ninth Annual Infectious Disease Symposium, Delaware Academy of Medicine, May 5, 1972.
2. Bowen, M. K., Thiele, L. C., Stearman, B. D., and Schaub, I. G.: The optochin sensitivity test; a reliable method for identification of pneumococci, J. Lab. Clin. Med. **49:**641-642, 1957.
3. Bowers, E. F., and Jeffries, L. R.: Optochin in the identification of *Str. pneumoniae*, J. Clin. Pathol. **8:**58, 1955.
4. Hall, C. T.: Summary analysis of results for the proficiency testing survey in bacteriology (Jan. 9, 1970), National Communicable Disease Center, May 7, 1970.
5. Lorian, V., and Popoola, B.: Pneumococci producing beta hemolysis on agar, Appl. Microbiol. **24:**44-47, 1972.
6. Lund, E., and Rasmussen, P.: Omniserum—a diagnostic pneumococcus serum reacting with the 82 known types of pneumococcus, Acta Pathol. Microbiol. Scand. **68:**458-460, 1966.
7. Merrill, C. W., Gwaltney, J. M., Jr., Hendley, J. W., et al.: Rapid identification of pneumococci. Gram stain vs. the quellung reaction, N. Engl. J. Med. **288:**510-512, 1973.
8. Washington, J. A., II. editor: Laboratory procedures in clinical microbiology, Boston, 1974, Little, Brown, and Co.

CHAPTER 19

Neisseria and Branhamella

Members of the family Neisseriaceae to be considered in this chapter, *Neisseria* and *Branhamella,* are **gram-negative cocci** occurring in pairs or in masses and are aerobic or facultatively anaerobic. These organisms may be found in the oropharynx or nasopharynx and the genitourinary tract of humans and animals.

Of the six recognized species in the genus *Neisseria,* two are well-known pathogens of humans. The remainder and the one *Branhamella* species are of doubtful pathogenicity, although certain species have been associated with respiratory disorders and meningeal infections.

NEISSERIA GONORRHOEAE (GONOCOCCUS)

The **gonococcus** is a gram-negative diplococcus in which the paired cells have flattened adjacent walls. The organism is fastidious, requiring enriched media, such as blood agar or chocolate agar, for cultivation. It is an obligate human parasite not found naturally in other animals, and it causes a number of human infections, including urethritis, prostatitis, epididymitis, cervicitis, salpingitis, proctitis, pharyngitis, perihepatitis, and other complications in adults, vulvovaginitis in children, and ophthalmia in newborns.

The disease is transmitted by sexual intercourse in most instances, and causes acute purulent (and sometimes asymptomatic) urethritis in men and cervicitis, (frequently asymptomatic) in women. Infection may also occur by extension to other sites, including the joints, rectal crypts, blood, and so forth. Less frequently, gonorrheal ophthalmia of the newborn may occur following passage through an infected birth canal; gonorrheal vulvovaginitis in the preadolescent female is usually the result of criminal assault or by contact with an infected fomite. Humans are the only known host. Careful attention to appropriate specimen collection and transport procedures should yield the organism from many of the above-mentioned sites. When there is reason to suspect homosexuality or participation in oral or anal sexual acts, urethral or endocervical swabs should be supplemented by oropharyngeal and anorectal swabs for culture. For obvious reasons, smears obtained from either the rectum or pharynx should not be relied on for diagnosis.

Isolation. Neisseria gonorrhoeae is a fastidious parasite, requiring an enriched culture medium and incubation under increased CO_2 (3% to 10%) for its recovery from clinical material. Luxuriant growth occurs on chocolate agar supplemented with yeast extract or a similar enrichment,* or on modified Thayer-Martin (MTM) medium, a modified chocolate agar made selective for the gonococcus and meningococcus by the addition of certain antibiotics.† MTM and supplemented chocolate agar without antibacterial agents added are the preferred media for isolation of *N. gonorrhoeae;* their use is described in detail in chapter 11, along with some of the newer developments regarding the transport of specimens and isolation of this organism.

On **moist** MTM medium after 20 hours'

---

*Available as IsoVitaleX (BBL), Supplement B (Difco).

†Available from Baltimore Biological Laboratory, Cockeysville, Md.; Difco Laboratories, Detroit; and others.

incubation in a candle jar at 35 C,* typical colonies of *N. gonorrhoeae* appear as small (1 to 2 mm), transluscent, raised, moist grayish white colonies with entire to lobate margins. They are usually mucoid and tend to come off as whole colonies when picked from the agar surface. Colony size varies, depending on age of the culture or crowding on the plate. Plates without growth should be returned to the incubator for 48 hours' incubation. Cultures should be processed **immediately** after removal from the incubator, as gonococci tend to autolyze in the absence of CO_2.

A recent study[2] notes that gonococci that require arginine, hypoxanthine, and uracil form small atypical colonies, are very sensitive to penicillin, and are associated with both asymptomatic urethritis and disseminated gonococcal infection.

Presumptive identification. All members of the genus *Neisseria*, as well as members of other genera (such as *Pseudomonas, Moraxella,* and *Aeromonas*) are

_____

*Many strains of gonococci will not grow well at 37 C.

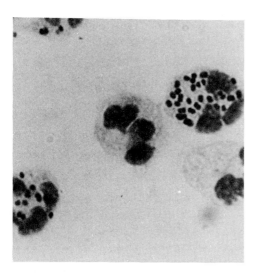

Fig. 19-1. *Neisseria gonorrhoeae.* Smear of exudate, showing intracellular gonococci. (Gram stain; 1200×.)

oxidase positive. In the oxidase reaction, a bacterial enzyme called indophenol oxidase will oxidize a redox dye, which results in a color change in the bacterial colony. In the oxidase test (described in Chapter 43), a freshly prepared or refrigerated (for no longer than 1 week) 1.0% solution of tetramethyl-p-phenylenediamine dihydrochloride is applied to a suspected colony on the MTM or chocolate agar plate; if **positive**, a color change in the colony will be observed—pink, progressing to maroon, to dark red, and finally to black. At this point (generally in 5 to 10 minutes) the organisms have been killed; however, they retain their gram-staining and FA-staining properties. Subcultures to a fresh MTM or chocolate agar plate should be made at the **pink color stage.** A reagent impregnated filter paper strip* also may be used for the oxidase test. If no characteristic colonies are observed, the MTM or chocolate agar plate may be flooded with the oxidase reagent to detect inapparent colonies. The oxidase test should not be used directly on Transgrow medium unless it has first been equilibrated to ambient atmosphere.

A thin smear of the oxidase-positive colony is then prepared, gram stained, and examined microscopically under the oil immersion lens. Typical neisseriae will appear as gram-negative diplococci with their **flattened sides adjacent;** the microscopist should note, however, that certain gram-negative diplobacilli (for example, *Moraxella osloensis*) may resemble gonococci morphologically.

A nationwide program has been under way to provide physicians with a rapid laboratory diagnosis of gonorrheal infection. The Center for Disease Control (CDC) has recommended that material obtained from the genitourinary tract, which has been inoculated to MTM (or Transgrow) medium and shows growth

_____

*Pathotec strips, General Diagnostics Division, Warner-Lambert Co., Morris Plains, N.J.

of typical oxidase-positive colonies consisting of gram-negative diplococci, provides sufficient criteria for the presumptive identification of *N. gonorrhoeae.*[1] This may be reported as: "Presumptive identification—*N. gonorrhoeae.*"* Isolation of neisseriae from other sites, such as a pharyngeal culture, or in special social or medicolegal situations, should be identified by the following, more definitive procedure.

Confirmatory identification. A presumptive identification of *N. gonorrhoeae* may be confirmed by carbohydrate fermentation reactions or less reliably by direct FA staining. *N. gonorrhoeae* **ferments glucose only,** producing acid but no gas (Table 19-1). The recommended base medium—cystine trypticase agar (CTA), pH 7.6—readily supports the growth of **fresh** isolates of gonococci and meningococci, although occasional strains of the former may grow poorly or not at all. The carbohydrates used (glucose, maltose, sucrose, and lactose) are added in 1% concentration to the sterilized medium (see Chapter 41 for preparation), which may be stored in the refrigerator.

CTA media are inoculated with a **heavy** suspension prepared from a subculture plate (taken from a single colony on the original plate) in about 0.5 ml saline or trypticase soy broth. Two to 3 drops of the suspension are deposited on the surface of the medium and then stabbed into the upper third of its depth using a sterile, cotton-plugged capillary pipet.

The screw caps of the tubes are then **tightened** and are incubated without added CO_2 at 35 C.* They are examined daily for evidence of growth and production of acid, indicated by turbidity and a **yellow color** of the upper layer of medium. After 48 to 72 hours' incubation, the tubes showing acidity should be checked for purity by microscopic examination of a Gram stain. If only glucose is fermented, report *"Neisseria gonorrhoeae isolated."* An occasional strain of gonococcus may fail to ferment glucose on the first attempt; repeat inoculation of CTA medium enriched with 10% ascitic fluid may

---

*Approximately 98% of these isolates from urogenital sites are confirmed as *N. gonorrhoeae* by carbohydrate fermentation or FA.

*Some workers prefer incubation in a candle jar, with loose screw caps. A wet paper towel should be included.

Table 19-1. Differentiation of selected neisseriae*

Organism	Carbohydrate fermentation reactions (CTA base, 1 to 4 days' incubation at 35 C)						Growth	
	Glucose	Maltose	Sucrose	Lactose	Fructose	Mannitol	On MTM	On nutrient agar at 22 C
Branhamella catarrhalis	−	−	−	−	−	−	−†	+
Neisseria gonorrhoeae	A‡	−	−	−	−	−	+	−
N. meningitidis	A	A	−	−	−	−	+	−
N. sicca	A	A	A	−	A	−	−†	+
N. lactamicus	A	A	−	A(slow)	−	−	+	
N. mucosa§	A	A	A	−	A			+
N. subflava‖	A	A	−	−	−	−	−†	+
N. flavescens‖	−	−	−	−	−	−	−†	+

*Modified from Dr. Charles T. Hall, Licensure and Proficiency Testing Branch, Center for Disease Control, Atlanta.
†A heavy inoculum can yield growth.
‡A, acid production (no gas).
§Not recognized in Bergey's Manual of determinative bacteriology (ed 8). Reduces nitrate to gas.
‖Bacterial growth shows a yellowish pigmentation on Loeffler's serum medium.

encourage fermentation by these strains. Preferably, use the "nongrowth carbohydrate degradation test" described in detail in Cumitech 4 by Kellogg and associates.[7] A rapid carbohydrate fermentation test using a barium hydroxide indicator (converted to the white precipitate barium carbonate by CO_2 produced during fermentation) also was recently described by Slifkin and Pouchet (J. Clin. Microbiol. **5:**15-19, 1977).

The direct FA staining procedure[11] is a less time-consuming and laborious procedure for confirming colonies of *N. gonorrhoeae*. However, because of cross reactions with group B meningococci and staphylococci, the FA procedure is less satisfactory for primary cultures and for accurately separating *N. meningitidis* from *N. gonorrhoeae* than are fermentation tests. The technique is described in Chapter 39.

Recently, a relatively rapid, highly specific identification procedure was described by Bawdon and co-workers (J. Clin. Microbiol. **5:**108-109, 1977). This employs DNA extracted from the culture to be identified (nonviable or mixed cultures or purulent discharge can be used) to transform genetically a uracil- and arginine-deficient auxotroph of *N. gonorrhoeae* to prototrophy.

NEISSERIA MENINGITIDIS (MENINGOCOCCUS)

The normal habitat of *Neisseria meningitidis* is the human nasopharynx; many persons may carry the organism for indefinite periods without symptoms. In susceptible persons, however, the meningococcus gains access to the central nervous system primarily through the hematogenous route (with a bacteremic precursor). A suppurative infection of the meninges then occurs, producing the characteristic syndrome of **bacterial meningitis.** The bloodstream invasion may result in an early petechial rash (smears of which may show meningococci) or take the form of septic shock accompanied by disseminated intravascular coagulation, with rapidly fatal outcome (Waterhouse-Friderichsen syndrome). Septic arthritis and endocarditis may also be seen.

The meningococcus is similar to the gonococcus in morphology and staining reactions. The paired cocci have a miniature coffee bean appearance. Some of the cells may be swollen and appear larger. The organism is fastidious in its growth requirements and may be cultivated on enriched media, such as chocolate agar or blood agar, on which it develops relatively large, smooth, raised colonies. No hemolysis is produced on blood agar. The colonies tend to autolyze fairly rapidly. The colonies may be identified as *Neisseria* by the oxidase test, which shows the same color changes observed with gonococcal colonies (see above).

The organism is **extremely sensitive** to temperature and dehydration; thus the same facilities provided for the culture of the gonococcus must be accorded the culturing of this species. Cultivation in a 3% to 10% carbon dioxide atmosphere greatly enhances growth.

N. meningitidis may be classified serologically into groups A, B, C, and D. Groups A and C are encapsulated, whereas group B strains are usually nonencapsulated. The former give rise to larger and more mucoid colonies than are observed with strains of group B, in which the colonies are smaller, rougher, and yellowish. Group A strains are generally involved in epidemics of meningitis, whereas group B strains are isolated in the majority of sporadic cases between outbreaks.

The majority of strains of *N. meningitidis* isolated from patients with meningococcal infections or from healthy carriers fall into serological groups B and C; group A and D strains are rarely isolated in the United States. In 1961, Slaterus[14] in Holland reported the isolation of meningococci that did not react with antisera prepared against group A, B, C, or D strains, and provisionally labeled these new types as X, Y, and Z. Meningococci of group Z (CDC group 29E), Boshard (Slaterus Y), and 135 have also been described.

Isolation. The organism may be isolated from cerebrospinal fluid, the nasopharynx, joint fluid, blood, transtracheal aspirates, skin petechiae, and from miscellaneous sites, such as the eye. Material from such sources is streaked on blood or chocolate agar plates or on modified Thayer-Martin (or Martin-Lester) medium if a mixed flora is anticipated, and incubated in a candle jar at 35 C. Strains of meningococci may grow on blood agar plates incubated in 5% to 10% CO_2. A positive oxidase test on the colonies is presumptive evidence of the presence of the organism. Gram stain indicates the typical morphology. Recognized colonies of the organism may then be transferred to obtain pure cultures for biochemical tests.

Identification. Inoculation of the recommended carbohydrate media—glucose, maltose, lactose, and sucrose—in either CTA or serum or ascitic fluid semisolid agar (see Chapter 41) will help to identify the meningococcus, which ferments only **glucose** and **maltose.** Occasional strains fail to produce acid from either carbohydrate. Repeated subculture or use of nongrowth carbohydrate tests[7] (see above) usually lead to typical reactions.

The serology of these organisms may be determined by the **capsular swelling** reaction, although it should be noted that only groups A and C possess capsular antigen; group B strains are identified by the agglutination reaction. Only cells from actively growing cultures, or cells directly from clinical material (for example, sediment from cerebrospinal fluid) showing a sufficient number of gram-negative diplococci, should be used for capsular swelling tests. The current method of choice in most diagnostic laboratories, however, is the **slide agglutination test,** using first polyvalent and then monovalent antisera. The reader is cautioned about the limitations of this technique, especially in reference to group B meningococci, since antigenic cross relationships exist between this organism and other neisseriae. There also appears to be some antigenic cross relationship between groups A and C and between *Escherichia coli* and groups B and C.

The direct FA staining procedure also may be used for the identification of meningococci in cerebrospinal fluid; fluorescein-labeled antimeningococcal conjugates are available commercially.*

Techniques for capsular swelling and slide agglutination tests are described in Chapter 37; the immunofluorescent procedure is described in Chapter 39.

ANTIBIOTIC SUSCEPTIBILITY OF PATHOGENIC NEISSERIAE

In a comparative study of the susceptibility of gonococci to penicillin in the United States, Martin and co-workers[9] reported that from 1955 to 1965 there was an increase from 0.6% to 42% in cultures requiring more than 0.05 unit per milliliter of penicillin to inhibit growth. Procaine penicillin G (or ampicillin), however, remains the antibiotic of choice in most patients; those allergic to the penicillin drugs may be treated effectively with spectinomycin or tetracycline.

There have been several recent reports, however, of the failure of penicillin G in the treatment of gonococcal disease related to production of penicillinase by the organism.[12,13] There is no evidence of reduced communicability or invasiveness in these strains, which accounted for 9% of the isolates in a Liverpool clinic.[12] This will clearly have an impact on current therapeutic recommendations. Spectinomycin was effective in 21 of 22 patients with such organisms, and certain penicillinase-resistant cephalosporins may also have promise.[12]

Present knowledge indicates that penicillin G or ampicillin in high dosage is

---

*Polyvalent antiserum (groups A through D) from Burroughs Wellcome Co., Research Triangle Park, N.C.; Difco Laboratories, Detroit.

the drug of choice in the treatment of meningococcal disease but is unsatisfactory for the prophylactic treatment of carriers.[4] Most of these strains can be eliminated by treatment with rifampin[3] or minocycline. Many or most are resistant to sulfadiazine, formerly the agent of choice for carriers.

OTHER NEISSERIA SPECIES AND BRANHAMELLA

Since other *Neisseriae* may be isolated from sputum, the throat, the nasopharynx, and occasionally from cerebrospinal fluid, they should be recognized and differentiated from the gonococcus and the meningococcus. For these reasons they are included in Table 19-1. *Branhamella catarrhalis* grows as a grayish white friable colony, granular and difficult to emulsify; *N. subflava* and *N. sicca* generally produce small yellowish to greenish colonies that may be smooth or hard and wrinkled and are considered to be nonpathogenic. *N. flavescens* and *N. lactamicus* produce colonies similar to *N. meningitidis,* but may have a yellowish pigment on primary isolation. These bacteria have been recovered from some pathologic processes, including endocarditis and meningitis. *B. catarrhalis* has been reported from otitis media and, rarely, from pneumonia and endocarditis. The organism is highly susceptible to penicillin.

Although previous authors reported the isolation of **lactose-utilizing** strains of neisseriae,[6,10] little attention was paid until Hollis and co-workers at CDC[5] published results of their study of organisms referred for confirmation as *N. meningitidis,* but which were subsequently found to be lactose-positive. This probably reflects those laboratories' not using lactose in their routine biochemical workup for neisseriae.[5] Since then, the species *N. lactamicus* has been characterized. The majority of strains have been recovered from pharyngeal and nasopharyngeal specimens, but occasionally strains have been recovered from sputum, tracheal aspirations, blood, amniotic and cerebro-

spinal fluids, lung tissue, and so forth. Most studies indicate that *N. lactamicus* is of little clinical significance. It has been most frequently misidentified as *N. meningitidis* or *N. subflava* (Table 19-1).

REFERENCES

1. Balows, A., and Printz, D. W.: CDC program for diagnosis of gonorrhea, Letter, J.A.M.A. **222:** 1557, 1972.
2. Crawford, G., Knapp, J., and Holmes, K. K.: Gonococci that require arginine, hypoxanthine and uracil (AHU) cause asymptomatic urethritis, Clin. Res. **25:**156A, 1977.
3. Deal, W. G., and Sanders, E.: Efficacy of rifampin in the treatment of meningococcal carriers, N. Engl. J. Med. **281:**641-649, 1969.
4. Dowd, J. M., Blink, D., Miller, C. H., Frank, P. F., and Pierce, W. E.: Antibiotic prophylaxis of carriers of sulfadiazine resistant meningococci, J. Infect. Dis. **116:**473-480, 1966.
5. Hollis, D. G., Wiggins, G. T., and Weaver, R. E.: *Neisseria lactamicus* sp. n.; a lactose-fermenting species resembling *Neisseria meningitidis,* Appl. Microbiol. **17:**71-77, 1969.
6. Jensen, J.: Studien uber gramnegative Kokken, Zentr. Bakteriol. Parasitenk. Abt. I. Orig. **133:** 75-88, 1934.
7. Kellogg, D. S., Jr., Holmes, K. K., and Hill, G. A.: Laboratory diagnosis of gonorrhea, Cumitech 4, Washington, D. C., 1976, American Society for Microbiology.
8. Knapp, J. K., and Holmes, K. K.: Disseminated gonococcal infections caused by *Neisseria gonorrhoeae* with unique nutritional requirements, J. Infect. Dis. **132:**204-208, 1975.
9. Martin, J. E., Jr., Lester, A., Price, E. V., and Schmale, J. D.: (Note) Comparative study of gonococcal susceptibility to penicillin in the United States, 1955-1969, J. Infect. Dis. **122:** 459-461, 1970.
10. Mitchell, M. S., Rhoden, D. L., and King, E. O: Lactose-fermenting organisms resembling *Neisseria meningitidis,* J. Bacteriol. **90:**560, 1965.
11. Peacock, W. L., Welch, B. G., Martin, J. E. Jr., and Thayer, J. D.: Fluorescent antibody technique for identification of presumptively positive gonococcal cultures, Public Health Rep. **83:** 337-339, 1968.
12. Percival, A., Corkill, J. E., Arya, O. P., Rowlands, J., Alergant, C. D., Rees, E., and Annels, E. H.: Penicillinase-producing gonococci in Liverpool, Lancet **2:**1379-1382, 1976.
13. Phillips, I. β-Lactamase–producing penicillin-resistant gonococcus, Lancet **2:**656-657, 1976.
14. Slaterus, K. W.: Serological typing of meningococci by means of micro-precipitation, Antonie van Leewenhoek J. Microbiol. Serol. **27:**304-315, 1961.

CHAPTER 20

Enterobacteriaceae

The Enterobacteriaceae are by far the most commonly encountered of the aerobic or facultative gram-negative bacilli in clinical specimens. Good perspective is furnished by the study of U. Blachman and M. J. Pickett (Unusual aerobic bacilli in clinical bacteriology, Scientific Developments Press, Los Angeles. In Press.) in which 768 clinical isolates were analyzed (see Diagram 1). Of these isolates, 78% were Enterobacteriaceae, 12% nonfermenters, 9% *Haemophilus* species, and 1% "unusual gram-negative bacilli" (*Actinobacillus; Brucella; Campylobacter; Cardiobacterium;* CDC Groups DF-1, DF-0-2, EF-4, and TM-1; *Eikenella, Haemophilus aphrophilus, Haemophilus vaginalis,* and *Streptobacillus*).

With the advent of the eighth edition of *Bergey's Manual of Determinative Bacteriology,*[6] the current classification of the family Enterobacteriaceae has become somewhat clouded and many clinical microbiologists will feel they are faced with something of a dilemma. This situation has been brought about primarily because of the impact the classification system of Edwards and Ewing[19] has had over the years in this country. Without any intent of passing judgment on the different schemes, the classification of Edwards and Ewing is used in this book.

Both systems are presented in Table 20-1. Table 20-2 shows specific changes in nomenclature relating to each classification scheme.

FAMILY ENTEROBACTERIACEAE

Members of the Enterobacteriaceae are gram-negative straight rods, some of which are motile and some nonmotile. The motile species possess **peritrichous**

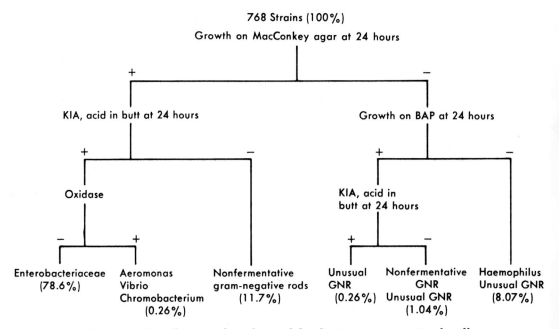

Diagram 1. Distribution of aerobic and facultative gram-negative bacilli.

Table 20-1. Classification of Enterobacteriaceae

Edwards and Ewing	Bergey's Manual (ed. 8)
FAMILY Enterobacteriaceae	FAMILY Enterobacteriaceae
TRIBE I Escherichieae	GENUS I Escherichia
GENUS I Escherichia	SPECIES *E. coli*
SPECIES *E. coli*	GENUS II Edwardsiella
GENUS II Shigella	SPECIES *E. tarda*
SPECIES *S. dysenteriae*	GENUS III Citrobacter
S. flexneri	SPECIES *C. freundii*
S. boydii	*C. intermedius*
S. sonnei	GENUS IV Salmonella
TRIBE II Edwardsielleae	SPECIES *S. cholerae-suis*
GENUS I Edwardsiella	*S. typhi*
SPECIES *E. tarda*	*S. enteritidis*
TRIBE III Salmonelleae	GENUS V Shigella
GENUS I Salmonella	SPECIES *S. dysenteriae*
SPECIES *S. cholerae-suis*	*S. flexneri*
S. typhi	*S. boydii*
S. enteritidis	*S. sonnei*
GENUS II Arizona	GENUS VI Klebsiella
SPECIES *A. hinshawii*	SPECIES *K. pneumoniae*
GENUS III Citrobacter	*K. ozaenae*
SPECIES *C. freundii*	*K. rhinoscleromatis*
C. diversus	GENUS VII Enterobacter
TRIBE IV Klebsielleae	SPECIES *E. cloacae*
GENUS I Klebsiella	*E. aerogenes*
SPECIES *K. pneumoniae*	GENUS VIII Hafnia
K. ozaenae	SPECIES *H. alvei*
K. rhinoscleromatis	GENUS IX Serratia
GENUS II Enterobacter	SPECIES *S. marcescens*
SPECIES *E. cloacae*	GENUS X Proteus
E. aerogenes	SPECIES *P. vulgaris*
E. hafnia	*P. mirabilis*
E. agglomerans	*P. morganii*
GENUS III Serratia	*P. rettgeri*
SPECIES *S. marcescens*	*P. inconstans*
S. liquefaciens	GENUS XI Yersinia
S. rubidaea	SPECIES *Y. enterocoliticia*
TRIBE V Proteeae	*Y. pseudotuberculosis*
GENUS I Proteus	*Y. pestis*
SPECIES *P. vulgaris*	GENUS XII Erwinia (plant pathogens)
P. mirabilis	SPECIES *E. herbicola* (has been considered
P. morganii	a human pathogen)
P. rettgeri	
GENUS II Providencia	
SPECIES *P. stuartii*	
P. alcalifaciens	
TRIBE VI Yersineae	
GENUS I Yersinia	
SPECIES *Y. enterocolitica*	
Y. pseudotuberculosis	
Y. pestis	
TRIBE VII Erwinieae (plant pathogens)	
GENUS I Erwinia	
GENUS II Pectobacterium	

Table 20-2. Specific changes in nomenclature

Edwards and Ewing	Bergey's Manual (ed. 8)
Citrobacter diversus	Citrobacter intermedius
Arizona hinshawii	Salmonella arizonae
Enterobacter agglomerans	Erwinia herbicola
Providencia stuartii/alcalifaciens	Proteus inconstans

flagella, differing from members of the Pseudomonadaceae, which have polar flagella. Several strains of *Salmonella, Shigella, Escherichia, Klebsiella, Enterobacter,* and *Proteus* possess fimbriae or pili.[5] The latter are not organs of locomotion, are considerably smaller than flagella, and bear no antigenic relationship to them.[15] They are readily observed under the electron microscope.

All species ferment glucose. Aerogenic and anaerogenic forms are found. The absence of gas in the fermentation of carbohydrates is characteristic of some genera. Nitrates are usually reduced to nitrites. Indophenol-oxidase is not produced.

The family is composed of a large and diverse group of organisms varying in antigenic structure and biochemical properties. The genera within the family have been established mainly on the basis of biochemical characteristics, whereas original species—the names of many of which still remain—were established on both biochemical and ecologic bases. The antigenic complexity of these bacteria has led to the development of antigenic schemes, patterned after the Kauffmann-White scheme for *Salmonella,* in which numerous serotypes are listed. Many of these serotypes are biochemically similar and can be distinguished only by serologic procedures.

The organisms are found in the intestine of humans and other animals, in the soil, and on plants. Many are parasites; others are saprophytes. Many species are pathogenic for humans, producing enteric and septicemic infections.

Culturally, these bacteria produce similar growth on blood agar, usually appearing as relatively large, shiny, gray colonies, which may or may not be hemolytic. Species that produce hydrogen sulfide show a definite greening around subsurface colonies in blood agar. On trypticase soy agar, nutrient agar, or meat infusion agar, the colonies may vary in size, depending on the genus, but they are usually grayish white, translucent, and slightly convex. Some colonies are large and mucoid, as in the case of *Klebsiella,* certain types of *Shigella,* and certain variants of *Salmonella,* especially S. *typhimurium.* Colony variation does occur, giving rise to smooth and rough forms. Individual species or type colony characteristics are described under each of the respective genera.

ISOLATION OF ENTEROBACTERIACEAE

Because this book is primarily confined to diagnostic procedures for **pathogenic** microorganisms, emphasis is on the pathogenic members of the family.

The general procedures described in Chapters 6 and 9 for isolation of the gram-negative enteric bacteria should be followed to obtain best results. Since these organisms may be isolated from various clinical sources, their isolation from fecal material, blood, urine, and other body fluids is presented again in this chapter. Laboratory personnel are reminded that clinical specimens may contain relatively few pathogens, and appropriate enrichment procedures are not only highly recommended but often necessary. It should be recalled that enrichment procedures satisfactory for *Salmonella* and *Shigella* are not applicable to coliforms and *Proteus,* since the latter are usually inhibited by their use.

Isolation from stools. The number of

pathogenic enteric organisms may decrease during the period between collection and processing in the laboratory. If any delay is anticipated in their arrival at the laboratory, specimens should be preserved by one of the procedures described earlier (see Chapter 9). Specimens from carriers may show very few pathogens, and in certain cases the appearance of these organisms may be only intermittent. Stuart's medium[83] is highly recommended as a transport medium for such specimens if they are to be shipped to a diagnostic laboratory. Ewing and coworkers[27] reported excellent recovery of *Salmonella typhi* from stools in transit in this medium after 1 week.

Because of the wide and varied flora present in fecal material, one or more of the following **enrichment media** inhibitory for the normal intestinal flora must be used. The enrichment media normally employed are **GN broth** (Hajna), the **selenite broth** of Liefson, and the **tetrathionate broth** of Mueller. Kauffmann's modification[44] of tetrathionate gives excellent recovery of salmonellae but inhibits most shigellae. It is recommended that these enrichment media be inoculated with one part stool specimen to 10 parts broth. If mucus is present, a portion of it should also be placed in the enrichment broth. If preserved diluted specimens are used (see Chapter 6), at least 2 ml should be placed in the enrichment broth. After 18 to 24 hours of incubation at 35 C, appropriate plating media should be streaked for isolation with inoculum from the enrichment broth. It can be beneficial to hold enrichment media for an additional 24 hours and, if the first plates are negative, to inoculate another set of plates.

Numerous **plating media** are in use today. Some of these are selective, whereas others are differential. Desoxycholate citrate agar, Salmonella-Shigella (SS) agar, Hektoen enteric agar (HEA), bismuth sulfite agar, brilliant green agar (BGA), eosin-methylene blue (EMB) agar, xylose lysine desoxycholate (XLD) agar, and MacConkey agar are among the most widely used.* Most laboratories prefer to employ one selective medium, such as SS or HEA, and one differential medium, such as MacConkey or EMB agar. Two procedures are followed: (1) the **direct** plating of the specimen on these media, and (2) the **indirect** procedure using enrichment first (see above). Bismuth sulfite is the medium of choice where *Salmonella typhi* is suspected, because it is still the most efficient in the isolation of this pathogen.

To isolate enterotoxic *Escherichia coli, Klebsiella, Enterobacter,* or *Citrobacter* from fecal material, tetrathionate and selenite enrichment broths are **not** recommended, since both are inhibitory for most strains of these genera. In these instances, the less inhibitory media (either MacConkey or EMB agar) are used for primary isolation by the direct plating procedure. Blood agar also is recommended by some investigators.

Some lactose-fermenting, gram-negative enteric bacteria can tolerate the inhibitory substances present in the enrichment broths and the selective media. These bacteria can be recognized readily by their appearance on selective plates. **Lactose-negative** bacteria, such as *Salmonella* and *Shigella,* give rise to small **colorless** colonies on desoxycholate citrate, MacConkey, EMB, XLD, and SS media. On HEA, salmonellae and shigellae appear **bluish green.** Colonies of *Proteus* may be confused with *Salmonella* and *Shigella,* especially on desoxycholate and also on EMB and MacConkey media containing 5% agar, because of their lactose-negative characteristic. Colonies of **lactose-fermenting** organisms on desoxycholate citrate agar (if not inhibited), on MacConkey agar, and on SS medium appear **red;** on EMB agar they appear **dark purple** to **black** and often have a **metallic**

---

*To inhibit the spreading of *Proteus* strains on MacConkey or EMB media, the agar concentration may be increased to 5%.

sheen. On HEA, they appear **salmon** to **orange** in color.

On bismuth sulfite agar, *Salmonella typhi* gives rise to **black** colonies with a metallic sheen if the colonies are well separated. A pour plate is recommended in those instances where only a few organisms are likely to be present. In this procedure a relatively large inoculum (3 to 5 ml) of fluid stool or preserved stool specimen is placed in a Petri dish, and approximately 15 ml of the melted medium is added and rotated thoroughly to mix. Subsurface colonies usually have a typical appearance. For example, subsurface colonies of *S. typhi,* if well separated, are circular, **jet black,** and well defined. Only those near the surface will exhibit the characteristic metallic sheen. Some strains of salmonellae (*S. enteritidis* and *S. enteritidis* serotype Paratyphi-B), as well as certain other members of Enterobacteriaceae, will give rise to black colonies on this medium. Thus, every black colony that appears should not be presumptively identified as *S. typhi.* Generally, salmonellae other than *S. typhi* (bioserotype Paratyphi-A, serotype Typhimurium, and *S. cholerae-suis)* grow as dark green, flat colonies or as colonies with black centers and green peripheries.

Isolation from blood. Blood culture specimens are usually collected from individuals with a febrile disease of unknown etiology, and therefore the media used is generally dictated by the needs of the suspected organism. General methods are discussed in Chapter 7, but if blood cultures are desired for *Salmonella* or *Shigella* **specifically** (shigellae are only rarely found in the blood), approximately 10 ml of blood should be placed in 90 to 100 ml of bile broth and incubated at 35 C. If the culture is negative after 24 hours, incubation should be continued for 10 to 14 days before a negative result is reported.

In **typhoid fever,** a positive blood culture is usually obtained during the first or second week after onset. In septicemias produced by other salmonellae, blood cultures* should be taken during the first week and, if negative, should be repeated during the second week and thereafter if considered necessary. Laboratory personnel are reminded that in typhoid fever a blood culture may be positive **before** stool cultures become positive. During the first week blood cultures are positive in about 90% of the cases, whereas stool cultures are positive in only about 10% of the cases during this period.

Subculture of blood cultures on the appropriate type of agar medium—bismuth sulfite, EMB, or MacConkey—is carried out as described for stool isolations. Many investigators find that a plate of either EMB or MacConkey agar will suffice at this phase of the procedure.

Isolation from urine. *Salmonella typhi* may be isolated from urine in about 25% of cases of typhoid fever. Other members of the Enterobacteriaceae may also be isolated from this source, including other species of *Salmonella* and, more commonly, certain members of the genera *Escherichia, Klebsiella,* and *Proteus.* Direct plating of the specimen on selective and differential media is recommended. Best results for isolating salmonellae from urine are usually obtained by centrifuging the specimen at 2,500 to 3,000 rpm for 20 to 30 minutes to sediment the bacteria. Several loopfuls of the sediment may then be plated on selective and differential media and the remainder added to enrichment broth. The enrichment procedure will aid in the recovery of *S. typhi.* Subculture from the enrichment broth is carried out on the appropriate plating media after 24 or 48 hours. If centrifugation is not employed for specimens from suspected *Salmonella* infections, 2 to 3 ml of the specimen should be added to the enrichment broth.

---

*Cultures of bone marrow also may be helpful in salmonelloses.

PRELIMINARY SCREENING OF CULTURES

Colonies of salmonellae and shigellae, the characteristics of which have been discussed earlier in this chapter, may be recognized by their **lactose-negative** appearance on isolation plates. Suspected colonies of these two genera should be picked carefully with an inoculating needle and transferred to the screening medium. If time permits, it is advisable to select two or three colonies from each plate, since mixed infections are not infrequent. **It is poor technique to touch the agar surrounding a colony to test the temperature of the needle, as members of the inhibited flora may be still present and viable although not visible. Only the center of the colony should be touched, using a cool needle.**

With the inoculating needle, a part of the colony should be stabbed first into the butt of a slant of either **triple sugar iron agar (TSI)** or **Kliger's iron agar (KIA)** and then streaked in a zigzag fashion over the slanted surface. The tube is always closed with a cotton plug or a **loose** closure, never with a tightly fitting rubber stopper or cork. The latter procedure can lead to a misinterpretation of the results, as is explained in the following section. The formulas for these screening media are given in Chapter 41.

Examination and interpretation of reactions. TSI agar contains the three sugars, glucose, lactose, and sucrose; phenol red indicator to indicate fermentation; and ferrous sulfate to demonstrate hydrogen sulfide production (indicated by blackening in the butt).The glucose concentration is one tenth of the concentration of lactose and sucrose in order that the fermentation of this carbohydrate **alone** may be detected. KIA is exactly like TSI agar except it lacks the carbohydrate, sucrose. The small amount of acid produced by fermentation of glucose is oxidized rapidly in the slant, which will remain or revert to an alkaline pH; in contrast, the acid reaction is maintained in the butt because it is under lower oxygen tension. To enhance the alkaline condition in the slant, **free exchange of air** must be permitted through the use of a loose closure, as stated earlier. If the tube is tightly closed with a stopper or screw cap, an acid reaction (caused solely by glucose fermentation) will also involve the slant. The reactions in TSI, which is basically **red** in color when uninoculated, are shown in Table 20-3. **It is to be emphasized that ideally the reactions should be read after 18 to 24 hours, and they cannot be properly interpreted if the slants are incubated for more than 48 hours.**

IDENTIFICATION OF ENTEROBACTERIACEAE

Since some colonies of *Proteus* species can be confused with salmonellae and other lactose-negative enterobacteria on initial isolation, all TSI (or KIA) cultures should be further screened on Christensen **urea agar slants,** or in Rustigian and Stuart tubed **urea broth** (weakly buffered). A heavy inoculum is prescribed, and this is spread over the surface of the agar slant or properly emulsified in the urea broth. Urease activity is observed by

Table 20-3. Reactions observed in TSI agar

Reaction	Explanation
Acid butt (yellow), alkaline slant (red)	Glucose fermented
Acid throughout medium, butt and slant yellow	Lactose or sucrose or both fermented
Gas bubbles in butt, medium sometimes split	Aerogenic culture
Blackening in the butt	Hydrogen sulfide produced
Alkaline slant and butt (medium entirely red)	None of the three sugars fermented

Table 20-4. Interpretation of reactions on TSI agar

Reaction	Carbohydrates fermented	Possible organisms
Acid butt Acid slant Gas in butt No H$_2$S	Glucose with acid and gas Lactose and/or sucrose with acid and gas	*Escherichia** *Klebsiella* or *Enterobacter* *Proteus* or *Providencia* Intermediate coliforms
Acid butt Alkaline slant Gas in butt H$_2$S produced	Glucose with acid and gas Lactose and sucrose not fermented	*Salmonella* *Proteus* *Arizona* (certain types) *Citrobacter* (certain types) *Edwardsiella*
Acid butt Alkaline slant No gas in butt No H$_2$S	Glucose with acid only Lactose and sucrose not fermented	*Escherichia* (anaerogenic biotypes) *Salmonella*† *Shigella* *Proteus* *Providencia* *Serratia*
Acid butt Acid slant Gas in butt H$_2$S produced	Glucose with acid and gas Lactose and/or sucrose with acid and gas	*Arizona* *Citrobacter*
Alkaline or neutral butt Alkaline slant No H$_2$S	None	*Alcaligenes*‡ *Pseudomonas*‡ *Acinetobacter*‡

*Rarely, a strain is H$_2$S positive.
†*Salmonella typhi* produces a small amount of H$_2$S but seldom gas.
‡Included here because colonies of these organisms may be frequently confused with lactose-negative members of Enterobacteriaceae and may be selected from isolation plates.

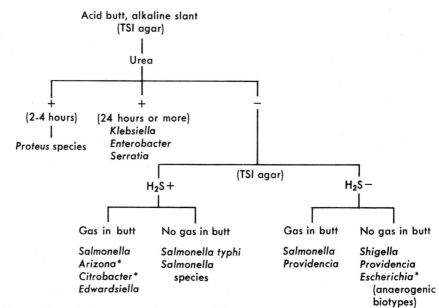

*Delayed lactose fermentation may not show up on TSI agar.

Diagram 2. Tentative differentiation of Enterobacteriaceae.

a change of color (red) in the indicator due to the production of ammonia. A positive test in 2 to 4 hours in the broth or on urea agar slants at 35 C indicates *Proteus*. A positive test after 24 hours might indicate a member of the genera *Klebsiella* or *Enterobacter*.

The reactions observed on TSI (or KIA) agar slants, **together with the effect on urea,** usually indicate the possible genus to which the isolate belongs (see reaction tree in Diagram 2). One should not draw premature conclusions, however, because final identification depends on biochemical and serologic confirmation. Many investigators use the growth from TSI (or KIA) agar slants for slide agglutination tests with polyvalent *Salmonella* or *Shigella* antisera. This is justified in cases of outbreaks or when rapid reporting is required. Cultures from these agar slants should be streaked on meat infusion agar, trypticase soy agar, or MacConkey agar for purity. A single colony may then be transferred to a tube of broth for culture, followed by a complete series of tests.

Another medium that has been used

Table 20-5. Typical reaction patterns of Enterobacteriaceae on TSI and LIA*

Organism	TSI				LIA		
	Slant	Butt	Gas	H$_2$S	Slant	Butt	H$_2$S
Arizona	K or A	A	+	+	K	K	+ or −
Citrobacter							
freundii	A or K	A	+	+	K	A	+ or −
diversus	A or K	A	+	−√	K	A	−
Escherichia coli†	A	A	+ or −	−	K	K or A	−
	K	A	−	−	K	K or A	−
	K	A	−	−	K	K√	−
	K	A	+√	−	K	K or A	−
Edwardsiella‡	K	A	+	+	K	K	+
Enterobacter§							
cloacae	A or K	A	+	−	K	A√	−
aerogenes	A	A	+	−	K	K√	−
hafniae	K	A	+	−	K	K	−
Klebsiella	A or K	A	+	−	K	K (or A‖)	−
Proteus							
vulgaris	K or A	A	+	+√	R	A	−
mirabilis	K	A	+	+√	R	A	−
morganii	K	A	+	−√	K¶ or R	A	−
rettgeri	K	A	+	−√	R	A	−
Providencia	K	A	+	−	R	A	−
Salmonella							
typhi	K	A	−	+ or −√	K	K⚡	+ or −
other	K	A	+	+⚡	K	K	+ or −
Serratia	K or A	A	−	−	K	K	−
Shigella	K	A	−√	−	K	A	−

*Modified from Washington, J. A. II, editor: Laboratory procedures in clinical microbiology, Boston, 1974, Little, Brown and Co.

†Mucoid colonies with negative motility and ornithine tests should be referred.

‡Refer for confirmation.

§DNase and oxidase tests must be done on all *Enterobacter* patterns. If negative, send out as *Enterobacter*. If DNase is positive, send out as *Serratia marcescens* and refer for confirmation. If oxidase is positive, refer. Any *Enterobacter* pattern with a green sheen must be referred. Refer all K/A TSIA and K/K LIA *Enterobacter* patterns without doing DNase or oxidase tests.

‖May be reported as *Klebsiella* if colonial morphology is mucoid and oxidase test is negative.

¶Refer for phenylalanine deaminase test.

⚡Rare exceptions.

A, acidic, K, alkaline; R, red; S, slant; √, key reaction.

Table 20-6. Differentiation of Enterobacteriaceae by biochemical tests*

Test	Escherichia	Shigella	Edwardsiella	Salmonella	Arizona	Freundii	Diversus	Klebsiella pneumoniae	Cloacae	Aerogenes	Hafniae	Agglomerans	Marcescens	Liquefaciens	Rubidaea	Vulgaris	Mirabilis	Morganii	Rettgeri	Alcalifaciens	Stuartii	Pestis†	Pseudotuberculosis†	Enterocolitica‡	Erwinia§	Pectobacterium†
Tribe / Genus	Escherichieae		Edwardsielleae	Salmonelleae		Citrobacter		Klebsielleae	Enterobacter				Serratia			Proteeae / Proteus				Providencia		Yersineae / Yersinia			Erwinieae	
Indole	+	− or +	+	−	−	−	+	− or +	−	−	−	− or +	−	−	−	+	−	+	+	+	+	−	−	=	−	− or +
Methyl red	+	+	+	+	+	+	+	− or +	−	−	− or +	− or +	− or +	+ or −	− or +	+	+	+	+	+	+	+	+ʷ 35 C / + 25 C	+	−	+ or −
Voges-Proskauer	−	−	−	−	−	−	−	+	+	+	+ or −	+ or −	+	− or +	+ or −	−	− or +	−	−	−	−	− 35 C / − 25 C	− 35 C / − 25 C	− 35 C / + 25 C	+	− 35 C / + 25 C
Simmons' citrate	−	−	−	d	+	+	+	+	+	+	d	d	+	+	+ or (+)	d	+ or (+)	−	+	+	+	−	−	−	+	d 35 C / + or (+) 25 C
Hydrogen sulfide (TSI)	−	−	+	+	+	+ or −	−	−	−	−	−	−	−	−	−	+	+ or (+)	−	−	−	−	−	−	−	−	−
Urease	−	−	−	−ᵉ	−	d*	d*	+	+ or −	−	−	d*	d*	d*	d*	+	+	+	+	−	+	−	+	+	−	d
KCN	−	−	−	−	−	+	+	+	+	+	+	− or +	+	+	− or +	+	+	+	+	+	+	−	−	−	−	+ or −
Motility	+ or −	−	+	+ or (+)	+ or (+)	+	+	−	− or (+)	+	+	+ or −	+	+	+ or −	+	+	+ or −	+	+	+	− 35 C / − 25 C	(+) or + 35 C / 25 C	− 35 C / + 25 C	+	+ or −
Gelatin (22 C)	−	−	−	−	(+)	−	−	−	(+) or −	− or (+)	−	d	+ or (+)	+ or (+)	+ or (+)	+	+	−	−	−	−	−	−	−	−	+ or (+)
Lysine decarboxylase	d	−	+	+	+	−	−	+	−	+	+	−	+	+	−	−	−	−	−	−	−	−	−	−	−	−
Arginine dihydrolase	d	d	−	+ or (+)	+ or (+)	d	+ or (+)	−	+	−	d	−	−	+ or (+)	−	−	−	−	−	−	−	−	−	−	−	−
Ornithine decarboxylase	d	d#	+	+	+	d	+	−	+	+	+	−	+	+	−	−	+	+	−	−	−	−	−	+	−	−
Phenylalanine deaminase	−	−	−	−	−	−	−	−	−	−	−	− or +	−	−	−	+	+	+	+	+	+	−	−	−	−	−
Malonate	−	−	−	−	+	− or +	− or +	+	+ or −	+ or −	+ or −	+ or −	−	+	+ or −	−	−	−	−	−	−	−	−	−	−	− or +

Gas from glucose	+	−#	+	+	+	+	+	+	+	+	+	− or +	+ or −**	+ or −	d	+ or −	+	+ or −	− or +	+ or −	−	−	−	−	−	− or + 35 C d 25 C
Lactose	+	−#	−	−	d	(+) or +	d	+	+ or (+)	+	d	d	−	d	+	−	−	−	−	−	−	−	−	−	−	d 35 C + or (+) 25 C
Sucrose	d	−#	−	−	−	d	− or +	+	+	+	d	d	+	+	+	+	d	−	d	d	(+) or +	−	−	+	+	+ or − 35 C + 25 C
Mannitol	+	+ or −	−	+	+	+	+	+	+	+	+	+	+	+	+	−	−	−	+ or −	−	d	+	+	+	−	+ or −
Dulcitol	d	d	−	d‡‡	−	d	+ or −	− or +	− or +	−	−	− or +	−	−	−	−	−		−	−	−		−	−	−	−
Salicin	d	−	−	−	−	d	(+) or +	+	+ or (+)	+	d	d	+	+	+ or (+)	d	d	−	d	−	−	+	(+)	d	+	d 35 C + 25 C
Adonitol	−	−	−	−	−	−	+	+ or −	− or +	+	−	−	d	d	+ or (+)	−	−	−	d	+	− or +	−	−	−	−	−
Inositol	−	−	−	d	−	−	−	+	d	+	−	d	d	+ or (+)	d	−	−	−	+	−	+		−	d	−	−
Sorbitol	d	d	−	+	+	+	+	+	+	+	−	d	+	+	−	−	−	−	d	−	d	− or +	−	+	d	−
Arabinose	+	d	− or +	+‡‡	+	+	+	+	+	+	+	+	−	+	+	−	−	−	−	−	−	+	+ or (+)	+	d	+ or − 35C + 25C
Raffinose	d	d	−	−	−	d	−	+	+	+	−	d	−	+	+	−	−	−	−	−	−	−	− or +	−	+	d 35C + or (+) 25C
Rhamnose	d	d	−	+	+	+	+	+	+	+	+	+ or (+)	−	d	−	−	−	−	+ or −	−	−	−	+	−	−	d

*Adapted from Edwards, P. R., and Ewing, W. H.: Identification of Enterobacteriaceae, ed. 3, Minneapolis, 1972, Burgess Publishing Co.

†Adapted from Sonnenwirth, A. C.: *Yersinia.* In Lennette, E. H., Spaulding, E. H., and Truant, J. P., editors: Manual of clinical microbiology, ed. 2, Washington, D.C., 1974, American Society for Microbiology.

‡Adapted from Darland, G., Ewing, W. H., and Davis, B. R.: The biochemical characteristics of *Yersinia enerocolitica* and *Yersinia pseudotuberculosis,* DHEW Publ. No. (CDC) 75-8294, Washington, D.C., 1974, Department of Health, Education, and Welfare.

§Adapted from Buchanan, R. E., and Gibbons, N. E., editors: Bergey's manual of determinative bacteriology, ed. 8, Baltimore, 1974, The Williams & Wilkins Co.

‖Majority of strains isolated in the United States are indole positive, whereas most of the strains isolated in Europe have been indole negative.

¶Rare exceptions.

#Certain biotypes of *S. flexneri* produce gas; cultures of *S. sonnei* ferment lactose and sucrose slowly and decarboxylate ornithine.

**Gas volumes produced by cultures of *Serratia, Proteus,* and *Providencia* are small.

‡‡*S. typhi, S. cholerae-suis, S. enteritidis* bioser. Paratyphi-A and Pullorum and a few others ordinarily do not ferment dulcitol promptly. *S. cholerae-suis* does not ferment arabinose.

+, 90% or more positive in 1 or 2 days; −, 90% or more negative; d, different biochemical types [+, (+), −]; (+), delayed positive (decarboxylase reactions, 3 or 4 days); + or −, majority of cultures positive; − or +, majority negative; w, weakly positive reaction.

NB. This chart is simply a guide. Users are urged to consult other publications, such as CDC publications entitled "Biochemical Reactions Given by Enterobacteriaceae in Commonly used Tests" and "Differentiation of Enterobacteriaceae by Biochemical Reactions" (W. H. Ewing, 1973), for percentage data, additional tests, and references.

with a good deal of success in conjunction with TSI or KIA agar is **lysine-iron-agar** (LIA).[20] Using a straight wire, inoculum is taken carefully from a well-selected colony and transferred to KIA or TSI agar in the usual way, and **without going back** to the colony on the plate, a tube of LIA is inoculated by stabbing the butt of the medium **twice** and then streaking the slant. If the spot where the stab was made in the TSI (or KIA) agar medium is touched with the tip of the wire, a suf-ficient number of organisms will be obtained for inoculation of the LIA medium. The combinations of reactions obtained by this[19,25] or other screening methods[25,86] yield much useful information at a relatively early time (Table 20-5).

Motility-indole-lysine (MIL) medium is a reliable medium except for occasional falsely weak or negative indole reactions. Used together with TSI and urea agar, MIL permits reliable early recognition of enteric pathogens of the Enterobacteriaceae.[63] (See Fig. 20-1 for motility test.)

Table 20-6 shows the tests on which differentiation of the Enterobacteriaceae may be made. Tests to which reference is made in Table 20-6 are given in Chapters 41 and 43.

Lactose-fermenting members of Enterobacteriaceae. A large number of organisms within the family Enterobacteriaceae ferment the carbohydrate **lactose,** and those that do so may be tentatively identified by the reactions shown in Diagram 3. In recent years, lactose-fermenting strains of *Salmonella typhi* (and other *Salmonella*) have been reported. Such strains are otherwise like *S. typhi* in possessing the Vi, O, and H antigens of this organism (antigenic structure is shown in Table 20-8). Some lactose-positive members of the family ferment the sugar promptly, whereas others exhibit a delayed reaction. In Diagram 2 the differ-

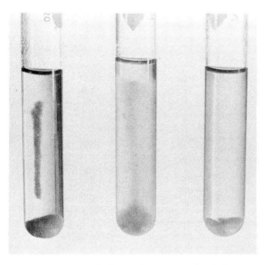

Fig. 20-1. Semisolid motility medium. Tube on left shows *Shigella* (nonmotile). Center tube is *E. coli* (motility results in diffuse growth pattern). Tube on right is uninoculated.

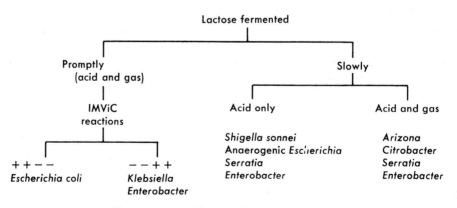

Diagram 3. Differentiation of lactose-fermenting Enterobacteriaceae.

entiation between these genera is shown.

The IMViC reaction is used primarily to distinguish between the coliform bacteria, but it also may be applied advantageously to other organisms in the family. The letters stand for **indole, methyl red, Voges-Proskauer,** and **citrate** reactions (the "i" is inserted for euphony). These may be carried out and interpreted as shown.

TEST FOR INDOLE. Inoculate tryptophan broth. Test after 48 hours by the addition of the Kovacs or Ehrlich reagent (see Chapter 43). A **red** color indicates production of indole from the amino acid.

METHYL RED TEST. Inoculate MR-VP medium (Clark and Lubs dextrose broth medium). Test after 48 to 96 hours by adding 5 drops of methyl red indicator (see Chapter 43 for more rapid procedures). A **red** color is read as positive. A yellow color is read as negative. Read the test immediately after adding the reagent.

VOGES-PROSKAUER TEST. Inoculate MR-VP medium (same as in previous test) and test for the production of acetylmethylcarbinol after 48 hours of incubation by adding to 1 ml of culture 15 drops of 5% alpha-naphthol in absolute ethyl alcohol and 10 drops of 40% potassium hydroxide (see Chapter 43 for more rapid procedures). The necessary volume of culture may be pipetted to a small tube before the MR test above is performed. A positive test is the development of a **red** color in 15 to 30 minutes. An alternative test is given in Chapter 43.

CITRATE TEST. Inoculate Simmons' citrate agar with a light inoculum. A positive test is indicated by the development of a **Prussian blue color** in the medium, showing that the organism can utilize citrate as a sole source of carbon.

ANTIGENIC COMPLEXITY OF ENTEROBACTERIACEAE

The members of the Enterobacteriaceae exhibit a mosaic of antigens that fall into three main categories as follows:

1. The **K** (from German *Kapsel*), or en-velope, antigens are those that, by concept, surround the cell. With certain exceptions these are heat labile. In the genus *Klebsiella* the subdivision into capsular types is based on the K antigens. K antigens mask the heat-stable somatic antigens of the cell and cause live cells to be inagglutinable in O antisera. Examples are the Vi antigen of *Salmonella typhi* and the B antigen found in certain types of *Escherichia coli.*

2. The **O** (from German *Ohne Hauch* nonspreading), or **somatic,** antigens, which are heat stable, are located primarily in the cell wall. Chemically they are polysaccharide in nature. The O complex of antigens determines the somatic sub-group to which the organism belongs, in the genera *Salmonella, Arizona, Citrobacter, Escherichia, Providencia, Serratia,* and others.

3. The **H** (from German *Hauch,* spreading) antigens are the **flagellar** antigens. These are located in the flagella, are protein in nature, and are heat labile. The serotypes within the somatic groups in *Salmonella* and some other genera in the Enterobacteriaceae are determined by the H antigens.

IDENTIFICATION OF GENERA WITHIN THE FAMILY ENTEROBACTERIACEAE

Genus Salmonella

Salmonella cholerae-suis is the type species of the genus. Salmonellae are usually motile, but nonmotile forms do occur.[45] With the exception of *S. typhi* and *S. enteritidis* serotype Gallinarum, they all produce **gas** in glucose. Suspected colonies of salmonellae on isolation media are inoculated to slants of TSI (or KIA) agar. Isolates that produce acid, gas, and hydrogen sulfide in the butt and an alkaline slant in this medium and are **urease negative** should be tested with *Salmonella* polyvalent antisera. Pure cultures should be tested for motility and inoculated to the biochemical media shown in Table 20-7. Typical salmonellae will give the reactions shown.

Table 20-7. Biochemical reactions of genus *Salmonella*

Test	Reaction	Test	Reaction
Adonitol	−	Methyl red	+
Dulcitol	+	Voges-Proskauer	−
Glucose	+ with gas	Simmons' citrate	+
Inositol	variable	KCN*	−
Lactose	−	Phenylalanine deaminase*	−
Mannitol	+	Sodium malonate*	−
Salicin	−	Lysine decarboxylase	+
Sucrose	−	Arginine dihydrolase*	+
Indole	−	Ornithine decarboxylase*	+

*Descriptions of these media and procedures for the tests performed may be found in Chapters 41 and 43. Although not used routinely in all laboratories, they aid in group differentiation.

Table 20-8. Some serotypes of the Kauffmann-White or *Salmonella* antigenic scheme

Bioserotypes and serotypes	O antigens	H antigens Phase 1	H antigens Phase 2
Group A			
Paratyphi-A	1, 2, 12	a	—
Group B			
Tinda	1, 4, 12, 27	a	e, n, z_{15}
Paratyphi-B	1, 4, 5, 12	b	1, 2
Typhimurium	1, 4, 5, 12	i	1, 2
Heidelberg	4, 5, 12	r	1, 2
Group C_1			
Paratyphi-C	6, 7, Vi	c	1, 5
Thompson	6, 7	k	1, 5
Group C_2			
Newport	6, 8	e, h	1, 2
Group D			
S. typhi	9, 12, Vi	d	—
S. enteritidis	1, 9, 12	g, m	—
Sendai	1, 9, 12	a	1, 5
Group E_1			
Oxford	3, 10	a	1, 7
London	3, 10	l, v	1, 6

Serologic identification. The Kauffmann-White or *Salmonella* antigenic scheme has a long list of serotypes that are arranged in O, or somatic, subgroups. The H, or flagellar, antigens, as previously stated, determine the type.

Table 20-8 shows an abbreviated example of a Kauffmann-White scheme, but it will help the student and laboratory worker to understand the schematic arrangement of the serotypes. There are more alphabetized somatic groups in the scheme than are shown (i.e., some 1,700 serotypes). The tabulated types shown are not necessarily the most common.

Because approximately 65 different somatic antigens have been recognized, serologic typing with polyvalent O antisera is essential. These sera are commercially available;* they will agglutinate the majority of strains found in the United States and Canada. **Group** identification is determined by O-typing sera, and **type** identification is determined by H-typing antisera. O-grouping sera for subgroups A, B, C_1, C_2, D, and E as well as H-typing sera for flagellar antigens a, b, c, d, i, 1, 2, 3, 5, 6, and 7 are available.* **Vi** antiserum is also available.

The serologic procedure usually employed is the **slide agglutination test,** in which a concentrated suspension of cells in saline is used. Details of the technique may be found in Chapter 37, but the reader's attention at this point is drawn to certain important considerations.

1. The culture to be tested must be **smooth** and not autoagglutinable in saline. A preliminary test with a 0.2% solution of acriflavine in 0.85% saline is an excellent indicator of smoothness. If a loopful of the test suspension, as a control, is mixed **gradually** on a slide with a loopful of acriflavine and the cells remain in homogeneous suspension, the culture may be considered smooth.

2. If the culture suspension fails to agglutinate in the O diagnostic sera, heat it at 100 C for 15 to 30 minutes, cool, and retest with the same sera. Some salmonellae possess K antigens, previously mentioned, and are inagglutinable in the live or unheated form in O antisera. Examples of envelope antigens are the Vi antigen of *S. typhi* and *S. enteritidis* ser. Paratyphi-C, the 5 antigen in somatic group B, and the M antigen in some serotypes. Vi-containing types will agglutinate in the live or unheated form in Vi antiserum.

3. O-inagglutinable cultures should be sent to a reference laboratory* for identification by complete serologic analysis. **These cultures should be sent through the state or provincial laboratories.**

4. The majority of the salmonellae are **diphasic;** that is, the motile types may exhibit two antigenic forms referred to as phases (see Edwards and Ewing[19] for a complete discussion of this subject). These phases share the same O antigens but possess different H antigens, as seen in Table 20-7, and to identify the serotype it is necessary to identify the specific H antigens in both phases. These may not always be in evidence, and phase-suppression procedures may be necessary to reveal the **latent** phase. This can be accomplished by inoculating the organism into a small Petri dish containing semisolid agar in which is incorporated specific antiserum against the antigen(s) of the one identified phase. The homologous phase will be arrested at the site of inoculation, whereas the other phase will develop (or express itself) and may be identified. Such procedures can be carried out only by properly equipped laboratories. It should be noted that some flagellated salmonellae are nonmotile, and special procedures are required for their identification.[2]

5. To completely identify the O antigen and H antigen complexes, the use of absorbed **single factor** typing sera is required. These antisera are prepared by adding a concentrated suspension of cells containing the appropriate antigens to a suitable dilution of the multifactor serum, incubating in a water bath for 2 hours at 50 C, and refrigerating overnight. After centrifugation, the supernate

---

*Lederle Laboratories, Pearl River, N.Y.; Difco Laboratories, Detroit; Baltimore Biological Laboratory, Cockeysville, Md.; Lee Laboratories, Grayson, Ga.

*Enteric Section, Center for Disease Control, Atlanta; Central Public Health Laboratory, Colindale, London.

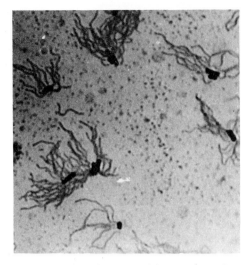

Fig. 20-2. *Salmonella typhi*, showing peritrichous flagellation using Gray's method. (1200×.)

will contain the unabsorbed and desired agglutinating antibodies. Such a procedure is referred to as **agglutinin absorption.**

Characteristics of Salmonella typhi. The typhoid organism exhibits its characteristic biochemical activity in the carbohydrates and produces small amounts of hydrogen sulfide in KIA or TSI agar. The organism is **anaerogenic,** a criterion that aids in its identification. A flagella stain of *S. typhi* is shown in Fig. 20-2.

This organism undergoes several different types of variation, among which is **H-to-O** variation, involving the loss of flagella. The H form is motile, the O form is nonmotile. There are two well-recognized variants of *S. typhi:* H901 (motile) and O901 (nonmotile), which are used widely in the preparation of H and O antigen suspensions, respectively, for the Widal test. The preparation of these antigens and the Widal test are discussed in Chapter 37.

Another type of variation exhibited by the typhoid organism is **V-to-W** variation, involving loss of the Vi antigen. In the **V** form the organism is virulent and inagglutinable in O antiserum. The **W** form readily agglutinates in O antiserum and

is avirulent. V colonies on nutrient or starch agar medium appear orange-red by oblique light, whereas W colonies appear greenish blue. Only the V form is typable by the typhoid Vi phages. If the phage type of the organism is required for epidemiologic purposes, a fresh culture in the V form should be sent to a reference laboratory.

The following criteria may serve to identify the organism. If a gram-negative isolate on TSI (or KIA) agar shows an acid butt, no gas, a small amount of hydrogen sulfide (may resemble a mustache), and an alkaline slant and also agglutinates in Vi typing serum, chances are good that this organism is *S. typhi*. Laboratory personnel should confirm this, however, with further biochemical studies and serology.

The reader is reminded of the existence, although rare, of lactose-positive variants of *S. typhi*.

Genus Arizona

Members of the genus *Arizona* are gram-negative, short, motile rods that show a close relationship to the salmonellae. The type species of the genus is *Arizona hinshawii*. Lactose may be fermented with acid and gas in 24 hours, but most strains ferment the carbohydrate after 7 to 10 days. The reaction on TSI agar slants closely resembles that of salmonellae, including hydrogen sulfide production. Gelatin is liquefied by these organisms in 7 to 30 days. They are sensitive to KCN[52] (negative) and do not produce urease.

These organisms[21,49] can be important in human infections. Many serotypes can cause disease in chickens, dogs, and cats. An antigenic scheme has been established that contains approximately 35 different O antigen groups with a total of about 350 serotypes.

The members of this genus show the biochemical reactions listed in Table 20-9.

Because of the similarity of arizonae to

Table 20-9. Biochemical reactions of genus *Arizona*

Test	Reaction	Test	Reaction
Adonitol	−	Methyl red	+
Dulcitol	−	Voges-Proskauer	−
Glucose	+ with gas	Simmons' citrate	+
Inositol	−	KCN	−
Lactose	+ or delayed	Phenylalanine deaminase	−
Mannitol	+	Sodium malonate	+
Salicin	−	Lysine decarboxylase	+
Sucrose	−	Arginine dihydrolase	+
Indole	−	Ornithine decarboxylase	+

Table 20-10. Biochemical reactions of genus *Citrobacter*

Test	Reaction		Test	Reaction	
	C. freundii	*C. diversus*		*C. freundii*	*C. diversus*
Adonitol	−	+	Voges-Proskauer	−	−
Dulcitol	+ or −	+ or −	Simmons' citrate	+	+
Glucose	+ with gas	+ with gas	Gelatin	−	−
Inositol	− or delayed	−	Urease	(+) or −	variable
Lactose	+ or delayed	variable	KCN	+	−
Mannitol	+	+	Phenylalanine deaminase	−	−
Salicin	variable	variable	Sodium malonate	−	+ or −
Sucrose	variable	− or +	Lysine decarboxylase	−	−
Indole	−	+	Arginine dihydrolase	+	+
Methyl red	+	+	Ornithine decarboxylase	variable	+

+ or −, majority are positive; − or +, majority are negative.

salmonellae, they are frequently mistaken for the latter in initial biochemical screening tests. However, fermentation of lactose, although usually delayed, failure to ferment dulcitol, growth in sodium malonate, and slow gelatin liquefaction help distinguish arizonae from salmonellae.

Genus Citrobacter

The genus *Citrobacter* includes the type species *C. freundii* and *C. diversus*.[19] The latter species comprises those strains of *Citrobacter* that are indole positive, KCN negative, adonitol positive, and positive for ornithine decarboxylase. The members of this genus are gram-negative motile rods that can ferment lactose. Because of their biochemical reactions, particularly during preliminary screening, they are often confused with *Salmonella* and *Arizona*. They are not truly pathogenic and are considered opportunists. Nevertheless, *Citrobacter* is found in a variety of infections, particularly urinary tract infection and bacteremia, on occasion. A summary of the biochemical reactions for the genus is shown in Table 20-10.

Certain *C. freundii* strains possess the Vi antigen found in *Salmonella typhi*. The KCN test is positive only for *C. freundii* (the organisms grow in this medium), whereas *Salmonella* and *Arizona* are inhibited.

Considerable work has been done on the development of a serologic test scheme for members of this genus, but to date none has evolved for routine use.

Genus Shigella

All members of the genus *Shigella* are **nonmotile;** they do not produce hydrogen sulfide; and with a few exceptions (bio-

Table 20-11. Biochemical reactions of genus *Shigella*

Test	Reaction	Test	Reaction
Adonitol	−	Methyl red	+
Dulcitol	variable	Voges-Proskauer	−
Glucose	+ no gas	Simmons' citrate	−
Inositol	−	KCN	−
Lactose	variable	Phenylalanine deaminase	−
Mannitol	variable	Sodium malonate	−
Salicin	−	Lysine decarboxylase	−
Sucrose	−	Arginine dihydrolase	−*
Indole	variable	Ornithine decarboxylase	−*

*S. sonnei and *S. boydii* 13 are usually ornithine decarboxylase positive and some strains are arginine dihydrolase positive. The other shigellae are negative for all three amino acids.

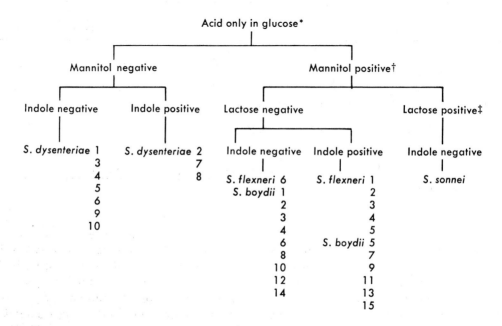

*S. flexneri 6 varieties may be aerogenic (Newcastle and Manchester).
†Certain cultures of *S. flexneri* 4, *S. flexneri* 6, and *S. boydii* 6 may not produce acid from mannitol.
‡Lactose fermentation is delayed with *S. sonnei* (usually 4 to 7 days). Closure of the fermentation tube with a tightly fitting stopper will hasten the reaction.

Diagram 4. Biochemical differentiation of genus *Shigella*. (Adapted from Edwards, P. R., and Ewing, W. H.: Manual for enteric bacteriology, Atlanta, 1951, Center for Disease Control.)

types of *S. flexneri* 6) they are anaerogenic. Although many shigellae produce catalase,[8] for example, *S. flexneri,* this is an inconsistent property within the genus. *S. dysenteriae* 1, the type species of the genus, is invariably negative.[7] Aside from bacillary dysentery, *Shigella* may occasionally cause bacteremia, pneumonia, or other infection.

Colonies of shigellae are usually smaller than those of salmonellae, but on occasion mucoid variants may be found in subgroup C. Lactose-negative colonies that resemble shigellae are picked from isolation plates to TSI (or KIA) agar slants. If these show an acid butt, no hydrogen sulfide, and an alkaline slant and prove to be urease negative, they may be tested with *Shigella* polyvalent typing sera. This procedure is recommended only for presumptive identification; cultures should always be checked for purity and motility and then confirmed with additional biochemical tests. The tests listed in Table 20-11 are recommended for screening of *Shigella* cultures. The reaction tree shown in Diagram 4 will serve as a guide in biochemical identification.

Serologic identification. The current antigenic scheme for *Shigella* was proposed by Ewing in 1949 and modified and extended by the Shigella Commission of the Enterobacteriaceae Subcommittee.[19] The scheme is based in part on biochemical characteristics, on antigenic relationships, and on tradition (the names of Shiga, Boyd, Flexner, and Sonne are obvious in the nomenclature). **Four** subgroups (species) are presently identified: subgroup A—*S. dysenteriae,* types 1 to 10; subgroup B—*S. flexneri* types 1 to 6; also X and Y variants; subgroup C—*S. boydii,* types 1 to 15; subgroup D—*S. sonnei.*

Subgroup A consists of those types that are nonfermenters of mannitol. Each type exhibits a type-specific antigen not related to other shigellae.

The types that make up **subgroup B** are usually fermenters of mannitol (mannitol-negative variants of *S. flexneri* 6 do exist) and in addition are interrelated through group or subsidiary antigens. Each type, however, possesses a type-specific antigen that differentiates it from other shigellae. Serotypes 1, 2, 3, and 4 also possess subtypes based on group antigen differences. The X and Y variants represent those forms that have lost the type-specific antigen.

In **subgroup C** all types are fermenters of mannitol and possess individual type-specific antigens not related significantly to other shigellae.

Subgroup D comprises only one type, *S. sonnei,* and this organism possesses a type-specific antigen that bears no significant relationship to other shigellae. There exist different "forms" of *S. sonnei* that differ antigenically.

In view of the number of serotypes, it is recommended that preliminary serologic screening be carried out with **polyvalent** typing antisera. Groups A, B, C, and D antisera for slide agglutination tests are available commercially.* The overall technique is quite similar to that used for the *Salmonella* O antigens. Since K antigens are present also in some shigellae, live bacteria **may fail to agglutinate** in any of the above-mentioned groups. Thus, one is advised to heat the test suspension in 0.5% NaCl at 100 C for 30 to 60 minutes, cool, and retest with the same antisera.

Type determination of shigellae can be carried out with monovalent antisera against the specific antigens. These are prepared from known types and are absorbed with suspensions of appropriate cultures to obtain type-specific diagnostic antisera. Agglutinin absorption is particularly necessary for most types in group B because of their group or minor antigens.

Colicin typing. In North America and in the United Kingdom, *S. sonnei* has become the most common etiologic agent in bacillary dysentery or shigellosis. In the United States, 85% of *Shigella* isolates are *S. sonnei* and 14% are *S. flexneri.*[61] Colicin typing of strains of *S. sonnei* is being performed in some reference laboratories. **Colicins,** which are produced by different gram-negative enteric bacteria, are antibioticlike substances that may have a lethal effect on other bacteria from the same habitat. Colicin typing of shigellae is based on colicinogeny rather than colicin susceptibility.

Fifteen colicin types of *S. sonnei* are

---

*Lederle Laboratories, Pearl River, N.Y.; Difco Laboratories, Detroit; Baltimore Biological Laboratory, Cockeysville, Md.; Lee Laboratories, Grayson, Ga.

currently recognized,[1] and these appear to be sufficiently stable to use in epidemiologic investigations. Carpenter[7] reported that during 1964 to 1969 approximately one third of the strains isolated in England were noncolicinogenic (CTO), and of the typable strains one third were of type 7 and the remainder comprised the other 14 types. The most prevalent of the latter in the United Kingdom were colicin types 1A, 2, 4, and 6.

In evaluating the colicin-typing method for differentiating epidemic strains of *S. sonnei* in the United States, Morris and Wells[54] reported that 40% of their strains were nontypable. The remaining 60% were distributed among all 15 established colicin types, with type 9 accounting for 22% of the total. These authors felt colicin typing offered a useful method of differentiating strains with reasonably good reproducibility, although it should not be interpreted with the same degree of confidence as some other characteristics, such as serotyping. They recommended that known type strains be included as controls in each run to ensure uniformity of results.

Genus Escherichia

The genus *Escherichia* includes the former *Alkalescens-Dispar* group, which although anaerogenic, are biochemically and antigenically related to the escherichiae. The type species of the genus is *Escherichia coli*.

Typical escherichiae are readily recognized and may be differentiated from other members of the Enterobacteriaceae by their rapid **fermentation of lactose** with acid and gas and their response to the classical **IMViC** tests (+ + − −). Some strains, however, either fail to ferment lactose or do so slowly. Both motile and nonmotile forms occur.

Most strains produce large, characteristic lactose-fermenting colonies on MacConkey agar and EMB agar. They are usually inhibited in enrichment broths and the more highly selective agar plating media, as stated earlier in this chapter. For direct isolation of *Escherichia*, the less inhibitory (differential) media, such as MacConkey agar or EMB agar, are recommended. Blood agar plates should also be used because certain enteropathogenic strains may not grow on MacConkey agar but will grow well on blood plates. Most strains are nonhemolytic.

Pure cultures of typical *E. coli* will give the reactions shown in Table 20-12.

E. coli may be associated with a number of disease syndromes. Among these are (1) often severe and sometimes fatal infections, such as cystitis, pyelitis, pyelonephritis, appendicitis, peritonitis, gallbladder infection, septicemia, meningitis, and endocarditis; (2) epidemic diarrhea of adults and children; and (3) "traveler's" diarrhea.

The antigenic complexity of the genus is very similar to that of *Salmonella* and other genera within the Enterobacteri-

Table 20-12. Biochemical reactions of genus *Escherichia*

Test	Reaction	Test	Reaction
Adonitol	−	Methyl red	+
Dulcitol	variable	Voges-Proskauer	−
Glucose	+ with gas	Simmons' citrate	−
Inositol	−	KCN	−
Lactose*	+	Phenylalanine deaminase	−
Mannitol	+	Sodium malonate	−
Salicin	variable	Lysine decarboxylase	−
Sucrose	+	Arginine dihydrolase	−
Indole	variable	Ornithine decarboxylase	−

*Some are late or nonlactose fermenting.

aceae. Approximately 160 O antigens are recognized. Many of these strains also possess K antigens, which now number 93. These may be of the L, A, or B type and thus exhibit various physical and immunologic properties. *E. coli* serotypes have been established on the basis of the O, K, and H antigens. K antigens are somatic antigens that occur as envelopes or as capsules. This antigen masks or otherwise makes inaccessible the O complex. The O inagglutinability due to K antigens is analogous to the O and Vi relationship in *Salmonella typhi.*

Enteropathogenic Escherichia coli

Fourteen O antigen groups of *E. coli* have been associated with epidemic diarrhea of the newborn. They may also cause diarrhea in adults. However, enterotoxin production and invasiveness correlate more with enteric disease than does serotype.

Toxigenicity. Recent studies of the pathogenesis of bacterial diarrhea have identified intestinal secretion of fluid and electrolytes in diseases associated with **enterotoxins.** The enterotoxins of both cholera and certain of the noninvasive (toxigenic) strains of *E. coli* cause diarrhea through a similar mechanism(s), that is, stimulation of adenylate cyclase in the small intestine (and in nonintestinal tissue), thereby giving rise to excessive secretion of fluid by the small intestine.[3,29] In *Shigella* and *Salmonella,* enterotoxins have been identified, but it is not known what role, if any, they play in enteritis.[65]

A large amount of evidence has accumulated in recent years to implicate the **enterotoxigenic** strains of *E. coli* as a cause of severe diarrheal disease in human beings.[12,16,32,33,66-68,72] There is not necessarily any correlation between toxigenicity and the classic enteropathogenic serotypes.

Two distinct enterotoxins, one heat stable and nonantigenic (**ST**) and the other heat labile, antigenic (**LT**), and choleralike (see below), have been characterized from *E. coli.*[23,74] LT has been definitely shown to cause diarrheal disease in humans,[32,69] and recent evidence implicates ST as well.[53]

For detecting these enterotoxins several assay methods are currently available; they vary in their ability to detect one or the other of the enterotoxins. When observed at 18 hours the rabbit ileal loop test[84] detects primarily LT, a positive test being dilatation of the ligated segment with accumulated fluid. The infant rabbit examined at 6 hours presumably indicates the presence of both enterotoxins.[23] Two tissue culture assays have been developed that are highly sensitive assays for LT only. One is the adrenal cell technic[13] and the other employs a Chinese hamster ovary tissue culture cell line as described by Geurrant and Dickens.[35] The infant mouse assay method detects ST.[12]

In an attempt to determine the *E. coli* serotypes of the enterotoxigenic strains obtained from diarrhea in adults and children, Ørskov and co-workers[56] examined 106 strains submitted to them from many parts of the world for O and H antigens (K antigens were not studied because a reliable procedure to test for these antigens was not yet available). These authors found some O:H types consistently from different geographic locations (i.e., O6:H-16, O8:H9, O15:H11; O25:H42; O78:H11 and O78:H12).

Invasive *E. coli* strains may cause enteric disease resembling shigellosis by invading the bowel epithelium. These strains are identified by the Sereny test, in which a positive result is the development of keratoconjunctivitis in the guinea pig eye.[53]

Genus Edwardsiella

The type species of this genus is *Edwardsiella tarda.* Members of this genus are motile, and their biochemical pattern conforms to that of the family Enterobacteriaceae. The first isolates were reported in 1959 and examined under biotype 1483-59 by various workers between 1962 and 1964. The generic term was suggested by Ewing and co-workers[26] in 1965.

These organisms have been isolated from a variety of human sources, including diarrheal disease; from normal stools, blood, wounds, visceral abscesses, and urine; and also from warm- and cold-blooded animals. The biochemical reactions are shown in Table 20-6.

Genus Klebsiella

The establishment of *Klebsiella* and *Enterobacter* as separate genera was effected in 1962, following the recommendations of the Enterobacteriaceae Subcommittee.[22,40,51] The tribe Klebsielleae currently comprises the genera *Klebsiella, Enterobacter,* and *Serratia.*

Members of the genus *Klebsiella* are gram-negative, **nonmotile,** encapsulated, short rods, which possess the characteris-

Table 20-13. Biochemical reactions of genus *Klebsiella*

Test	Reaction	Test	Reaction
Adonitol	+ or −	Voges-Proskauer	+
Dulcitol	− or +	Simmons' citrate	+
Glucose	+ with gas	KCN	+
Inositol	+ with gas	Urease	+ (slow)
Lactose	+	Gelatin liquefaction	−
Mannitol	+	Phenylalanine deaminase	−
Salicin	+	Sodium malonate	+
Sucrose	+	Lysine decarboxylase	+
Indole	− (+)	Arginine dihydrolase	−
Methyl red	−	Ornithine decarboxylase	−

Table 20-14. Biochemical reactions of *Enterobacter* species

Test	E. cloacae	E. aerogenes	E. hafniae 35 C	E. hafniae 22 C	E. agglomerans
Adonitol	− or +	+	−	−	−
Dulcitol	− or +	−	−	−	− or +
Glucose (+ gas)	+	+	+	+	− or +
Inositol (+ gas)	d	+	−	−	d
Lactose	+	+	− or (+)	− or (+)	d
Mannitol	+	+	+	+	+
Salicin	+ or (+)	+	d	d	d
Sucrose	+	+	d	d	d
Indole	−	−	−	−	− or +
Methyl red	−	−	+ or −	−	− or +
Voges-Proskauer	+	+	+ or −	+	+ or −
Simmons' citrate	+	+	+ or −	d	d
KCN	+	+	+	+	− or +
Urease	+ or −	−	−	−	d
Gelatin liquefaction	(+) or −	− or (+)		−	+
Phenylalanine deaminase	−	−	−	−	− or +
Sodium malonate	+ or −	+ or −	+ or −	+ or −	+ or −
Arginine dihydrolase	+	−	−	−	−
Lysine decarboxylase	−	+	+	+	−
Ornithine decarboxylase	+	+	+	+	−

d, different biochemical types; + or −, majority are positive; − or +, majority are negative; (+), delayed reaction.

tics shown in Table 20-13. The type species is *K. pneumoniae*. The klebsiellae can cause severe enteritis in infants* and pneumonia, septicemia, meningitis, wound infection, peritonitis, and an increasing number of hospital-acquired urinary tract infections in adults. On blood agar, EMB agar, MacConkey agar, trypticase soy agar, and other routine plating media, they generally give rise to large **mucoid** colonies that have a tendency to coalesce and usually string out when touched with a needle. Growth in broth also can be very stringy and difficult to break up when making transfers.

Some 72 capsular types have been identified. Cross-reactions between types are known to occur; in the preparation of pooled typing sera, related types are placed in the same pool. Epidemiologic significance of these bacteria, particularly in hospital-acquired infections, is perhaps best served by the methods of serotyping. It should be emphasized that greater accuracy in identification of the *Klebsiella* types is obtained by use of the quellung reaction rather than reliance on agglutinating antisera.

K. rhinoscleromatis is the cause of scleroma of the nose, pharynx, and other respiratory tract structures. This disease is endemic in Central and South America and North Africa but is seen rarely in the United States.

Genus Enterobacter

According to current classification (Table 20-14), the genus *Enterobacter* contains four species—*E. cloacae*, *E. aerogenes*, *E. hafniae* (the former *Hafnia* group), and *E. agglomerans*,[24] which includes the former Herbicola-Lathyri bacteria. The clinical significance of *E. agglomerans* is presently undetermined. However, in 1972, Pien and associates[60] examined 56 strains of *E. agglomerans* isolated from 51 patients during an 18-month period at the Mayo Clinic. Infections occurred primarily in accident victims during the summer and autumn months. *E. agglomerans* was considered to be directly pathogenic in only six wound infections and one urinary tract infection. No cases of septicemia were seen. Practically all strains isolated were sensitive to the antibiotics tested except that some were relatively resistant to ampicillin and cephalothin.

The species of *Enterobacter* are found in soil, water, dairy products, and the intestines of animals, including humans. Considered as opportunists, they are now being isolated more frequently in urinary tract infection, in septicemia, and from other clinical sites, particularly in hospitalized patients who are quite ill.

Genus Serratia

Members of the genus *Serratia* are gram-negative motile rods; only a small percentage of strains are **chromogenic.** Three species currently exist: *S. marcescens*, *S. liquefaciens*, and *S. rubidaea*.[28] Those strains of *S. marcescens* that are chromogenic produce a red, non-water-soluble pigment at room temperature, but seldom at 35 C or above. Considered for many years as innocuous, these bacteria are primarily found in water and in soil. Only in recent years have they been directly implicated in human infection, including pulmonary infection, urinary tract infection, and septicemia.[58,70,71,91,93] This again reflects the vulnerability of the compromised patient to opportunistic pathogens. The indiscriminate use of antibiotics also must be considered a contributory factor, especially since *Serratia* is typically quite resistant.

Fifteen somatic antigens for *Serratia* have been previously identified serologically, and there is good evidence that serotyping can be of value in studying nosocomial infections. The biochemical pattern exhibited by the genus is shown in Table 20-15; the biochemical speciation is shown in Table 20-6.

---

*Enterotoxin-producing klebsiellae have been isolated from the small intestine of individuals with tropical sprue.

Table 20-15. Biochemical reactions of genus *Serratia*

Test	Reaction	Test	Reaction
Arabinose	−	Voges-Proskauer	+
Adonitol	variable	Simmons' citrate	+
Dulcitol	−	Gelatin (22 C)	+
Glucose	+ gas variable	KCN	+
Inositol	variable	Phenylalanine deaminase	−
Lactose	− or delayed	DNase	+
Mannitol	+	Sodium malonate	−
Salicin	+	Lysine decarboxylase	+
Sucrose	+	Arginine dihydrolase	−
Indole	−	Ornithine decarboxylase	+
Methyl red	− or +		

− or +, majority are negative.

Genus Proteus

The tribe Proteeae consists of two genera, *Proteus* and *Providencia*.

Proteus species are gram-negative, somewhat pleomorphic motile rods. The type species is *Proteus vulgaris*. They are typically **lactose negative** and often rapidly **urease positive**. The species are actively motile at 25 C but often weakly motile at 35 C. On moist agar or agar of 1.5% concentration (usual strength for plating media), *P. vulgaris* and *P. mirabilis* tend to swarm, producing a bluish gray confluent surface growth at both 25 and 35 C. Swarming may be inhibited on MacConkey and on EMB plates if the agar concentration is increased to 5%.

The appearance of the spreading growth will vary from a rippled form of growth that is easily recognized to one that is smooth and almost transparent. This growth may sometimes go unobserved in examining plates containing mixed cultures, but a loop drawn over a seemingly colony-free area readily reveals its presence. Because of careless technique, *Proteus* is often the contaminant in "pure" cultures obtained from selected colonies.

Due to their lactose-negative characteristic, discrete *Proteus* colonies are sometimes selected from differential or selective media as suspicious salmonellae, shigellae, or other gram-negative

organisms. This is one of the reasons why screening on TSI (or KIA) agar, followed by a rapid urease test, is highly recommended, as was previously discussed.

Proteus organisms can frequently be found in large numbers in the stools of individuals undergoing oral antibiotic therapy. They are frequently responsible for infections of the urinary tract and may be seen in bacteremia, intraabdominal infection, wound infection, and so forth.

The biochemical reactions of the *Proteus* species are given in Table 20-16. Species identification of *Proteus* isolates is certainly to be encouraged, since *P. mirabilis* responds more readily to antimicrobial therapy (the other species are more resistant).[85]

Genus Providencia

Formerly known as the 29911 paracolon bacterium,[81,82] the genus *Providencia* currently consists of two species: *P. alcalifaciens* and *P. stuartii*. The former is the type species. They are gram-negative motile rods that are lactose negative and grow well on most enteric isolation media. Because they are hydrogen sulfide negative and may or may not produce gas in glucose, they often resemble shigellae on TSI (or KIA) agar. If sucrose is fermented, the reaction is usually delayed, and thus fermentation of this carbohydrate will not be detected in 48 hours in

Table 20-16. Biochemical characteristics of *Proteus* species*

	P. mirabilis	*P. vulgaris*	*P. morganii*	*P. rettgeri*
Adonitol	−	−	−	+
Dulcitol	−	−	−	−
Glucose	+ (gas)	+ (gas)	+ (gas)	+
Inositol	−	−	−	+
Lactose	−	−	−	−
Mannitol	−	−	−	+
Salicin	variable	+	−	variable
Sucrose	+ (3-8 days)	+	−	+ (delayed)
Xylose	+	+	−	−
Gelatin	+	+	−	−
H₂S (TSI)	+	+	−	−
Indole	−	+	+	+
Methyl red	+	+	+	+
Voges-Proskauer	− or +	−	−	−
Simmons' citrate	+ (variable)	− (variable)	−	+
Phenylalanine deaminase	+	+	+	variable
Ornithine decarboxylase	+	−	+	−

*All *Proteus* species are urease positive.

Table 20-17. Biochemical reactions of *Providencia* species

	P. alcalifaciens	*P. stuartii*		*P. alcalifaciens*	*P. stuartii*
Adonitol	+	−	H₂S (TSI)	−	−
Arabinose	−	−	Indole	+	+
Dulcitol	−	−	Methyl red	+	+
Glucose (gas)	+ or −	−	Voges-Proskauer	−	−
Inositol	−	+	Urease	−	−
Lactose	−	−	Simmons' citrate	+	+
Sodium malonate	−	−	KCN	+	+
Mannitol	−	d	Phenylalanine deaminase	+	+
Salicin	−	−	Lysine decarboxylase	−	−
Sucrose	d	d	Arginine dihydrolase	−	−
Gelatin	−	−	Ornithine decarboxylase	−	−

d, different biochemical types; + or −, majority are positive.

the TSI slant. *Providencia* may be distinguished from the shigellae, however, by their motility and utilization of citrate. Negative urease reactions differentiate *Providencia* from *Proteus*. *Providencia* may be distinguished also from other genera within the Enterobacteriaceae by their biochemical characteristics, indicated in Table 20-17.

The organisms (particularly *P. stuartii*) have been incriminated in urinary tract infection, especially in patients with underlying urologic disorders[30,90] and in various infections of patients in a burn unit.[57] *P. stuartii*, in particular, not uncommonly

is resistant to multiple antibiotics, like so many other nosocomial pathogens. *Providencia* has been felt to cause diarrhea,[39] but this is not well established.

An antigenic scheme has been established for the genus, in which there are 62 O antigen groups and about 175 serotypes.

Recently, some investigators have questioned the current classification of urease-positive and urease-negative strains of *Proteus rettgeri* and *Providencia stuartii*, respectively, based on observed variation in urease activity of endemic hospital strains of these orga-

Table 20-18. Distinguishing properties of *Yersinia* species

Property	*Y. pestis*	*Y. enterocolitica*	*Y. pseudotuberculosis*
Colony forms	Two	Two	Two
Optimal growth temperature	25 to 30 C	25 to 30 C	25 to 30 C
Motility at 25 C	−	+	+
Serogroups	One	Two	Five
Catalase	+	+	+
Oxidase	−	−	−
Coagulase	+*		−
Fibrinolysin	+		−
Hemolysis on blood agar	−	alpha	−
Medium containing bile salts	+	+	+
H₂S	−	−	−
Indole	−	−	−
Methyl red	+	+	+
Voges-Proskauer	−	+	−
Urease	−	+	+
Nitrate'	Variable	Reduced	Reduced
Cellobiose	−	+	−
Glucose	+	+	+
Glycerol	Variable	+	+
Lactose	−	−	−
Melibiose	−	−	+
Maltose	+	+	+
Mannitol	+	+	+
Rhamnose	−	−†	+
Salicin	+	−	+
Sucrose	−	+	−

*Using rabbit plasma.
†Occasional strains are rhamnose positive.

nisms. Penner and associates[59] reported that all of their isolates gave the same reactions in 17 biochemical tests and possessed O antigens characteristic of *Providencia* O-type strains 4 or 17. Study of the isolates indicated that their endemic strains were capable of undergoing variation in urease activity and that all strains should be accommodated in a single group.

Genus Yersinia

Recently incorporated into the family Enterobacteriaceae, this genus consists of the species *Yersinia pestis, Y. enterocolitica,* and *Y. pseudotuberculosis.* The distinguishing properties of these species are shown in Table 20-18.

Yersinia pestis

Yersinia pestis is the etiologic agent of **plague,** a disease with a high mortality in human beings and rats, and is infectious for mice, guinea pigs, and rabbits. Three clinical forms of plague are recognized in humans: **bubonic, pneumonic,** and **septicemic.** The main vector for transmission is the rat flea. Humans are accidental hosts of this flea. Plague bacilli are short, plump, nonmotile, gram-negative rods, sometimes elongated and pleomorphic, usually appearing singly or in pairs and occasionally in short chains. Bipolar staining can be demonstrated with polychrome stains, such as the Giemsa or Wayson stains, but **not** by the Gram stain. Coccoid, round, filamentous, elongated, and other forms commonly occur, especially in old cultures. A capsule can be demonstrated in animal tissue and in young cultures. The latter will give a positive fluorescent-antibody reaction, by which a presumptive diagnosis may be made.

Y. pestis grows slowly on nutrient agar and fairly rapidly on blood agar, producing small, nonhemolytic, round, transparent, glistening, colorless colonies with an undulate margin.[79] Older colonies enlarge, becoming opaque with yellowish centers and whitish edges, which develop a soft, mucoid consistency due to capsular material. A characteristic type of growth occurs in old broth cultures overlaid with sterile oil, in which a pellicle forms with "stalactite" streamers. Biochemical properties are shown in Table 20-18.

Y. pestis is pathogenic for rats and guinea pigs. After subcutaneous inoculation, the animals die within 2 to 5 days and show certain postmortem characteristics, including ulceration at the site of injection, regional adenopathy, congested spleen and liver, and pleural effusion. The bacilli may be demonstrated in splenic smears; they may also be recovered by culture. Cultures may be identified by specific bacteriophage typing, by the fluorescent-antibody test, or by agglutination with specific antisera.

Great care must be taken by the laboratory worker when handling suspect cultures or pathologic materials. A biologic safety hood should be used. The worker should be masked and wearing rubber gloves. The creation of aerosols should be avoided, and all contaminated materials must be autoclaved.

Yersinia enterocolitica

This organism appears to be closely related to *Y. pseudotuberculosis*. *Y. enterocolitica* has been implicated in human disease with a variety of clinical syndromes; these include gastroenteritis, bacteremia, peritonitis, cholecystitis, visceral abscesses, and mesenteric lymphadenitis.*[80] Septicemia, with a 50% mortality, has been reported in the compromised host.[62]

---

*Recent outbreaks of illness due to *Y. enterocolitica* have been traced to milk (CDC Morbidity and Mortality Weekly Report **26:**53-54, 1977).

Y. enterocolitica is not easily isolated and identified in the microbiology laboratory.[79] The organism grows slowly at 35 C. Biochemical reactions for the most part resemble those of many members of the family Enterobacteriaceae; thus, these organisms may be frequently overlooked. The organism is a gram-negative coccobacillus occasionally showing bipolar staining. Table 20-18 shows biochemical and other properties of this microorganism.

The serologic and biochemical characteristics of 24 human isolates of *Y. enterocolitica* submitted to the California Department of Health from 1968 through 1975 were reported by Bissett.[4] Nine different serotypes were represented, with the majority of strains being serotype O:8 (6 strains) and O:5 (5 strains). Sources of the isolates included feces (12 cases), blood (3), sputum or throat (3), bile or bowel drainage (2), wounds (2), breast abscess (1), and skin abscess (1). Underlying medical conditions existed in 13 patients.

Y. enterocolitica is frequently resistant to penicillins and cephalosporins. Aminoglycosides, cotrimoxazole, and chloramphenicol are the drugs most active against this organism (Bottone, E. J.: CRC Crit. Rev. Microbiol. **5:**211-241, 1977).

Yersinia pseudotuberculosis

Y. pseudotuberculosis causes disease primarily in rodents, particularly guinea pigs, but it also causes two recognized forms of disease in human beings. Of these the most serious is a **fulminating septicemia,** which is usually fatal, while the more common form is a **mesenteric lymphadenitis,** which may simulate appendicitis. The organism has been responsible for many infections in Europe.[47] In recent years, cases have been reported in the United States by Weber and co-workers.[92]

The bacterium is a small gram-negative coccobacillus, exhibiting pleomorphism, found singly, in short chains, and in small clusters. It grows well on blood agar and on media containing bile salts, such as MacConkey agar.[79] Two colonial forms may be observed: a smooth grayish yellow translucent colony in 24 hours at 25 to 30 C, and a raised colony, with an opaque center and a lighter margin with serrations, that develops with continued incubation. A rough variant with an irregular outline may also develop. The characteristics of the species are shown in Table 20-18.

Tribe Erwinieae

Although a certain amount of confusion still exists with regard to the taxonomy of these organisms, it is reasonably well established that these bacteria, noted primarily as plant pathogens, are members of the family Enterobacteriaceae, hence their inclusion here. At this writing, there is very little else to add other than to point out that Ewing's current classification consists of the tribe Erwinieae and the two genera *Erwinia* and *Pectobacterium* (see Table 20-1). Biochemical reactions of a representative species for each genus are shown in Table 20-6. Readers interested in this group of organisms should consult Chapter 15 in Edwards and Ewing's text.[19]

IDENTIFICATION OF ENTEROBACTERIACEAE BY RAPID METHODS

Rapid procedures designed to reduce the amount of time for identifying members of the Enterobacteriaceae have been developed commercially and have been compared with conventional methods used in the identification of gram-negative bacterial isolates from clinical material. The results have been evaluated and reported by various investigators. There are currently in use at least six different systems, and a brief description of these is given in alphabetical order. In the performance of each, however, emphasis is placed on the competence and technical skill of the person or persons carrying out the tests, and it should be recognized that

the **results will only be as good as the laboratory expertise shown.** Complete information on the use of any of these test systems can be obtained from the manufacturer. All systems have been rated as good, with an accuracy correlation ranging from 87% to 96% or higher, depending on the number and variety of cultures tested. In addition, many of these systems are in their third or fourth generation, which should increase their accuracy even more.

API System (Analytab Products Inc.*). The API 20 enteric system utilizes **22** biochemical tests that can produce results from a single bacterial colony in 18 to 24 hours. The colony is emulsified in about 5 ml distilled water to supply the inoculum, and inoculation is carried out with a Pasteur pipet. Viable cells are introduced into the small plastic cupules and tubes arranged on a plastic-covered strip that is incubated in a plastic tray, to which water has been added to provide humidity, for 18 hours. This system has been evaluated by several investigators, and correlation with conventional tests has been reported as high as 96.4%.[36,78,88] Washington[87] feels that the API strip is currently the most complete and accurate kit for speciation of Enterobacteriaceae.

Enterotube System (Roche Diagnostics†). The new improved tube permits simultaneous inoculation and performance of **11** biochemical tests from a single colony. A well-isolated colony is usually selected from MacConkey, EMB, or Hektoen agar plates that have been inoculated with a clinical specimen. The multitest tube is self-inoculated by touching the needle to a single colony and drawing it through all media in the tube. It has been reported that the citrate and urease reactions tend to present the greatest problems in the interpretation of the test. Reports by Morton and Monaco,[55] Dou-

---

*Plainview, N.Y.
†Division of Hoffman-LaRoche, Inc., Nutley, N.J.

glas and Washington,[14] and Smith[75] provide evidence of the practical use of this system. Washington[87] notes that this is the simplest and most convenient of all kits to use.

Inolex (Auxotab) Enteric 1 System (Colab Laboratories, Inc.*). This system consists of a card with **10** capillary units containing 10 different reagents designed to differentiate between the genera of the Enterobacteriaceae on a reduced time schedule and to differentiate the species of *Enterobacter* and *Proteus*. The reliability of this test system in regard to accuracy of results and laboratory safety has been reported.[64,89] In his latest appraisal, Washington[87] notes conflicting reports in the literature and no recent evaluations and therefore reserves judgment on the system for the present.

Inoculum is prepared from a single colony selected from a primary isolation plate, and from this suspension the battery of tests is set up. Readings can be recorded in 7 hours at 35 C.

Minitek System (BBL).† This test system utilizes paper disks impregnated with individual substrates. These disks are placed in wells in a plastic plate and inoculated with a broth suspension of the isolate. Subsequent identification is based on color reactions occurring in the disks following overnight incubation. There are presently **35** different disks available, permitting the user to select those tests most appropriate for identification.

Several reports have attested to the reliability and accuracy of the Minitek system in the identification of the Enterobacteriaceae.[31,37,46] However, Washington's experience was less favorable.[87] He feels that at this point the accuracy of identification with Minitek is not of the same order as that achievable with the API and Enterotube systems.

*PathoTec Rapid I-D System.** This system consists of a set of **12** test strips impregnated with various biochemical reagents in carefully measured concentrations. Selected test strips have been reported to identify approximately 95% of the Enterobacteriaceae isolated in approximately 4 hours after initial isolation.[76] The strips are added to prepared test tubes (13 × 100 mm). Two of the strips, cytochrome oxidase and esculin hydrolysis, are inoculated by rubbing inoculum from selected colonies on the designated areas, while the remaining strips are placed in tubes to which a measured amount of cell suspension has been added. The cytochrome oxidase test is read after 30 seconds, while the remaining tests are recorded after approximately 4 hours at 35 C. It is the only kit that can be regarded as truly "rapid."

R/B System.† The basic system contains **eight** biochemicals in two tubes: phenylalanine deaminase, hydrogen sulfide, indole, motility, lysine and ornithine decarboxylase, gas from glucose, and lactose. Two additional tubes have been introduced, namely, Cit/Rham (citrate and rhamnose) and Soranase (sorbitol, deoxyribonuclease, raffinose, and arabinose), permitting an expansion to **14** biochemicals in four tubes. The addition of these tubes permits differentiation of *Enterobacter* species and a differentiation of other genera.[42,50,77] The tests are read after overnight incubation at 35 C.

This system has been evaluated favorably by several investigators.[42,50]

Other approaches. Several systems employ mathematical analysis of data from biochemical reactions for rapid identification of Enterobacteriaceae. Thus, API has a Profile Register for computer use. Encise II is a large data base for use with the Enterotube; it can be approached by a binomial computer system or a four-

---

*Glenwood, Ill.
†Division of Becton, Dickenson and Co., Cockeysville, Md.

*General Diagnostics Division, Warner-Lambert Co., Morris Plains, N.J.
†Corning Medical, Medfield, Mass.

digit number reference. Still another such unit is the Enteric Analyzer (Diagnostic Research Inc.).

Computer-assisted bacterial identification utilizing antimicrobial susceptibility patterns has also been done.[9,73]

Other approaches to rapid identification of the Enterobacteriaceae include the use of bile-esculin agar to detect esculin hydrolysis in 4 hours,[48] a 3-hour deoxyribonuclease test,[34] and direct identification of colonies of *Salmonella* and *Shigella* by coagglutination of protein A–containing staphylococci sensitized with specific antibody.[17] Von Graevenitz (Mt. Sinai J. Med. **43:**727-735, 1976) found that use of a DNase-indole medium for non-lactose-fermenting gram-negative rods often led to rapid identification of a variety of Enterobacteriaceae and other gram-negative bacilli. An interesting simplified approach to prompt lactose-fermenting Enterobacteriaceae strains is presented by Hicks and Ryan.[38] Flat, spot indole-positive colonies were identified as *Escherichia coli*. Spot indole-negative organisms forming mucoid colonies were identified as *Klebsiella* sp. or *Enterobacter* sp. on the basis of semisolid agar motility and ornithine decarboxylase tests. This very simple scheme yielded 97.4% accuracy as compared with conventional and API identifications.

ANTIMICROBIAL SUSCEPTIBILITY

Antimicrobial susceptibility patterns may vary widely in different institutions. It is helpful to the clinician to receive periodic summaries of specific susceptibility patterns to guide decisions regarding therapy until studies have been done on isolates from patients. In general, the organisms most commonly encountered that are frequently resistant are *Pseudomonas, Klebsiella,* and *Serratia.* Resistance to cephalosporins and aminoglycosides is not uncommon in many institutions. Amikacin is presently the most active of all antimicrobials against the Enterobacteriaceae.

REFERENCES

1. Abbott, J. D., and Shannon, R.: A method for typing *Shigella sonnei* using colicin production as a marker, J. Clin. Pathol. **11:**71-75, 1958.
2. Bailey, W. R.: Studies on the transduction phenomenon I. Practical applications in the laboratory, Can. J. Microbiol. **2:**549-553, 1956.
3. Banwell, J. G., and Sherr, H.: Effect of bacterial enterotoxins on the gastrointestinal tract, Gastroenterol. **65:**467-497, 1973.
4. Bissett, M. L.: *Yersinia enterocolitica* isolates from humans in California, 1968-1975, J. Clin. Microbiol. **4:**137-144, 1976.
5. Brinton, C. C.: Non-flagellar appendages of bacteria, Nature **183:**782-786, 1959.
6. Buchanan, R. E., and Gibbons, N. E., editors: Bergey's manual of determinative bacteriology, ed. 8, Baltimore, 1974. The William & Wilkins Co.
7. Carpenter, K. P.: Personal communication; 1969.
8. Carpenter, K. P., and Lachowicz, K.: The catalase activity of *Sh. flexneri,* J. Pathol. Bact. **77:** 645-648, 1959.
9. Darland, G.: Discriminant analysis of antibiotic susceptibility as a means of bacterial identification, J. Clin. Microbiol. **2:**391-396, 1975.
10. Darland, G., Ewing, W. H., and Davis, B. R.: The biochemical characteristics of *Yersinia enterocolitica* and *Yersinia pseudotuberculosis,* DHEW Publ. No. (CDC) 75-8294, Washington, D.C., 1974, Department of Health, Education, and Welfare.
11. De, S. N., Bhattacharya, K., and Sarker, J. K.: A study of the pathogenicity of strains of *B. coli* from acute and chronic enteritis, J. Path. Bact. **71:**201-209, 1956.
12. Dean, A. G., Ching, Y. C., Williams, G., and Barden, B.: Test for *Escherichia coli* enterotoxin using infant mice: application in a study of diarrhea in children in Honolulu, J. Infect. Dis. **125:**407-411, 1972.
13. Donta, S. T., Moon, H. W., and Whipp, S. C.: Detection of heat-labile *Escherichia coli* enterotoxin with the use of adrenal cells in tissue culture, Science **183:**334-336, 1974.
14. Douglas, G. W., and Washington, J. A. II: Identification of Enterobacteriaceae in the clinical laboratory, Atlanta, 1970, Center for Disease Control.
15. Duguid, J. P., Smith, I. W., Dempster, G., and Edmonds, P. N.: Non-flagellar filamentous appendages ("fimbriae") and haemagglutinating activity in *Bacterium coli,* J. Path. Bact. **70:** 335-348, 1955.
16. DuPont, H. L., Formal, S. B., Hornick, R. B., Snyder, M. J., Libonati, J. P., Sheahan, D. G., LaBrec, E. H., and Kalas, J. P.: Pathogenesis of *Escherichia coli* diarrhea, N. Engl. J. Med. **258:** 1-9, 1971.

17. Edwards, E. A., and Hilderbrand, R. L.: Method for identifying *Salmonella* and *Shigella* directly from the primary isolation plate by coagglutination of protein A–containing staphylococci sensitized with specific antibody, J. Clin. Microbiol. **3:**339-343, 1976.

18. Edwards, P. R., and Ewing, W. H.: Manual for enteric bacteriology, Atlanta, 1951, National Communicable Disease Center.

19. Edwards, P. R., and Ewing, W. H.: Identification of Enterobacteriaceae, ed. 3, Minneapolis, 1972, Burgess Publishing Co.

20. Edwards, P. R., and Fife, M. A.: Lysine-iron agar in the detection of *Arizona* cultures, Appl. Microbiol. **9:**478-480, 1961.

21. Edwards, P. R., Kauffmann, F., and van Oye, E.: A new diphasic *Arizona* type, Acta Pathol. Microbiol. Scand. **31:**5-9, 1952.

22. Enterobacteriaceae Subcommittee: Third report, Int. Bull. Bact. Nomenclat. Taxon. **8:**25-70, 1958.

23. Evans, D. G., Evans, D. J., Jr., and Pierce, N. F.: Differences in the response of rabbit small intestine to heat-labile and heat-stable enterotoxins of *Escherichia coli*, Infect. Immun. **7:**873-880, 1973.

24. Ewing, W. H., and Fife, M. A.: Biochemical characterization of *Enterobacter agglomerans*, DHEW Publ. No. (HSM) 73-8173, Washington, D.C., 1972, Department of Health, Education, and Welfare.

25. Ewing, W. H., and Martin, W. J.: Enterobacteriaceae. In Lennette, E. H., Spaulding, E. H., and Truant, J. P., editors: Manual of clinical microbiology, ed. 2, Washington, D.C., 1974, American Society for Microbiology.

26. Ewing, W. H., McWhorter, A. C., Escobar, M. R., and Lubin, A. M.: *Edwardsiella*, a new genus of *Enterobacteriaceae* based on a new species *E. tarda*, Int. Bull. Bact. Nomencl. Taxon. **15:**33-38, 1965.

27. Ewing, W. H., McWhorter, A. C., and Montague, T. S.: Transport media in the detection of *Salmonella typhi* in carriers, J. Conf. State Prov. Public Health, Lab. Directors **24:**63-65, 1966.

28. Ewing, W. H., Davis, B. R., Fife, M. A., and Lessel, E. F.: Biochemical characterization of *Serratia liquefaciens* (Grimes and Hennerty), Bascomb et al. (formerly *Enterobacter liquefaciens* and *Serratia rubidaea* (Stapp) comb. nov. and designation of type and neotype strains, Int. J. Syst. Bacteriol. **23:**217-225, 1973.

29. Field, M.: Intestinal secretion, Gastroenterol. **66:**1063-1084, 1974.

30. Fields, B. N., Uwaydah, M. M., Kunz, L. J., and Swartz, M. N.: The so-called "paracolon" bacteria. A bacteriologic and clinical reappraisal, Am. J. Med. **42:**89-106, 1967.

31. Finklea, P. J., Cole, M. S., and Sodeman, T. M.: Clinical evaluation of the Minitek differential system for identification of *Enterobacteriaceae*, **4:**400-404, 1976.

32. Gorbach, S. L., Kean, B. H., Evans, D. G., Evans, D. J., Jr., and Bessudo, D.: Travelers diarrhea and toxigenic *Escherichia coli*, N. Engl. J. Med. **292:**933-936, 1975.

33. Gorbach, S. L., and Khurana, C. N.: Toxigenic *Escherichia coli* as a cause of infantile diarrhea in Chicago, N. Engl. J. Med. **287:**791-795, 1972.

34. Greenwood, J. R., and Pickett, M. J.: Deoxyribonuclease: detection with a three-hour test, J. Clin. Microbiol. **4:**453-454, 1976.

35. Guerrant, R. L., and Dickens, M. D.: Toxigenic bacterial diarrhea: a nursery outbreak involving multiple strains, Fourteenth Interscience Conference on Antimicrobial Agents and Chemotherapy, Abstr. 130, 1974.

36. Guillermet, F. N., and Desbresles, A. M. B.: A propos de l'utilisation d'une micromethode d'identification des Enterbacties, Revue de l'Instit Pasteur de Lyon **4:**71-78, 1971.

37. Hanson, S. L., Hardesty, D. R., and Myers, B. M.: Evaluation of the BBL Minitek system for the identification of *Enterobacteriaceae*, Appl. Microbiol. **28:**798-801, 1974.

38. Hicks, M. J., and Ryan, K. J.: Simplified scheme for identification of prompt lactose-fermenting members of the *Enterobacteriaceae*, J. Clin. Microbiol. **4:**511-514, 1976.

39. Hobbs, B. C., Thomas, M. E. M., and Taylor, J.: School outbreak of gastro-enteritis associated with a pathogenic paracolon bacillus, Lancet **257:**530-532, 1949.

40. Hormaeche, E., and Edwards, P. R.: A proposed genus *Enterobacter*, Int. Bull. Bact. Nomenclat. Taxon. **10:**71-74, 1960.

41. Hormaeche, E., and Munilla, M.: Biochemical tests for the differentiation of Klebsiella and Cloaca, Int. Bull. Bact. Nomenclat. Taxon. **7:**1-20, 1957.

42. Isenberg, H. D., and Painter, B. G.: Comparison of conventional methods, the R/B System, and modified R/B System as guides to major divisions of Enterobacteriaceae, Appl. Bact. **22:**1126-1134, 1971.

43. Julianelle, L. A.: A biological classification of *Encapsulatus pneumoniae* (Friedlander's bacillus), J. Exp. Med. **44:**113-128, 1926.

44. Kauffmann, F.: Ein kombiniertes Anreicherungsverfahren fur Typhus und Paratyphusbazillen, Zbl. Bakt. **119:**148-152, 1930.

45. Kauffmann, F.: Die Bakteriologie der Salmonella-Gruppe, Copenhagen, 1941, Einar Munksgaard.

46. Kiehn, T. E., Brennan, K., and Ellner, P. D.: Evaluation of Minitek system for identification of *Enterobacteriaceae*, Appl. Microbiol. **28:**668-671, 1974.

47. Knapp, W., and Masshoff, W.: Zur Ätiologie der abszedierenden retikulozytären Lymphadenitis: einer praktisch wichtigen, vielfach unter dem Bilde einer aktuen Appendizitis verlaufenden Erkrankung, Deutsch Med. Wschr. **79:** 1266-1271, 1954.

48. Lindell, S. S., and Quinn, P.: Use of bile-esculin agar for rapid differentiation of *Enterobacteriaceae*, J. Clin. Microbiol. **1:**440-443, 1975.

49. Martin, W. J., Fife, M. A., and Ewing, W. H.: The occurrence and distribution of the serotypes of *Arizona*, Atlanta, 1967, National Communicable Disease Center.

50. McIlroy, G. T., Yu, P. K. W., Martin, W. J., and Washington, J. A. II: Evaluation of modified R-B system for identification of members of the family Enterobacteriaceae, Appl. Microbiol. **24:** 358-362, 1972.

51. Minutes of the Enterobacteriaceae Subcommittee meeting, Montreal, 1962; report of the Subcommittee on taxonomy of the Enterobacteriaceae, Int. Bull. Bact. Nomenclat. Taxon. **13:**69-93, 139, 1963.

52. Moeller, V.: Diagnostic use of the Braun KCN test within Enterobacteriaceae, Acta Pathol. Microbiol. Scand. **34:**115-126, 1954.

53. Morris, G. K., Merson, M. H., Sack, D. A., Wells, J. G., Martin, W. T., DeWitt, W. E., Feeley, J. C., Sack, R. B., and Bessudo, D. M.: Laboratory investigation of diarrhea in travelers to Mexico: evaluation of methods for detecting enterotoxigenic *Escherichia coli*, J. Clin. Microbiol. **3:**486-495, 1976.

54. Morris, G. K., and Wells, J. G.: Colicin typing of *Shigella sonnei*, Appl. Microbiol. **27:**312-316, 1974.

55. Morton, H. E., and Monaco, M. A. J.: Comparison of enterotubes and routine media for the identification of enteric bacteria, Am. J. Clin. Pathol. **56:**64-66, 1971.

56. Ørskov, F., Ørskov, I., Evans, D. J., Jr., Sack, R. B., Sack, D. A., and Wadström, T.: Special *Escherichia coli* serotypes among enteropathogenic strains from diarrhea in adults and children, Med. Microbiol. Immunol. **162:**73-80, 1976.

57. Overturf, G. D., Wilkins, J., and Ressler, R.: Emergence of resistance of *Providencia stuartii* to multiple antibiotics: speciation and biochemical characterization of *Providencia*, J. Infect. Dis. **129:**353-357, 1974.

58. Patterson, R. H., Banister, G. B., and Knight, V.: Chromobacterial infection in man, Arch. Intern. Med. **90:**79-86, 1952.

59. Penner, J. L., Hinton, N. A., Whiteley, G. R., and Hennessy, J. N.: Variation in urease activity of endemic hospital strains of *Proteus rettgeri* and *Providencia stuartii*, J. Infect. Dis. **134:**370-376, 1976.

60. Pien, F. D., Martin, W. J., Hermans, P. E., and Washington, J. A. II: Clinical and bacteriologic observations on the proposed species, *Enterobacter agglomerans* (the Herbicola-Lathyri bacteria), Mayo Clin. Proc. **47:**739-745, 1972.

61. Pruneda, R. C., and Farmer, J. J. III: Bacteriophage typing of *Shigella sonnei*, J. Clin. Microbiol. **5:**66-74, 1977.

62. Rabson, A. R., Hallett, A. F. and Koornhof, H. J.: Generalized *Yersinia enterocolitica* infection, J. Infect. Dis. **131:**447-451, 1975.

63. Reller, L. B., and Mirrett, S.: Motility-indole-lysine medium for presumptive identification of enteric pathogens of Enterobacteriaceae, J. Clin. Microbiol. **2:**247-252, 1975.

64. Rhoden, D. L., Tomfohrde, K. M., Smith, P. B., and Balows, A.: Auxotab—a device for identifying enteric bacteria, Appl. Microbiol. **25:**284-286, 1973.

65. Roat, W. R., Formal, S. B., Dammin, G. J., and Giannella, R. A.: Pathophysiology of Salmonella diarrhea in the rhesus monkey: intestinal transport, morphological and bacteriological studies. Gastroenterol. **67:**59-70, 1974.

66. Rowe, B., Taylor, J., and Bettelheim, K. A.: An investigation of travellers diarrhea, Lancet **1:**1-5, 1970.

67. Sack, R. B., Gorbach, S. L., Banwell, J. G., Jacobs, B., Chatterjee, B. D., and Mitra, R. C.: Enterotoxigenic *Escherichia coli* isolated from patients with severe cholera like disease, J. Infect. Dis. **123:**378-385, 1971.

68. Sack, R. B., Hirschhorn, N., Brownlee, I., Cash, R. A., Woodward, W. E., and Sack, D. A.: Enterotoxigenic *Escherichia coli*-associated diarrheal disease in Apache children, N. Engl. J. Med. **292:**1041-1045, 1975.

69. Sack, R. B., Jacobs, B., and Mitra, R.: Antitoxin responses to infections with enterotoxigenic *Escherichia coli*, J. Infect. Dis. **129:**330-335, 1974.

70. Sanders, C. V., Jr., Luby, J. P., Johanson, W. G., Jr., Barnett, J. A., and Sanford, J. P.: *Serratia marcescens* infections from inhalation therapy medications, nosocomial outbreak, Ann. Intern. Med. **73:**15-21, 1970.

71. Schaberg, D. R., Alford, R. H., Anderson, R., Farmer, J. J., III, Molly, M. A., and Schaffner, W.: An outbreak of nosocomial infection due to multiple resistant *Serratia marcescens:* evidence of inter-hospital spread, J. Infect. Dis. **134:**181-188, 1976.

72. Shore, E. G., Dean, A. G., Holik, K. J., and Davis, B. R.: Enterotoxin producing *Escherichia coli* and diarrheal disease in adult travellers; a prospective study, J. Infect. Dis. **129:**577-582, 1974.

73. Sielaff, B. H., Johnson, E. A., and Matsen, J. M. Computer-assisted bacterial identification utilizing antimicrobial susceptibility profiles gen-

erated by Autobac 1, J. Clin. Microbiol. **3:** 105-109, 1976.

74. Smith, H. W., and Gyles, C. L.: The relationship between two apparently different enterotoxins produced by enteropathogenic strains of *Escherichia coli* of porcine origin, J. Med. Microbiol. **3:**387-401, 1970.

75. Smith, P. B.: Roundtable on *Enterobacteriaceae*, American Society for Microbiology Meeting, Miami Beach, May 1973.

76. Smith, P. B., Rhoden, D. L., and Tomfohrde, K. M.: Evaluation of the Pathotec rapid I-D system for identification of *Enterobacteriaceae*, J. Clin. Microbiol. **1:**359-362, 1975.

77. Smith, P. B., Tomfohrde, K. M., Rhoden, D. L., and Balows, A.: Evaluation of the modified R/B system for identification of Enterobacteriaceae, Appl. Microbiol. **22:**928-929, 1971.

78. Smith, P. B., Tomfohrde, K. M., Rhoden, D. L., and Balows, A.: API system: a multitube micromethod for identification of Enterobacteriaceae, Appl. Microbiol **24:**449-452, 1972.

79. Sonnenwirth, A. C.: *Yersinia.* In Lennette, E. H., Spaulding, E. H., and Truant, J. P., editors: Manual of clinical microbiology, ed. 2, Washington, D.C., 1974, American Society for Microbiology.

80. Sonnenwirth, A. C., and Weaver, R. E.: *Yersinia enterocolitica*, N. Engl. J. Med. **283:**1468, 1970.

81. Stuart, C. A., Wheeler, K. M., Rustigian, R., and Zimmerman, A.: Biochemical and antigenic relationships of the paracolon bacteria, J. Bacteriol. **45:**101-119, 1943.

82. Stuart, C. A., Wheeler, K. M., and McGann, V.: Further studies of one anaerogenic paracolon organism, type 29911, J. Bacteriol. **52:**431-438, 1946.

83. Stuart, R. D.: Transport medium for specimens in public health bacteriology, Public Health Rep. **74:**431-438, 1959.

84. Taylor, J., Maltby, M. P., and Payne, J. M.: Factors influencing the response of ligated rabbit-gut segments to injected *Escherichia coli,* J. Pathol. Bacteriol. **76:**491-499, 1958.

85. Waisbren, B. A., and Carr, C.: Penicillin and chloramphenicol in the treatment of infections due to *Proteus* organisms, Am. J. Med. Sci. **223:** 418-421, 1952.

86. Washington, J. A. II: Laboratory procedures in clinical microbiology, Boston, 1974, Little, Brown and Co.

87. Washington, J. A. II: Laboratory approaches to the identification of Enterobacteriaceae, Human Pathol. **7:**151-159, 1976.

88. Washington, J. A. II, Yu, P. K., and Martin, W. J.: Evaluation of accuracy of multitest micromethod system for identification of Enterobacteriaceae, Appl. Microbiol. **22:**267-269, 1971.

89. Washington, J. A. II, Yu, P. K., and Martin, W. J.: Evaluation of the Auxotab Enteric 1 System for identification of Enterobacteriaceae, Appl. Microbiol. **23:**298-300, 1972.

90. Washington, J. A. II, Senjem, D. H., Haldorson, A., Schutt, A. H., and Martin, W. J.: Nosocomially acquired bacteriuria due to *Proteus rettgeri* and *Providencia stuartii,* Am. J. Clin. Pathol. **60:**836-838, 1973.

91. Wassermann, M. M., and Seligmann, E.: *Serratia marcescens* bacteriophages, J. Bacteriol. **66:** 119-120, 1953.

92. Weber, J., Finlayson, N. B., and Mark, J. B. D.: Mesenteric lymphadenitis and terminal ileitis due to *Yersinia pseudotuberculosis,* N. Engl. J. Med. **283:**172-174, 1970.

93. Wilfert, J. N., Barrett, F. F., and Kass, E. H.: Bacteremia due to *Serratia marcescens,* N. Engl. J. Med. **279:**286-289, 1968.

Vibrionaceae

Vibrio · Aeromonas · Plesiomonas

GENUS VIBRIO

This genus has been assigned to the family *Vibrionaceae* in the eighth edition of Bergey's manual. The species of medical importance are *V. cholerae* and *V. parahaemolyticus.* The El Tor vibrio is not considered a separate species but as a biotype of *V. cholerae* (this assignment was adopted by the International Subcommittee on Cholera). Both the El Tor and *V. cholerae* biotypes produce **cholera,** but because of their biochemical and other differences, it is of epidemiologic significance to distinguish between them. The other species, *V. parahaemolyticus,* is perhaps the most important member of this genus because of its pathogenic potential in this hemisphere.

Vibrio cholerae

This species and its biotype El Tor are gram-negative, actively motile rods possessing a single polar flagellum. With careful preparation of stained smears, one may observe the slightly curved rods. Most investigators feel the El Tor biotype is less susceptible to environmental changes than the *V. cholerae* type, and for this reason is more readily recovered from clinical specimens (usually stool) submitted to the laboratory. The noncholera vibrios are similar biochemically to *V. cholerae* but fail to agglutinate in *V. cholerae* antiserum.

As has been pointed out by Balows and co-workers,[1] the most effective and rapid bacteriologic diagnosis of cholera is accomplished only by **proper communication** between the clinician and the laboratory. As advocated by these investigators,

a liquid stool is best collected by rectal catheter, and formed stools should be collected in disinfectant-free containers. Rectal swabs are most effective when inserted beyond the anal spincter. In all cases, specimens should be taken prior to any antimicrobial therapy and inoculated to appropriate media immediately at the laboratory. The Center for Disease Control (CDC) recommends the use of **two plating** media and **one enrichment** broth. Thiosulfate-citrate-bile salt-sucrose (TCBS) agar at pH 8.6 and gelatin agar, the latter a noninhibitory medium, give good results. The selective TCBS medium may be inoculated quite heavily.

On TCBS at 35 C after 18 to 24 hours, *V. cholerae* appears as medium-sized, smooth, **yellow colonies** with opaque centers and transparent periphery. On gelatin medium, the colonies are somewhat flattened and transparent, surrounded by a cloudy halo. Refrigeration tends to accentuate this characteristic, which demonstrates gelatin liquefaction.

The enrichment broth medium is alkaline peptone water, which, by virtue of its high alkalinity, tends to suppress other intestinal bacteria.[1] Culturing in this medium should not be extended beyond 18 to 20 hours because suppressed forms may begin to develop. Agar media may be inoculated from the enrichment broth after 6 to 8 hours of incubation. Suspicious colonies are selected and inoculated to appropriate media or tested with O typing sera by slide test. The young enrichment broth culture also may be examined for the characteristic darting motility by darkfield microscopy. The motility test, if

Table 21-1. Distinguishing characteristics of *Vibrio* species

Test	V. cholerae	V. cholerae El Tor biotype	V. para- haemolyticus
String test after 45 to 60 seconds	+	+	−
Hemagglutination test	−	+	−
Polymyxin B susceptibility test	+*	−	
Phage IV susceptibility test	+*	−	
VP	−	+	−
Hemolysis of sheep RBC	−	+†	−
Sucrose	+	+	−
Salt-free broth	+	+	−
Broth containing 7% to 10% NaCl	−	−	+
Cholera red test	+	+	−
Agglutination in O group serum	+	+	−

*+indicates susceptibility.
†Considerable variability in hemolytic activity has been reported.

positive, can be extended by carrying out the immobilization test with *V. cholerae* pooled O antiserum. If this is positive, it may be used as presumptive evidence of identity.

Several tests will help to distinguish *V. cholerae* from the noncholera vibrios and also differentiate between *V. cholerae* and its biotype El Tor. Among these are the "string test,"[16,21] the hemagglutination test,[8] the polymyxin B susceptibility test,[9] the phage IV susceptibility test,[15] the VP test, and the hemolysis test.[13]

The **string test** consists of testing for the viscid character of a cholera culture in 0.5% sodium desoxycholate on a slide. The "string" is detected by lifting a loop of the mixture from the slide. The **hemagglutination test** is performed by mixing a loopful of washed chicken RBC with a heavy suspension of a pure culture of the organism on a slide. Visible clumping of the RBC is a positive test. The **polymyxin B susceptibility test** determines the inhibitory effect of the antimicrobic on the organism by the appearance of a zone of inhibition around polymyxin B disks on a seeded plate. The **phage IV susceptibility test** aids in differentiating *V. cholerae* from El Tor. The **hemolysis test,** performed with washed sheep RBC, is

claimed by some investigators also to be effective in differentiating these biotypes (Table 21-1).

Vibrio parahaemolyticus

This species is perhaps more important in the United States, since *V. cholerae* and the El Tor biotype are rarely encountered. *V. parahaemolyticus* can cause gastroenteritis or food poisoning, associated with the consumption of contaminated seafood, and more serious infection, such as septicemia with shock, hemolytic anemia, disseminated intravascular coagulation, and other manifestations.[19,25] Infection with this organism is most prevalent in the Orient but has been reported from other countries, and the organism is present in most, if not all, of the coastal marine environment of the United States.[4] The organism is resistant to penicillin, colistin, and polymyxin; variable in sensitivity to ampicillin, erythromycin, and kanamycin; and sensitive to chloramphenicol and gentamicin.[3,22] It is relatively sensitive to tetracycline (maximum MIC 6 μg/ml) and borderline in susceptibility to streptomycin.

The organism is **halophilic** and grows well in peptone water medium containing 7% to 8% NaCl, but it does not normally utilize citrate. For the characteristics of this species see Table 21-1.

Other Vibrios

V. alginolyticus is closely related to *V. parahaemolyticus* (it is considered a biotype of it in Bergey's Manual, ed. 8). It is differentiated from *V. parahaemolyticus* by growth in 10% NaCl, positive VP test, acid from sucrose, and negative methyl red. *V. alginolyticus*, *V. parahaemolyticus*, and other unclassified marine vibrios have been isolated from extraintestinal sources such as blood, spinal fluid, otitis externa, and infected extremity wounds.[7,12,20]

GENUS AEROMONAS

Members of the genus *Aeromonas* are gram-negative, motile rods with a single polar flagellum. They utilize carbohydrates **fermentatively,** with the production of acid or acid and gas, and are oxidase positive. Their normal habitat appears to be natural bodies of water, nonfecal sewage, marine life, and foods; they have also been isolated from fecal specimens of asymptomatic persons. *A. hydrophila* has been suggested as the type species of the genus.*

An increasing incidence of human infections caused by *A. hydrophila* has been reported[6,10,11,14,17,18,23,24] and includes infected traumatic wounds (some with a history of exposure to soil or water), septicemia, meningitis, osteomyelitis, postoperative wound infections (usually in mixed culture), and so forth. A unique presentation is a relatively benign or slow-moving myonecrosis, which may cause significant liquefaction of muscle. Patients with hematologic and other malignancies appear to be particularly susceptible to infection with this organism.

A. hydrophila grows well on routine laboratory media, producing on blood agar small (1 to 3 mm), smooth, convex **beta-**hemolytic colonies that become dark green after 3 to 5 days.[11] Good growth also can occur on enteric isolation media, including EMB and MacConkey agar, SS agar, and others, both at 25 and 35 C. On TSI medium, an alkaline slant over an acid and gas butt is observed; indole is usually produced from tryptophane, which helps to separate the aeromonads from pseudomonads, *Alcaligenes*, flavobacteria, and others. Demonstration of a **positive oxidase reaction** (Kovacs' method) and **polar flagellation** are most useful in distinguishing these organisms from other facultative gram-negative bacilli of clinical importance. The usefulness of these two test procedures in the identification of members of the genus *Aeromonas* cannot be overemphasized, since these organisms may be misidentified as members of the Enterobacteriaceae. They show similarity to *Escherichia coli*, *Enterobacter* sp., and *Providencia*, particularly on conventional biochemical test media used in the early stages of the identification process. At the Mayo Clinic, isolates of *A. hydrophila* were most often confused with *Enterobacter*.[24]

Aeromonas hydrophila is highly susceptible to the aminoglycosides gentamicin and kanamycin and also to chloramphenicol and tetracycline; little activity is demonstrated by penicillin, carbenicillin, ampicillin, or cephalothin.[24]

GENUS PLESIOMONAS

The one species in this genus, *P. shigelloides*, previously was in the genus *Aeromonas*.[5] It is **oxidase positive** and produces acid without gas from carbohydrates. It gives a positive arginine dehydrogenase reaction, is usually lysine decarboxylase positive, and may be ornithine decarboxylase positive. It is lipase negative. Also, in contrast to *A. hydrophila*, it ferments inositol and does not ferment mannitol or sucrose. It does not break down gelatin. Most strains are sensitive to the vibriostatic agent, 2,4-diamino-6,7-diisopropyl pteridine.[2] Like

---

*The reader is referred to an excellent monograph on the genus *Aeromonas* by Ewing and co-workers.[6]

Aeromonas, it may grow on a number of enteric media.

P. shigelloides has been isolated from feces of humans and from blood and spinal fluid cultures. It has been thought to be a cause of acute gastroenteritis.

REFERENCES

1. Balows, A., Hermann, G. J., and DeWitt, W. E.: The isolation and identification of *Vibrio cholerae*—a review, Health Lab. Sci. **8:**167-175, 1971.
2. Buchanan, R. E., and Gibbons, N. E., editors: Bergey's manual of determinative bacteriology, ed. 8, Baltimore, The Williams & Wilkins Co., 1974.
3. Chatterjee, B. D., Neogy, K. N., and Chowdhury, B. R. R.: Drug-sensitivity of *Vibrio parahaemolyticus* isolated in Calcutta during 1969, Bull. W.H.O. **42:**640-641, 1970.
4. Dadisman, T. A., Jr., Nelson, R., Molenda, J. R., and Garber, H. J.: *Vibrio parahaemolyticus* gastroenteritis in Maryland, Am. J. Epidemiol. **96:**414-426, 1973.
5. Ewing, W. H., and Hugh, R.: *Aeromonas*. In Lennette, E. H., Spaulding, E. H., and Truant, J. P., editors: Manual of clinical microbiology, ed. 2, Washington, D. C., 1974, American Society for Microbiology.
6. Ewing, W. H., Hugh, R., and Johnson, J. G.: Studies on the Aeromonas group, Atlanta, 1961, National Communicable Disease Center.
7. Fernandez, C. R., and Pankey, G. A.: Tissue invasion by unnamed marine vibrios, J.A.M.A. **233:**1173-1176, 1975.
8. Finkelstein, R. A., and Mukerjee, S.: Hemagglutination; a rapid method for differentiating *Vibrio cholerae* and El Tor vibrios, Proc. Soc. Exp. Biol. Med. **112:**355-359, 1963.
9. Gangarosa, E. S., Bennett, J. V., and Boring, J. R. III: Differentiation between *Vibrio cholerae* and *Vibrio* biotype El Tor by the polymyxin B disc test: comparative results with TCBS, Monsur's, Mueller-Hinton and nutrient agar media, Bull. W.H.O. **35:**987-990, 1967.
10. Gifford, R. R. M., Lambe, D. W., Jr., McElreath, S. D., and Vogler, W. R.: Septicemia due to *Aeromonas hydrophila* and *Mima polymorpha* in a patient with acute myelogenous leukemia, Am. J. Med. Sci. **263:**157-161, 1972.
11. Gilardi, G. L., Bottone, E., and Birnbaum, M.: Unusual fermentative gram-negative bacilli isolated from clinical specimens. II. Characteristics of *Aeromonas* species, Appl. Microbiol. **20:** 156-159, 1970.
12. Hollis, L. G., Weaver, R. E., Baker, C. N., and Thornsberry, C.: Halophilic *Vibrio* species isolated from blood cultures, J. Clin. Microbiol. **3:**425-431, 1976.
13. Hugh, R.: A comparison of *Vibrio cholerae*, Pacini and *Vibrio* El Tor, Int. Bull. Bact. Nomenclat. Taxon. **15:**61-68, 1965.
14. Ketover, B. P., Young, L. S., and Armstrong, D.: Septicemia due to *Aeromonas hydrophila:* clinical and immunologic aspects, J. Infect. Dis. **127:**284-290, 1973.
15. Mukerjee, S.: The bacteriophage susceptibility test in differentiating *Vibrio cholerae* and *Vibrio* El Tor, Bull. W.H.O. **28:**333-336, 1963.
16. Neogy, K. N., and Mukherji, A. C.: A study of the string test in *Vibrio* identification, Bull. W.H.O. **42:**638-641, 1970.
17. Nygaard, B. S., Bissett, M. L., and Wood, R. M.: Laboratory identification of aeromonads from men and other animals, Appl. Microbiol. **19:** 618-620, 1970.
18. Qadri, S. M. H., Gordon, L. P., Wende, R. D., and Williams, R. P.: Meningitis due to *Aeromonas hydrophila*, J. Clin. Microbiol. **3:**102-104, 1976.
19. Roland, F. P.: Leg gangrene and endotoxin shock due to *Vibrio parahaemolyticus*—an infection acquired in New England coastal waters, N. Engl. J. Med. **282:**1306, 1970.
20. Rubin, S. J., and Tilton, R. C.: Isolation of *Vibrio alginolyticus* from wound infections, J. Clin. Microbiol. **2:**556-558, 1975.
21. Smith, H. L.: A presumptive test for vibrios: the "string" test, Bull. W.H.O. **42:**817-818, 1970.
22. von Graevenitz, A., and Carrington, G. O.: Halophilic vibrios from extraintestinal lesions in man, Infection **1:**54-58, 1973.
23. von Graevenitz, A., and Mensch, A. H.: The genus *Aeromonas* in human bacteriology, N. Engl. J. Med. **278:**245-249, 1968.
24. Washington, J. A. II: *Aeromonas hydrophila* in clinical bacteriological specimens, Ann. Intern. Med. **76:**611-614, 1972.
25. Zide, N., Davis, J., and Ehrenkranz, N. J.: Fulminating *Vibrio parahaemolyticus* septicemia, Arch. Intern. Med. **133:**479-481, 1974.

CHAPTER 22

Nonfermentative, gram-negative bacilli

Pseudomonas · Comamonas · Alcaligenes · Achromobacter · Acinetobacter Moraxella · Kingella · Flavobacterium · Eikenella

Predominantly opportunistic in nature, this group of organisms owes its invasiveness or infectivity to an altered or already debilitated host, who has been compromised by potent medications, varied instrumentation, or the dramatic and prolonged surgical procedures recently developed.

In the past few years major advances have been made in the characterization and taxonomy of these nonfermentative bacteria; excellent descriptions of this group can be found in several sources.[1,9,28,34] No doubt one of the most important tests in this area was introduced by Hugh and Leifson in their fermentation (O-F) medium,[22] enabling the bacteriologist to determine whether an isolate was oxidative, fermentative, or inactive with respect to carbohydrate metabolism. These workers recognized that conventional fermentation media contained a high content of peptone (1%) and, when attacked, gave rise to alkaline amines capable of neutralizing any acidity formed during the fermentation. This is not a problem when large amounts of acid are produced by active "fermenters," but the oxidative bacteria produce **low levels** of acidity, which can be masked due to the accumulation of the alkaline amines. With a carbohydrate medium low in peptone (0.2%), the oxidative, fermentative, or inactive properties of gram-negative bacilli can be determined.

The base medium* with bromthymol blue indicator is sterilized by autoclaving (see Chapter 41 for preparation); after it is cooled, a filter-sterilized solution of the desired carbohydrate is added to give a final concentration of 1%. The recommended carbohydrates include glucose, lactose, sucrose, maltose, mannitol, and xylose. For each isolate, **two** tubes of glucose O-F medium are inoculated with a light stab from a young culture, and one of the tubes is overlaid with at least ¼ inch of sterile, stiff petrolatum or vaspar. Sterile mineral oil is not recommended for this purpose. They are then incubated at 35 C for several days and examined daily. An acid reaction is indicated by a color change from the uninoculated blue-green color to a **yellow** color. Reactions of characteristic groups of organisms are shown on opposite page.

Generally, these nonfermentative bacteria are gram-negative, nonsporulating, obligately aerobic bacilli that produce no change on TSI agar (occasionally an alkaline slant) and are indole (except the flavobacteria) and ornithine decarboxylase negative. For purposes of simplicity,

---

*Available from Baltimore Biological Laboratory, Cockeysville, Md.; Difco Laboratories, Detroit. Test systems similar to those used in the identification of the Enterobacteriaceae are now available for identification of nonfermentative gram-negative bacilli (API, Plainview, N.Y.; Corning Medical, Medfield, Mass.; and Roche Diagnostics, Nutley, N.J.).

Organism	Glucose Open	Covered	Group
Alcaligenes faecalis	—	—	I* Nonoxidizers Nonfermenters
Pseudomonas aeruginosa	A	—	II Oxidizers Nonfermenters
Shigella dysenteriae	A	A	IIIa Fermenters (Anaerogenic)†
Salmonella enteriditis	A	AG	IIIb Fermenter (Aerogenic)†

these bacteria may be classified into several major groups:

1. Oxidase-positive motile rods: *Pseudomonas, Alcaligenes*
2. Oxidase-negative nonmotile diplobacilli: *Acinetobacter*
3. Oxidase-positive nonmotile diplobacilli and bacilli: *Moraxella, Flavobacterium*

Oberhofer and associates[31] have recently described a medium for rapid detection of lysine and ornithine decarboxylase and arginine dihydrolase, which gave results in 4 to 24 hours and generally agreed with results on the Moeller medium, which often requires 3 to 7 days' incubation.

Pickett provides important perspective regarding the incidence of nonfermentative gram-negative rods in clinical laboratory practice.[34] Among 1,032 strains of aerobic or facultatively anaerobic gram-negative bacilli recovered from clinical specimens in the UCLA Clinical Laboratories in June 1976, 705 (68%) were enteric bacilli and 169 (16%) were nonfermenters. Among 486 strains of nonfermentative bacilli recovered from clinical specimens in a 7-month period in 1968, 322 (66%) were *Pseudomonas aeruginosa*. Other *Pseudomonas* species and

Pseudomonas-like organisms accounted for 77 strains (16%). *Acinetobacter* strains totaled 45 (9%); *Flavobacterium* strains, 22 (4%); *Moraxella* strains 11, (2%); and *Alcaligenes* strains, 6 (1%). There were two isolates of *Bordetella bronchiseptica* (see Chapter 23).

Only those organisms of proved clinical significance will be discussed in detail; the interested reader is referred to authors already cited for further information.

PSEUDOMONAS SPECIES
Pseudomonas aeruginosa

Pseudomonas aeruginosa is the most frequently implicated member of the genus in human infections; it may infect burn sites, wounds, the urinary tract, and the lower respiratory tract, particularly in patients whose defenses have been compromised. Infection also may result in a serious septicemia. Although *P. aeruginosa* may be isolated from the skin and feces of normal humans, most of the infections are exogenous in origin.[41,43] Since the organism is part of the hospital environment—it can survive and even multiply in moist environments with minimal amounts of organic matter—it has been incriminated in from 5% to 15% of all hospital-acquired infections.[8]

P. aeruginosa is a polar **monotrichous,** gram-negative rod occurring singly, in pairs, or in short chains. On blood agar, the organism grows as a large, flat colony with a ground-glass appearance, and produces a zone of hemolysis. In patients with cystic fibrosis, **mucoid** strains are frequently isolated from the sputum. The colonies tend to spread and give off a characteristic **grapelike odor.** Most strains excrete pyocyanin and fluorescein (pyoverdin), giving the colony a characteristic **blue-green** color; approximately 4% are apyocyanogenic.

P. aeruginosa is **oxidase positive** by Kovacs' method[27] (described in Chapter 43) and utilizes glucose **oxidatively** in O-F medium; gluconate is oxidized to

---

*Adapted from Hugh and Leifson.[22]

†The criterion of gas production applies to vaspar-sealed tubes only.

ketogluconate* (but not by other pseudomonads). *P. aeruginosa* is lysine and ornithine decarboxylase negative and arginine dihydrolase positive. Most strains grow at 42 C on trypticase agar slants; they also grow on a selective agar medium containing cetyltrimethylamine bromide (Cetrimide).† Most other members of the genus are generally inhibited on the latter medium.

Because *P. aeruginosa* grows on EMB or MacConkey agar as a nonlactose-fermenting organism, it is frequently selected and transferred to TSI (or Kligler's) agar as a suspicious colony from stool cultures and may be incorrectly identified because of the alkaline slant and butt reaction, which is characteristic for this organism. One other pseudomonad, *P. putrefaciens*, also can be misidentified as an enteric organism because it produces an appreciable amount of H_2S in these media.[46]

An identification system that uses a standardized technique of bacteriophage typing has been recommended for tracing *Pseudomonas* strains during an epidemiologic investigation.[42] In addition, a system of typing by pyocin production[17] and a serologic typing system[21,45] have been used as markers to trace the sources of infections due to this organism.

P. aeruginosa is resistant to kanamycin but susceptible to the aminoglycosides gentamicin, tobramycin, and amikacin, the drugs of choice in the treatment of serious *Pseudomonas* infections. Unfortunately, in several centers significant resistance to the first two agents listed has been noted. Many strains are also sensitive to carbenicillin and ticarcillin, semisynthetic penicillins that are recommended for therapy primarily as an adjunct to aminoglycosides in serious infections. These organisms are also inhibited by

*Gluconate tablets, from Key Scientific Products Co., Los Angeles.
†Available as Pseudosel agar, Baltimore Biological Laboratory, Cockeysville, Md.; Cetrimide agar; Difco Laboratories, Detroit.

the polymyxin antibiotics, polymyxin B and colistin, which are much less useful because of poor distribution in the body and inactivation by pus and other organic matter.

Pseudomonas maltophilia

This pseudomonad is ubiquitous in nature. Like *P. aeruginosa*, it is being increasingly isolated from blood, cerebrospinal and other body fluids, sputum, urine, and abscesses.

P. maltophilia is multitrichous, with tufts of two or more flagella per pole. It produces a yellow to tan pigment on trypticase soy agar and an ammoniacal odor. On glucose O-F medium, it produces an early (overnight) alkaline reaction, becoming weakly acid on further incubation; on **maltose** O-F medium, acidity is promptly produced oxidatively.

P. maltophilia is **oxidase negative,** ONPG and DNase positive; it does not reduce nitrate to nitrogen gas. It is lysine decarboxylase positive and arginine dihydrolase negative. Most strains are susceptible to polymyxin and colistin, chloramphenicol, and cotrimoxazole[30]; varying results occur with other antimicrobial agents.

Other opportunistic pseudomonads

P. fluorescens and *P. putida* are occasionally isolated from humans; sources have included respiratory tract and urinary tract infections, wounds, and contaminated blood-bank blood. They fail to grow at 42 C (25 C optimal). *P. fluorescens* will grow at refrigerator temperatures. *P. fluorescens* can be differentiated from *P. putida* by the former's ability to liquefy gelatin and produce lecithinase. Both produce fluorescein, but not pyocyanin or pyorubrin; they are generally resistant to carbenicillin and sensitive to kanamycin,[33] gentamicin, and polymyxin and often to tetracycline.[29]

P. stutzeri is a polar, monotrichous, nonfluorescent pseudomonad that produces a yellow, wrinkled colony resembling that of *P. pseudomallei*. Oxidative in glucose O-F medium, it reduces nitrate to nitrogen gas and grows in 6.5% NaCl broth, which aids in differentiating it from other pseudomonads. It is lysine and ornithine decarboxylase negative and arginine dihydrolase negative. It is usually susceptible to aminoglycosides and often to carbenicillin, ampicillin, and tetracycline.

P. cepacia (synonyms: *P. multivorans*, *P. kingii*, EO-1) has occasionally been recovered from clinical material, including blood cultures.[13] It also has been isolated from contaminated detergent solutions in urinary catheter kits,[18] as well as from hospital water supplies, and has been implicated in several outbreaks of infections in hospitals. Included in its characteristics are a variable oxidase reaction, lack of growth on Salmonella-Shigella agar, presence of lysine decarboxylase, and resistance to polymyxin and colistin.[13] Chloramphenicol is the drug most consistently active against it, with two thirds of strains susceptible to kanamycin. All strains were shown recently to be susceptible to trimethoprim alone and to cotrimoxazole.[30] Many strains produce a nonfluorescent yellow pigment (particularly on TSI) that diffuses into the agar medium.

P. alcaligenes and *P. putrefaciens* (the latter produces H_2S on TSI or Kligler's agar) are examples of rarely isolated but potentially pathogenic pseudomonads. A description of these and related pseudomonads may be found in references 15, 28, 37, and 39.

Pseudomonas pickettii, a recently described species (Riley, P. S., and Weaver, R. E.: J. Clin. Microbiol. **1**:61-64, 1975), which probably includes strains previously identified as VA-2, has been isolated from such clinical specimens as blood, urine, wounds and abscesses, and spinal fluid. It is oxidase and catalase positive and urea positive, produces acid oxidatively from glucose but not mannitol, produces gas during nitrate reduc-

tion, and is arginine dihydrolase negative. It does not grow on SS or cetrimide agars. It is motile by means of one to two polar flagella.

CDC groups VE-1 and VE-2[16] resemble the pseudomonads but have not yet been classified. They have occasionally been isolated from clinical material, particularly wounds and abscesses. They are oxidase negative, catalase positive, and produce yellow pigment. The ends of the cells are slightly tapered. Colonies may be wrinkled or semirough on blood agar, and the medium is discolored green or lavender-green. The organisms are motile by means of either one polar flagellum or a tuft of more than three polar flagella. The slant of TSI (Kligler's) is alkalinized, and no H_2S is produced. The organisms are oxidative. The VE group has a unique pattern of antimicrobial agent susceptibility; they are sensitive to aminoglycosides (including kanamycin), carbenicillin, ampicillin, polymyxin, tetracycline, chloramphenicol, and erythromycin.

Pseudomonas pseudomallei

Pseudomonas pseudomallei is the causative agent of **melioidosis** in humans. This disease presents a varying clinical picture, ranging from unsuspected asymptomatic infection to acute, severe pneumonia or overwhelming and highly fatal septicemia. The organism has been isolated frequently from moist soil, market fruits and vegetables, and from well and surface waters in Southeast Asia and other tropical areas. Although apparently rare in natives, the disease was an important and sometimes fatal infection in the U.S. Armed Forces in Vietnam. A closely related organism was recovered from an infection in the United States (McCormick, J. B., et al.: J. Infect. Dis. **135**:103-107, 1977). *P. pseudomallei* is multitrichous, with a tuft of three or more flagella per pole, and morphology similar to *P. aeruginosa*, but the colonies frequently are **wrinkled** and on prolonged incubation become umbonate in character, particularly on blood agar. An oxidative acidity is produced in glucose O-F medium; the oxidase reaction is positive, and growth occurs at 42 C, but pyocyanin or fluorescent pigment are not produced. The organism is arginine dihydrolase positive and lysine decarboxylase negative.

These pseudomonads may be cultivated on most laboratory media, growing well on trypticase soy agar, on blood agar, and on MacConkey agar, but not on Salmonella-Shigella (SS) agar or cetrimide agar. A selective medium for isolating *P. pseudomallei* from contaminated clinical material has been described.[10]

P. pseudomallei is generally resistant to most antimicrobials, with chloramphenicol and tetracycline being the drugs of choice and kanamycin showing good activity. Cotrimoxazole also offers promise.[25]

Thomason and associates[44] have reported that identification of *P. pseudomallei* in mixed culture on a slide may be made by the fluorescent-antibody technique.

Characteristics useful in identification of some *Pseudomonas* species are noted in Table 22-1.

Pseudomonas mallei

Formerly classified as *Actinobacillus mallei*, this organism is the causative agent of **glanders,** an infectious disease of horses that has occasionally been transmitted to humans by direct contact through trauma or inhalation. *P. mallei* is a **nonmotile,** coccoid- to rod-shaped organism, sometimes occurring in filaments or, under special conditions, with branching involution forms. Colonies on infusion agar (especially when glycerol is added) appear in 48 hours as grayish white and translucent, later becoming yellowish and opaque. *P. mallei* oxidizes glucose, fails to grow at 42 C, and may be weakly oxidase positive. It is lysine decarboxylase negative. Male guinea pigs injected intraperitoneally with culture

Table 22-1. Some characteristics of species of *Pseudomonas**

Organism	Pyocyanin CHCl₃ soluble)	Oxidase (Kovacs')	Growth at 42 C	NO₃ reduction	Growth on SS	Maltose oxidation	Lysine decarboxylase	Ornithine decarboxylase	Arginine dihydrolase
P. aeruginosa	+ (−)	+	+	+†	+	−	−	−	+
P. fluorescens	−	+	−	−†	+	var	−	−	+
P. putida	−	+	−	−	+	var	−	−	+
P. cepacia	var	var	−	+/−	− (+)	+	+	− (+)	−
P. stutzeri	−	+	+	+†	− (+)	+	−	−	−
P. maltophilia	−	−	+	− (+)	−	+	− (+)	− (+)	−
P. pseudomallei	−	+	+	+†	−	+	−	−	+

*+, positive reaction; −, negative reaction; ( ), occasional strain; var, variable.
†Some strains reduce NO₃ to gas.

material containing *P. mallei* develop a tender and swollen scrotum 2 to 4 days later (Straus test).

COMAMONAS

Comamonas terrigena[1,9] is the only species in the genus; it includes organisms previously known as *P. acidovorans* and *P. testosteroni. C. terrigena* is not commonly encountered in the clinical laboratory but has been recovered from blood culture, from spinal and pleural fluids, and from abscesses, urine, and the respiratory tract, as well as feces. It is oxidase positive and produces an alkaline reaction in O-F glucose medium. It is motile, with a tuft of one to six flagella at one pole (usually three or more). It does not produce H_2S on Kligler's (or TSI) agar.

ALCALIGENES AND ACHROMOBACTER

In the eighth edition of Bergey's Manual, the genus *Alcaligenes* is listed under genera of uncertain affiliation and *Achromobacter* is not listed as a genus. These groups have long been poorly defined and, accordingly, are difficult to identify.

The CDC group[28] feels that *Alcaligenes* should be defined as follows: gram-negative rods, motile with **peritrichous flagella, oxidase positive, nonsaccharolytic,** and **urease negative.** Growth occurs on MacConkey agar but is inhibited on SS agar. The three most important species of *Alcaligenes* are *A. faecalis, A. odorans,* and *A. denitrificans. A. odorans* produces a pronounced dark green color on blood agar, whereas the other species vary from indeterminate lysis to greenish brown. *A. odorans* produces a **sweet odor** similar to that of peeled apples. The most frequent clinical sources of these organisms are the ear, urine, blood, spinal fluid, pleural fluid, wounds and abscesses, and feces. Susceptibility patterns vary considerably. With *A. faecalis*, the most active drugs are kanamycin, tetracycline, and polymyxin. With *A. odorans*, polymyxin, carbenicillin, and gentamicin are most active. Tetracycline and carbenicillin are most active against *A. denitrificans.*

Achromobacter is also **peritrichously flagellated** and **oxidase positive** but attacks **carbohydrates oxidatively.** This genus is strictly **aerobic** and **fails to produce 3-ketolactonate.** There are two biotypes each of *A. xylosoxidans* and of *Achromobacter* species. All grow on MacConkey and SS agars, and *A. xylosoxidans* usually grows on cetrimide agar. *Achromobacter* species is urease positive (Christensen's agar); *A. xylosoxidans* is not.

Achromobacter is most often isolated from blood, spinal fluid, urine, the respiratory tract, wounds, and feces. Only polymyxin and carbenicillin show significant activity against this group, with chloramphenicol active against two thirds of strains.

ACINETOBACTER SPECIES

There is a single species, *Acinetobacter calcoaceticus*. Some prefer to recognize two varieties, *A. calcoaceticus* var. *anitratus* (formerly *Herellea vaginicola* or *B. anitratus*) and var. *lwoffii* (*Mima polymorpha* of the oxidase-negative type). *A. calcoaceticus* var. *anitratus* is the third most frequently isolated nonfermenter, after *P. aeruginosa* and *P. maltophilia.*[1]

These organisms are **oxidase negative, catalase positive, nonmotile** diplococcoid to coccobacillary forms that **do not reduce nitrate** (some may show activity). They grow well on MacConkey agar but poorly or not at all on SS agar. *A. calcoaceticus* var. *anitratus* produces acidity in the open tube of glucose, 10% lactose, and other carbohydrate O-F media. *A. calcoaceticus* var. *lwoffii* is negative in glucose and lactose. Both species give strongly positive tests for catalase but fail to demonstrate decarboxylase, dihydrolase, or deaminase; neither grows on cetrimide agar. A characteristic reaction on Sellers medium (blue slant, yellow band, green butt) is produced by *A. calcoaceticus* var.

anitratus, which readily differentiates it from other nonfermentative bacilli.

A. calcoaceticus var. *anitratus* has been recovered from a wide variety of clinical sources, including the central nervous system, the upper and lower respiratory tracts, urinary tract, wounds, nosocomial infections, and bacteremia secondary to intravenous catheterization. Its role has been primarily one of an opportunistic pathogen, generally occurring in mixed cultures from low-grade infections, although occasional cases of septicemia or pneumonia have been reported in debilitated hospital patients.[11]

A. calcoaceticus var. *lwoffii's* role in human infections is more difficult to assess; most isolates are of questionable clinical significance. It is part of the normal flora of the skin, external genitalia, and other sites, and may be recovered as a commensal from these areas.[33] It is probably best considered as an occasionally opportunistic agent, particularly in superficial infections.

Both organisms are susceptible in vitro to the aminoglycosides kanamycin and gentamicin, as well as to carbenicillin, cotrimoxazole, colistin, and polymyxin B[14,33,47]; *A. calcoaceticus* var. *lwoffii* appears to be susceptible to a wider variety of antimicrobial agents. Both are resistant to penicillin.

MORAXELLA SPECIES

Moraxellae are **nonmotile** bacilli or coccobacilli that are **oxidase positive** and **penicillin sensitive;** most strains are **biochemically inactive** with respect to carbohydrate oxidation, denitrification and the production of deaminase, decarboxylase, and dihydrolase.[12] Some strains are fastidious, requiring enriched media or increased humidity for growth. Several species are generally recognized by medical microbiologists and include the following:

1. *M. lacunata* (incorporates *M. liquefaciens*)
2. *M. osloensis*

3. *M. nonliquefaciens*
4. *M. phenylpyruvica*

M. lacunata (Morax-Axenfeld bacillus) was originally described as the etiologic agent in chronic conjunctivitis. It is infrequently isolated; characteristically, it causes **pitting,** or lacunae, on the surface of Loeffler slants, with subsequent digestion of the medium. Serum is required for its growth, preferably under increased CO_2.

M. osloensis[2] was previously classified as *Mima polymorpha* var. *oxidans* (De-Bord); the latter was considered an illegitimate epithet.[35] Although this organism's involvement in human infections is comparatively rare, *M. osloensis* is easily confused with *N. gonorrhoeae* from genitourinary sources, since it is also **oxidase positive** and appears as gram-negative coccobacillary forms resembling gonococci. *M. osloensis* **does not ferment** CTA glucose medium and grows on nutrient agar, whereas *N. gonorrhoeae* ferments CTA glucose and does not grow on nutrient agar. *M. osloensis* may be isolated from blood and spinal fluid, among other sources, and therefore may also be confused with the meningococcus; differentiation should not be difficult.

M. nonliquefaciens resembles *M. osloensis* but is somewhat more fastidious. The former reduces nitrate to nitrite, but only about one fourth of *M. osloensis* strains do this. *M. nonliquefaciens* occurs more frequently in the respiratory tract than any other *Moraxella* species.

M. phenylpyruvica is characterized chiefly by hydrolysis of urea and deamination of phenylalanine or tryptophan. Major clinical sources of this species have been urine, blood, spinal fluid, and the urethra and vagina.

M. atlantae, recently proposed as a new species,[3] resembles *M. phenylpyruvica* but lacks urease and phenylalanine and tryptophan deaminase activities. Four of the five known strains were isolated from the blood.

Still another proposed new species is

M. urethralis.[38] It resembles both *M. osloensis* and *M. phenylpyruvica*, but it reduces nitrites, possesses phenylalanine deaminase, utilizes citrate, and is urease negative. Whether this organism is of pathologic significance remains to be determined. It has been recovered primarily from urine and from female genital samples.

Moraxella species are uniformly susceptible to penicillin, ampicillin, tetracycline, chloramphenicol, aminoglycosides, and erythromycin in vitro.[14]

KINGELLA

A new genus was proposed recently[20] to accommodate the organism originally known as *Moraxella kingii*[19] and subsequently *M. kingae;* this organism would thus become *K. kingae*. The new genus was created primarily because the organism in question differed strikingly from other moraxellae in several respects: production of acid from certain sugars (glucose and maltose especially) and lack of catalase activity, in particular. The organism is oxidase positive. Growth on agar spreads and corrodes the agar and produces beta hemolysis; there is a tendency to pellicle formation in fluid media. The cells show polar fimbriation and twitching motility. Poor growth is obtained on nonenriched media. The organism has been isolated from the blood, nose, throat, joints, and bone lesions. It is very sensitive to penicillin, streptomycin, chloramphenicol, oxytetracycline, and erythromycin.

Two additional species recently have been proposed for this genus: *K. indologenes* (isolated from eye infections) and *K. denitrificans* (isolated from the pharynx).[40] It is not clear that they are pathogenic.

FLAVOBACTERIUM

Members of the genus *Flavobacterium* are gram-negative, nonmotile, proteolytic bacilli that can weakly ferment carbohydrates. Presently two species are recognized, *Flavobacterium meningosepticum*, which was first characterized by King,[26] and another unnamed *Flavobacterium* called Group IIb. Biochemical tests and other characteristics readily permit differentiation of these two species.[28] They grow on MacConkey agar (usually) or blood agar, usually producing a **lavender-green** color on the latter due to extensive proteolytic enzymatic activity. Although these organisms are fermentative, they are often inactive during the first 24 to 48 hours of incubation. If they are inoculated into sealed glucose O-F medium, acid production is slight and often delayed. If inoculated in a liquid peptone medium with carbohydrates, reactions may not be recognized for 14 to 21 days. For this reason the flavobacteria are best treated like oxidizers of carbohydrates, using unsealed O-F media for carbohydrate reactions.

Pigment production by *F. meningosepticum* can be either **beige** or a **light yellow**.[28] Group IIb, however, is markedly yellow to orange. TSI (or Kligler's) reaction is alkaline over alkaline or alkaline over neutral for both organisms. They are oxidase positive and catalase positive, and both form indole. In gelatin, strains of *F. meningosepticum* are mostly positive; most IIb strains are also positive, but the reaction is delayed. On O-F media, the reactions generally are as follows for *F. meningosepticum:* glucose, mannitol, and maltose positive and xylose, lactose, and sucrose negative. Except for glucose, the O-F reactions are all negative for the IIb organism. On Gram stain, the flavobacteria are long, thin, gram-negative bacilli that usually appear with slightly swollen ends.

Flavobacterium is widely distributed in nature, particularly in water, soil, and any moist area.[35,37] It does not readily colonize in adult patients with intact host defenses. In hospitals the organisms have been isolated from water fountains, faucets, sinks, water baths, air conditioners, humidifiers, ice machines, and so forth.[6,32,36] The organism grows best in a cool environment and does not remain viable at 35 C when stored for 5 to 7 days. Organisms isolated from patients during an outbreak of postoperative fever did not survive at 38 C.[32] However, *Flavobacterium* will survive at 0 C.[35,37] Group IIb flavobacteria are only rarely pathogenic in infant or adult patients.[6,28] *Flavobacterium meningosepticum*, however, may cause a variety of nosocomial infections. Most important is the organism's ability to produce outbreaks of **neonatal meningitis.**[5,26,28] In these outbreaks, many infants had positive nasal and throat cultures for this organism but remained healthy. In those who developed meningitis, mortality was in excess of 50%. Epidemiologic investigation failed to delineate a definite source or mode of transmission in the outbreaks. Only infrequently is *F. meningosepticum* pathogenic in adult patients. Sporadic cases of bacteremia, subacute bacterial endocarditis, pneumonia, and meningitis have been described.[7,26] The mode of transmission in such sporadic infections is not well understood. There are six serotypes of *F. meningosepticum*, A through F, and although serotyping may be of assistance during epidemiologic investigations, it is not widely employed or available. No typing system exists for Group IIb. *Flavobacterium* species are frequently resistant to many antimicrobial agents.[6] Among the more active compounds are novobiocin, chloramphenicol, and erythromycin.

EIKENELLA

Eikenella corrodens, originally known as *Bacteroides corrodens*[23] and clearly distinct from that obligately anaerobic organism, is a well-documented pathogen that is not at all uncommon.[4,24] CDC has characterized close to 600 strains referred there in recent years. It is usually recovered in mixed culture, primarily along with aerobic gram-positive cocci. Many or most *Eikenella* infections originate from the oral cavity or the bowel.

Sources of cultures positive for *E. corrodens*, then, include abscesses of the face and neck, sputum, pleural fluid, and abdominal wounds, as well as blood, spinal fluid, brain abscess, bone, and so forth. Infections with *Eikenella* are seen with high frequency in association with methylphenidate (Ritalin) abuse.

E. corrodens is a straight gram-negative rod that is **nonmotile** and **nonsaccharolytic, oxidase positive,** usually catalase negative, lysine and ornithine decarboxylase positive, arginine dihydrolase negative, and reduces nitrate to nitrite only. Colonies are small to tiny and frequently are situated in shallow **craters in the agar.** A pale **yellow pigment** is produced. Optimal growth is obtained under increased CO_2 tension. An **odor** similar to hypochlorite bleach is produced. Growth in broth is usually granular, with the granules adhering to the side of the tube.

Disk susceptibility tests are not reliable, but the **resistance** of *E. corrodens* to clindamycin is so striking that the organism grows right to the disk margin (this may facilitate recovery of the organism from mixed culture). *Eikenella* organisms are susceptible to penicillin, ampicillin, carbenicillin, and tetracycline and usually to chloramphenicol and colistin. They are resistant to methicillin, relatively resistant to aminoglycosides, and of variable susceptibility to cephalothin.

REFERENCES

1. Blazevic, D. J.: Current taxonomy and identification of non-fermentative gram negative bacilli, Human Pathol. **7:**265-275, 1976.
2. Bøvre, K., and Henriksen, S. D.: A new *Moraxella* species, *Moraxella osloensis*, and a revised description of *Moraxella nonliquefaciens*, Intern. J. Syst. Bact. **17:**127-135, 1967.
3. Bøvre, K., Fuglesang, J. E., Hagen, N., Jantzen, E., and Froholm, L. O.: *Moraxella atlantae* sp. nov. and its distinction from *Moraxella phenylpyrouvica*, Intern. J. Syst. Bacteriol. **26:**511-521, 1976.
4. Brooks, G. F., and White, A.: *Eikenella corrodens* comes of age, South. Med. J. **69:**533-534, 1976.
5. Cabrera, H. A., and Davis, G. H.: Epidemic meningitis of the newborn caused by Flavo-

bacteria. I. Epidemiology and bacteriology, Am. J. Dis. Child. **101**:289-295, 1961.

6. Center for Disease Control: National nosocomial infections study quarterly report, third and fourth quarters, 1973, Atlanta, March 1975.
7. DuPont, H. L., and Spink, W. W.: Infections due to gram negative organisms: an analysis of 860 patients with bacteremia at the University of Minnesota Medical Center, Medicine **48**:307-332, 1969.
8. Eickhoff, T. C.: Hospital infections, disease-a-month, Chicago, September 1972, Year Book Medical Publishers, Inc.
9. Elliott, T. B., Gilardi, G., Hugh, R., and Weaver, R. E.: ASM Workshop on identification of glucose non-fermenting gram negative rods, Washington, D.C., 1975, American Society for Microbiology.
10. Farkas-Himsley, H.: Selection and rapid identification of *Pseudomonas pseudomallei* from other gram-negative bacteria, Am. J. Clin. Pathol. **49**:850-856, 1968.
11. Gardner, P., Griffin, W. B., Swartz, M. N., and Kunz, L. J.: Nonfermentative gram-negative bacilli of nosocomial interest, Am. J. Med. **48**: 735-749, 1970.
12. Gilardi, G. L.: Diagnostic criteria for differentiation of pseudomonads pathogenic for man, Appl. Microbiol. **16**:1497-1502, 1968.
13. Gilardi, G. L.: Characterization of EO-1 strains *(Pseudomonas kingii)* isolated from clinical specimens and the environment, Appl. Microbiol. **20**:521-522, 1970.
14. Gilardi, G. L.: Antimicrobial susceptibility as a diagnostic aid in the identification of nonfermenting gram-negative bacteria, Appl. Microbiol. **22**:821-823, 1971.
15. Gilardi, G. L.: Characterization of nonfermentative nonfastidious gram-negative bacteria encountered in medical bacteriology, J. Appl. Bacteriol. **34**:623-644, 1971.
16. Gilardi, G. L., Hirschl, S., and Mandel, M.: Characteristics of yellow-pigmented nonfermentative bacilli (groups VE-1 and VE-2) encountered in clinical bacteriology, J. Clin. Microbiol. **1**:384-389, 1975.
17. Gillies, R. R., and Govan, J. R. W.: Typing of *Pseudomonas pyocyanea* by pyocine production, J. Pathol. Bacteriol. **91**:339-345, 1966.
18. Hardy, P. C., Ederer, G. M., and Matsen, J. M.: Contamination of commercially packaged urinary catheter kits with the pseudomonad EO-1, N. Engl. J. Med. **282**:33-35, 1970.
19. Henriksen, S. D., and Bøvre, K.: *Moraxella kingii* sp. nov., a hemolytic, saccharolytic species of the genus *Moraxella*, J. Gen. Microbiol. **51**:377-385, 1968.
20. Henriksen, S. D., and Bøvre, K.: Transfer of *Moraxella kingae* Henriksen and Bøvre to the genus *Kingella* gen. nov. in the family *Neisser-*

iaceae, Intern. J. Syst. Bacteriol. **26**:447-450, 1976.
21. Homma, J. Y., Kim, K. S., Yamada, H., Ito, M., Shionoya, H., and Kawabe, Y.: Serological typing of *Pseudomonas aeruginosa* and its cross-infection, Jap. J. Exp. Med. **40**:347-359, 1970.
22. Hugh, R., and Leifson, E.: The taxonomic significance of fermentative versus oxidative metabolism of carbohydrates by various gram-negative bacteria, J. Bacteriol. **66**:24-26, 1953.
23. Jackson, F. L., and Goodman, Y. E.: Transfer of the facultatively anaerobic organism *Bacteroides corrodens* Eiken to a new genus, *Eikenella*, Intern. J. Syst. Bacteriol. **22**:73-77, 1972.
24. Jackson, F. L., Goodman, Y. E., Bel, F. R., Wong, P. C., and Whitehouse, R. L. S.: Taxonomic status of facultative and strictly anaerobic "corroding bacilli" that have been classified as *Bacteroides corrodens*, J. Med. Microbiol. **4**:171-184, 1971.
25. John, J. F., Jr.: Trimethoprim-sulfamethoxazole therapy of pulmonary melioidosis, Am. Rev. Respir. Dis. **114**:1021-1025, 1976.
26. King, E. O.: Studies on a group of previously unclassified bacteria associated with meningitis in infants, Am. J. Clin. Pathol. **31**:241-247, 1959.
27. Kovacs, N.: Identification of *Pseudomonas pyocyanea* by the oxidase reaction, Nature **178**:703, 1956.
28. Lennette, E. H., Spaulding, E. H., and Truant, J. P., editors: Manual of clinical microbiology, ed. 2, Washington, D.C., 1974, American Society for Microbiology, Chapter 24.
29. Martin, W. J., Maker, M. D., and Washington, J. A. II: Bacteriology and *in vitro* antimicrobiol susceptibility of the *Pseudomonas fluorescens* group isolated from clinical specimens, Am. J. Clin. Pathol. **60**:831-835, 1973.
30. Moody, M. R., and Young, V. M.: In vitro susceptibility of *Pseudomonas cepacia* and *Pseudomonas maltophilia* to trimethoprim and trimethoprim-sulfamethoxazole, Antimicrob. Ag. Chemother. **7**:836-839, 1975.
31. Oberhofer, T. R., Rowen, J. W., Higbee, J. W., and Johns, R. W.: Evaluation of the rapid decarboxylase test for the differentiation of nonfermentative bacteria, J. Clin. Microbiol. **3**:137-142, 1976.
32. Olsen, H.: *Flavobacterium meningosepticum* isolated from outside hospital surroundings and during routine examination of patient specimens, Acta Pathol. Microbiol. Scand. **75**:313-322, 1969.
33. Pedersen, M. M., Marso, M. A., and Pickett, M. J.: Nonfermentative bacilli associated with man. III. Pathogenicity and antibiotic susceptibility, Am. J. Clin. Pathol. **54**:178-192, 1970.
34. Pickett, M. J.: Nonfermentative gram-negative bacilli, Current Concepts in Clinical Micro-

biology Course, UCLA Extension Division, August 1976, 10 pp.

35. Pickett, M. J., and Manclark, C. R.: Nonfermentative bacilli associated with man. I. Nomenclature, Am. J. Clin. Pathol. **54:**155-163, 1970.

36. Pickett, M. J., and Pedersen, M. M.: Characterization of saccharolytic non-fermentative bacteria associated with man, Can. J. Microbiol. **16:**351-362, 1970.

37. Pickett, M. J., and Pedersen, M. M.: Nonfermentative bacilli associated with man. II. Detection and identification, Am. J. Clin. Pathol. **54:**164-177, 1970.

38. Riley, P. S., Hollis, D. G., and Weaver, R. E.: Characterization and differentiation of 59 strains of *Moraxella urethralis* from clinical specimens, Appl. Microbiol. **28:**355-358, 1974.

39. Riley, P. S. Tatum, H. W., and Weaver, R. E.: *Pseudomonas putrefaciens* isolates from clinical specimens, Appl. Microbiol. **24:**798-800, 1972.

40. Snell, J. J. S., and Lapage, S. P.: Transfer of some saccharolytic *Moraxella* species to *Kingella* Henriksen and Bøvre 1976, with descriptions of *Kingella indologenes* sp. nov. and *Kingella denitrificans* sp. nov, Int. J. Syst. Bacteriol. **26:**451-458, 1976.

41. Sutter, V. L., and Hurst, V.: Sources of *Pseudomonas aeruginosa* infection in burns: study of wound and rectal cultures with phage typing, Ann. Surg. **163:**596-603, 1966.

42. Sutter, V. L., Hurst, V., and Fennell, J.: A standardized system for phage typing *Pseudomonas aeruginosa*, Health Lab. Sci. **2:**7-16, 1965.

43. Sutter, V. L., Hurst, V., Grossman, M., and Calonje, R.: Source and significance of *Pseudomonas aeruginosa* in sputum, J.A.M.A. **197:**854-858, 1966.

44. Thomason, B. M., Moody, M. D., and Goldman, M.: Staining bacterial smears with antibody. II. Rapid detection of varying numbers of *Malleomyces pseudomallei* in contaminated materials and infected animals, J. Bacteriol. **72:**362-367, 1956.

45. Verder, E., and Evans, J.: A proposed antigenic schema for the identification of strains of *Pseudomonas aeruginosa*, J. Infect. Dis. **109:**183-193, 1961.

46. von Graevenitz, A., and Simon, G.: Potentially pathogenic, nonfermentative, H_2S-producing gram-negative rod (1 b), Appl. Microbiol. **19:**176, 1970.

47. Washington, J. A. II: Antimicrobial susceptibility of enterobacteriaceae and nonfermenting gram-negative bacilli, Mayo Clin. Proc. **44:**811-824, 1969.

CHAPTER 23

Gram-negative coccobacillary facultative bacteria

Pasteurella · *Francisella* · *Bordetella* · *Brucella* · *Haemophilus*

The bacteria that comprise the genera of this chapter are, for the most part, small gram-negative rods, occurring singly, in pairs, in short chains, and in other arrangements. Encapsulation may occur, and some may show bipolar staining. Others exhibit pleomorphism. The organisms are aerobic to facultatively anaerobic. Carbon dioxide in excess of normal atmospheric concentration may favor the growth of some species, whereas serum or blood as culture medium enrichment will enhance the growth of others. X and V factors (p. 199) are required for the cultivation of certain fastidious species.

Some species have the capacity to invade living tissue, gaining entrance through the mucous membranes of the skin. Several zoonotic species are transmissible to humans, producing diseases such as brucellosis, tularemia, respiratory illness, meningitis, and others. Many species are obligate animal parasites.

GENUS PASTEURELLA

Organisms of the genus *Pasteurella* are oxidase positive and include *Pasteurella haemolytica*, *P. multocida*, *P. pneumotropica*, and *P. ureae*. As was described in Chapter 20, members of the genus *Yersinia* are oxidase negative and include *Yersinia pestis*, *Y. enterocolitica*, and *Y. pseudotuberculosis*. The organism formerly known as *P. tularensis* is currently the type species of the genus *Francisella*.[17] In the genus *Pasteurella*, the only species of clinical significance is *P. multocida* (*P. septica*).

Pasteurella multocida

Primarily an animal pathogen, *Pasteurella multocida* causes a form of hemorrhagic septicemia in lower animals and cholera in chickens. Humans become infected through a **bite or scratch** from a cat or dog[9] or through contact with a diseased carcass, which may occur in abattoir workers and veterinarians. Pulmonary infections also may occur, particularly in patients suffering from bronchiectasis. Other infections described with this organism include bacteremia, meningitis, brain abscess, septic arthritis, osteomyelitis, appendiceal abscess, and liver abscess.[10] Clinical specimens include sputum, pus, blood, spinal fluid, and tissues.

P. multocida is a small, coccoid, nonmotile, gram-negative rod often showing bipolar staining. It grows well at 35 C on chocolate agar or blood agar, where it produces small, nonhemolytic, translucent colonies with a characteristic musty odor. Several colony forms are recognized. Many strains isolated from the respiratory tract or from chronic infections produce mucoid (M), relatively avirulent colonies. This form is highly pathogenic for animals, however. Highly virulent strains for humans produce smooth (S), fluorescent colonies. Nonfluorescent, smooth, transitional forms are weakly virulent, and the R form, which is granular and dry, is avirulent. Colonies may be blue, iridescent, or punctiform. *P. multocida* is inhibited on bile-containing media, such as SS, XLD, or Hektoen enteric agar. The

Table 23-1. Characteristics of *Pasteurella multocida*

Property	Observation	Property	Observation
Colony forms	Several	Voges-Proskauer	−
Optimal growth temperature	35 to 37 C	Urease	−
Motility at 25 C	−	Nitrates	Reduced
Serotypes	16	Glucose	+
Catalase	+	Glycerol	−
Oxidase	+	Lactose	−
Coagulase	−	Melibiose	−
Fibrinolysin	−	Maltose	−
Hemolysis on blood agar	−	Mannitol	+ or −
Medium containing bile salts	No growth	Rhamnose	−
H₂S	+	Salicin	−
Indole	+	Sucrose	+
Methyl red	−		

biochemical properties and other characteristics are shown in Table 23-1.

Serologically, *P. multocida* has been found to have 16 serotypes.[7] Animal pathogenicity tests and serologic tests with specific typing sera are of value. In vitro, the organism is **very susceptible to penicillin.** The zone of inhibition around a two-unit disk can frequently lead one to suspect its presence on routine culture plates streaked with sputum or bronchoscopic secretions. Tetracycline, erythromycin, and chloramphenicol are also effective therapeutic agents.

GENUS FRANCISELLA
Francisella tularensis

This organism is the etiologic agent of **tularemia,** a disease of rodents (particularly rabbits) that is directly transmissible to humans through the handling of infected animals or indirectly transmissible by blood-sucking insects (chiefly ticks and deer flies in the United States). Tularemia may also be spread by water and by the aerosol route. In culture, *Francisella tularensis* is a minute, highly pleomorphic, nonmotile, gram-negative rod with capsules occurring in vivo. The characteristic bipolar staining may be seen on Gram stain or preferably on Giemsa stain.

The organism reproduces by different methods, including budding,[8] binary fission, and the production of filaments. The organism exhibits a filterable phase, and in this respect it resembles members of the pleuropneumonia mycoplasmas. Clinical specimens include blood (first week), sputum, tissue, pleural fluid, and conjunctival scrapings.

F. tularensis requires enriched media, such as blood-cystine-glucose agar (see Chapter 41). On this medium, minute, transparent, droplike, mucoid, readily emulsifiable colonies are formed after 2 to 5 days of incubation at 35 C. Colonies may take as long as 10 to 14 days to develop. The organism is an obligate aerobe and grows optimally at 35 C. Glucose, maltose, and mannose are fermented without gas; other carbohydrates are attacked irregularly.

Further identification procedures include the testing for susceptibility to specific bacteriophages, agglutination by specific antisera, direct or indirect FA staining, and the demonstration of virulence by intraperitoneal inoculation of guinea pigs. The danger of handling infected animals and virulent cultures of *F. tularensis* cannot be overemphasized. **Many laboratory workers have become infected, and some have died of tularemia.** It is advised that specimens be sent to a laboratory equipped for handling *F.*

tularensis, such as a reference laboratory or state health laboratory. Streptomycin or tetracycline is the drug of choice therapeutically.

GENUS BORDETELLA

The genus *Bordetella* consists of three species that are minute, gram-negative, motile or nonmotile coccobacilli. Some require complex media for primary isolation, and all three species have been implicated in whooping cough, or an infection clinically resembling it, in humans.

Bordetella pertussis

Bordetella pertussis, the type species, is the etiologic agent of **whooping cough,** or **pertussis.** For initial isolation, it requires Bordet-Gengou agar (potato-blood-glycerol agar; see Chapter 41). The addition of 0.25 to 0.5 unit of penicillin per milliliter is recommended for reducing overgrowth of gram-positive organisms. On this medium, small, smooth, convex colonies with a pearllike luster (resembling **mercury droplets**) develop in 3 to 4 days. The colonies are mucoid and tenacious and are surrounded by a zone of hemolysis. The organism is **nonmotile** and may occur singly, in pairs, and occasionally in short chains. The cells tend to show bipolar staining and may be encapsulated.

Indole is not produced by the organism, and citrate is not utilized. Nitrates are not reduced, nor is urea hydrolyzed. X and V factors (p. 199) are not required; catalase is produced.

When isolated from patients with pertussis (a nasopharyngeal swab is the specimen of choice and should be plated immediately), the organism gives rise to smooth, encapsulated phase I colonies on Bordet-Gengou medium. The other phases of the organism (II, III, and IV) may be determined by antigenic analysis. Identification of *B. pertussis* is further confirmed by a slide agglutination test with specific antiserum.* Immunofluo-

---

*Difco Laboratories, Detroit.

rescent procedures also are useful as an aid in identification.[12]

Bordetella parapertussis

Occasionally isolated from patients with an acute respiratory tract infection resembling mild whooping cough, *B. parapertussis* is morphologically and colonially similar to *B. pertussis. B. parapertussis* develops a large colony on Bordet-Gengou agar and produces a **brown** pigment in the underlying medium. It is **nonmotile** and does not require X or V factors for growth. Indole is not produced, and carbohydrates are not fermented. The organism is normally urease and catalase positive and utilizes citrate.

Although it is serologically homogeneous, *B. parapertussis* shares common somatic antigens with *B. pertussis* and *B. bronchiseptica* and may cross-agglutinate with these organisms. An absorbed high-titer antiserum is available for the serologic identification of *B. parapertussis* (see Chapter 37).

Bordetella bronchiseptica

Bordetella bronchiseptica has been isolated occasionally from patients with a pertussis-like disease. It differs from *B. pertussis* in that it is **motile** and possesses peritrichous flagella. The organism grows readily on blood agar, producing smooth, raised, glistening colonies with hemolytic zones. Indole is not produced, and none of the carbohydrates is fermented. Urease is positive within **4 hours,** catalase is formed, nitrates are often reduced, and citrate is utilized as a source of carbon. Cross-agglutination occurs with *B. pertussis* and *B. parapertussis.* Once thought to cause canine distemper, *B. bronchiseptica* is a common cause of bronchopneumonia in guinea pigs and rabbits; it may also occur in these animals as normal flora of the respiratory tract.

GENUS BRUCELLA

The genus *Brucella* consists of four main species of nonmotile, gram-negative, rod- to coccoid-shaped cells; they are pathogenic for humans and a variety of domestic animals.

The brucellae are **obligate parasites,** characterized by their intracellular existence, and are capable of invading all animal tissue, where they cause various brucelloses, including contagious abortion in goats, cows, and hogs, as well as undulant fever in humans.

A definitive diagnosis of brucellosis is established by the isolation and identification of the organism from clinical specimens. **Blood** is the material most frequently found to be positive on culture, particularly when drawn during the febrile period of illness. Brucellae may be recovered occasionally from cultures of bone marrow, from biopsied lymph nodes and other tissue, and also from urine and cerebrospinal fluid. The use of the modified Castañeda bottle (see Chapter 41) is strongly recommended for culturing blood from multiple specimens. Incubation of these cultures in a candle jar or CO_2 incubator (3% to 10% carbon dioxide) is imperative, along with high humidity; cultures should be retained for a minimum of 21 days before being discarded as negative. Blood cultures, as a rule, are negative after the acute symptoms have subsided, which generally coin-

cides with the development of humoral antibodies in the patient.

The agglutination test (see Chapter 38), which uses a standardized, heat-killed, smooth *Brucella* antigen, is the most reliable of the serologic tests. The standard agglutination test does not detect antibodies to *B. canis*, but there is a specific serologic test for this agent.[15] An indirect FA test has been reported successful for detecting antibody in human sera.[2] The opsonocytophagic test is of doubtful value.

The four important species may be readily differentiated by their susceptibility to certain bacteriostatic dyes, by their reaction in carbohydrates, and by their requirements for additional concentrations of carbon dioxide for primary isolation on laboratory media, and by H_2S production (Table 23-2).

Guinea pigs are susceptible to all species and will develop an infection within 30 days after injection of primary smooth (S) isolates.

Brucellae are aerobic and grow best at 35 C. *B. abortus* requires increased carbon dioxide tension (3% to 10%) on primary isolation, although many strains lose this requirement on subculture. The nutrition of these organisms is complex. Best growth is obtained primarily on en-

Table 23-2. Differential characteristics of four species of the genus *Brucella*

Species	CO₂ requirement (5%)	H₂S production	Growth* Thionine A	B	C	Basic fuchsin B	C
B. melitensis	−	− to +/− (throughout 4 days)	−	+	+	+	+
B. abortus	+	+ (first 2 days only)	−	+	+†	+†	+†
B. suis	−	+ (throughout 4 days)	+	+	+	−†	−†
B. canis	−	−	+	+	+	−	−

*Dye concentration: A, 1:25,000; B, 1:50,000; C, 1:100,000.
†There is some strain variation; these reactions are the most typical.

riched media, such as liver infusion tryptose,* trypticase,† or brucella‡ agar at pH 7 to 7.2 (pH 7.5 to 7.8 in 3% carbon dioxide). After 24 to 48 hours' incubation, small, convex, smooth, translucent colonies appear, which become brownish with age. Brucellae may also be cultivated on synthetic media containing amino acids, vitamins, mineral salts, and glucose.

The three major species reduce nitrates, with *B. abortus* and *B. suis* carrying the reduction to nitrogen gas. **Urea** is rapidly hydrolyzed by *B. suis* but slowly, if at all, by *B. melitensis* and *B. abortus*. All strains are catalase positive, with *B. suis* the most active. *B. suis* strains isolated in the United States are active producers of hydrogen sulfide (lead acetate paper), whereas the other species produce only small amounts or none. Brucellae do not liquefy gelatin or produce indole and are MR and VP negative.

Species of *Brucella* show a **differential sensitivity** to a number of aniline dyes, such as thionine, basic fuchsin, crystal violet, pyronin, azure A, and so forth. These may be incorporated in the agar medium in concentrations of 1:25,000 to 1:100,000, depending on the dye content of each lot of dye and on the medium used (Table 23-2). Inhibition of growth may also be determined by the use of dye tablets§ similar to antibiotic disks. An inhibition zone of 4 mm or more in diameter around the disks is interpreted as susceptibility.

B. melitensis also can be differentiated from *B. abortus* and *B. suis* by the use of the agglutination test, using absorbed monospecific antisera. The latter two species, however, cannot be so differentiated, since they share an equal concentration of two identical antigens.

*Difco Laboratories, Detroit.
†Baltimore Biological Laboratory, Cockeysville, Md.
‡Pfizer Laboratories, Flushing, N. Y.; Difco Laboratories, Detroit; BBL, Cockeysville, Md.
§Medical Research Specialties, Loma Linda, Calif.

B. canis has occasionally been implicated in human disease, usually a relatively mild illness.[15] Persons exposed to infected dogs have a low risk of disease, but infection has been transmitted to humans from the dog. Routine *Brucella* agglutinin tests do not detect antibody to *B. canis;* special serologic studies are available through CDC.

Tetracycline for 3 weeks is the therapy of choice for brucellosis. Some workers feel that the addition of streptomycin is desirable.

GENUS HAEMOPHILUS

Haemophilus species are small, nonmotile, gram-negative rods that require hemoglobin in the culture medium or are stimulated by its presence. Whole blood contains the following two factors that are necessary for the growth of the type species of the genus *H. influenzae:*

1. **X factor:** a **heat-stable** substance, hemin, associated with hemoglobin.
2. **V factor:** a **heat-labile** substance, which is coenzyme I, nicotinamide-adenine-dinucleotide (NAD), supplied by yeast, potato extract, and certain bacteria, in addition to that found in blood.

Table 23-3 shows the characteristics of the pathogenic species.

Trypticase soy agar or brain-heart infusion agar plates used with commercially available filter paper strips or disks* containing factors X and V is a convenient means for testing the requirements of *Haemophilus* for these factors. To prevent carryover of X factor, which may be present in trace amounts in blood agar, it is recommended that colonies be inoculated into nutrient broth.[20] After thorough mixing, this broth suspension is inoculated to a trypticase soy agar plate by streaking with a sterile cotton swab. Place paper strips (or disks) containing X factor, V factor, and both X and V factors (not too close together) on the inoculated plate

*Difco Laboratories, Detroit; BBL, Cockeysville, Md.

Table 23-3. Hemolytic activity and the X and V requirements of members of the genus *Haemophilus*

Organism	Infection site	X factor	V factor	Hemolysis
H. influenzae	Respiratory tract, meninges, blood, and other areas	+	+	−
H. aegyptius	Conjunctiva	+	+	−
H. haemolyticus	Respiratory tract (not pathogenic)	+	+	+
H. parainfluenzae	Respiratory tract (rarely pathogenic)	−	+	−
H. parahaemolyticus	Respiratory tract (not pathogenic)	−	+	+
H. ducreyi	Genital region	+	−	+/−
H. aphrophilus	Respiratory tract, blood, brain, others	+/−	−	−

and incubate at 35 C under an atmosphere of 3% to 10% CO_2. The presence or absence of growth around each strip determines the species of *Haemophilus* (Table 23-3).

The hemolytic properties of members of the genus *Haemophilus* are readily determined by streaking a loopful of the above broth suspension to a rabbit blood agar plate and incubating it overnight.

Although all members of the genus are parasitic in nature and require growth factors, they may or may not be pathogenic for humans. Some are members of the normal flora of the respiratory tract; others are important human pathogens, capable of causing severe respiratory tract disease, meningitis, pyogenic arthritis, subacute bacterial endocarditis, and other types of infection.

Haemophilus influenzae

Originally regarded by Pfeiffer and others as being the causative agent of influenza, *H. influenzae* is found primarily in the respiratory tract of humans. It plays an important etiologic role in acute respiratory tract infections and conjunctivitis and may also cause septicemia, subacute bacterial endocarditis, septic arthritis, and purulent meningitis in children. This form of meningitis occurs only rarely in adults.[14] *H. influenzae* also can produce a characteristic obstructive epiglottitis or laryngotracheal infection that may prove fatal in children 2 to 5 years of age.

H. influenzae is a fastidious organism, requiring an infusion medium containing X and V factors. Luxuriant growth occurs on **chocolate agar** (see Chapter 41). Growth on this medium appears in 18 to 24 hours as small (1 to 2 mm), colorless, transparent, moist colonies with a distinct "mousy" odor. On transparent media with Fildes enrichment, colonies are translucent and bluish; on richer media, such as Levinthal's transparent agar, the colonies may be larger. On sheep blood agar the organism grows poorly, if at all, producing only tiny colonies. Colonies are large and characteristic on blood agar, however, when they are growing **near** colonies of staphylococci, neisseriae, pneumococci, and other organisms capable of synthesizing V factor. This factor diffuses into the surrounding medium and stimulates growth of *H. influenzae* in the vicinity of such colonies. This phenomenon is known as "**satellitism.**"

All strains of *H. influenzae* reduce nitrates to nitrites and are soluble in sodium deoxycholate*; indole is produced by the encapsulated organisms. Fermentation reactions are variable. Glucose and other carbohydrates are utilized by some strains but not by others. Six serologic types (**a, b, c, d, e,** and **f**) are recognized by

---

*To 0.8 ml of a young broth culture, add 0.2 ml of a 10% solution of sodium deoxycholate. Incubate for 2 hours at 35 C. The tube should become clear except for a slight opalescence.

the quellung* method or by the precipitin reaction. Most meningeal infections
are caused by **type b,** but the majority of
respiratory strains are not type specific
and are apparently less virulent. Virulent
H. influenzae strains appear to be immunologically related to the pneumococcus;
cross-reactions occur between the capsular substances of these two organisms.
For example, type **b** cross-reacts with
pneumococcus types 6 and 29.

Recently, ampicillin resistance due to
beta-lactamase production has been a
problem.[11] Spurious ampicillin resistance may be noted due to an ampicillin
antagonist in certain lots of Difco Supplement C (Washington, J. A. II, et al.:
Antimicrob. Ag. Chemother. **9:**199-200,
1976). Resistance to tetracycline and
chloramphenicol is also seen occasionally.
Cotrimoxazole is active against this organism clinically.

Haemophilus aegyptius (Koch-Weeks bacillus)

Haemophilus aegyptius is associated
with the highly communicable form of
conjunctivitis known as **pink eye.** It was
recently recovered from patients with
pneumonia, as well (Marraro, R. V., et al.:
J. Clin. Microbiol. **6:**172-173, 1977). Resembling *H. influenzae* morphologically,
it also requires both X and V factors for its
growth, and it produces small, transparent, nonhemolytic colonies on blood
agar. Satellitism occurs with neighboring
Staphylococcus colonies. On transparent
agar the colonies show a bluish sheen
with transmitted light. Indole is not produced, and the reaction in carbohydrates
is inconsistent. Nitrates are reduced to
nitrites, and the organism is bile soluble.
H. aegyptius is serologically related to
and possibly identical with *H. influenzae.*

Haemophilus haemolyticus

Haemophilus haemolyticus is found
normally in the upper respiratory tract of
humans. It requires both X and V factors
for growth. Colonies on blood agar resemble *H. influenzae* but are surrounded by a
wide zone of **beta hemolysis.**

In examining throat cultures on blood
agar* plates, it is important that colonies
of *H. haemolyticus* be differentiated from
those of beta-hemolytic streptococci,
since both are small colonies surrounded
by a zone of clear hemolysis. Colonies of
H. haemolyticus are generally soft,
pearly, and translucent, in contrast to the
firm, white, opaque colonies of group A
streptococci. Gram stain of the colony will
readily differentiate the two.

Haemophilus parainfluenzae and Haemophilus parahaemolyticus

H. parainfluenzae and *H. parahaemolyticus* are found in the normal respiratory
tract of humans and are rarely associated
with infections. Isolated case reports, however, periodically implicate *H. parainfluenzae* in upper respiratory tract infections
and subacute bacterial endocarditis. Both
require **only the V factor** for growth. *H.
parainfluenzae* resembles *H. influenzae*
both morphologically and colonially;
it is not hemolytic, and it exhibits satellitism around staphylococcal colonies. *H.
parahaemolyticus* resembles *H. haemolyticus,* although it produces somewhat
larger colonies on blood agar. These are
surrounded by a zone of beta hemolysis.
Biochemically, both species are similar to
other members of the genus.

H. paraphrophilus, an organism that is
readily confused with *H. parainfluenzae,*
has recently been reported as a cause of
endocarditis (Geraci, J. E., Wilkowske, C.
J., Wilson, W. R., and Washington, J. A. II:
Mayo Clin. Proc. **52:**209-215, 1977).

---

*Polyvalent and type-specific antisera are available
from Hyland Laboratories, Los Angeles, and Difco
Laboratories, Detroit.

---

*Rabbit and horse blood do not contain inhibitory
substances against *Haemophilus* species; human,
and especially sheep, blood greatly inhibit the
growth of these organisms.

Haemophilus ducreyi

Haemophilus ducreyi is the causative agent of an ulcerative venereal disease known as **chancroid** (soft chancre) in humans. (This is one of the classic five venereal diseases.) This small gram-negative rod occurs in long strands in smears obtained from the genital ulcer, where it is usually associated with other pyogenic bacteria. *H. ducreyi* is grown with considerable difficulty and grows best in fresh clotted rabbit, sheep, or human blood heated to 55 C for 15 minutes. Smears made after 1 to 2 days of incubation will show the tangled chains of *H. ducreyi*, if present.

Patients infected with this microorganism can develop a hypersensitivity reaction, which can be detected by the intradermal injection of heat-killed cells. Since the test is positive 1 to 2 weeks after infection, this can be useful as a diagnostic aid.

Haemophilus aphrophilus

Haemophilus aphrophilus is a small, gram-negative coccobacillus, which does not require V factor and is variable in X factor requirement. Fastidious in its growth requirements, this organism appears to grow best in humid air with 10% CO_2.[18,22] Colonies on blood agar plates are similar in appearance to other species of *Haemophilus* that grow on this medium except that colonies of *H. aphrophilus* tend to be more opaque.[22] Easily confused with *Actinobacillus actinomycetem-comitans*, differentiation can be readily obtained by the catalase test; *H. aphrophilus* does not produce catalase. Other biochemical tests of value include fermentation by *H. aphrophilus* of lactose, sucrose, and trehalose but not mannitol and xylose. Conversely, *A. actinomycetem-comitans* does not ferment lactose, sucrose, and trehalose but does utilize mannitol and xylose.

Infections due to *H. aphrophilus* are infrequent but are often very severe and can occur in both adults and children. They include endocarditis, septicemia, brain abscess, meningitis, and others.[13,16,18,21] Specimens include blood, spinal fluid, pus, and sputum.[16,18]

Haemophilus vaginalis (Corynebacterium vaginale)

The organism known as *Haemophilus vaginalis* is frequently recovered from genital tract specimens of women with clinically diagnosed vaginitis. Its role in vaginitis has been questioned, however. *H. vaginalis* has also been reported to be etiologically significant in cases of neonatal sepsis, postpartum bacteremia, and nonspecific urethritis.[19]

Gardner and Dukes[5] originally placed the organism in the genus *Haemophilus*. Later work showed that neither X nor V factor is required for growth. Zinneman and Turner[23] recommended placing *H. vaginalis* in the genus *Corynebacterium* because it stained gram positive. However, cell wall analysis, guanine plus cytosine ratios, and electron microscopy have shown *H. vaginalis* to be distinct from *Corynebacterium* and to have the cellular morphology and biochemical composition of a true gram-negative rod.[3,4] It is suggested that the name *Haemophilus vaginalis* be retained until taxonomic studies determine the proper genus of the organism.

While several methods have been described for isolation of *H. vaginalis*, the medium recently described by Greenwood et al.[6] seems to provide efficient isolation, rapid screening of plates, and quantification of this organism from clinical material. Colonies on this "vaginalis agar" are used as inoculum for buffered single substrates (glucose, maltose, and starch).[6] *H. vaginalis* is glucose, maltose, and starch positive. Biochemical characteristics and other properties are shown in Table 23-4.

The finding of so-called "clue cells" (vaginal epithelial cells covered with bac-

Table 23-4. Characteristics of
Haemophilus vaginalis
(Corynebacterium vaginale)

Property	Observation
Motility	−
Oxidase	−
Catalase	−
Indole	−
Gelatinase	−
Urease	−
Hemolysis on human blood agar	+
Hemolysis on sheep blood agar	−
Hippurate hydrolysis	+
Anaerobic growth	+
Nitrate to nitrite	−
Voges-Proskauer	−
B-galactosidase (ONPG)	majority +
Glucose	+
Maltose	+
Dextrin	+
Starch	+
Sorbitol	−
Inulin	−
Salicin	−

teria) in wet mounts has been frequently stated to be diagnostic for the presence of *H. vaginalis*. Recent evidence, however, has questioned this method of identification in that it lacks both sensitivity and specificity.[1]

REFERENCES

1. Akerlund, M., and Mårdh, P.-A.: Isolation and identification of *Corynebacterium vaginale (Haemophilus vaginalis)* in women with infections of the lower genital tract, Acta Obstet. Gynecol. Scand. **53**:85-90, 1974.
2. Biegeleisen, J. Z., Jr., Bradshaw, B. R., and Moody, M. D.: Demonstration of *Brucella* antibodies in human serum, a comparison of the fluorescent antibody and agglutination techniques, J. Immunol. **88**:109-112, 1962.
3. Criswell, B. S., Marston, J. H., Stenback, W. A., Black, S. H., and Gardner H. L.: *Haemophilus vaginalis* 594, a gram-negative organism? Can. J. Microbiol. **17**:865-869, 1971.
4. Criswell, B. S., Stenback, W. A., Black, S. H., and Gardner, H. L.: Fine structure of *Haemophilus vaginalis*, J. Bacteriol. **109**:930-932, 1972.
5. Gardner, H. L., and Dukes, C. D.: *Haemophilus vaginalis* vaginitis, Am. J. Obstet. Gynecol. **69**: 962-976, 1955.
6. Greenwood, J. R., Pickett, M. J., Martin, W. J., and Mack, E. G.: *Haemophilus vaginalis (Corynebacterium vaginale):* Method for isolation and rapid biochemical identification, Health. Lab. Sci. **14**:102-106, 1977.
7. Heddleston, K. L., and Wessman, G.: Characteristics of *Pasteurella multocida* of human origin, J. Clin. Microbiol. **1**:377-383, 1975.
8. Hesselbrook, W., and Foshay, L.: The morphology of *Bacterium tularense*, J. Bacteriol. **49**: 209-231, 1945.
9. Holloway, W. J., Scott, E. G., and Adams, Y. B.: *Pasteurella multocida* infection in man, Am. J. Clin. Pathol. **51**:705-708, 1969.
10. Johnson, R. H., and Rumans, L. W.: Unusual infections caused by *Pasteurella multocida*, J.A.M.A. **237**:146-147, 1977.
11. Kammer, R. B., Preston, D. A., Turner, J. R., and Hawley, L. C.: Rapid detection of ampicillin-resistant *Haemophilus influenzae* and their susceptibility to sixteen antibiotics, Antimicrobiol. Ag. Chemother. **8**:91-94, 1975.
12. Kendrick, P. L., Eldering, G., and Eveland, W. C.: Application of fluorescent antibody techniques; methods for the identification of *Bordetella pertussis*, Am. J. Dis. Child. **101**:149-154, 1961.
13. King, E. O., and Tatum, H. W.: *Actinobacillus actinomycetemcomitans* and *Hemophilus aphrophilus*, J. Infect. Dis. **111**:85-94, 1962.
14. Merselis, J. G., Sellers, T. F., Jr., Johnson, J. E., and Hook, E. W.: *Hemophilus influenzae* meningitis in adults, Arch. Intern. Med. **110**: 837-846, 1962.
15. Munford, R. S., Weaver, R. E., Patton, C., Feeley, J. C., and Feldman, R. A.: Human disease caused by *Brucella canis*, J.A.M.A. **231**: 1267-1269, 1975.
16. Page, M. I., and King, E. O.: Infection due to *Actinobacillus actinomycetemcomitans* and *Haemophilus aphrophilus*, N. Engl. J. Med. **275**: 181-188, 1966.
17. Philip, C. B., and Owen, C. R.: Comments on the nomenclature of the causative agent of tularemia, Internat. Bull. Bact. Nomen. Taxon. **11**:67-72, 1961.
18. Sutter, V. L., and Finegold, S. M.: *Haemophilus aphrophilus* infections: clinical and bacteriologic studies. Ann. N. Y. Acad. Sci. **174**:468-487, 1970.
19. Venkataramani, T. K., and Rathbun, H. K., *Corynebacterium vaginale (Hemophilus vaginalis)* bacteremia: clinical study of 29 cases, Johns Hopkins Med. J. **139**:93-97, 1976.
20. Washington, J. A. II, editor: Laboratory procedures in clinical microbiology, Boston, 1974, Little, Brown and Co.

21. Witorsch, P., and Gorden, P.: *Hemophilus aphrophilus* meningitis, Ann. Intern. Med. **60:**957-961, 1964.

22. Young, V. M.: *Haemophilus.* In Lennette, E. H., Spaulding, E. H., and Truant, J. P., editors: Manual of clinical microbiology, ed. 2, Washington, D.C., 1974, American Society for Microbiology.

23. Zinneman, K., and Turner, G. C.: The taxonomic position of *"Haemophilus vaginalis" (Corynebacterium vaginale),* J. Pathol. Bacteriol. **85:**213-219, 1963.

CHAPTER 24

Spirochetes and curved rods

Campylobacter · Spirillum · Borrelia · Treponema · Leptospira

Most of the curved bacteria encountered as pathogens in humans belong to two families, the Spirillaceae and the Spirochaetaceae.

Members of the Spirillaceae are rigid, **helically curved** rods with from less than one turn to many turns. They are motile with a corkscrew motion, by means of **polar flagella.** Pathogenic forms stain readily with aniline dyes (they are gram-negative) and Giemsa or Wright's stains. The two genera in this family are *Campylobacter* and *Spirillum.*

The Spirochaetaceae are **helically coiled** organisms. They consist of a protoplasmic cylinder intertwined with one or more **axial fibrils;** both of these are enclosed by an outer envelope. Although gram-negative, only *Borrelia* stain well with aniline dyes; Giemsa or silver impregnation are best for staining. The three genera containing organisms pathogenic for humans are *Borrelia, Treponema,* and *Leptospira.*

CAMPYLOBACTER

One species, *C. fetus (Vibrio fetus),* contains organisms pathogenic for humans: two of the three subspecies, *C. fetus* ss. *intestinalis* and ss. *jejuni* (the latter had been called "related vibrio" by King).[7]

The manifestations of *Campylobacter* infection in humans are variable and include fever alone (which may be relapsing), thrombophlebitis (involving both upper and lower extremities), bacteremia, bacterial endocarditis, septic arthritis, meningoencephalitis (sometimes chronic and indolent and sometimes fulminant and lethal), pleuropulmonary infection, diarrhea, and fever and abortion in pregnant women. Many patients have underlying disease, such as malignancy. *C. fetus* ss. *jejuni* may be found in humans (in the intestinal tract) in the absence of disease; this has not been true for ss. *intestinalis.* The organism is not uncommon in infections in cattle, sheep, and goats and may be found in poultry. However, many cases in humans have occurred in the absence of animal contact.

The organism is an **obligate microaerophile,** growing under reduced oxygen tension (as in 3% to 10% CO_2) but not aerobically or anaerobically. As a result, it may be more difficult to initiate growth than is true for many other organisms. It is motile with a **single polar flagellum** at one or both ends of the cell. It is **nonfermentative, oxidase positive,** and catalase positive. It reduces nitrate to nitrite.

A sensitive indirect hemagglutination test is available.

The organism is relatively susceptible to antimicrobial agents. Representative MICs are chloramphenicol 4 μg/ml, tetracycline 1 μg/ml, gentamicin < 1 μg/ml, streptomycin 2 to 4 μg/ml, erythromycin 2 to 8 μg/ml, and penicillin 32 to 64 units/ml. It is resistant to bacitracin, polymyxin, and novobiocin.

SPIRILLUM MINOR (MINUS)

This organism, already discussed in Chapter 7 on blood cultures, is one of the causes of **rat bite fever** (sodoku), a disease characterized by an initially inflamed (occasionally ulcerating) wound associated with lymphadenopathy and an erythematous rash. The organism may be visualized by darkfield examination in material from the bite wound, affected lymph node, or in the blood. Blood films also should be stained by the Wright or Giemsa technique and examined microscopically. The organism has not been cultivated to date.

When the microscopic examination is unrevealing, one must resort to animal inoculation, using at least several mice and a guinea pig. These are injected intraperitoneally with 1 to 2 ml of the patient's blood, and the peritoneal fluid (mice) or defibrinated blood (guinea pig) is examined by darkfield microscopy or by Giemsa or Wright's stain weekly for 4 weeks for the presence of short, thick, actively motile spiral forms of two to three spirals and **bipolar polytrichous tufts of flagella.** Penicillin and streptomycin, and probably tetracycline, are effective therapeutically.

BORRELIA SPECIES

These spirochetes normally are parasitic for several species of arthropods, including body lice and ticks; human beings or other animals acquire infection (**relapsing fever**) by the bite of the infected vector. *Borrelia recurrentis* is the only species transmitted by lice; this species is presently confined primarily to Eastern Africa. There are nine species of *Borrelia* transmitted by ticks (various species of Ornithodoros); the principal species found in the United States are *B. hermsii, B. parkeri,* and *B. turicatae.*[4,6] The tick bite is usually painless, and the tick drops off the host after 30 to 60 minutes, so the subject may not be aware of tick contact.

Borreliae are best demonstrated in the blood of an infected individual early in the course of a febrile period by direct examination of stained blood films. When this is unrewarding, a similar specimen of blood is inoculated intraperitoneally into mice, and the animals are examined daily for *Borrelia* organisms in films of tail blood. The organisms are 10 to 20 μm long, with five to seven open spirals, and

are actively motile in a corkscrew fashion; they stain relatively well with Giemsa or Wright's stain, particularly with prolonged staining. Counterstaining of blood films, stained in the above manner, with 1% crystal violet for 30 seconds, is effective. **Direct darkfield examination** of blood is recommended. A drop of blood is placed on a slide, a coverslip is placed over it, and the edges of the coverslip are sealed with lanolin. Under 400X to 500X magnification (high dry), movement of erythrocytes (occasioned by the movement of the spirochetes) is readily noted and the organisms themselves may then be detected. They exhibit forward and backward motion, with bending and looping. The organism may also be cultured; the interested reader is referred to Kelly's description of the technique.[6]

The drug of choice in therapy is tetracycline.

TREPONEMA SPECIES

There are numerous species in the genus *Treponema,* and they comprise two major groups: those normally present in the mouth, urogenital tract, and gastrointestinal tract of humans and other animals, and the several pathogenic species. In the former group are found *T. macrodentium* and *T. orale* from the oral cavity, *T. refringens,* from the genital area, and so forth. In the latter group are *T. pallidum,* the causative agent of **syphilis,** *T. pertenue,* from the tropical disease **yaws;** and *T. carateum,* which causes a chronic skin disease, **pinta,** endemic in Central and South America.

All the treponemes are **obligate anaerobes;** they are actively motile and contain numerous tight, rigid coils. They are difficult to stain and are best observed by **darkfield microscopy** (see Chapter 2). This is an important laboratory procedure, since a diagnosis of syphilis in the early stages can be immediately confirmed by demonstrating *T. pallidum* in material from suspected lesions (usually anogenital) or affected regional lymph nodes. *T.*

pallidum has not been cultivated in vitro.

Darkfield examination for T. pallidum

The following procedure is recommended for darkfield examination:

1. After donning rubber gloves, thoroughly cleanse the suspected lesion with gauze sponge, removing crusts if present (after soaking in saline), and abrading it to produce serous fluid.
2. Pinch the lesion with gloved fingers so that a drop of fluid is expressed from its border or surface; if none is obtained (avoid blood), moisten with a drop of saline and after a minute or two, repeat the attempt to obtain fluid.
3. Touch a coverslip to the fluid, invert it over a microscope slide so as to eliminate air bubbles, seal the edges with lanolin, and immediately examine for characteristic forms of motile *T. pallidum*, using darkfield microscopy and an oil-immersion objective.
4. Look for a thin, tightly wound, corkscrew-shaped organism of eight to 14 uniform, rigid spirals, slightly longer than the size of an average red blood cell. Characteristic movement is slowly backward or forward or a corkscrew rotation about the long axis, sometimes with slight flexion.
5. Cautiously interpret a positive preparation from a lesion in the mouth—the commensal treponemes may be confused with *T. pallidum*.

A direct FA staining procedure for delayed examination of fluids to detect the presence of *T. pallidum* is of diagnostic value.[3]

It is beyond the scope of this text to discuss the clinical manifestations of primary, secondary, or tertiary syphilis. The serologic tests available for laboratory diagnosis of syphilis are discussed in Chapter 38. (See also references 2, 8, and 9.)

GENUS LEPTOSPIRA

The principal sources of leptospirae infecting humans are urine and tissues of **infected animals.** The disease is acquired by direct contact or indirectly through contact with contaminated water. The organisms enter through abrasions of the skin or through the mucosal surfaces of the body; occupational exposure is a prime factor in acquiring the infection. Clinical manifestations range from a mild illness to a severe infection with liver, kidney, and central nervous system involvement.

The species of *Leptospira* are also thin, flexible, tightly coiled spirals of approximately the same size as the species of *Treponema*. One or both ends of the spirochete, being more flexible than the center portion, may be bent to form a hook. Spinning takes place on the long axis of the cell. If one end only is hooked, forward movement is in the direction of the straight end. With both ends hooked, a lashing effect may be noticed as the cell spins.

The organism is **aerobic** and may be cultivated in a supplemented ascitic fluid medium or in the medium recommended by Fletcher (see Chapter 41), consisting of salts, amino acids, and rabbit albumin.

All pathogenic leptospires are placed in one species, *Leptospira interrogans*, with many serotypes. Serotypes encountered in the United States include icterohemorrhagiae, canicola, ballum, grippotyphosa, bataviae, autumnalis, and pomona.[5] The manifestations of the disease and its severity are **not** related to the specific serotype involved.

The spirochetes can be observed in blood or spinal fluid drawn from a patient during the first week of the disease (see Chapter 7 for discussion of blood culture of *Leptospira*) when the symptoms are chills, fever, muscle pains, headache, and abdominal pain. During the second stage, manifested by jaundice and skin lesions, the spirochetes may be readily observed in the urine by the darkfield technique.[1]

The **serologic** detection of leptospiral antibodies in humans and animals is an invaluable adjunct in the diagnosis of leptospirosis. Various techniques have been used, including the microscopic agglutination–lysis test, the genus-specific hemagglutination test, and the FA test. Stable, formalized suspensions of *Leptospira* serotypes are available* and can be used in a rapid macroscopic slide agglutination test for serologic diagnosis of the infection.

The more serious symptoms of leptospirosis are related to the immune response to the organism. Antimicrobial therapy is generally felt to be of no value, particularly after the first 48 hours of illness. Corticosteroids are contraindicated.

ANAEROBIC VIBRIOS

Certain obligately anaerobic vibrios cause disease in humans on rare occasion. These organisms are discussed in Chapter 27.

---

*Difco Laboratories, Detroit.

REFERENCES

1. Alston, J. M., and Broom, J. C.: Leptospirosis in man and animals, London, 1958, E. & S. Livingston, Ltd.
2. Center for Disease Control: The laboratory aspects of syphilis, Atlanta, 1971, The Center.
3. Daniels, K. C., and Ferneyhough, H. S.: Specific direct fluorescent antibody detection of *Treponema pallidum*, Health Lab. Sci. **14**:164-171, 1977.
4. Felsenfeld, O.: *Borrelia*: Strains, vectors, human and animal borreliosis, St. Louis, 1971, Warren H. Green, Inc.
5. Finegold, S. M., and Meyer, R. D.: Leptospirosis. In Tice's practice of medicine, vol. 3, New York, 1975, Harper & Row, Publishers.
6. Kelly, R. T.: *Borrelia*. In Lennette, E. H., Spaulding, E. H., and Truant, J. P., editors: Manual of clinical microbiology, ed. 2, Washington, D.C., 1974, American Society for Microbiology.
7. King, E. O.: The laboratory recognition of *Vibrio fetus* and a closely related *Vibrio* isolated from cases of human vibriosis, Ann. N.Y. Acad. Sci. **98**:700-711, 1962.
8. Manual of tests for syphilis, Public Health Service Pub. No. 411, Washington, D.C., 1969, U.S. Government Printing Office.
9. Rohde, P., editor: BBL manual of product and laboratory procedures, Cockeysville, Md., 1969, Baltimore Biological Laboratory, Division of BioQuest, Becton, Dickinson and Co.

CHAPTER 25

Aerobic or facultative gram-positive spore-forming bacilli

In this chapter, discussion of the clinical significance of the gram-positive spore-forming bacilli is limited to certain species of *Bacillus*. The genus contains a large number of species that are aerobic or facultative, usually gram-positive, and spore forming. They are widely distributed in nature and are therefore frequent contaminants in laboratory cultures from clinical specimens. They may or may not grow on eosin methylene blue (EMB) agar. Because they may vary in Gram stain and oxidase reaction and spores may not be evident, they may resemble gram-negative bacilli. Sporulation is often stimulated on esculin agar.

It should be noted that certain species of *Bacillus*, other than the well-recognized pathogen *B. anthracis*, can be involved, though infrequently, in human disease processes. For example, *B. cereus* has been implicated in food-borne illness both in Europe and in the United States and may be isolated from wounds. *B. cereus*, *B. circulans*, *B. pumilus*, *B. sphaericus*, and *B. subtilis* have also been incriminated in cases of meningitis, pneumonia, and septicemia, as reported by several authors.[1,3,5,8,10] Sepsis due to *B. licheniformis* has also been described recently (Sugar, A. M., and McCloskey, R. V.: J.A.M.A. **238**:1180-1181, 1977). Fatal pneumonia with bacteremia due to *B. cereus* in a patient with subacute lymphocytic leukemia has been reported by Coonrod and co-workers.[4] These authors further reported on the antimicrobial susceptibility of several *Bacillus* species in which they found that all were susceptible to tetracycline, kanamycin, gentamicin, and chloramphenicol.

Susceptibility to penicillin G, ampicillin, methicillin, and cephalothin was species related, being high for *B. subtilis*, intermediate for *B. pumilus*, and low for *B. cereus*.

B. cereus colonies on laboratory media vary from small, shiny, and compact to the large, feathery, spreading type. A **lavender-colored** colony with beta hemolysis is seen on sheep blood agar. *B. subtilis* colonies are normally large, flat, and dull, with a ground-glass appearance.

BACILLUS ANTHRACIS

Bacillus anthracis is the primary human pathogen in the genus. However, it is seldom encountered in the average hospital or public health laboratory. Nevertheless, because of its importance in some areas, its identification and pathogenicity should be discussed. Cases in the United States are most often related to handling of imported wool or goat hair, animal hides, shaving brushes, and so forth (chiefly from Asia and Africa).

The organism is a facultative, large, square-ended, nonmotile rod, with an ellipsoidal to cylindrical centrally located spore. The sporangium is usually **not swollen**. The cells frequently occur in long chains, giving a **bamboo** appearance, especially on primary isolation from infected tissue or discharge. The chains of virulent forms are usually surrounded by a **capsule**. Encapsulation will occur also in enriched media and when grown on sodium bicarbonate agar under 5% CO_2. Encapsulated strains may be used for fluorescent-antibody staining, an important ancillary test. Smears may be sent to the Center for Disease Control for the FA

209

test. Avirulent forms are usually non-encapsulated. Sporulation occurs in the soil and on inanimate media but not in living tissue.

The colonies of *B. anthracis* are normally large (4 to 5 mm), opaque, raised, and irregular, with a **curled** margin. When the margin of the colony is pushed inward and then lifted gently with an inoculating needle, the disturbed portion of the colony stands up like beaten egg whites. Comma-shaped colony outgrowths are common. On sheep blood agar, the colonies are invariably **nonhemolytic**. Smooth and rough colony forms may be observed, and both of these may be virulent.

Extreme caution must be used when working with suspected *B. anthracis.* Avoid creating aerosols and decontaminate all areas thoroughly.

The optimal temperature for growth is 35 C, and when grown at 42 to 43 C, the organism becomes **attenuated** or avirulent. This was shown by Louis Pasteur years ago. The loss of virulence is attributed to loss of the capsule. In broth, the bacillus produces a heavy pellicle with little if any subsurface growth. Biochemically the organism is characterized as follows:

Carbohydrate fermentation: Glucose, fructose, maltose, sucrose, and trehalose fermented with acid only; arabinose, xylose, galactose, lactose, mannose, raffinose, rhamnose, adonitol, dulcitol, inositol, inulin, mannitol, and sorbitol not fermented.
Gelatin: Inverted pine-tree growth; liquefaction (slow)
Nitrates: Reduced to nitrites
Starch: Hydrolyzed
Voges-Proskauer: Positive

A simple presumptive test for identification of *B. anthracis* has been proposed recently (Bailie, W. E., and Stowe, E. C.: Abstr. Ann. Mtg. Am. Soc. Microbiol. 1977, Abstr. C80, p. 48).

Specific identification may be made by use of a gamma bacteriophage (at CDC or state health department laboratories).

Pathogenicity of Bacillus anthracis.

The organism is the cause of **anthrax,**[*] which in humans may be manifested in three forms:

1. **Cutaneous anthrax** (malignant pustule), the most common form in the United States. Infection is initiated by the entrance of bacilli through an abrasion of the skin. A pustule usually appears on the hands or forearms. The bacilli are readily recognized in the serosanguineous discharge.

2. **Pulmonary anthrax,** or woolsorters' disease. The bacilli may be found in large numbers in the sputum. Spores are inhaled during shearing or sorting of animal hair. If not properly treated, this form can readily progress to fatal septicemia.

3. **Gastrointestinal anthrax,** the most severe and rarest form. The bacilli or spores are swallowed, thus initiating an intestinal infection. The organisms may be isolated from the stools. This form is also usually fatal if not treated.

The pathogenicity of the organism is determined most efficiently by injecting each of 10 white mice (2 to 3 weeks of age) subcutaneously with 0.2 ml of a saline suspension of the organism. Rabbits or guinea pigs may also be used. Details of the technique are given by Feeley and Brachman.[6] Animals usually die 2 to 5 days after inoculation, but they may survive 10 days. Death is due to **septicemia,** and the organism is readily recovered from the heart, blood, spleen, liver, and lungs of the animal.

Ascoli test. The Ascoli test is a **precipitin test** used to diagnose anthrax in dead animals and to determine whether hides for industrial use have been removed from infected animals. The antigen is prepared by boiling a small piece of spleen or hide in 5 to 10 ml of physiologic saline for 15 minutes. This is cooled and filtered

---

*Much valuable information on the clinical picture of anthrax may be gained from a 1954 symposium.[7]

to clarify. One milliliter of the filtered antigen is then layered carefully over an equal volume of antianthrax serum, which has been prepared in rabbits against the encapsulated anthrax organism, using an agglutinating or capillary tube. A positive reaction is recognized by the formation of a ring of **precipitate** at the interface of the two reacting substances. The antigenic material is probably a high molecular weight polypeptide that is found in the capsule and has diffused through the tissues. The antiserum is not commercially available in the United States.

REFERENCES

1. Allen, T. B., and Wilkinson, H. A.: A case of meningitis and generalized Shwartzman reaction caused by *Bacillus sphaericus*, Johns Hopkins Med. J. **125**:8-13, 1969.
2. Ascoli, A.: Der Ausbau meiner Präzipitinreaktion zur Milzbranddiagnose, Z. Immunitaetsforsch, Exper. Ther. **11**:103, 1911.
3. Curtis, J. R., Wing, A. J., and Coleman, J. C.: *Bacillus cereus* bacteremia; a complication of intermittent haemodialysis, Lancet **1**:136-138, 1967.
4. Coonrod, J. D., Leadley, P. J., and Eickhoff, T. C.: Antibiotic susceptibility of *Bacillus* species, J. Infect. Dis. **123**:102-105, 1971.
5. Farrar, W. E., Jr.: Serious infections due to "nonpathogenic" organisms of the genus *Bacillus;* review of their status as pathogens, Am. J. Med. **34**:134-141, 1963.
6. Feeley, J. C., and Brachman, P. S.: *Bacillus anthracis*. In Lennette, E. H., Spaulding, E. H., and Truant, J. P., editors: Manual of clinical microbiology, ed. 2, Washington, D.C., 1974, American Society for Microbiology.
7. Hospital of the University of Pennsylvania: A symposium on anthrax in man, Philadelphia, 1954, University of Pennsylvania Press.
8. Leff, A., Jacobs, R., Gooding, V., Hauch, J., Conte, J., and Stulberg, M.: *Bacillus cereus* pneumonia, Am. Rev. Resp. Dis. **115**:151-154, 1977.
9. Pennington, J. E., Gibbons, N. D., Strobeck, J. E., Simpson, G. L., and Myerowitz, R. L.: *Bacillus* species infection in patients with hematologic neoplasia, J.A.M.A. **235**:1473-1474, 1976.
10. Stopler, T. V., Cāmuescu, V., and Voiculescu, M.: Bronchopneumonia with lethal evolution determined by a microorganism of the genus *Bacillus (B. cereus)*, Romanian Med. Rev. **19**:7-9, 1969.

CHAPTER 26

Gram-positive non-spore-forming bacilli

Corynebacterium · Listeria · Erysipelothrix · Nocardia

Although there is a degree of pleomorphism within these four genera, this characteristic is perhaps most prominent with members of the genus *Corynebacterium.*

GENUS CORYNEBACTERIUM

The corynebacteria (Gk. *koryne*, a club) are gram-positive, nonsporulating, nonmotile (with occasional exceptions) rods. They are often club shaped and frequently banded or beaded with irregularly staining granules (Fig. 26-1). These bacteria frequently exhibit characteristic arrangements, resembling Chinese letters and palisades. The corynebacteria are generally aerobic or facultative, but microaerophilic species do occur. Anaerobic organisms formerly in this genus are now placed in the genus *Propionibacterium* (see Chapter 30). They have a wide distribution in nature. Some species are parasites or pathogens of plants and domestic animals; others are part of the normal human respiratory flora. The type species, *Corynebacterium diphtheriae,* produces a powerful exotoxin that causes **diphtheria** in humans.

Corynebacterium diphtheriae

Corynebacterium diphtheriae (Klebs-Loeffler bacillus) is a facultative, nonmotile, slender, gram-positive rod that is highly pleomorphic. In addition to straight or slightly curved bacilli, club-shaped and branching forms are seen. The rods generally do not stain uniformly with methylene blue but show alternate bands of stained and unstained areas as well as deeply stained granules (metachromatic granules) that give the organism a beaded appearance. Individual cells tend to lie parallel or at acute angles to each other, resulting in V, L, or Y shapes. Although the appearance of *C. diphtheriae* in stained smears is highly characteristic, it **should not be identified by morphology alone;** many diphtheroids and actinomyces stain in the same irregular fashion and are also pleomorphic.

On primary isolation, *C. diphtheriae* should be cultivated on enriched media, such as infusion agar with added blood; the most characteristic morphologic forms, however, are found in smears made from a 12- to 18-hour culture on Loeffler serum medium (Chapter 41), which seems to enhance pleomorphism. The addition of potassium tellurite to blood or chocolate agar provides both a differential and selective medium. Bacterial contaminants are inhibited, and after 1 day *C. diphtheriae* appears as gray or

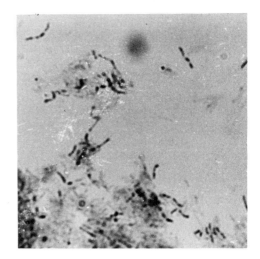

Fig. 26-1. *Corynebacterium diphtheriae,* granule stain. (1000×.)

black colonies. This characteristic aids in distinguishing the organism in mixed culture and helps to differentiate the three types: *gravis, mitis,* and *intermedius.* Although the growth of *C. diphtheriae* is not distinctive on blood agar, specimens should be streaked on this medium (in addition to a tellurite plate) because a few strains are very sensitive to potassium tellurite.[13] Furthermore, the use of a blood agar plate will permit isolation of group A beta-hemolytic streptococci, which may be responsible for the throat lesion or which may be part of a dual infection. The streptococci will not grow in the presence of the concentration of tellurite used for isolation of *C. diphtheriae.* The characteristics of *C. diphtheriae* are summarized in Table 26-1. In general, the diphtheria bacilli on tellurite media are shorter, staining is more uniform, and the granules are less readily

seen than when grown on Loeffler medium. For these reasons most laboratories use **both** media.

In nature, *C. diphtheriae* occurs only in the nasopharyngeal area (rarely on the skin or in wounds) of infected persons or healthy carriers. *C. diphtheriae* endocarditis has been described (Davidson, S., et al.: Am. J. Med. Sci. **271**:351-353, 1976). The organisms are spread to susceptible individuals by droplets or direct contact. The **primary focus** of growth is on the mucous membrane of the respiratory tract, where the exotoxin is produced. Absorption of the toxin leads to the pathologic syndrome that characterizes the disease.

Virulence test. Not all strains of *C. diphtheriae* elaborate exotoxin; however, it is now well established that infection of this bacterium by prophage B or a closely related lysogenic bacteriophage is re-

Table 26-1. Characteristics of *Corynebacterium diphtheriae* types and other members of the genus

Organism	Morphology	Colonies on blood-tellurite agar	Hemo-lysis	Starch and gly-cogen	Glucose	Sucrose	Toxi-genic
C. diphtheriae type *gravis**	Short, evenly staining	Large, dark gray centrally, matt, striated ("daisy-head"), irregular; brittle	−	+	+	−	+
C. diphtheriae type *mitis*	Long, curved, with many metachromatic granules	Small, black, shiny, convex, entire, soft	+	−	+	−	+
C. diphtheriae type *intermedius*	Long, barred forms with clubbed ends, few granules	Very small, flat, dry, gray, or black	−	−	+	−	+
C. pseudo-diphtheriticum†	Short, thick, with 1 or 2 unstained septa; granules absent	Resembles *mitis* type; not as dark	−	−	−	−	−
C. xerosis†	Rods showing polar staining; occasionally clubbed; resembles *C. diphtheriae*	Resembles *mitis* type; not as dark	−	−	+	+	−

*There is apparently no correlation between type and severity of disease in the United States.
†Diphtheroids, normal inhabitants of human beings; not associated with disease.

sponsible for the production of diphtheria toxin. In this regard, the one reliable criterion for identifying this organism is its **ability to produce exotoxin.** There are two methods for determining the toxigenicity of a suspect strain of *C. diphtheriae,* one in vivo and the other in vitro.

The **in vivo** test that yields satisfactory results for most workers is the one described by Fraser and Weld.[9] This is outlined below and may be applied to either a guinea pig or a white rabbit. The latter will permit a greater number of tests.

1. Inoculate a Loeffler slant and a tube of sugar-free infusion broth from a suspect colony of each culture to be tested. Incubate both at 35 C. Check the slant culture for purity by staining after 24 hours. Use the broth culture after 48 hours of incubation for the test.

2. Prepare the test animal by clipping the back and sides closely. Disinfect the clipped area with alcohol or some suitable disinfectant. Make a line over the backbone with an indelible pen or pencil and mark off a series of 2-cm-square areas on either side of the line.

3. Using a 2-ml syringe graduated in tenths of a milliliter and fitted with a 24-gauge needle, draw up 1 to 2 ml of each 48-hour culture and inject 0.2 ml intracutaneously into the marked squares, leaving in each case the square immediately adjacent (below) for the control. Refrigerate the syringe and contents for later use. **Inject a similar amount of a known toxigenic strain into one square as a positive control.**

4. After 5 hours, inject 500 units of diphtheria antitoxin intravenously into the rabbit's ear and wait 30 minutes. If a guinea pig is used, inject the antitoxin intraperitoneally.

5. Using the refrigerated syringe cultures, inject 0.2 ml of each culture into the corresponding square imme-

diately adjacent (**below**) the original test site. This provides the **postantitoxin control.**

6. Read at 24 and 48 hours because the lesions are usually fully developed after the longer interval. A **toxigenic strain** will produce a **central necrotic area,** about 5 to 10 mm in diameter, surrounded by a somewhat larger erythematous zone. The corresponding **control** should show only a **pinkish swollen area** of 5 to 10 mm in diameter without any necrosis.

7. If both test sites show necrosis, toxigenicity of the test culture cannot be confirmed. This may be due to the culture not being toxigenic or impure, or the antitoxin was inadequate or mislabeled, or the test culture was some species other than *C. diphtheriae.* Some strains of *C. ulcerans,* a closely related organism, produce diphtheria toxin as well as an unrelated toxin that is not neutralized by diphtheria antitoxin.[13]

The **in vitro** test, first reported by Elek,[8] has been modified by a number of workers. The procedure recommended here is the modification of Hermann and colleagues.[14] To an autoclaved and cooled basal medium of proteose peptone agar* is added an enrichment of Tween 80, glycerol, and casamino acids* plus potassium tellurite. This is then thoroughly mixed in a Petri dish. Before the agar hardens, a 1- by 8-cm sterile paper strip saturated with diphtheria antitoxin* is placed on the agar surface and gently pressed below the surface with sterile forceps. The plate is dried at 35 C to ensure a moisture-free surface. A loopful of the culture of suspected *C. diphtheriae* is then streaked across the plate perpendicular to the paper strip in a single line, as shown in Fig. 26-2. Four to five cultures may be tested on a single plate. The plate is incubated at 35 C for 1 to 3 days and

---

*Difco Laboratories, Detroit.

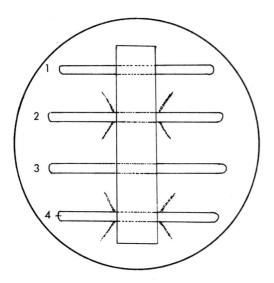

Fig. 26-2. Elek method for in vitro demonstration of toxigenicity of *Corynebacterium diphtheriae*. Filter paper strip impregnated with antitoxin (from top to bottom). *1*, known nontoxigenic strain (negative control); *2*, known toxigenic strain (positive control); *3*, unknown culture that is nontoxigenic; *4*, unknown culture that is toxigenic.

examined. The antitoxin from the paper strip and any toxin produced by the growing diphtheria cultures will diffuse through the agar medium and react in optimal proportions to produce thin lines of precipitate, as shown. These lines arise at angles of approximately 45° from those strains that are toxigenic. The test should include known **positive** and **negative** controls.

Brief mention should be made of the application of the FA test for the detection of *C. diphtheriae* in suspected cases of diphtheria. The test is considered useful in the rapid **presumptive** diagnosis of diphtheria when properly titered reagents are employed, but should be used in conjunction with conventional cultural procedures.[6]

Diphtheroid bacilli

As their collective name implies, diphtheroids are morphologically similar and sometimes indistinguishable from the diphtheria bacilli (Table 26-1). *Corynebacterium pseudodiphtheriticum* occurs in the normal throat and is not pathogenic for human beings or laboratory animals. *C. xerosis* also occurs on mucous membranes of humans; it is basically nonpathogenic but has been recovered from patients with endocarditis, as have a number of other types of diphtheroids (not all classifiable).[7,12] Bacteremia due to a previously undescribed *Corynebacterium* that produces colonies with a distinctive metallic sheen has been described by Hande and associates.[12] Other types of infection involving diphtheroids include meningitis, brain abscess, pneumonia and empyema, osteomyelitis, and soft-tissue and wound infection.[12,17] Animal strains, such as *C. bovis* and *C. equi*, have also been isolated from infection in humans[3,10] *C. haemolyticum* has occasionally been isolated from serious infections. (Ceilley, R.: Ann. Intern. Med. **86:**239, 1977). In general, however, diphtheroids are relatively nonpathogenic and are primarily opportunistic pathogens, causing disease in persons with underlying valvular heart disease or with implanted foreign bodies or who are immunosuppressed.

Diphtheroids may be remarkably resistant to antimicrobial agents.[7,12] Vancomycin appears to be the only agent that is consistently active.

Two poorly described species of *Corynebacterium—C. minutissimum* and *C. tenuis*—are said to be responsible for minor superficial infections, erythrasma and trichomycosis axillaris, respectively.

GENUS LISTERIA

The genus *Listeria* contains the clinically significant species *L. monocytogenes* plus the species *L. grayi*, *L. murrayi*, and *L. denitrificans*. *L. monocytogenes* is a small, nonsporulating, nonencapsulated, gram-positive rod, 0.5 µm by 1 to 2 µm in size, which exhibits distinctive **tumbling motility**. Isolated originally from rabbits with a disease characterized by a great increase in circulating mononuclear leukocytes, it usually does not cause monocytosis in humans. Listeria have been isolated from blood, cerebrospinal fluid, meconium, and from various visceral and cutaneous lesions of humans. *L. monocytogenes* is responsible for an acute meningoencephalitis in infants and adults. It also is implicated in abortion and stillbirth, in bacteremia, endocarditis, disseminated abscesses or granulomata in infants, spontaneous peritonitis in patients with cirrhosis, oculoglandular infection, and cutaneous infection. It is an important opportunistic pathogen in patients with hematologic and other malignancies and patients receiving immunosuppressive therapy. Special **cold enrichment** may be needed to effect recovery of the organism from sources with a heavily mixed flora.[15]

L. monocytogenes is a facultative organism that grows well on infusion media, on tryptose agar with or without added glucose and blood, or on modified McBride medium (Chapter 41). It is not always isolated from tissue or other clinical material on the first attempt. The bacterium grows well at 25 and 35 C but poorly at 42 C and slowly (4 to 7 days) at 4 C. On sheep blood agar incubated at 35 C for 24 hours, it produces small (1 to 2 mm), translucent, gray to white, **beta-hemolytic** colonies; hemolysis is slow to appear on sheep or rabbit blood agar. When cultivated on a clear medium, such as McBride agar, for 18 to 24 hours and examined by unfiltered, oblique illumination with a scanning microscope,* colonies appear characteristically **blue green** in color. They are 0.2 to 0.8 mm in diameter, translucent, round, slightly raised, and watery in consistency.

Indole and hydrogen sulfide are not produced by this organism; nitrate is not reduced; citrate is not utilized; urease is not formed; gelatin and coagulated serum are not liquefied. It is catalase positive and oxidase negative. The methyl red test is positive; the Voges-Proskauer reaction is negative with the O'Meara test but positive with Coblentz and other methods using alpha-naphthol, potassium hydroxide, and creatine. The organism hydrolyzes esculin, as demonstrated by the **blackening** of bile esculin agar after overnight incubation.

The **motility** of *L. monocytogenes* is best demonstrated by the stab inoculation of two tubes containing semisolid motility medium (see Chapter 41), incubating one at room temperature (20 to 25 C) and the other at 35 C. Motility is more pronounced at room temperature than at the higher temperature, and a motile culture shows spread of growth from the line of the stab and the development of an "umbrella" 3 to 5 mm below the surface.

In meat extract broth with carbohydrates and bromcresol purple indicator (see Chapter 41), cultures of *Listeria* grown at 35 C for 1 week will produce acid in glucose, levulose, rhamnose, maltose, and salicin; will show irregular acidity in lactose, sucrose, xylose, arabinose, and dextrin; and will rarely, if ever, produce acid in mannitol, inositol, sorbitol, dulcitol, or raffinose.

Test for pathogenicity. The **ocular** test (Anton) is reliable for pathogenicity. It is carried out by introducing two to three drops of an overnight tryptose agar slant culture, suspended in 5 ml of distilled water, into the conjunctival sac of a rabbit. A purulent conjunctivitis will develop within several days, which eventually heals completely.

Constantly mounting evidence indicates the important role *L. monocytogenes* plays in both human and animal infections. The medical microbiologist should be fully aware that small, gram-positive, diphtheroidlike organisms isolated from cerebrospinal fluid, blood, vaginal swabs, and so forth are not always

---

*The plate may also be placed on a laboratory tripod and examined with a hand lens, with 45-degree oblique illumination.

contaminants. Some may prove to be *L. monocytogenes.* If facilities are not available for carrying out this identification, it is strongly recommended that a subculture be sent to a reference laboratory. For further information on listeriosis, the reader is referred to the excellent review by Gray and Killinger[11] and to Killinger's work in the *Manual of Clinical Microbiology.*[15] Ampicillin and tetracycline are effective therapeutically in listeriosis.

GENUS ERYSIPELOTHRIX

Erysipelothrix rhusiopathiae (synonym, *E. insidiosa*) is the single member of the genus *Erysipelothrix.* It is a nonsporulating, nonencapsulated, nonmotile, gram-positive rod with a tendency to form long filaments. It causes swine erysipelas and septicemia in mice. In humans, it usually causes a self-limited infection of the fingers or hand (**erysipeloid**), but may cause bacteremia with arthritis or endocarditis. An erythematous, elevated, spreading lesion develops at the point of entrance of the organism; the disease generally can be traced to contact with animals or animal products. On blood agar medium, smooth (S) colonies are clear and very small (0.1 mm) and contain small slender rods. Rough (R) colonies are 0.2 to 0.4 mm in diameter and contain long filamentous forms, showing beading and swelling. The latter colonies can spread out to simulate miniature anthrax-type colonies. Both forms stain evenly and may show deeply stained granules.

The organism is microaerophilic and facultatively anaerobic and grows best at 30 to 35 C. Hemolysis on blood agar occurs when 10% horse blood is used. A "test tube brush" type of growth (lateral, radiating projections) characteristically occurs in gelatin stab cultures at room temperature after 48 hours. Carbohydrate reactions are variable; most strains produce acid from glucose and lactose but not from sucrose and other carbohydrates. Hydrogen sulfide is produced;

nitrate is not reduced; indole is not formed; and catalase is negative. *E. rhusiopathiae* is susceptible to penicillin and erythromycin, among other drugs.

The diagnosis rests on the isolation of the organism from a skin biopsy or blood culture; the skin fragment is incubated in thioglycollate broth overnight and then subcultured to blood agar. Inoculation of white mice may be used to confirm the identification.[19]

GENUS NOCARDIA

Nocardiosis is a bacterial infection caused by species of *Nocardia* in humans and the lower animals. It may result in chronic suppuration and draining sinuses of the subcutaneous tissue (mycetoma) or in a primary pulmonary infection resembling tuberculosis, which may spread to the pleural space and chest wall or metastasize to other organs, especially the brain and meninges.

Nocardia asteroides is an important opportunistic pathogen in patients with malignancy or receiving immunosuppressive therapy and is usually responsible for a pulmonary infection. *N. brasiliensis*, on the other hand, is the primary pathogen in mycetoma but may occasionally be found in disseminated disease, particularly in patients with poor host defense mechanisms.[5] *N. brasiliensis* also may cause a lymphocutaneous infection resembling sporotrichosis, as may *N. asteroides.*[16] Bacterial endocarditis and other infections have been produced by a nocardioform organism, *Oerskovia turbata.*[18] (See also Sottnek, F. O., et al.: Int. J. Syst. Bacteriol. **27**:263-270, 1977.) Rarely, *Nocardia caviae* may cause disseminated infection (Arroyo, J. C., et al.: Am. J. Med. **62**:409-412, 1977).

N. asteroides is a **partially acid-fast** aerobic to microaerophilic or capnophilic (CO_2 requiring) bacterium composed of branched filaments (Fig. 26-3) that fragment readily into bacillary and coccoid forms. The organism grows as a saprophyte in the soil, and systemic infection

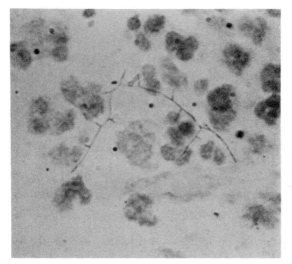

Fig. 26-3. *Nocardia* in Gram stain of sputum. Note thin, branching filamentous organism with irregular staining. (1000×.)

follows inhalation (**exogenous**) or introduction through skin abrasions, especially on the feet. *N. asteroides* may be demonstrated in direct smears of pus (opaque or pigmented sulfur granules are present occasionally) or sputum and in the sediment of centrifuged cerebrospinal fluid. Gram-stained smears reveal thin, gram-positive, branching filaments or coccoid and diphtheroid forms. An acid-fast stain* will show that some filaments retain the carbolfuchsin; young cultures are usually more strongly acid fast, whereas older cultures or subcultures are less so.

All infected material in which delicate, branching, gram-positive filaments have been demonstrated should be inoculated heavily on two plates or tubes each of infusion blood agar and Sabouraud agar **without antibiotics**†; both media should be incubated under 10% CO_2 at room temperature and at 35 C.‡ Cultures should also be set up for *Actinomyces* and *Arachnia* (see Chapter 30).

*The Kinyoun stain is recommended, followed by **light** decolorization with acid alcohol or, preferably, 0.5% to 1% aqueous sulfuric acid.
†*N. asteroides* will not grow on media containing chloramphenicol.
‡Growth on primary culture is much more rapid in an atmosphere of 10% CO_2.[4]

It should be pointed out that *N. asteroides* frequently survives the sodium hydroxide or other decontamination procedures used in preparing sputum specimens for the isolation of *Mycobacterium tuberculosis* and will grow well on the usual isolation media.[2] *N. asteroides* appears as a moist glabrous colony on the tuberculosis media and grows out within 1 to 2 weeks. The colonies resemble those of the saprophytic or other mycobacterial species; slide cultures will reveal the branching acid-fast mycelium characteristic of *N. asteroides.*

On Sabouraud agar and blood agar, growth may appear as early as 3 days as **small yellow colonies** resembling those of *M. tuberculosis.* After 5 to 10 days the colonies become waxy, cerebriform, irregularly folded, and yellow to deep orange in color. Microscopically, the colony is composed of delicate **branching** filaments that break up into bacillary forms. These are gram positive and acid fast.* Cultures should be held for 30 days before discarding.

Because of the similarity between some species of *Nocardia* and species of *Streptomyces,* a variety of biochemical

*Pure cultures grown in litmus milk for several weeks will show a strong acid-fastness when stained as described above.

tests have been devised to differentiate them. The following tests are used by the Mycology Unit of the Center for Disease Control (CDC).

Demonstration of branching. Demonstration of branching is best done in slide culture. *N. asteroides*, *N. brasiliensis*, and *Streptomyces* species branch; filamentous bacteria do not.

Acid fastness. As shown by the Kinyoun stain with 1% sulfuric acid decolorization, *N. asteroides*, and *N. brasiliensis* are partially acid fast; spores of *Streptomyces* may be acid fast.

Hydrolysis of casein. See Chapter 43. *N. asteroides* does **not** hydrolyze casein, but casein is readily hydrolyzed by *N. brasiliensis* and *Streptomyces* species.

Growth in gelatin. See Chapter 43. *N. asteroides* either fails to grow or grows poorly, with a small, thin, flaky, white growth. *N. brasiliensis* grows well, forming discrete, compact, round colonies. *Streptomyces* species may grow well; growth is flaky or stringy.

Animal pathogenicity. Most isolates of *N. asteroides* are pathogenic for white mice and guinea pigs inoculated intraperitoneally; *N. brasiliensis* is usually not pathogenic. *Streptomyces* is not pathogenic, although some strains may cause a toxic death within 24 hours after inoculation.

As a test of pathogenicity for laboratory animals, a heavy suspension of the suspected culture should be prepared by grinding the growth from several slants of Sabouraud agar (incubated 1 to 2 weeks) with sterile saline, using a sterile mortar and pestle. This is combined with an equal amount of 5% hog gastric mucin, and 1 ml is injected intraperitoneally into a small (200 gm) guinea pig. The animal will generally die in 7 to 14 days, revealing a considerable amount of purulent peritoneal exudate at autopsy. Smears of this will reveal the characteristic gram-positive, acid-fast elements of *N. asteroides*. Since, however, there is considerable variation in strain virulence

and guinea pig susceptibility to *N. asteroides*, animal injection is not reliable unless positive. *N. asteroides* is best identified by colonial and biochemical characteristics.

An early diagnosis of pulmonary nocardiosis, combined with vigorous treatment, is important to prevent metastasis to the brain; sulfonamides appear to be the drugs of choice. It has been suggested that a sulfonamide combined with ampicillin or trimethoprim may be more effective.[1] Minocycline is very active in vitro, and limited clinical experience with cycloserine is encouraging. Amikacin is also very active in vitro (Wallace, R. J., Jr., et al.: J. Infect. Dis. **135**:568-576, 1977).

REFERENCES

1. Adams, A. R., Jackson, J. M., Scopa, J., Lane, G. K., and Wilson, R.: Nocardiosis, Med. J. Aust. **1**:669-674, 1971.
2. Ajello, L., Grant, V. Q., and Gutzke, M. A.: The effect of tubercle bacillus concentration procedures on fungi causing pulmonary mycoses, J. Lab. Clin. Med. **38**:486-491, 1951.
3. Bolton, W. K., Sande, M. A., Normansell, D. E., Sturgill, B. C., and Westervelt, F. B., Jr.; Ventriculojugular shunt nephritis with Corynebacterium bovis, Am. J. Med. **59**:417-423, 1975.
4. Boncyk, L. H., Millstein, C. H., and Kalter, S. S.: Use of CO_2 for more rapid growth of the *Nocardia* species, J. Clin. Microbiol. **3**:463-464, 1976.
5. Causey, W. A., and Sieger, B.: Systemic nocardiosis caused by Nocardia brasiliensis, Am. Rev. Resp. Dis. **109**;134-137, 1974.
6. Cherry, W. B., and Moody, M. D.: Fluorescent-antibody techniques in diagnostic bacteriology, Bacteriol. Rev. **29**:222-250, 1965.
7. Davis, A., Binder, M. J., Burroughs, J. T., Miller, A. B., and Finegold, S. M.: Diphtheroid endocarditis after cardiopulmonary bypass surgery for the repair of cardiac valvular defects. Antimicrobial agents and chemotherapy—1963, Ann Arbor, Mich., 1964, American Society for Microbiology, pp. 643-656.
8. Elek, S. D.: The plate virulence test for diphtheria, J. Clin. Pathol. **2**:250-258, 1949.
9. Fraser, D. T., and Weld, C. B.: The intracutaneous virulence test for C. diphtheriae, Trans. R. Soc. Can. Sect. **20**:343-345, 1926.
10. Gardner, S. E., Pearson, T., and Hughes, W. T.: Pneumonitis due to *Corynebacterium equi*, Chest **70**:92-94, 1976.
11. Gray, M. L., and Killinger, A. H.: *Listeria*

monocytogenes and *Listeria* infections, Bacteriol. Rev. **30:**309-382, 1966.

12. Hande, K. R., Witebsky, F. G., Brown, M. S., Schulman, C. B., Anderson, S. E., Jr., Levine, A. S., Mac Lowry, J. D., and Chabner, B. A.: Sepsis with a new species of *Corynebacterium*, Ann. Intern. Med. **85:**423-426, 1976.

13. Hermann, G. J., and Bickham, S. T.: *Corynebacterium.* In Lennette, E. H., Spaulding, E. H. and Truant, J. P., editors: Manual of clinical microbiology, ed. 2, Washington, D.C., 1974, American Society for Microbiology.

14. Hermann, G. J., Moore, M. S., and Parsons, E. I.: A substitute for serum in the diphtheria in vitro test, Am. J. Clin. Pathol. **29:**181-183, 1958.

15. Killinger, A. H.: *Listeria monocytogenes.* In Lennette, E. H., Spaulding, E. H., and Truant, J. P., editors: Manual of clinical microbiology, ed. 2, Washington, D.C., 1974, American Society for Microbiology.

16. Mitchell, G., Wells, G. M., and Goodman, J. S.: Sporotrichoid *Nocardia brasiliensis* infection, Am. Rev. Resp. Dis. **112:**721-723, 1975.

17. Nazemi, M. M., and Musher, D. M.: Empyema due to aerobic diphtheroids following dental extraction, Am. Rev. Resp. Dis. **108:**1221-1223, 1973.

18. Reller, L. B., Maddoux, G. L., Eckman, M. R., and Pappas, G.: Bacterial endocarditis caused by *Oerskovia turbata*, Ann. Intern. Med. **83:**664-666, 1975.

19. Weaver, R. E.: *Erysipelothrix.* In Lennette, E. H., Spaulding, E. H., and Truant, J. P., editors: Manual of clinical microbiology, ed. 2, Washington, D.C., 1974, American Society for Microbiology.

CHAPTER 27

Anaerobic gram-negative non-spore-forming bacilli

Bacteroides · *Fusobacterium* · Anaerobic vibrios

Five organisms or groups of organisms account for about two thirds of all clinically significant anaerobic infections.[5] These are *Bacteroides fragilis, B. melaninogenicus, Fusobacterium nucleatum, Clostridium perfringens,* and the anaerobic cocci. Thus, the anaerobic gram-negative rods are very important. They are the most commonly encountered of all anaerobes in infections.

The identification of the anaerobic species of *Bacteroides* and *Fusobacterium* is based on a number of morphologic, physiologic, and genetic characteristics much too detailed to be considered satisfactorily in a text of this type. The interested reader is strongly advised to consult the excellent anaerobic bacteriology manuals currently available for full descriptions of technical procedures and methods of identification, including gas-liquid chromatography.[3,8,13]

Most of the anaerobes commonly encountered in clinically significant infections can be identified definitively, or with reasonable accuracy, by means of a relatively small number of tests that are easy to carry out in the clinical laboratory setting. It is not essential for all laboratories to do gas chromatography. These tests, noted in Tables 27-1 through 27-3, are described in Chapter 43.

Preliminary **grouping procedures** utilizing observations on colonial and cellular morphology, susceptibility to antibiotic disks, and the results of a few simple biochemical tests are described in Chapter 13. These data may provide early and useful information to the clinician await-

ing more definitive identification. Each "preliminary group" is studied by means of additional tests appropriate to definitive identification or further breakdown of identification within the group.

GENUS BACTEROIDES
Bacteroides fragilis

Bacteroides fragilis is the anaerobe **most frequently isolated from clinical infections**[5]; it is also the predominant organism of the normal human intestinal tract. Its importance is further underlined by the fact that it is more resistant to antimicrobial agents than is any other anaerobe. The group has been further divided into five species, including *B. fragilis, B. distasonis, B. ovatus, B. thetaiotaomicron,* and *B. vulgatus. B. fragilis* and *B. thetaiotaomicron* are the species most commonly found in infections.

On blood agar, *B. fragilis* grows as a 1- to 3-mm smooth, white to gray, nonhemolytic, translucent, glistening colony. Gram stain shows a pale-staining gram-negative bacillus with rounded ends; some are pleomorphic, with filaments and irregular staining. Bile is not inhibitory (it may be stimulatory); catalase production is usually positive. *B. fragilis* is uniformly resistant to colistin, kanamycin, and vancomycin by the grouping technique described in Chapter 13, is usually resistant to penicillin, but is sensitive to erythromycin and rifampin. Speciation of *B. fragilis* is done as noted in Table 27-1, using the indole test and three carbohydrate fermentations (see Chapter 43). Miniaturized systems in-

Table 27-1. Typical characteristics of members of *Bacteroides fragilis* group

Characteristic	Species				
	Distasonis	*Fragilis*	*Vulgatus*	*Ovatus*	*Theta-iotaomicron*
Indole production	−	−	−	+	+
Fermentation of:					
Mannitol	−	−	−	+	−
Rhamnose	V	−	+	V	V
Trehalose	+	−	−	V	+

V, variable.

corporating these and other tests are available commercially; preliminary studies indicate they offer some promise.[6,9,11]

Bacteroides melaninogenicus

This species of *Bacteroides* has been divided into three subspecies: *B. melaninogenicus* ss. *melaninogenicus, asaccharolyticus,* and *intermedius.* However, ss. *asaccharolyticus* has been proposed as a separate species—*Bacteroides asaccharolyticus* (Finegold, S. M., and Barnes, E. M.: Int. J. Syst. Bacteriol. **27:**388-391, 1977). Normally part of the microflora of the oropharynx and upper respiratory tract, gastrointestinal tract, and genitourinary tract, it is less frequently isolated from human infections than is *B. fragilis* but is an important pathogen nonetheless. Characteristically it produces a **brown to black colony** on blood agar after 5 to 7 days; it grows more rapidly and pigments much earlier on media containing **laked** blood. Young colonies on blood agar (prior to pigmentation) will show a **brick-red fluorescence** when examined under ultraviolet light.* By Gram stain, the organism appears coccobacillary and slightly pleomorphic in broth media. Bile inhibits its growth, catalase production is negative, and indole is variable; the organism is sensitive to erythromycin, penicillin (usu-

ally), and rifampin, but is resistant to kanamycin and is variable to vancomycin by the disk identification technique described. Further biochemical tests are required for subspeciation.[7] These are described in Chapter 43. Table 27-2 shows the characteristics of the three subspecies.

Other Bacteroides species

B. oralis and *B. corrodens* (the obligately anaerobic variety)* are occasionally isolated from oral and pleuropulmonary or other infections[1] and, rarely, are isolated from blood cultures.[4] Colonies of *B. oralis* are similar to those of *B. fragilis,* while *B. corrodens* characteristically produces a small, translucent nonhemolytic colony that is depressed in the agar or **pits** the agar around the colony. *B. ochraceus* is closely related to *B. oralis,* but can grow in 5% to 10% CO_2 in air. Characteristics of these two organisms are shown in Table 27-2. All three species are inhibited by bile; indole and catalase are not produced. *B. oralis* and *B. ochraceus* are resistant to kanamycin in the disk identification test but sensitive to erythromycin, penicillin, and rifampin; variable reactions occur with colistin and vancomycin. *B. corrodens* is sensitive to all of the above except vancomycin. *B. corrodens* produces nitrite and hydrolyzes urea.

---

*"Blak-Ray" ultraviolet lamp and viewbox from Ultra-Violet Products, San Gabriel, Calif.

---

*Suggested new nomenclature for aerobic strains: *Eikenella corrodens* (Int. J. Syst. Bact. **22:**73-77, 1972, and **23:**75-76, 1973).

Table 27-2. Characteristics of *Bacteroides melaninogenicus-oralis-ochraceus* group

| | B. melaninogenicus | | | | |
	ss. Asaccharo-lyticus	ss. Inter-medius	ss. Melanino-genicus	B. oralis	B. ochraceus
Growth in 5% to 10% CO$_2$ in air	−	−	−	−	+
Indole production	+	+	−	−	−
Nitrate reduction	−	−	−	−	−
Fermentation of:					
Esculin	−	−	−	−	−
Glucose	−	+	+	+	+
Starch	−	+	+	+	+
Hydrolysis of:					
Esculin	−	−	+	+	+
Starch	−	+	+	+	+
Effect of bile on growth	I	I	I	I	I
Lipase	−	+/−	−	−	−

I, inhibition.

FUSOBACTERIUM SPECIES

A number of species belong to the genus *Fusobacterium;* included among those isolated from clinical specimens are *F. necrophorum* (formerly *Sphaerophorus necrophorus*), *F. nucleatum* (formerly *F. fusiforme*), *F. mortiferum* (formerly *S. mortiferus*), and *F. varium* (*S. varius*). Normally present in the upper respiratory tract, gastrointestinal tract, and genitourinary tract, the fusobacteria are responsible for serious pleuropulmonary, bloodstream, or metastatic suppurative infections (for example, brain abscess, lung abscess, septic arthritis, and so forth).

Morphologically, members of this genus characteristically appear as long, slender, spear-shaped, pale staining gram-negative bacilli with **tapered ends** (Fig. 27-1); some species, such as *F. mortiferum,* show a bizarre pleomorphism, with spheroid swellings along irregularly stained filaments and free round bodies. Colonies on blood agar are generally nonhemolytic (some are alpha or slightly beta hemolytic) and vary from flat to convex, with opaque centers and translucent irregular margins (for example, "fried egg" appearance of *F. mortiferum*). Convex, glistening alpha-hemolytic colonies with a flecked internal

Fig. 27-1. *Fusobacterium nucleatum.* Note pointed ends. (1000×.)

structure, or bread crumb colonies, are characteristic of *F. nucleatum.*

Bile is generally inhibitory and indole is produced (*F. mortiferum* is indole-negative and neither *F. mortiferum* nor *F. varium* are affected by bile); *F. necrophorum* often produces lipase on egg yolk agar. The fusobacteria are usually sensitive to colistin, kanamycin, and penicillin; are resistant to vancomycin; but show variability with respect to erythromycin and rifampin on the disk identification test. Other biochemical

Table 27-3. Characteristics of certain *Fusobacterium* species

Characteristic	F. gonidia-formans	F. navi-forme	F. necro-phorum	F. nucle-atum	F. morti-ferum	F. varium
Indole production	+	+.	+	+	−	+
Nitrate reduction	−	−	−	−	−	−
Fermentation						
Glucose	+	−	−	−	+	+
Levulose	−	−	−	+		
Mannose	−	−	−	−	+	+
Hydrolysis						
Esculin	−	−	−	−	+	−
Starch	+	−	−	−	V	−
Effect of bile on growth	I	I	V	I	No effect or stimulation	No effect or stimulation
Lipase production	−	−	+/−	−	−	−

I, inhibition; V, variable.

tests are required for precise identification. Certain key characteristics are shown in Table 27-3.

ANAEROBIC VIBRIOS

Anaerobic vibrios, usually not further identified, have been recovered, rarely, from central nervous system infection, bacteremia, pulmonary infection, and soft-tissue infections.[5] *Butyrivibrio fibrisolvens* has been recovered from a patient with endophthalmitis and *Succinivibrio* from a patient with bacteremia.[5]

CLINICAL EFFECTIVENESS OF ANTIMICROBIAL AGENTS

The in vitro sensitivity testing of bacteroides and fusobacteria and other anaerobes to various antimicrobial agents is discussed in Chapter 36. Typical patterns of susceptibility to 23 antimicrobial agents are given elsewhere.[12]

Penicillin G is the drug of choice for infections due to most of the gram-negative anaerobic rods **other than** *B. fragilis*, which is typically resistant. Resistance to penicillin G is also seen in some strains of *B. melaninogenicus* and *F. varium*. Chloramphenicol is consistently active against all anaerobes and penetrates the central nervous system well. Clindamycin is also very effective,[2] but it does not cross the blood-brain barrier well and a

few strains of *B. fragilis* and up to one third of *F. varium* strains are resistant. Metronidazole, not yet approved by the FDA, is consistently active against all gram-negative anaerobic rods[14]; it gets into the central nervous system well. It is the only agent consistently bactericidal against all susceptible anaerobes, including *B. fragilis*.[10] It should therefore be excellent in endocarditis. Surgical drainage and debridement are essential in the therapy of most anaerobic infections.

REFERENCES

1. Bartlett, J. G., and Finegold, S. M.: Anaerobic pleuropulmonary infections, Medicine **51**:413-450, 1972.
2. Bartlett, J. G., Sutter, V. L., and Finegold, S. M.: Treatment of anaerobic infections with lincomycin and clindamycin, N. Engl. J. Med. **287**:1006-1010, 1972.
3. Dowell, V. R., Jr., and Hawkins, T. M.: Laboratory methods in anaerobic bacteriology, CDC laboratory manual, DHEW Publ. No. (CDC) 74-8272, Washington, D.C., 1974, U.S. Government Printing Office.
4. Felner, J. M., and Dowell, V. R., Jr.: "Bacteroides" bacteremia, Am. J. Med. **50**:787-796, 1971.
5. Finegold, S. M.: Anaerobic bacteria in human disease, New York, 1977, Academic Press, Inc.
6. Hansen, S. L., and Stewart, B. J.: Comparison of API and Minitek to Center for Disease Control methods for the biochemical characterization of anaerobes, J. Clin. Microbiol. **4**:227-231, 1976.
7. Harding, G. K. M., Sutter, V. L., Finegold, S. M., and Bricknell, K. S.: Characterization of

Bacteroides melaninogenicus, J. Clin. Microbiol. **4:**354-359, 1976.

8. Holdeman, L. V., and Moore, W. E. C., editors: Anaerobe laboratory manual, ed. 3, Blacksburg, Va., 1975, Staff of the Anaerobe Laboratory, Virginia Polytechnic Institute and State University.

9. Moore, H. B., Sutter, V. L., and Finegold, S. M.: Comparison of three procedures for biochemical testing of anaerobic bacteria, J. Clin. Microbiol. **1:**15-24, 1975.

10. Nastro, L. J., and Finegold, S. M.: Bactericidal activity of five antimicrobial agents against *Bacteroides fragilis.* J. Infect. Dis. **126:**104-107, 1972.

11. Stargel, M.D., Thompson, F. J., Phillips, J. E., Lombard, G. L., and Dowell, V. R., Jr.: Modification of the Minitek miniaturized differentiation system for characterization of anaerobic bacteria, J. Clin. Microbiol. **3:**291-301, 1976.

12. Sutter, V. L., and Finegold, S. M.: Susceptibility of anaerobic bacteria to 23 antimicrobial agents, Antimicrob. Ag. Chemother. **10:**736-752, 1976.

13. Sutter, V. L., Vargo, V. L., and Finegold, S. M.: Wadsworth anaerobic bacteriology manual, ed. 2, Los Angeles, 1975, Dept. of Continuing Education and Health Sciences, School of Medicine, UCLA.

14. Tally, F. P., Sutter, V. L., and Finegold, S. M.: Treatment of anaerobic infections with metronidazole, Antimicrob. Ag. Chemother. **7:**672-675, 1975.

CHAPTER 28

Anaerobic cocci

Peptostreptococcus · Peptococcus · Veillonella ·
Acidaminococcus · Megasphaera

Next to the anaerobic gram-negative bacilli, the anaerobic gram-positive cocci are the anaerobes most commonly encountered in clinically significant infections.[6] Reflecting differences in their presence as normal flora, the **gram-positive** anaerobic cocci are relatively more prevalent in respiratory tract and related infections. They are seen in female genital tract infections and are less commonly encountered in intraabdominal processes than are the gram-negative anaerobic bacilli.[6,14] As with the gram-negative anaerobic rods, however, they may be found in virtually all types of infection. The **gram-negative** anaerobic cocci are not significant pathogens as a rule, but may be found in mixed infections and, rarely, even in pure culture.[2]

Unfortunately, classification and characterization of the anaerobic cocci is somewhat complicated. For example, organisms formerly classified as *Peptococcus morbillorum* and *Peptostreptococcus intermedius* actually belong to the genus *Streptococcus,* since they produce lactic acid as a metabolic end product without significant amounts of other products. These organisms become aerotolerant after one or more subcultures. On the other hand, some strains of *Streptococcus* are obligately anaerobic. The so-called **microaerophilic** streptococci have occasioned the greatest confusion. These organisms have often been included with the anaerobic cocci because they are readily overlooked if one does not use good anaerobic transport and culture techniques;

thus, workers interested in anaerobes have recovered these organisms much more commonly than have other investigators. However, the microaerophilic streptococci are actually members of the genus *Streptococcus* on the basis of metabolic end-product analysis (Harder, E . J., Sutter, V. L., and Finegold, S. M.: Unpublished data). They can often be speciated and include, for example, S. *mutans, S. mitis, S. anginosus, S. sanguis,* S. *MG,* and S. *salivarius.* Although "viridans streptococci" have been said to be nonpathogenic except in subacute bacterial endocarditis, it is clear that these organisms are not infrequently involved (and not uncommonly in pure culture) in such serious infections as brain abscess, necrotizing pneumonia, liver abscess, postabortal sepsis, and so forth.[6] Anaerobic cocci, particularly *Peptococcus,* may also be found as (skin) contaminants in blood cultures.

Definitive identification of the anaerobic cocci depends on gas chromatographic analysis of end products; nevertheless, as shown in Table 28-1, the use of a number of simple tests permits speciation of many of these organisms. Certainly characterization adequate for the usual clinical purposes (immediate treatment of the infected patient) is possible by these means. Reference laboratories may be used for definitive identification where this is of interest.

Preliminary grouping of these organisms is discussed in Chapter 13. The tests noted in Table 28-1, for further char-

Table 28-1. Characteristics of some anaerobic cocci found in human infection

	Inhibition by SPS	Catalase	Indole production	Nitrate reduction	Gelatin liquefaction	Fermentation						Esculin hydrolysis	Growth stimulation with Tween 80
						Cellobiose	Glucose	Lactose	Levulose	Maltose	Sucrose		
Gram-negative cocci													
Acidaminococcus fermentans	–	–	–	–	–	–	–	–	–	–	–	–	–
Megasphaera elsdenii	–	+	–	–	–	–	+	–	+	+	–	–	–
Veillonella alcalescens	–	+	–	+	–	–	–	–	–	–	–	–	–
V. parvula	–	–	–	+	–	–	–	–	–	–	–	–	–
Gram-positive cocci													
Peptococcus asaccharolyticus	–	–	+	–	–	+	+	–	+	–	–	–	V
P. constellatus	–	–	–	V	–	+	+	+	+	+	+	+	–
P. magnus*	–	–	–	–	+	–	–	–	–	–	–	–	V
P. prevotii†	–‡	–	–	–	+	–	–	–	–	–	–	–	V
P. variabilis*	–	–	–	–	–	–	+	–	+	–	–	–	V
Peptostreptococcus anaerobius	+	–	–	–	–	–	–	–	–	–	–	–	V
P. micros	–‡	–	–	–	+	–	+	+	–	–	–	–	+
P. parvulus	–	–	–	–	–	–	+	+	–	–	–	–	V
P. productus	–	–	–	–	–	+	+	–	+	+	+	+	–

Note: *Peptococcus morbillorum* and *Peptostreptococcus intermedius* are not included, since they become aerotolerant after one or more subcultures and produce lactic acid without significant amounts of other acids or gas. They should be identified as *Streptococcus* species. These organisms are commonly designated "microaerophilic streptococci." *P. magnus* and *P. variabilis* are probably the same as *Peptococcus anaerobius* (Bergey's, ed. 8). However, these names are retained until there is clarification of their status and the status of *P. anaerobius.*
†*P. prevotii* is not recognized in Bergey's (ed. 8) and may be a variant of *P. asaccharolyticus.*
‡Occasional strains have a small zone of inhibition about an SPS disk.

acterization, are described in Chapter 43. Additional data are found in three reference manuals.[5,7,13]

PEPTOSTREPTOCOCCUS

Material suspected of harboring anaerobic streptococci, particularly in **foul-smelling** pus, should be plated **without delay** on fresh blood agar and enriched thioglycollate medium, placed immediately into an **anaerobic** atmosphere (see Chapter 13 for anaerobic methods), and examined after 48 hours' incubation. Peptostreptococci appear as minute, shiny, smooth colonies. Hemolysis is variable; in thioglycollate medium, gas and a foul odor are generally apparent. Subcultures incubated aerobically and anaerobically should reveal only anaerobic growth of a catalase-negative, gram-positive streptococcus; this may sometimes exhibit a sharp pungent odor.[10] See Table 28-1 for characterization tests. In addition, it has been noted that DNase is produced by *P. anaerobius* and *P. intermedius* (the latter is a *Streptococcus*).[9] However, occasional strains of other anaerobic gram-positive cocci also produce DNase. The anaerobic coccus most commonly encountered in infections, *P. anaerobius*

(Fig. 28-1), is readily identified by the SPS disk test[15] (see footnote, p. 108).

A Gram stain of a **direct smear** of the purulent material also should be examined; a report of the presence of **tiny** gram-positive cocci in chains may alert the clinician to the possibility of a peptostreptococcal infection.

The drug of choice in the treatment of infections due to anaerobic streptococci is generally benzyl penicillin (penicillin G), although occasional strains require as much as 32 units per milliliter[12] for inhibition and therefore are probably best treated with another agent. Clindamycin and chloramphenicol have also proved effective, especially in mixed anaerobic infections.[1] Metronidazole is active against 97% of strains.[12] There is evidence that the peptostreptococci may act synergistically with *Bacteroides* or *S. aureus* to produce tissue necrosis in mixed infections.[8]

PEPTOCOCCUS

The species of this genus are **obligate anaerobes,** and although they are found normally in the respiratory and genital tracts and on the skin of humans, they have been isolated from a variety of in-

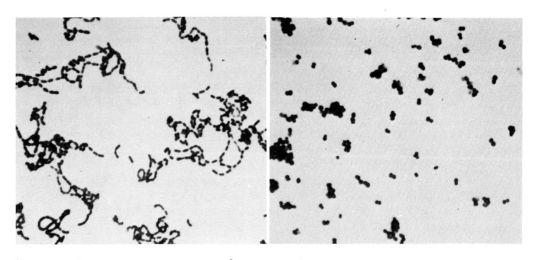

Fig. 28-1. *Peptostreptococcus anaerobius.* (1000×.)

Fig. 28-2. *Peptococcus prevotii.* (1000×.)

fections, usually abscesses, and often in mixed culture. They occur in irregular masses, and some resemble *S. aureus* microscopically (Fig. 28-2). Such an appearance on direct Gram stain, with no staphylococci recovered on aerobic culture after 18 to 24 hours, should suggest the possibility of anaerobic micrococci.

Peptococci are incriminated in anaerobic infections less frequently than peptostreptococci, and they may be distinguished from the latter in that peptococci are sometimes **catalase positive** and do not produce a sharp, pungent odor. In general, however, there is no simple way to distinguish between *Peptococcus* and *Peptostreptococcus*. The tests noted in Table 28-1 will often provide this differentiation and speciation as well.

Peptococci are routinely very sensitive to penicillin G, and this drug would be the agent of choice in *Peptococcus* infections. Chloramphenicol is also always effective. All but 2% of strains are susceptible to metronidazole. However, in the case of clindamycin, significant resistance has become apparent recently; at present, 17% of strains are resistant.[12]

GRAM-NEGATIVE ANAEROBIC COCCI

Organisms in the genus *Veillonella* occur in pairs, short chains, and irregular clumps. They are gram negative and strictly **anaerobic.** The cells are smaller than neisseriae and are less fastidious in their nutritional requirements. They occur normally in the respiratory tract, the intestinal tract, and the genitourinary tract of humans and animals. Two species have been described, *V. alcalescens* and *V. parvula.* Normally *Veillonella* is not considered pathogenic, but it is found in mixed infection and, rarely, as a single infecting organism.[2] Chow and associates[4] report that *Veillonella* strains fluoresce **red** under long-wave ultraviolet light. Unlike *B. melaninogenicus,* fluorescence does not depend on blood in the medium and is lost rapidly after exposure of colonies to air.

Acidaminococcus and *Megasphaera* differ from *Veillonella* in being nitrate negative. All three genera are distinct in terms of metabolic end products.[3] *Megasphaera* is fermentative, whereas *Acidaminococcus* is not (Table 28-1). *M. elsdenii* commonly stains gram positive, although it is really gram negative on the basis of cell-wall composition. Both *Acidaminococcus* and *Megasphaera* are found normally in the lower intestinal tract of humans.[11] *M. elsdenii* presumably is indigenous to the oral cavity as well, since it has been recovered as part of the flora of a putrid lung abscess.[11] *A. fermentans* was found in a mixed aerobic-anaerobic intraabdominal abscess.[11]

REFERENCES

1. Bartlett, J. G., Sutter, V. L., and Finegold, S. M.: Treatment of anaerobic infections with lincomycin and clindamycin, N. Engl. J. Med. **287**-1006-1010, 1972.
2. Borchardt, K. A., Baker, M., and Gelber, R.: *Veillonella parvula* septicemia and osteomyelitis, Ann. Intern. Med. **86**:63-64, 1977.
3. Buchanan, R. E., and Gibbons, N. E., editors: Bergey's manual of determinative bacteriology, ed. 8, Baltimore, 1974, The Williams & Wilkins Co.
4. Chow, A. W., Patten, V., and Guze, L. B.: Rapid screening of *Veillonella* by ultraviolet fluorescence, J. Clin. Microbiol. **2**:546-548, 1975.
5. Dowell, V. R., Jr., and Hawkins, T. M.: Laboratory methods in anaerobic bacteriology, CDC laboratory manual, DHEW Publ. No. (CDC) 74-8272, Washington, D.C., 1974, U.S. Government Printing Office.
6. Finegold, S. M.: Anaerobic bacteria in human disease, New York, 1977, Academic Press, Inc.
7. Holdeman, L. V., and Moore, W. E. C., editors: Anaerobe laboratory manual, ed. 3, Blacksburg, Va., 1975, Staff of the Anaerobe Laboratory, Virginia Polytechnic Institute and State University.
8. Pien, F. D., Thompson, R. L., and Martin, W. J.: Clinical and bacteriologic studies of anaerobic gram-positive cocci, Mayo Clin. Proc. **47**:251-257, 1972.
9. Porschen, R. K., and Sonntag, S.: Extracellular deoxyribonuclease production by anaerobic bacteria, Appl. Microbiol. **27**:1031-1033, 1974.
10. Rogosa, M.: Peptococcaceae, a new family to include the gram-positive, anaerobic cocci of the genera *Peptococcus, Peptostreptococcus,* and *Ruminococcus,* Int. J. Syst. Bacteriol. **21**: 234-237, 1971.

11. Sugihara, P. T., Sutter, V. L., Attebery, H. R., Bricknell, K. S., and Finegold, S. M.: Isolation of *Acidaminococcus fermentans* and *Megasphaera elsdenii* from normal human feces, Appl. Microbiol. **27**:274-275, 1974.

12. Sutter, V. L., and Finegold, S. M.: Susceptibility of anaerobic bacteria to 23 antimicrobial agents, Antimicrob. Ag. Chemother. **10**:736-752, 1976.

13. Sutter, V. L., Vargo, V. L., and Finegold, S. M.: Wadsworth anaerobic bacteriology manual, Los Angeles, 1975, Dept. of Continuing Education and Health Sciences, School of Medicine, UCLA.

14. Thomas, C. G. A., and Hare, R.: The classification of anaerobic cocci and their isolation in normal human beings and pathological processes, J. Clin. Pathol. **7**:300-304, 1954.

15. Wideman, P. A., Vargo, V. L., Citronbaum, D. M., and Finegold, S. M.: Evaluation of the sodium polyanethol sulfonate disk test for the identification of *Peptostreptococcus anaerobius,* J. Clin. Microbiol. **4**:330-333, 1976.

CHAPTER 29

Anaerobic gram-positive spore-forming bacilli

Clostridium

The genus *Clostridium* contains the **anaerobic, spore-forming** bacteria. A large number of species are involved, and the majority are **obligate anaerobes.** Some species are aerotolerant, showing growth on enriched media on aerobic incubation. The sporangia are often characteristically **swollen** (Fig. 29-1), showing spindle, drumstick, and "tennis racket" forms and containing central, subterminal, and terminal spores. With rare exceptions in the aerotolerant forms, the enzymes catalase, cytochrome oxidase, and peroxidase are **not** produced. Many attack carbohydrates, and some are proteolytic.

True **exotoxins** are produced by the pathogenic clostridia, several species of which are important in medicine. The organisms are widely distributed in soil, dust, and water and are common inhabitants of the intestinal tract of animals, including humans. They are often found in unclean wounds and wound infections. The spores of some pathogenic species (*C. botulinum*) may appear in improperly home-canned produce, in which they can develop vegetatively under normal domestic conditions, in inefficiently sterilized surgical dressings and bandages, in plaster of Paris for casts, and on the clothing and skin of humans.

GENERAL METHODS OF ISOLATION AND CULTIVATION OF PATHOGENIC CLOSTRIDIA

The procedures for anaerobic culture discussed in Chapter 13 are applicable to the clostridia. It should be reemphasized here that any material for anaerobic cul-

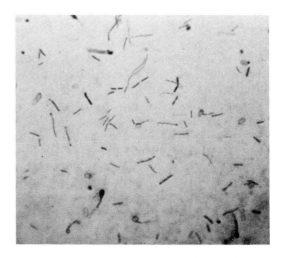

Fig. 29-1. *Clostridium septicum.* Note subterminal and free spores. (1000×.)

tivation should be inoculated **immediately,** using fresh media, without awaiting the results of aerobic culture. Egg yolk medium[4] should be used, in addition to the others mentioned in Chapter 13, when clostridia are anticipated to be present. Media with 5% agar may also be useful to minimize swarming of colonies.

In most instances, the clostridia occur in **mixed** culture with gram-negative bacteria, such as coliforms, *Proteus*, or *Pseudomonas*, and with various non–spore-forming anaerobes. The anaerobic plates, therefore, may be overgrown with such organisms, and isolation of the clostridia becomes difficult or impossible. Either of the two following methods may be used to combat this problem:

1. The original culture, shown to contain gram-positive sporulating rods, may be **heat shocked** at 80 C for 10

minutes and then streaked on blood agar and egg yolk agar plates for anaerobic cultivation. Alternatively, a fresh tube of enriched thioglycollate medium or of starch broth may be inoculated from the original culture, heated immediately, and incubated for 24 to 48 hours. From this, the plates may be streaked.

2. Incorporate 100 μg/ml neomycin in blood agar plates or egg yolk agar plates. This compound can be autoclaved.

Cooked-meat medium also may be used for cultivation of the spore-forming anaerobes and is preferred by some workers. Tubes of this medium require a petrolatum seal if obligate anaerobes are being cultivated, unless the medium is prereduced and anaerobically sterilized.

The **Nagler** reaction[5] is recommended for the rapid identification of *C. perfringens* and other alpha-toxin producers. The medium for this reaction, 10% egg yolk in blood agar base (see Chapter 41 for preparation), is placed in a Petri dish, and one half the surface is smeared with a few drops of *C. perfringens* type A antitoxin*(antilecithinase). The culture is then streaked in a single line across the plate at a right angle to the antitoxin. The toxin lecithinase produces a precipitate (opalescence) about the growth in the line of streak in the absence of antitoxin, but it is inhibited on the half of the plate with antitoxin.†

---

*Available from Wellcome Reagents Division, Burroughs Wellcome Co., Research Triangle Park, N.C.
†*C. bifermentans, C. sordellii,* and *C. paraperfringens (C. barati)* also produce alpha toxin and give a positive Nagler reaction. See Table 29-1 for means of differentiating these species.

IDENTIFICATION OF THE PATHOGENIC CLOSTRIDIA

By virtue of the disease syndromes they produce, the pathogenic species may be placed in five categories or groups:

Group I, the **gas gangrene** group, includes *C. perfringens* (type A), *C. novyi, C. septicum, C. sporogenes, C. bifermentans, C. sordellii, C. histolyticum,* and others. Of these, the three species first named are most important.

Group II, *C. tetani.*

Group III, the *C. botulinum* group.

Group IV, the miscellaneous infection group (wound infection, abscesses, bacteremia, and so forth). This group includes *C. perfringens, C. ramosum, C. bifermentans, C. sphenoides, C. sporogenes,* and a number of others.

Group V, *C. difficile,* responsible for antibiotic-induced colitis.*

In general, *C. perfringens* and *C. ramosum* are the most commonly isolated clostridia. Group IV infections are the most commonly encountered by far. However, clostridia are found in such infections only about one tenth as often as non–spore-forming anaerobes. *C. ramosum* (Fig. 29-2), although not as virulent as *C. perfringens,* takes on added importance by virtue of its **resistance** to antimicrobial agents. *C. ramosum* is second

---

*Bartlett, J. G., Onderdonk, A. B., Cisneros, R. L., and Kasper, D. L.: Clindamycin-associated colitis due to a toxin-producing species of *Clostridium* in hamsters, J. Infect. Dis. **136:**701-705, 1977; and George, W. L., Sutter, V. L., and Finegold, S. M.: Antimicrobial agent induced diarrhea—a bacterial disease. Editorial, J. Infect. Dis. **136:**822-828, 1977.

Table 29-1. Characteristics of lecithinase, α-toxin–producing *Clostridium* species

	Motility	Fermentation of lactose	Gelatin	Urease
C. perfringens	−	+	+	
*C. paraperfringens**	−	+	−	
C. bifermentans	+	−	+	−
C. sordellii	+	−	+	+

C. paraperfringens was formerly called *C. barati.*

Table 29-2. Characteristics of some non-α-toxin–producing *Clostridium* species

	Spores*	Aerobic growth	Motility	Lecithinase	Lipase	Indole production	Gelatin liquefaction	Meat digestion	Milk†	Fermentation				Esculin hydrolysis	Principal acid fermentation products‡
										Glucose	Lactose	Maltose	Mannitol		
C. botulinum (A,B,F,G)§	OS	−	V	−	+	−	+	+	CD	+	−	+	−	V	A IB B IV (P, V, IC, F)
C. botulinum (C,D)	OS	−	+	−	+	−	+	−	CD	+	−	+	−	V	APB (F)
C. botulinum (B,E,F)	OS	−	+	V	+	−	+	−	C	+	−	V	−	V	AB (F)
C. butyricum	OS	−	+	−	−	−	−	−	CG	+	+	+	+	+	ABF
C. difficile	ST	−	+	−	−	−	+	−	−	+	−	−	+	+	A, IB, B, IV, V, IC, F, L
C. histolyticum	OS	V	+	−	−	−	+	+	CD	−	−	−	−	−	A
C. innocuum	OT	−	−	−	−	−	−	−	−	+	−	−	+	+	ABL (F)
C. novyi A	OS	−	+	+	+	−	+	−	C	+	−	V	−	−	APBV
C. novyi B	OS	−	+	+	−	−	+	−	C	+	−	−	−	−	APBV
C. ramosum	R/OT	−	−	−	−	−	−	−	C	+	+	+	V	+	ALF
C. septicum	OS	−	+	−	+	−	+	−	CG	+	+	+	−	+	ABF
C. sporogenes	OS	−	+	−	+	−	+	+	CD	+	−	V	−	+	A, P, IB, B, IV, V, IC
C. subterminale	OS	−	+	−	−	−	+	+	CD	−	−	−	−	−	A, IB, B, IV
C. tertium	OT	+	+	−	−	−	−	−	C	+	+	+	+	+	ABLF
C. tetani	RT	−	+	−	−	+	+	V	−	−	−	−	−	−	APB

Adapted from Holdeman, L. V., and Moore, W. E. C., editors: Anaerobe laboratory manual, ed. 3, Blacksburg, Va., 1975, Virginia Polytechnic Institute and State University.

*O, oval; R, round; S, subterminal; T, terminal.

†C, clot; D, digested; G, gas; V, variable.

‡A, acetic; B, butyric; F, formic; IB, isobutyric; IC, isocaproic; IV, isovaleric; L, lactic; P, propionic; V, valeric.

§Letters refer to types.

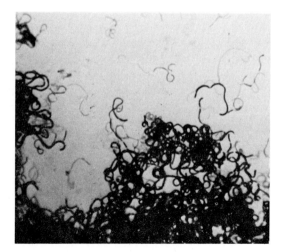

Fig. 29-2. *Clostridium ramosum.* (1000×.)

mice and guinea pigs are normally used in the laboratory. Susceptibility and rapidity of death may vary according to the virulence of the strain. This type of testing is beyond the scope of the clinical laboratory. Cultivation of *C. botulinum* should only be attempted by reference laboratories. Determination of toxigenicity, where indicated, should be done by reference laboratories.[1]

For a further study of the clostridia, the reader is referred to some of the sources listed in this chapter.[1-3,6-8]

only to *B. fragilis*, among the anaerobes, in such resistance. About 15% of strains are highly resistant to clindamycin, and many strains are resistant to tetracycline and erythromycin. Penicillin is the drug of choice against *C. ramosum* and clostridia in general, but MICs of *C. ramosum* are as high as 8 units per milliliter. Certain other clostridia (other than *C. perfringens*) are also resistant to clindamycin. Chloramphenicol and metronidazole are universally active against clostridia.

Most of the human pathogenic clostridia are pathogenic also for guinea pigs, mice, rabbits, and pigeons. To establish the toxigenicity of isolated strains, white

REFERENCES

1. Dowell, V. R., and Hawkins, T. M.: Detection of clostridial toxins, Toxin neutralization tests, and pathogenicity tests, Atlanta, 1968, Center for Disease Control.
2. Finegold, S. M.: Anaerobic bacteria in human disease, 1977, New York, Academic Press, Inc.
3. Gorbach, S. L., and Thadepalli, H.: Isolation of *Clostridium* in human infections: evaluation of 114 cases, J. Infect. Dis. **131:** S 81-S 85, 1975.
4. McClung, L. S., and Toabe, R.: The egg yolk plate reaction for the presumptive diagnosis of *Clostridium sporogenes* and certain species of gangrene and botulinum, J. Bacteriol. **53:**139-147, 1947.
5. Nagler, F. P. O.: Observations on a reaction between the lethal toxin of *Cl. welchii* (type A) and human serum, Br. J. Exp. Pathol. **20:**473, 1939.
6. Smith, L. DS.: The pathogenic anaerobic bacteria, ed. 2, Springfield, Ill., 1975, Charles C Thomas, Publisher.
7. Sterne, M., and van Heyningen, W. E.: The clostridia. In Dubos, R. J., and Hirsch, J. G.: Bacterial and mycotic infections of man, ed. 4, Philadelphia, 1965, J. B. Lippincott Co.
8. Willis, A. T.: Clostridia of wound infection, London, 1969, Butterworth & Co. (Publishers) Ltd.

CHAPTER 30

Anaerobic gram-positive non-spore-forming bacilli

Bifidobacterium · Propionibacterium Eubacterium
Lactobacillus · Actinomyces · Arachnia

The identification of this group of anaerobes requires the use of **gas chromatography** (or a substitute such as silicic acid chromatography), with rare exception. The catalase-positive forms, *P. acnes* and *A. viscosus*, may be recognized without it. Fortunately, only *Actinomyces, Arachnia,* and *Bifidobacterium eriksonii* are major pathogens and none of these is encountered often. Laboratories not equipped for gas chromatography will need to send selected cultures to reference laboratories.* In the case of actinomycosis, however, clinical and pathologic features may be very distinctive or diagnostic. A serologic test for actinomycosis is available.* The interested reader is referred to other sources for details of gas chromatographic and other identification procedures. Significant members of each genus are described on the following pages. Table 30-1 outlines identifying characteristics.

BIFIDOBACTERIUM ERIKSONII

This organism is part of the normal human oral and intestinal microflora, and occurs principally in mixed pulmonary infections. *B. eriksonii* grows as a white, convex, shiny colony with an irregular edge. In thioglycollate medium, growth is diffuse, and the Gram stain shows a diphtheroid to filamentous bacillus, branched or bifurcated. Catalase and in-

dole are not produced, nitrate is not reduced; gelatin is not liquefied, but esculin is hydrolysed. Carbohydrate fermentation reactions and metabolic end products are characteristic.[6,8]

PROPIONIBACTERIUM ACNES

This anaerobe* produces propionic acid by fermentation of glucose and is part of the resident flora of **normal skin;** consequently, it is the most frequent contaminant of blood cultures and often contaminates other cultures as well. It occasionally causes infection, especially endocarditis and ventricular shunt infections. *P. acnes* characteristically grows as a small, white to pinkish, shiny to opaque colony with entire margins. A Gram stain of the colony shows slender, slightly curved rods, sometimes with false branching or a beaded appearance. Tests for catalase and indole production are positive; nitrates are reduced, gelatin is liquefied, but esculin is not hydrolyzed. Characteristic patterns of carbohydrate fermentation and organic acid end products are obtained.[6,8] *P. acnes* is very sensitive to penicillin G and most other antimicrobials except aminoglycosides.[9]

EUBACTERIUM SPECIES

Eubacterium species are isolated from wound and other infections and are almost always associated with other an-

---

*For example, the Center for Disease Control, Atlanta, through the referring state laboratory.

*Some strains are microaerophilic.

235

Table 30-1. Characteristics of some gram-positive, non-spore-forming bacilli found in human infection*

	Oxygen tolerance	Catalase	Indole production	Nitrate reduction	Gelatin liquefaction	Esculin hydrolysis	Starch hydrolysis	Fermentation										Principal acid fermentation products
								Arabinose	Erythritol	Glucose	Inositol	Lactose	Mannitol	Raffinose	Sorbitol	Trehalose	
Actinomyces israelii	A,M†	−	−	−	−	+	−	−	−	+	+	+	+	+	−	+	AFLS‡
A. odontolyticus	A,M	−	−	+	−	V	−	V	−	+	+	+⁻	−	V	−	+⁻	AFLS
A. naeslundii	M,F	−	−	+	−	+	−	−	−	+	V	+	−	+	−	+	AFLS
A. viscosus	F	+	−	+	−	+	V	−	−	+	−	+⁻	−	+	−	V	AFLS
Arachnia propionica	A,M	−	−	+	V	−	−	+	−	+	−	V	V	V	−	+	APS (L)
Propionibacterium acnes	A,M	+	+	+	+	−	−	−	−	+	−	−	V	V	−	−	AP (IVFLS)
Bifidobacterium eriksonii§	A	−	−	−	−	+	V	+	−	+	−	+	+	+	−	+	AL (FS)
Lactobacillus catenaforme	A	−	−	−	−	+	+	−	−	+	−	+	−	+	−	−	L (AFS)
Eubacterium alactolyticum	A	−	−	−	−	−	−	−	−	+	−	−	+	−	−	−	ABH (CFS)
E. lentum	A	−	−	+	−	−	−	+	+	+	−	−	−	−	−	−	(AFLS)
E. limosum	A	−	−	−	−	+	−	+	+	+	−	−	+	−	−	−	ABL (S)

Adapted from Holdeman, L. V., and Moore, W. E. C., editors: Anaerobe laboratory manual, ed. 3, Blacksburg, Va., 1975, Virginia Polytechnic Institute and State University.

*Characteristics used to differentiate commonly encountered species after generic identification provided by fatty acid end-product analysis. See VPI manual[6] for other characteristics of these species and characteristics of other species.

†A, anaerobic; M, microaerophilic; F, facultative.

‡A, acetic; B, butyric; C, caproic; F, formic; H, heptanoic; IV, isovaleric; L, lactic; P, propionic; S, succinic.

§This organism is listed as *Actinomyces eriksonii, species incertae sedis* in Bergey's manual, ed. 8.

aerobes or facultative bacteria; they are part of the normal fecal microflora. They are not particularly pathogenic but may cause endocarditis. Growing as raised to convex, translucent to opaque colonies, they appear microscopically as pleomorphic bacillary to coccobacillary forms, occurring in pairs and short chains. *E. lentum* is relatively inactive biochemically; *E. limosum* hydrolyzes esculin and ferments several carbohydrates. Characteristic metabolic end products are observed.

LACTOBACILLUS SPECIES

These anaerobic members of the genus *Lactobacillus* produce primarily lactic acid from the fermentation of glucose and are only occasionally involved in human infections, usually pleuropulmonary. One species, *L. catenaforme,* was formerly classified as a member of the genus *Catenabacterium;* however, some strains of the latter were shown to produce spores and therefore belong to the clostridia.[6] *Lactobacillus catenaforme* grows as a convex, translucent colony; microscopic examination reveals pleomorphic gram-positive bacilli, sometimes in chains. Terminal swellings are also observed.

ACTINOMYCES AND ARACHNIA

The etiologic agents of **human actinomycosis** include the most common and important organism, *A. israelii,* and several other species, *A. naeslundii, A. odontolyticus, A. viscosus,* and *Arachnia propionica.*[2,4] These organisms are a part of the normal microbiota of the mouth and produce infection primarily as endogenous opportunists. While *A. israelii, A. odontolyticus,* and *Arachnia* grow best under anaerobic conditions, *A. naeslundii* and *A. viscosus* strains are either microaerophilic or facultative. *A. israelii* produces a characteristically heaped, rough, lobate colony resembling a **molar tooth;** microscopically it appears as long, filamentous, gram-positive bacilli, some

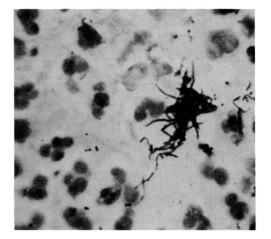

Fig. 30-1. *Actinomyces israelii,* crushed sulfur granule, showing bacterial forms and pus cells. (Gram strain; 1000×.)

of which may show branching. A Gram stain of "**sulfur granules**" (Fig. 30-1) from actinomycotic pus reveals a similar morphology, often with diphtheroid and coccal forms. Colonies of the other organisms generally are smooth, flat to convex, whitish, and transparent, with entire margins. *A. odontolyticus* colonies on blood agar may develop a red color after several days. Microscopically, the other causes of actinomycosis are gram-positive bacilli consisting of diphtheroidal forms and branched elements; some filaments may show clubbed ends.[3]

A. viscosus, being catalase positive, is often overlooked because it is assumed to be a *Propionibacterium* or diphtheroid and therefore not likely to be important. It differs from *P. acnes* by growing equally well aerobically and anaerobically, hydrolyzing esculin, not producing indole, not being proteolytic, not producing a pink sediment in thioglycollate broth, and fermenting melibiose, raffinose, sucrose, and salicin.[5] Direct identification of *Actinomyces* and of *Arachnia* by immunofluorescent techniques has proved successful (available from CDC, Atlanta).[7]

Penicillin G remains the drug of choice

in the treatment of actinomycosis; tetracycline and clindamycin also demonstrate activity against these organisms.[1]

REFERENCES

1. Bartlett, J. G., and Finegold, S. M.: Anaerobic pleuropulmonary infections, Medicine **51**:413-450, 1972.
2. Brock, D. W., Georg, L. K., Brown, J. M., and Hicklin, M. D.: Actinomycosis caused by *Arachnia propionica;* report of 11 cases, Am. J. Clin. Pathol. **59**:66-77, 1973.
3. Coleman, R. M., Georg, L. K., and Rozzell, A. R.: *Actinomyces naeslundii* as an agent of human actinomycosis, Appl. Microbiol. **18**:420-426, 1969.
4. Georg, L. K.: Diagnostic procedures for the isolation and identification of the etiologic agents of actinomycosis. In Proceedings of the International Symposium on Mycoses, Washington, D.C., 1970, Pan American Health Organization Scientific Publ. No. 205.
5. Gerencser, M. A., and Slack, J. M.: Identification of human strains of *Actinomyces viscosus*, Appl. Microbiol. **18**:80-87, 1969.
6. Holdeman, L. V., and Moore, W. E. C., editors: Staff of the Anaerobe Laboratory, Anaerobe laboratory manual, ed. 3, Blacksburg, Va., 1975, Virginia Polytechnic Institute and State University.
7. Lambert, Jr., R. F., Brown, J. M., and Georg, L. K.: Identification of *Actinomyces israelii* and *Actinomyces naeslundii* by fluorescent antibody and agar-gel diffusion techniques, J. Bacteriol. **94**:1287-1295, 1967.
8. Sutter, V. L., Vargo, V. L., and Finegold, S. M.: Wadsworth anaerobic bacteriology manual, ed. 2, Los Angeles, 1975, Dept. of Continuing Education and Health Sciences, School of Medicine UCLA.
9. Wang, W. L. L., Everett, E. D., Johnson, M., and Dean, E.: Susceptibility of *Propionibacterium acnes* to seventeen antibiotics, Antimicrob. Ag. Chemother. **11**:171-173, 1977.

CHAPTER 31

Mycobacteria

The mycobacteria may vary morphologically from the coccobacillary form to long, narrow, rod-shaped cells ranging from 0.8 to 5 μm in length and about 0.2 to 0.6 μm in thickness. They do not stain readily but, once stained, will resist decolorization with acid-alcohol and are therefore called **acid-fast bacilli.** Occurring as single bacilli or in small irregular clumps, mycobacteria are sometimes beaded, banded, or pleomorphic in stained smears. In addition to saprophytic species, the group includes numerous organisms pathogenic for humans, the most important of which is *Mycobacterium tuberculosis.*

LABORATORY DIAGNOSIS OF TUBERCULOSIS AND RELATED MYCOBACTERIOSES
General considerations

Tuberculosis is an infectious disease of a persistent and chronic nature that is usually caused by *Mycobacterium tuberculosis* but occasionally by other species, such as *M. bovis* and *M. kansasii.* Although capable of involving almost any organ of the body, tuberculosis is most commonly associated with the lungs, from which it spreads from person to person through coughing or expectoration. Tuberculosis is generally considered to be the most socioeconomically important specific communicable disease in the world today. Although it no longer ranks as the most common cause of death in nations with a high standard of living, it still remains a leading killer.

Only within the last two decades has it become generally accepted that mycobacteria other than *M. tuberculosis* can be the cause of human infections. Although *M. tuberculosis* is still the mycobacterium most frequently isolated from clinical specimens, *M. kansasii* and the *M. avium* complex are also well-established human pathogens, and their recognition becomes equally important. Mycobacteria other than *M. tuberculosis* may account for as much as 10% of all human mycobacterial infections.[2]

Definitive proof of a tuberculous infection is provided only by the demonstration of *M. tuberculosis* (or *M. bovis*) in clinical specimens obtained from the patient. The laboratory procedures used include (1) examination of a stained smear, (2) isolation by cultural procedures, and (3) identification.

Characteristics of the tubercle bacillus that differentiate it from other microorganisms are:

1. The resistance of stained tubercle bacilli to decolorization with strong decolorizing agents. Acid-alcohol (3% hydrochloric acid in 95% ethanol) is the usual agent.
2. The resistance of acid-fast bacilli to digesting agents, such as strong acids and alkalis. The use of these agents on material that may contain contaminating bacteria will destroy most contaminants without decreasing greatly the viability of any tubercle bacilli that may be present.

In 1967, Kubica and Dye[21] proposed a "Levels of Laboratory Service" program for identification of mycobacterial disease. This approach, which we consider a very important concept that might well be applied to other areas of the microbiology laboratory, such as mycology and anaerobic bacteriology, is detailed further in two recent publications,[26,52] each with a slightly different approach. Under this concept, a laboratory decides how far it can reasonably go **reliably** and then depends on reference or regional laboratories for more sophisticated or complex procedures that may be beyond the scope of a small facility. Of course, a laboratory can always extend its level of service if conditions permit or warrant this. In the case of mycobacterial infection, Wayne and associates[52] suggest four levels of service (or areas of proficiency):

1. Collection and transport of specimens; preparation and examination of smears for acid-fast bacilli
2. Detection, isolation, and identification of *M. tuberculosis*
3. Determination of drug susceptibility of mycobacteria
4. Identification of mycobacteria other than *M. tuberculosis*

Collection of clinical specimens

In suspected mycobacterial infection, as in all other diseases of microbial origin, the **diagnostic procedure begins not in the laboratory but at the bedside of the patient.** In the collection of clincial specimens there should be the same careful attention to detail on the part of the attending physician, nurse, and ward personnel as is required of the bacteriologist in carrying out the cultural procedures.

Secretions from the **lung** may be obtained by any one of the following methods: spontaneous or induced expectoration of sputum, aspiration of secretions during bronchoscopy, transtracheal aspiration, aspiration of gastric contents that contain swallowed sputum, or swabbing of the larynx.

Sputum

Since it may not be practical to provide constant supervision during the collection of a sputum specimen, it is necessary to elicit the intelligent cooperation of patients by giving them detailed instructions for the collection and explaining the purpose and importance of the test. The importance of sputum expectoration "from deep down in the lungs" should be emphasized as opposed to the expectoration of saliva or nasopharyngeal secretions. The specimen should not be collected immediately after the patient has used a mouth wash.

A clean, sterile container must be provided for collection; the Falcon sputum collection kit* appears to be ideally suited for this purpose. It consists of a plastic disposable 50-ml graduated conical centrifuge tube with an aerosol-free screw cap, which is fitted inside a funnellike disposable plastic outer container in such a manner that the risk of accidental contamination by handling is almost eliminated. Not more than 5 to 10 ml of sputum are collected in this container, which is then labeled with the patient's name, date, and other data.

A 4- to 6-ounce, clean, sterile widemouth glass jar with a screw cap and rubber or Teflon liner or a disposable sterile plastic cup with tightly fitted lid may also be used. However, specimens in these containers must be transferred to an appropriate aerosol-free centrifuge tube in the laboratory, a procedure both unpleasant and hazardous to the laboratory worker.

For optimal recovery of mycobacteria, a series of three to five single fresh **early morning** sputum specimens, not exceeding one fifth (10 ml) of the volume of the centrifuge tube, should be collected on successive days.[17] If the amount of this specimen is insufficient, sputum may be collected over a 24-hour period. Extending the collection of a single specimen over a longer period or the pooling of several days' accumulation is **not** recommended, since a reduced number of positive isolations may result, along with an increased rate of contaminated cultures.[25]

Nebulized and heated hypertonic saline may be used to induce sputum production in patients unable to raise a

---

*Falcon Plastics, No. 9002, Division of Becton, Dickinson, and Co. Cockeysville, Md., and Los Angeles.

satisfactory coughed specimen. The specimen is obtained (preferably by a member of the respiratory therapy service) 10 to 15 minutes after inhalation of aerosolized warmed (45 C) 10% sodium chloride solution; it appears to give a higher yield of positive cultures than those obtained by gastric lavage and is generally more acceptable to both patients and personnel. These specimens will appear watery. They should be labeled "Induced sputum" so the laboratory will not mistake them for saliva.

Sputum, and other specimens, should be delivered to the laboratory with minimal delay; specimens that cannot be delivered or processed immediately should be refrigerated. Despite this, a recent study showed 90% agreement between fresh specimens examined after a delay of 1 to 8 days in the mail (Dutt, A. K., et al.: J.A.M.A. **238**:886-887, 1977).

Gastric contents

If a patient is unable to raise a sufficient amount of sputum, is uncooperative, or cannot expectorate because he or she is comatose, a specimen of gastric contents is necessary. Gastric lavage is frequently required in young children from whom it is difficult to obtain a sputum sample. A minimum of three specimens should be submitted.

The gastric lavage should be performed before the patient gets out of bed in the morning, after having fasted for at least 8 hours prior to the collection. A disposable plastic gastric tube* is used. It is first moistened with sterile water and then inserted into the tip of a nostril or into the mouth. As the tube is gently advanced, the patient is instructed to swallow small sips of sterile water to assist him in swallowing the tube. After the gastric tube is properly in the stomach, gastric contents may be aspirated with a sterile 50-ml syringe and transferred to a sterile flask.

The patient should then be given 20 to 30 ml of sterile water (commercially distilled water for parenteral administration), either by mouth or by injection through the gastric tube.* The gastric washings are again aspirated and added to the first collection. The specimen is then delivered **immediately** to the laboratory, where prompt processing (within 4 hours) must be performed to neutralize the adverse effects of gastric acid on the tubercle bacillus. If this is not practical, some means of neutralizing the gastric acid must be used, such as the addition of 10% sodium carbonate until a pH of 7 is achieved, using phenol red as an indicator.

Urine

A minimum of three early morning midstream voided ("clean catch") or catheterized urine specimens is recommended; the entire volume of voided urine is collected in a sterile container. A 24-hour pooled specimen is sometimes used, although it is not recommended because it is likely to be contaminated and to contain fewer viable tubercle bacilli than a first-voided specimen.[15]

Other materials

Since tuberculosis may occur in almost any site of the body, types of clinical material other than those previously mentioned may occasionally be examined. Specimens of cerebrospinal fluid, pleural and pericardial fluid, pus, joint fluid, bronchial secretions, feces, and resected lung tissue† and autopsy material may be submitted for study. Collection of these specimens does not generally require supervision by the bacteriologist; an adequate supply of sterile specimen containers must be available, however.

---

*Falcon Plastics, Los Angeles.

*Water should not be injected until one is certain that the tube is in the stomach and not the trachea.
†Tissue specimens may be frozen when a delay in processing is necessary.

Processing of clinical specimens
Examination of stained smears

Although the demonstration of acid-fast bacilli in stained smears of sputum or other clinical material is only **presumptive** evidence of tuberculosis, the speed and ease of performance make the **stained smear** an important diagnostic aid, since it may be the first indication of a mycobacterial infection.

Because mycobacteria stain poorly by the Gram method, the conventional Ziehl-Neelsen carbolfuchsin stain or the fluorochrome staining technique, using auramine and rhodamine, is a required procedure. In the latter method, smears may be screened at 100× magnification, permitting a larger area of the slide to be examined in the same time as compared to 1,000× magnification using the conventional technique. By use of the Truant[39] technique (described in Chapter 42) and a properly adjusted fluorescence optical system, the mycobacteria and acid-fast nocardia are readily discerned as bright, **yellow fluorescent bacilli** against a dark background. It should be noted that an antigen-antibody reaction is **not** involved in the fluorochrome procedure. It should not be considered, therefore, as an FA or immunofluorescence test. Some laboratories confirm all positive fluorescent smears by the more specific Ziehl-Neelsen or Kinyoun techniques.

Preparation of the smear

Sputum specimen—direct smear
1. In a hood or biological safety cabinet, transfer a portion of the sputum to a disposable Petri dish and, with a wooden applicator stick broken in half, tease out a small portion of caseous, purulent, or bloody material and transfer it to a clean **new** slide.
2. Press another slide on top of it, squeeze the slides together, then pull them apart. This should result in two preparations of the proper thinness for staining and microscopy.
3. Label both slides, air dry, and flame immediately two or three times.
4. Stain according to the Ziehl-Neelsen or Kinyoun acid-fast or Truant fluorochrome[39] methods (see Chapter 42). Do not use staining dishes, as they permit transfer of mycobacteria between slides.
5. Dry in air or under an infrared lamp; do not blot.
6. Examine under oil immersion lens.

Sputum specimen—concentration. In the concentration method[31] the specimen is digested by sodium hypochlorite, and any mycobacteria present are concentrated in the sediment by centrifugation. This procedure will generally increase the number of positive smears, and is best carried out on 24-hour specimens. Since sodium hypochlorite is actively tuberculocidal as well as an excellent digestant, the resulting sediments are considered nonviable and will be useful **only for stained smears** and **not** for subsequent cultural procedures or animal inoculation.
1. Mix an equal volume (5 to 10 ml) of sputum and 5% sodium hypochlorite (Clorox household bleach is recommended) in a collection container of ample size; stir with a wooden applicator stick to ensure complete mixing. This should be done in an appropriately vented work area.
2. Shake for 2 to 3 minutes and keep at room temperature for 10 minutes or until complete digestion has taken place.
3. Transfer a portion to a 15-ml centrifuge tube and centrifuge for 10 minutes at 3,000 rpm.
4. Decant the supernatant fluid and allow the centrifuge tube to stand in an inverted position on a paper towel to drain the sediment; no neutralization is necessary.
5. Transfer the creamy white sediment to a slide with a cotton-tipped applicator stick. Air dry (no fixing is required) and stain by usual method.

Reporting and interpreting the microscopic examination

All sputum smears stained by conventional methods should be examined by carefully scanning the long axis of the smear three times, right-left-right, before reporting as negative.[40] Typical acid-fast bacilli are **red stained**, slender, slightly curved, long or short rods (2 to 8 μm), sometimes beaded or granular. Atypical forms are unusually thick or diphtheroid, very long, and sometimes coccoid. The presence of bacilli in sputum that are longer, broader, and more conspicuously banded than tubercle bacilli should lead to the suspicion that *Mycobacterium kansasii* may be present.*

Results of microscopic examination should be reported simply "**Positive for acid-fast bacilli**" or "**No acid-fast bacilli found**"; a positive finding should be based only on typical forms, but atypical rods should also be noted. When large numbers of typical acid-fast bacilli are found, it is reasonable to assume they are *M. tuberculosis;* when atypical rods are seen, they may represent other pathogenic or nonpathogenic mycobacteria or nocardia. It is desirable to report the number of acid-fast bacilli seen; the following criteria recommended by the American Lung Association may be used[2]:

Number of organisms seen	Report to read
1 to 2 in entire smear	Report number found and request another specimen
3 to 9 per slide	Rare (+)
10 or more per slide	Few (++)
1 or more per oil immersion field	Numerous (+++)

When only one or two acid-fast bacilli are seen in the **entire** smear, it is also rec-ommended[21] that they not be reported until confirmation is obtained by examining other smears from the same or another specimen. One study found that when **more than six** acid-fast organisms were present per high-power field, with either sputum or gastric contents, culture of the same material always yielded a pathogenic mycobacterium.[38]

As noted above, the demonstration of mycobacteria in sputum or other clinical material should be considered only as **presumptive** evidence of tuberculosis, since it does not specifically identify *M. tuberculosis.* For example, *M. gordonae*, a nonpathogenic scotochromogen commonly found in tap water, has been a particular problem when tap water or deionized water has been used in the preparation of smears or even when patients rinsed their mouths briefly with tap water prior to the use of aerosolized saline solution for inducing sputum.[7,9]

The examination of stained smears is considered the least sensitive of the diagnostic methods for tuberculosis; **cultures** should be performed on all specimens examined microscopically.* Because of its simplicity and speed, however, the stained smear is an important and useful test, since smear-positive patients ("infectious reservoirs") are the greatest risk to others in their environment.

Cultural methods
Culture of sputum and bronchial secretions by N-acetyl-L-cysteine-alkali method[36]

It has long been recognized that the conventional chemical methods of digestion and decontamination of sputum for the cultivation of tubercle bacilli result in the destruction of a large percentage of these organisms. For this reason, a milder de-

---

*It is important to wipe the objective thoroughly with lens paper **after every positive smear** to prevent the transfer of acid-fast organisms to the next slide by way of oil remaining on the objective.

---

*Authorities estimate that it requires from 10^4 to 10^5 organisms per milliliter of sputum to yield a positive direct smear, of which 10 to 100 organisms per milliliter are recoverable by culture.

contamination and digestion procedure, using the mucolytic agent N-acetyl-L-cysteine (NALC), was advocated by Kubica and associates.[22] (The preparation of the required reagents is described in Chapter 43.) The procedure is as follows:

1. Collect the sputum specimen as previously described and transfer approximately 10 ml to a 50-ml sterile, disposable, aerosol-free plastic centrifuge tube with a screw cap (Falcon Plastics, No. 2070).
2. Add an equal volume of NALC–sodium hydroxide solution.
3. Tighten screw caps; mix well in a Vortex mixer* for 5 to 20 seconds or until digested. Violent agitation is not recommended because denaturation of the NALC may take place.
4. Allow to stand at room temperature for **15 minutes** to effect decontamination.†
5. Fill the tube within ½ inch of the top with either sterile M/15 phosphate buffer, pH 6.8 (preferred), or sterile distilled water. This dilution acts to minimize the action of the sodium hydroxide as well as to reduce the specific gravity, thereby facilitating centrifugation.
6. Centrifuge at or near 3,000 rpm (1,800 to 2,400 *g*) for **15 minutes** and carefully decant the supernatant fluid into a splash-proof can containing a phenolic disinfectant. Wipe the lip of the tube with a cotton ball soaked with 5% phenol. Retain the sediment.
7. If direct drug susceptibility tests are to be performed (see step 12), prepare a smear of the sediment by using a sterile applicator stick or 3-mm-diameter wire loop to spread one drop of the material over an area approximately 1 by 2 cm. Stain by the Ziehl-Neelsen or fluorochrome method and determine the approximate number of AFB (acid-fast bacilli) per oil-immersion field.
8. Add 1 ml of 0.2% bovine albumin* to the sediment, using a sterile pipet. Shake gently by hand to mix. No neutralization is required, since the dilution of the sodium hydroxide by the phosphate buffer wash in step 5 and the strong buffering capacity of the bovine albumin makes this unnecessary. Refrigerate these albumin-suspended sediments overnight if inoculation to media is not practical at this time. It is recommended that bovine albumin be added **before** smears are made if the volume of sediment is small.
9. Make a 1:10 dilution of this sediment by adding 10 drops of it to 4.5 ml of sterile water.
10. Inoculate the following media, using 0.1 ml of the sediment for each tube or plate:
    a. **Löwenstein-Jensen (L-J) slants**
       One tube with undiluted sediment
       One tube with 1:10 dilution
    b. **7H10-Oleate-albumin-dextrose-catalase (OADC) biplates**
       One-half plate with undiluted sediment
       One-half plate with 1:10 dilution
    Spread with a glass spreader, using the 1:10 dilution first.
11. If the specimen shows a **positive** smear (step 7), dilute the concentrated sediment as follows:

AFB per oil-immersion field (avg. 20 fields)
  Less than 1
  1 to 10
  More than 10

Dilution of inoculum for control quadrants
  Undiluted, 1:100

---

*Vortex-type, Junior model, available from most laboratory supply houses.
†If contamination is expected to be heavy, the concentration of NaOH may be increased to 4%, but the exposure time must **not** be extended.

---

*Bovine albumin fraction V, adjusted to pH 6.8 with 4% NaOH; from Pentex, Inc., Kankakee, Ill.

1:10, 1:1,000
1:100, 1:10,000
Dilution of inoculum for drug quadrants
Undiluted
1:10
1:100

Dilution is necessary to ensure an inoculum size that will yield at least 40 to 50 colonies on the control plate when performing drug-susceptibility tests but not large enough to permit overgrowth of drug-resistant mutants, which can occur spontaneously in drug-susceptible populations.

12. **Direct** drug susceptibility testing* is encouraged for all specimens with **positive** smears. This may be carried out by inoculating the following media in Felsen quadrant 7H10 plates† (see Chapter 41 for preparation), using sterile, disposable capillary pipets and adding to each quadrant 3 drops (0.15 ml) of the two dilutions selected in step 11:

Plate number	Quadrant number	Drug	Amt (μg) per disk	Final drug concentration (μg/ml)
1	I	(Control no. 1)	—	0
	II	Isoniazid	1	0.2
	III	Isoniazid	5	1.0
	IV	Ethambutol	25	5.0
2	I	(Control no. 2)	—	0
	II	Streptomycin	10	2.0
	III	Streptomycin	50	10.0
	IV	Rifampin	5	1.0
3	I	p-Aminosalicylic acid	10	2.0
	II	p-Aminosalicylic acid	50	10.0
	III	—	—	—
	IV	—	—	—

Tests involving the secondary antituberculous drugs, including kanamycin, viomycin, ethionamide, cycloserine, and others, should be done only by reference laboratories.[52]

Paper disks impregnated with antituberculous drugs* are placed in sectors of quadrant plates that are then filled with 7H10 agar and incubated overnight to allow diffusion of the drug. They are then inoculated with the test strain as previously described.

13. Incubate L-J slants (steps 10 and 11) in a horizontal position for 1 to 2 days at 35 C in the dark. Examine weekly for 6 to 10 weeks. An atmosphere of **carbon dioxide** is beneficial for the growth of mycobacteria.[53] Thus, Lowenstein-Jensen slants (loosen screw caps) should be incubated in a carbon dioxide incubator for the first 2 weeks. Incubation in a candle jar is **not** recommended, since an increase in contamination and a decrease in the amount of growth may occur, apparently as a result of oxygen depletion within the jar.

14. Incubate 7H10 plates (steps 10 and 12) right side up in permeable polyethylene bags,† closed by stapling, **in the dark** in a carbon dioxide incubator (5% to 10% CO_2) at 35 C for 3 weeks.

15. Examine 7H10 plates for growth both macroscopically and microscopically (a dissecting microscope is recommended) after 5 to 7 days' incubation and weekly thereafter.‡ The transparent 7H10 medium permits good differentiation

---

*These tests should be performed on all suspected isolates of *Mycobacterium tuberculosis;* if not available, the culture should be referred to a state laboratory or other specialized facility for susceptibility tests or speciation.
†Falcon Plastics, Los Angeles, Calif.

*Antimicrobial drugs for use in culture media, Baltimore Biological Laboratory, Cockeysville, Md.
†Falcon Plastics, Los Angeles, or Baggies, Colgate-Palmolive Co., New York.
‡Runyon has written an excellent discussion of the identification of mycobacteria by microscopic examination of colonies.[34]

between corded and noncorded colonies under low-power magnification. **Group II scotochromogens** (see next section) are obvious by the presence of yellow, **noncorded** colonies on removal from the dark incubator; the development of **pigment** in colorless, noncorded colonies after exposure to light is characteristic of **Group I photochromogens.**

16. **Positive** cultures are reported as soon as growth is noted, and the final report is made after 6 to 8 weeks' or more incubation of 7H10 plates and L-J slants. All cultures, of any type, positive for mycobacteria should be saved at 5 C for 6 months in case special studies are required. Drug susceptibility tests are read as soon as possible, with the reservation that changes can occur in the final reading because of the slow growth of some initially susceptible strains in the presence of certain agents, particularly streptomycin.

In reporting results of the drug susceptibility tests, the report should include:

1. Type of test—direct or indirect
2. Number of colonies on control quadrant
3. Number of colonies on the drug quadrant
4. Concentration of the drug in each quadrant

From these data, a rough approximation of the percentage of organisms resistant to the drug may be calculated as follows:

$$\frac{\text{No. of colonies on drug quadrant}}{\text{No. of colonies on control quadrant}} \times 100 =$$

% resistance at that drug concentration

Refer to previously cited manual[40] and texts[36,42] for examples and photographs of these drug-susceptibility tests.

Culture of sputum by other methods

In addition to the previously described NALC-alkali method for digestion and de-

contamination, two other methods, if properly performed, are acceptable.

Trisodium phosphate-benzalkonium chloride method.[49] The trisodium phosphate-benzalkonium chloride method requires digestion with trisodium phosphate for only 1 hour. Due to the low survival rate of mycobacteria, the 12- to 24-hour exposure of the original method is no longer an acceptable procedure. The preparation of the required reagents is described in Chapter 43. The technique is as follows:

1. Mix equal volumes of sputum and trisodium phosphate-benzalkonium chloride* in a 50-ml disposable, leak-proof centrifuge tube†; shake on a shaking machine‡ for 30 minutes.
2. Allow to stand at room temperature for 20 to 30 minutes.
3. Centrifuge at 3,000 rpm for 20 minutes.
4. Decant supernatant fluid into a disinfectant, observing necessary precautions.
5. Resuspend sediment in 10 to 20 ml of M/15 sterile phosphate buffer, pH 6.6; recentrifuge for 20 minutes.
6. Again decant supernatant fluid and inoculate the sediment to egg media. If non-egg media, such as 7H10 agar, are used, residual benzalkonium chloride may inhibit the growth of mycobacteria. This inhibitory effect may be neutralized by adding 10 mg/100 ml of lecithin to the sterile buffer[21] or by employing an additional wash with sterile buffer. This is not necessary when using egg-containing media, since neutralizing phospholipid compounds are already present.

Sodium hydroxide method[32]

1. Mix equal volumes of sputum and 3% to 4% sodium hydroxide con-

---

*Zephiran, Winthrop Chemical Co., New York.
†Falcon Plastics, No. 2070, Los Angeles.
‡Paint conditioner, laboratory model No. 34, Red Devil Tools, Union, N.J.

taining 0.004% phenol red in a disposable leak-proof centrifuge tube; homogenize on a shaking machine for 10 minutes.

2. Centrifuge at 3,000 rpm for 20 minutes.
3. Decant supernatant fluid into a disinfectant, observing bacteriologic precautions.
4. Using a sterile capillary pipet, add 2 N hydrochloric acid one drop at a time until a definite yellow endpoint is obtained.
5. Back titrate with 4% sodium hydroxide to a **faint pink** endpoint (neutrality).
6. Inoculate sediment to L-J medium and 7H10 agar. The use of L-J medium containing penicillin and nalidixic acid (Gruft medium[10]), along with digestion with 2% NaOH, has been found effective in the recovery of mycobacteria from heavily contaminated specimens.

When sputum specimens are consistently contaminated with *Pseudomonas* species and other similar organisms, the method of Corper and Uyei[6] is recommended. It makes use of the decontaminating effect of oxalic acid as follows:

1. Mix equal volumes of sputum and 5% oxalic acid in a leak-proof centrifuge tube; homogenize in a Vortex mixer and allow to stand at room temperature for 30 minutes, shaking occasionally.
2. Add sterile physiologic saline to within 1 inch of the top; this lowers the specific gravity, permitting better concentration of bacilli in the sediment.
3. Centrifuge at 3,000 rpm for 15 minutes.
4. Decant supernatant fluid into a disinfectant, using bacteriologic precautions.
5. Neutralize sediment with 4% sodium hydroxide containing phenol red indicator; inoculate the desired media.

Cetylpyridinium method

For specimens that will be in transport more than 24 hours, CDC (Smithwick et al.: J. Clin. Microbiol. 1:411-413, 1975) recommends mixing equal volumes of sputum and a solution of 1% cetylpyridinium chloride and 2% sodium chloride. This serves to decontaminate, liquefy, and concentrate the sputum. Tubercle bacilli remain viable for 8 days. This method is **not** suitable for fungi.

Culture of gastric specimens

1. Add a pinch of NALC powder to about 25 ml of specimen in a 50-ml centrifuge tube.
2. Mix in Vortex as with sputum.
3. Centrifuge at 3,000 rpm for 30 minutes.
4. Aseptically decant supernatant fluid and resuspend sediment in 2 to 5 ml sterile distilled water.
5. Add an equal volume of NALC-alkali reagent and proceed as with sputum.
6. If the gastric specimen is quite fluid, centrifuge directly and proceed with step 5.

Culture of urine

Clean-voided, early morning urine specimens are handled as follows:

1. Pour approximately 50-ml portions into one or more centrifuge tubes; centrifuge at 3,000 rpm for 30 minutes.
2. Aseptically decant supernatant fluid, resuspend sediment in 2 to 5 ml sterile water, and handle as described under gastric specimens.

An alternate method described in the CDC manual[40] is as follows:

1. Centrifuge and combine sediments as described above.
2. Add to sediment an equal volume of 4% H_2SO_4.
3. Mix in Vortex; let stand for 15 minutes.
4. Fill tube to nearly the 50-ml mark with sterile distilled water, mix, and centrifuge as before.

5. After decanting supernatant fluid, add 0.2% bovine albumin to sediment and inoculate media as in sputum cultures.

Note: Since prolonged exposure to urine may be toxic to mycobacteria, urine should be decontaminated as soon as possible after receipt in the laboratory.

Culture of cerebrospinal fluid and other body fluids

The isolation of *M. tuberculosis* from cerebrospinal fluid and other body fluids is generally more difficult than from sputum and gastric secretions. Since **very few** organisms may be present, success will often depend on the amount of specimen available for culture—the larger the volume the greater the chance of recovering the organism. At least 10 ml of fluid should be submitted.

The fluid specimen should be centrifuged for 30 minutes at 3,000 rpm and the supernatant fluid discarded. (Cerebrospinal fluid supernate may be saved for other examinations, such as serologic or chemical tests.) Half of the sediment is used for inoculating media and preparing a smear; the other half should be used for guinea pig injection.* Because these specimens will generally contain no other bacteria, the decontaminating procedure is neither necessary nor desirable. It could result in the loss of the few organisms present. If the fluid specimen contains a clot, this is cut into small pieces with sterile scissors and subsequently accorded the same treatment as a sputum specimen. Purulent material is also handled in the manner used for sputum.

Kubica and Dye[21] suggest that specimens that have been obtained aseptically, as cerebrospinal fluid, synovial fluid, pleural fluid, and small bits of biopsied tissue, be inoculated directly to a fluid medium, such as Middlebrook 7H9, Tween-albumin medium, or Proskauer

and Beck medium, in a volume ratio of 1:5. These media are incubated at 35 C in a carbon dioxide incubator, and acid-fast stained smears are prepared and examined weekly. If mycobacteria are seen, the medium is inoculated to egg slants or 7H10 agar and incubated as previously described. If smears are negative after 4 weeks, the media are again subcultured at weekly intervals to L-J slants or 7H10 agar for 4 more weeks before discarding.

Specimens from superficial areas, such as skin lesions, should be set up in an additional culture to be incubated at **30 to 33 C** to permit detection of *M. marinum.*

Culture of feces

The examination of fecal material for *M. tuberculosis* is not very rewarding and its use should be limited to special **unusual** circumstances. The presence of tubercle bacilli in the feces does not necessarily indicate intestinal tuberculosis; it more likely indicates sputum swallowed by a patient with pulmonary disease.

In the culture of fecal material, suspend about 5 g in 20 to 30 ml of distilled water and mix well. Add sufficient sodium chloride to make a saturated solution and allow to stand for 30 minutes. Skim off the film of bacteria that forms on the surface with a sterile spoon and transfer to a sterile centrifuge tube. Add an equal volume of 4% sodium hydroxide, homogenize with vigorous shaking, incubate at 35 C for 3 hours, shaking at intervals, neutralize with 2 N hydrochloric acid, and centrifuge for 20 minutes at 3,000 rpm. Inoculate five tubes of media with the **top layer** of the supernatant fluid. Save a portion of the supernatant fluid, discarding the rest, and combine with a portion of the sediment, to be later inoculated into a guinea pig if desired. The remaining sediment is inoculated to five additional tubes of media (**ten** tubes in all). This inoculation of ten tubes is advised because of the increased likelihood of subsequent

---

*Wayne[43] has recommended the use of membrane filters for spinal fluid culture for mycobacteria.

loss from contamination in culturing fecal material.

Culture of tissue removed at operation or necropsy

Tissue removed surgically (lymph nodes, resected lung, and so forth) generally is not contaminated and does not require sodium hydroxide treatment. The specimen is finely minced with sterile scissors and transferred to a sterile tissue grinder,* where it is ground to a pasty consistency with sterile alundum and a small amount of sterile saline or 0.2% bovine albumin V. After settling, 0.2-ml portions of the supernatant fluid are removed and inoculated directly to egg media and 7H10 agar (liquid 7H9 medium also may be used). Also, a smear is made.

Tissue that is obviously contaminated (such as tonsils and autopsy tissue) can be handled as just described, except that the supernatant fluid obtained after grinding is treated as is sputum (decontamination with NALC—alkali solution, homogenization, centrifugation, neutralization, and so forth) before inoculation to culture media.

ANIMAL INOCULATION TESTS

The inoculation of laboratory animals (generally guinea pigs) has been accepted in the past as definitive in determining the pathogenicity of an acid-fast bacillus. This is no longer true for human isolates, since some organisms, notably the mycobacteria other than tubercle bacilli and some isoniazid-resistant strains of *Mycobacterium tuberculosis*, do not produce progressive disease in the injected animal.[21]

Furthermore, the development of excellent cultural procedures and the more frequent use of **multiple specimens** from a suspected case of tuberculosis make it generally unnecessary for the hospital laboratory to inoculate laboratory animals.*

The procedure, however, is a useful diagnostic tool in the detection of **small numbers** of tubercle bacilli, as in cerebrospinal fluid or in specimens that are consistently contaminated on culture. For these reasons, the technique has been included.

1. If clinical specimens are to be injected, they should be decontaminated and concentrated as described in previous sections under culture methods.

2. After a portion of the concentrated and neutralized sediment has been inoculated onto appropriate media, suspend the remainder in 1 ml of sterile physiologic saline and inject either subcutaneously in the groin or intraperitoneally into a guinea pig.

3. If one is using pure cultures of mycobacteria, a dose consisting of approximately 0.1 mg (moist weight) is the usual inoculum. This is suspended in 1 ml of physiologic saline and injected subcutaneously into the right groin of each of **two** guinea pigs, the animals having been pretested and found negative to 0.1 ml of 5% old tuberculin (OT) or a satisfactory substitute injected intracutaneously. A subculture is also made of the suspension for control purposes.

4. Animals inoculated either with clinical material or with cultures should be examined weekly to detect the presence of enlarged lymph nodes draining the site of inoculation. Tubercle bacilli of human or bovine origin usually give rise to progressive disease in the guinea pig, with development of caseous nodes and involvement of the spleen and usually the liver and lungs. **Actual invasion of deep tissue** must be demon-

---

*Ten Broeck tissue grinder, small size, heavy-walled Pyrex glass, Bellco Glass, Inc., Vineland, N.J.

*Specimens for animal inoculation are best handled in a mycobacteriology reference laboratory.

strated—the microbiologist must not rely on the simple production of a local lesion alone.

5. If two guinea pigs have been injected, 0.5 ml of OT is administered to one at the end of 4 weeks. In most instances an animal eventually proved tuberculous is so sensitized to tuberculin by this time that it will die within 1 to 2 days of OT injection. If the tuberculin reaction is negative, the animals are held another 4 weeks before autopsy is performed.
6. Examine tissue showing gross pathology, using the acid-fast stain.
7. Involvement of only the regional nodes is considered a **doubtful** test, and the test should be repeated.
8. If only extensive lung tuberculosis is observed, **spontaneous** infection should be suspected and the test repeated; this also applies to animals that die in less than 4 weeks and show negative autopsy findings.

CULTURAL CHARACTERISTICS OF MYCOBACTERIA

It should be emphasized that the **acid-fast** nature of all colonies growing on culture media must be confirmed before they can be identified as mycobacteria. Cultures for the isolation of *M. tuberculosis* on L-J media should be incubated at 35 to 36 C for a total of 6 to 10 weeks and examined at weekly intervals; 7H10 media are held for at least 6 weeks. **Positive** cultures should be reported as soon as identification has been completed.

Mycobacterium tuberculosis

Colonies of human tubercle bacilli generally appear on egg media after 2 to 3 weeks at 35 C; no growth occurs at 25 or 45 C. Growth first appears as small (1 to 3 mm), dry, friable colonies that are rough, warty, granular, and buff in color. After several weeks these increase in size (up to 5 to 8 mm); typical colonies have a flat irregular margin and a "cauliflower"

center. Because of their luxuriant growth these mycobacteria are termed **eugonic.** Colonies are easily detached from the medium's surface but are difficult to emulsify. After some experience, one can recognize typical colonies of human-type tubercle bacilli without great difficulty. However, final confirmation by biochemical tests must be carried out.

Virulent strains tend to orient themselves in tight, **serpentine cords,** best observed in smears from the condensation water or by direct observation of colonies on a cord medium.[28] Catalase is produced in moderate amounts but **not** after heating at 68 C for 20 minutes in pH 7 phosphate buffer; human strains resistant to isoniazid (INH) are frequently catalase negative and yield smooth colonies on egg media. Nitrate reduction is positive. The niacin test (p. 256) is useful in that most **niacin-positive** strains encountered in the diagnostic laboratory will prove to be *M. tuberculosis.* Susceptibility to antituberculous drugs is characteristically high.

Mycobacterium bovis

Bovine tubercle bacilli are rarely isolated in the United States but remain significant pathogens in other parts of the world. They require a longer incubation period—generally 3 to 6 weeks—and appear as tiny (less than 1 mm), translucent, smooth, pyramidal colonies when grown at 35 C. They adhere to the surface of the medium but are emulsified easily. On the basis of these characteristics, their growth is termed **dysgonic.** On 7H10 the colonies are rough and resemble those of *M. tuberculosis.*

M. bovis will grow only at 35 C. It forms serpentine cords in smears from colonies on egg media; the niacin reaction and nitrate reduction tests are negative. *M. bovis* is also susceptible to thiophen-2-carboxylic acid hydrazide (TCH), a useful test to differentiate it from other mycobacteria.[51] Its susceptibility to the primary antituberculous drugs is similar to that of *M. tuberculosis.*

Other mycobacteria

Although it has long been recognized that acid-fast bacilli other than *M. tuberculosis* are occasionally associated with both pulmonary and extrapulmonary disease, it has only been within the last two decades that these mycobacteria, variously called atypical, anonymous, or unclassified acid-fast bacilli, have been associated with pulmonary and other disease clinically diagnosed as tuberculosis.[4] Evidence, obtained primarily from skin testing, suggests that many persons in the United States may become naturally infected with these mycobacteria, even though most of them do not show clinical evidence of disease.[55]

In 1959, Runyon[33] proposed a scheme for separation of the medically significant unclassified mycobacteria, dividing them into four large groups. This scheme served as an initial classification system until more precise speciation could be established for members within each group. For example, studies by individual workers[16] and cooperative groups have demonstrated fundamental differences between various clinically significant and other mycobacteria through the use of easily performed metabolic and biochemical tests, many of which are described in a following section.

Group I photochromogens

Group I mycobacteria possess the outstanding characteristic of **photochromogenicity**—the capacity to develop pigment when exposed to light. A young, actively growing culture on L-J slant medium that has been exposed to light for as little as 1 hour and reincubated in the dark will produce a bright lemon yellow pigment within 6 to 24 hours.

Three well-defined photochromogenic mycobacteria comprise Runyon Group I: *M. kansasii*, *M. marinum*, and *M. simiae*.

Mycobacterium kansasii. This organism is responsible for pulmonary disease in humans, often appearing in white, emphysematous men over 45 years of age.

Table 31-1. Potential clinical significance of various mycobacteria*

Always or sometimes significant	Runyon group	Never or rarely significant
M. tuberculosis *M. bovis*		
M. kansasii	I	*M. kansasii* (low-catalase variety)
M. marinum (balnei) *M. simiae*		
M. scrofulaceum *M. szulgai*	II	*M. gordonae* *M. flavescens*
M. avium complex† *M. xenopi* *M. ulcerans*	III	*M. gastri* *M. terrae* complex *M. triviale*‡
M. fortuitum *M. chelonei*	IV	*M. smegmatis* *M. phlei* *M. vaccae*
M. rhodochrous§		

*Modified from Current Item No. 165 (updated), Laboratory Program, 1968, National Communicable Disease Center. Courtesy George Kubica, former Chief, Mycobacteriology Unit, NCDC, Atlanta.
†Includes *M. intracellulare.*
‡One human case of septic arthritis has been reported.
§This organism's taxonomic position is uncertain, but it is probably not a mycobacterium. It has been recovered from a variety of infections.

The disease with its complications is indistinguishable from that caused by *M. tuberculosis*, but it follows a more chronic and indolent course. Other types of disease, including disseminated infection, may occur.[27] The disease is not communicable, in distinct contrast to that caused by the tubercle bacillus.

Optimal growth of *M. kansasii* occurs after 2 to 3 weeks at 35 C (slower at 25 C); growth does not occur at 45 C. Colonies are generally smooth, although there is a tendency to develop roughness. They are cream colored when grown in the dark and become a bright **lemon yellow** if exposed to light.

A practical procedure for photochromogenicity testing has been suggested by Kubica[20]:

1. Two slants of L-J medium are inoculated with a barely turbid suspension of the culture; one tube is shielded from the light by wrapping with black x-ray paper or aluminum foil.
2. Both tubes are incubated at 35 C (32 to 33 C is recommended for suspected cultures of *M. marinum*) until visible growth occurs on the uncovered slant.
3. The shield is removed from the covered tube and any pigment is noted; if none is observed, wrap only **one half** of the tube with the shield, while exposing the other half to either a 60-watt bulb placed 8 to 10 inches from the tube or to bright daylight for 1 hour. Loosen cap during exposure.
4. Replace shield, reincubate both culture tubes overnight with the caps loose. Pigments are compared in colonies grown (a) unshielded, (b) shielded, and (c) exposed to light for an hour. Three classes of pigments can then be established:
   a. **Scotochromogens**—pigmented in dark
   b. **Photochromogens**—pigmented after exposure to light
   c. **Nonphotochromogens**—nonpigmented either in dark or light

If cultures are grown continuously under light (2 to 3 weeks), bright orange crystals of beta carotene will form on the surface of colonies, especially where growth is heavy. Nonphotochromogenic and scotochromogenic variants of *M. kansasii* occur very rarely.

Clinically significant strains of *M. kansasii* are strongly catalase positive, even after heating at 68 C at pH 7 (especially when using the semiquantitative test of Wayne,[45] described on p. 257). Low-catalase strains of *M. kansasii* have been described[24,45]; these were not associated with human pathogenicity. Most strains do not produce niacin, although aberrant strains have been noted; nitrates are re-

duced; mature colonies show loose cords.

Stained preparations of *M. kansasii* show characteristically **long, banded, and beaded** cells that are strongly acid fast. Such conspicuous cells in sputum smears should lead one to suspect *M. kansasii* infection. There is variable susceptibility to INH and streptomycin, moderate susceptibility to rifampin, and resistance to para-aminosalicylic acid (PAS), but *M. kansasii* infections typically respond to conventional antituberculous therapy.[3]

Mycobacterium marinum. Primarily associated with granulomatous lesions of the skin, particularly of the extremities, *M. marinum* infection usually follows exposure of the abraded skin to contaminated water. Known best for causing "swimming pool granuloma," this organism also has been implicated in infections related to home aquariums,[1] bay water, and industrial exposures involving water. It grows best at **25 to 32 C,** with sparse to no growth at 35 C (corresponding to the reduced skin temperatures of the extremities), and is never isolated from sputum, It may be distinguished from *M. kansasii* by its source, its more rapid growth at 25 C, negative nitrate reduction, and weaker catalase production.

M. simiae. First isolated from monkeys, *M. simiae* has subsequently been recovered from humans with pulmonary disease. Pigmentation may be erratic. It has a positive niacin reaction, a high catalase activity, a negative nitrate test, and is resistant to all first-line antituberculous drugs.

Group II scotochromogens

The **scotochromogens** are **pigmented in the dark** (Gk. *scotos,* dark), usually a deep yellow to orange, which darkens to an orange or dark red when the cultures are exposed to continuous light for 2 weeks. This pigmentation in the dark occurs on nearly all types of media at all stages of growth—characteristics that clearly aid in their identification.

The Group II scotochromogens are generally divided into two subgroups: the potential pathogens, *M. scrofulaceum* and *M. szulgai,* and the so-called "tap water scotochromogen," isolated from laboratory water stills, faucets, soil, and natural waters.[47] The latter organism is now classified as *M. gordonae.*

Since the tap water scotochromogen is not associated with human disease, it is important to differentiate it from the potentially pathogenic mycobacteria. The former may contaminate equipment used in specimen collection, as in gastric lavage.

M. scrofulaceum is a slow-growing organism, producing smooth, domed to spreading, **yellow** colonies in both the light and dark. When exposed to continuous light, the colonies may increase in pigment to an **orange** or **brick red;** the paper shield should remain in place (see under photochromogens) until visible growth occurs in the unshielded tube, then colonies in the shielded tube should be exposed to continuous light. Initial growth may be inhibited by too much light.[40] The hydrolysis of Tween 80 and urease activity separates *M. scrofulaceum* from the tap water organisms in that the latter will hydrolyze it within 5 days, while *M. scrofulaceum* remains negative up to 3 weeks. *M. scrofulaceum* is a cause of cervical adenitis and bone and other infections, particularly in children. It is often resistant to INH and PAS.

M. szulgai has been associated with pulmonary disease, cervical adenitis, and olecranon bursitis. It gives a positive nitrate reduction test. It is relatively susceptible to ethionamide, rifampin, ethambutol, and higher levels of INH.

Growth of *M. gordonae* appears late on L-J and 7H10 media, usually after 2 weeks and frequently after 3 to 6 weeks, as scattered small yellow-orange colonies in both the light and dark. Hydrolysis of Tween 80 characteristically occurs. *M. flavescens,* another nonpathogen Group II organism, also produces a yellow-pigmented colony in both the light and dark,

but is considerably more rapid in growth (usually within 1 week) and reduces nitrate as well as hydrolyzing Tween 80.

Group III nonphotochromogens

Runyon Group III is made up of a heterogeneous variety of both pathogenic and nonpathogenic mycobacteria that **do not** develop pigment on exposure to light. The following species are recognized:

1. *M. avium* complex (including Battey bacilli and *M. intracellulare*). These bacilli grow slowly at 35 C (10 to 21 days) and 25 C and produce characteristically thin, translucent, radially lobed to smooth, cream-colored colonies; rough variants* may show cording. These organisms are niacin negative (with rare exceptions[56]), do not reduce nitrate, and produce only a small amount of catalase. Tween 80 is not hydrolyzed in 10 days, but most of these organisms **reduce tellurite** within 3 days, a useful test to differentiate these potential pathogens from clinically insignificant members of Group III.

The *M. avium* complex organism causes serious tuberculosislike disease that is most difficult to treat. However, the organism may also occur in clinical specimens as a nonpathogen. Multiple drug regimens, usually including both isoniazid and rifampin, may be effective. Surgical resection of localized lesions may be necessary.

2. *M. xenopi* has been isolated from the sputum of patients with pulmonary disease.[29] The optimal temperature for its growth is 42 C, and it fails to grow at 22 to 25 C. Four to 5 weeks' incubation is required to produce tiny dome-shaped colonies of a characteristic yellow color; branching **filamentous extensions** are seen around colonies on 7H10 agar, resembling a miniature bird's nest. Tween 80 hydrolysis and tellurite reduction tests

---

*Most strains show a small proportion of rough colonies, which may resemble colonies of tubercle bacilli.

are negative; resistance is high to anti-tuberculous drugs.

3. *M. ulcerans* is also associated with skin lesions and is considered to be the causative agent of Buruli ulceration, a necrotizing ulcer found in African natives.[5] The organism requires several weeks' incubation at 32 C. Unlike *M. marinum*, this organism is **nonphotochromogenic** and is in Group III.

4. *M. terrae* complex organisms have been called "radish" bacilli. A number of these mycobacteria have been isolated from soil and vegetables as well as from humans, where their pathogenicity remains questionable. The slow-growing (35 C) colonies may be circular or irregular in shape and smooth or granular in texture. They actively hydrolyze Tween 80, reduce nitrate, and are strong catalase producers, but they do not reduce tellurite in 3 days. *M. terrae* is resistant to INH.

5. *Mycobacterium gastri* ("J" bacillus) has been described as occurring primarily as single colony isolates from gastric washings. However, it has not been associated with disease in humans. It is closely related to the low catalase-producing strains of *M. kansasii*, from which it may be readily differentiated by the photochromogenic ability of the latter. *M. gastri* may be differentiated from other members of Group III mycobacteria by its ability to hydrolyze Tween 80 rapidly, its loss of catalase activity at 68 C, and nonreduction of nitrate.

6. *Mycobacterium triviale* ("V" bacilli)[12] occurs on egg media as rough colonies that may be confused with *M. tuberculosis* or rough variants of *M. kansasii*. These bacilli have been recovered from patients with previous tuberculous infections but are considered to be unrelated to human infections. There is one report of human infection with this organism (Dechairo, D. C., et al.: Am. Rev. Resp. Dis. **108**:1224-1226, 1973). They are nonphotochromogenic, moderate to strong nitrate reducers, and maintain a high catalase activity at 68 C. They hydrolyze Tween 80 rapidly and do not reduce tellurite.

Group IV rapid growers

The Runyon Group IV mycobacteria are characterized by their ability **to grow in 3 to 5 days** on a variety of culture media, incubated either at 25 or 35 C. Two members, *M. fortuitum* and *M. chelonei*, are associated with human pulmonary infection, although *M. fortuitum* is also a common soil organism and may frequently be recovered from sputum without necessarily being implicated in a pathologic process.

M. smegmatis, *M. phlei*, and *M. vaccae* are considered **saprophytes** and are nonpathogenic. *M. smegmatis* and *M. phlei* produce pigmented colonies and show filamentous extensions from colonies growing on cornmeal-glycerol agar. The ability of *M. phlei* ("hay bacillus") to produce large amounts of carbon dioxide has been utilized to stimulate primary growth of *M. tuberculosis* on 7H10 agar plates incubated in carbon dioxide–impermeable (Mylar) bags.

M. fortuitum has been incriminated in progressive pulmonary disease,[8] usually with a severe underlying complication, and has resulted in death. This potential pathogen also grows rapidly (2 to 4 days), is generally nonchromogenic, and may be readily separated from the rapidly growing saprophytes by its **positive 3-day arylsulfatase** reaction and by growing on MacConkey agar within 5 days, producing a change in the indicator. Both rough and smooth colonies are produced, with increased dye absorption (greening) on L-J medium. The niacin test is negative.

M. fortuitum is usually resistant to PAS, streptomycin, and INH, but is considered susceptible to the tetracyclines.

M. chelonei (formerly *M. borstelense*) comprises two distinct subspecies: *M. chelonei*, which fails to grow on 5% NaCl medium, and *M. chelonei* ss. *abscessus*, which grows on the NaCl medium. This subspecies is considered a significant but rarely isolated pulmonary pathogen, whereas *M. chelonei* is considered a saprophyte.[36,40] *M. chelonei*, along with unidentified mycobacteria, has been isolated from preimplantation cultures of porcine heart valve prostheses; there is evidence of infection subsequently in some patients receiving these implants. One report (Arroyo, J., and Medoff, G.: Antimicrob. Ag. Chemother. **11:**763-764, 1977) indicates susceptibility of this organism to erythromycin, streptomycin, and rifampin.

It should be noted that the majority of Group IV mycobacteria are not stained by the fluorochrome (auramine-rhodamine) stain.[14] All, however, are stained by the Ziehl-Neelsen technique.

Aids in identifying mycobacteria*

As soon as visible nonpigmented growth (5 to 6 days) is observed on any media, they should be exposed to light (previously described). After overnight incubation, *M. kansasii* strains will become pigmented. All cultures of mycobacteria that are suspected of being members of Runyon's four groups (because of rapid growth, pigmentation, colonial characteristics, microscopic appearance, and so forth) should be subjected to the following procedures, as recommended by Wolinsky[54]:

1. Subculture three L-J slants to obtain isolated colonies.
2. Incubate one at room temperature (20 to 25 C).
3. Incubate the second at 35 C in light, either continuously exposed to fluorescent light or frequently exposed to bright light at hourly intervals after growth appears.
4. Incubate the third at 35 C in the dark, wrapped in aluminum foil or black paper and placed next to the second tube.
5. When slant No. 2 shows good growth, remove foil or paper from slant No. 3 and compare them.
6. Next, loosen cap of slant No. 3 (grown in the dark) and expose to bright light and observe for development of yellow pigment the next day.
7. Perform a niacin test on one of the original slants if growth is sufficient; subculture and reincubate if growth is scanty.
8. Other tests, such as cording, catalase, nitrate reduction, Tween 80 hydrolysis, arylsulfatase activity, ability to grow at 45 C, and so forth, may be required. See following section for a description of these procedures.

Interpretation. *M. tuberculosis* and *M. bovis* will not grow at room temperature; rapid growers will produce full-grown colonies in several days, even at room temperature. *M. kansasii* will show yellow colonies in the light and white colonies when grown in the dark; the latter will turn yellow overnight after exposure to light. Group II scotochromogens will be yellow to orange in both the light and dark tubes, the pigment generally being deeper in the light tube. *Mycobacterium avium* complex colonies will be cream colored at first, becoming a deeper yellow with aging. Light exposure has no effect on pigment production. *M. tuberculosis* and a few others will give a positive niacin test; a negative test does not preclude the possibility of *M. tuberculosis*, and it should be repeated at weekly intervals for a total of 6 weeks.

PROCEDURES USEFUL IN DIFFERENTIATING THE MYCOBACTERIA (INCLUDING CYTOCHEMICAL TESTS)

In considering the following, it should be emphasized that if facilities for carrying out these special procedures are lacking, the cultures should be **referred**

---

*The reader is referred to an excellent summary by Wayne and Doubek[46] of aids in the identification of most mycobacteria encountered in the clinical laboratory and to two publications by Wayne et al.[50,51]

promptly to a mycobacteriologic reference laboratory. In the meantime, an **interim report,** based on the available information (result of smear, growth on culture, and so forth) **should be sent to the clinician promptly.**

Niacin test

Strong niacin production by an acid-fast bacillus isolated from a clinical specimen is strong evidence of its possible identity as a **human** tubercle bacillus. Conversely, an accurately performed test resulting in a negative reaction generally indicates another species of *Mycobacterium. M. marinum* and *M. chelonei* may produce niacin and give doubtful or weakly positive results. *M. simiae* is also niacin positive.

The niacin test devised by Konno[19] and modified by Runyon and co-workers[35] depends on the formation of a complex color compound when a pyridine compound (niacin) from the organism reacts with cyanogen bromide (CNBr) and a primary or secondary amine.

If water of condensation is present on a culture slant of the organism to be tested (at least 3-week-old cultures **on egg media** should be used and have at least 100 colonies), it may be used for the niacin test. If not, the niacin may be extracted by adding a few drops of water or saline to the egg medium and placing the tube so that the liquid remains in contact with the colonies for 15 minutes. This serves to extract niacin if it is present. Puncturing the medium with the tip of a pipet aids in extracting niacin from the medium, especially if growth is confluent. One or 2 drops of the extract are then transferred to a white porcelain spot plate, and 2 drops of each of the following reagents are added: (1) 4% aniline in 95% ethanol, which should be nearly colorless, and (2) 10% aqueous CNBr. When not in use, both reagents are stored in the refrigerator in brown dropper bottles; they are made up fresh each month. A known positive strain of *M. tuberculosis* should always be included as a control.

Caution: The tests must be carried out under a chemical fume hood, since tear gas forms from CNBr. The production of an almost immediate **yellow color** indicates the presence of niacin. If a weak test is noted it should be repeated, using water in place of CNBr, and the results compared. On completion of the tests, several drops of 4% sodium hydroxide or 10% ammonia are added to the spot plate to destroy the residual CNBr and arrest tear gas formation. The plate may then be reconditioned by placing in boiling water for several minutes. Commercial niacin paper test strips are available and are strongly recommended.*

The only niacin-producing acid-fast organisms likely to be encountered in the clinical laboratory are those of *M. tuberculosis.* Most strains are niacin positive on L-J medium in 3 to 4 weeks; others may take up to 6 weeks. Some BCG strains (currently used in therapy of certain malignancies and sometimes causing disseminated infection in such patients) may give a weakly positive niacin test. Resistance to 1 μg/ml of thiophene-2-carboxylic acid[51] should also be tested in this situation. BCG strains are nitrate negative.

Nitrate reduction test

The test for reduction of nitrate[41] (nitroreductase) is helpful in differentiating the slower-growing organisms, *M. tuberculosis* and *M. kansasii,* from members of Group II mycobacteria and the clinically significant *Mycobacterium avium* complex organisms. The former are strong nitrate reducers, whereas the latter are generally negative. In Group II, *M. szulgai* and *M. flavescens* are also positive. Other Group III bacilli *(M. terrae, M. triviale),* which are not considered clinically significant, are also strongly positive.

---

*Bacto TB Niacin Test Strips, Difco Laboratories, Detroit.

Nitrate broth,* in 2-ml amounts, is heavily inoculated (spadeful) from a 4-week slant culture and incubated in a 36 C water bath for 2 hours. The suspension is then acidified with 1 drop of a 1:2 dilution of hydrochloric acid, followed by 2 drops of sulfanilamide solution and 2 drops of the coupling reagent. (See Chapter 43 for preparation of these reagents.)

A **positive** test is indicated by the immediate formation of a **bright red** color (as compared with the reagent control). A **negative** test should always be confirmed by the addition of a small amount of zinc dust; a red color, due to the reduction of nitrate by the zinc, confirms the negative test. An uninoculated reagent control and a negative and a positive control with a known strain of *M. tuberculosis* also should be included. Color standards ($\pm$ to 5+) may be prepared from dilutions of sodium nitrate but are stable for only 10 to 15 minutes. (See Chapter 43 for preparation.)

Catalase activity

Acid-fast bacilli produce the enzyme **catalase,** which is detected by the breakdown of hydrogen peroxide and the subsequent active ebullition of gas (oxygen) bubbles. Detection of catalase activity, which should be tested routinely, has proved useful in several ways: (1) tubercle bacilli that become resistant to INH will lose or show reduced catalase activity, (2) the catalase activity of human or bovine strains may be selectively inhibited by heat,[24] and (3) a Group III mycobacterium *(M. gastri)* and isolates of clinically insignificant *M. kansasii* also lose their catalase activity at pH 7 and at 68 C.

Catalase activity may be determined in three ways:

1. At **room temperature,** add 1 drop of a 1:1 mixture of 10% Tween 80* and 30% hydrogen peroxide† directly to a slant culture of modified L-J medium or the control quadrant of a 7H10 agar plate. A **positive** catalase test is indicated by grossly visible gas bubbles (nascent O_2 from breakdown of H_2O_2 by catalase) within 2 minutes.

2. To determine the effect of pH and temperature on catalase activity, scrape several loopfuls or spadefuls of growth from a slant culture and suspend in 0.5 ml phosphate buffer pH 7 (M/15) in a screw-capped tube and place in a **68 C** water bath for 20 minutes. After cooling to room temperature, add 0.5 ml of the Tween 80–peroxide mixture and observe the reaction for 15 or 20 minutes for the formation of gas bubbles.

3. A semiquantitative test, utilizing butt tubes (20 × 150 mm) of L-J medium (available commercially‡) and a measurement of the height of the column of gas bubbles is performed as follows.[23,45] Inoculate surface of the L-J medium with an actively growing egg slant or 7H9 broth culture and incubate the tube, with loosened cap, at 35 C for 2 weeks. Add 1 ml of the Tween 80–peroxide mixture and after 5 minutes measure the height of the gas bubbles; record as follows:

Negative = no bubbles
≤ 45 = less than 45 mm bubbles (low)
≥ 45 = more than 45 mm bubbles (high)

Interpretation. Except for *M. tuberculosis, M. gastri,* and low-catalase *M. kansasii* strains, all other mycobacteria commonly isolated from human sources

---

*Cater, J. C., and Kubica, G. P. (unpublished data) have shown that Difco nitrate broth is a satisfactory substitute for the buffered nitrate substrate originally described.

*Atlas Chemicals Division, ICI America, Inc., Wilmington, Del.
†Superoxol, Merck & Co., Rahway, N.J.; must be refrigerated when not in use.
‡Lowenstein-Jensen Medium Deeps, Difco Laboratories, Detroit.

retain their ability to produce catalase after heating at 68 C at pH 7.

With the semiquantitative test, most Group II scotochromogens, Group IV rapid growers (including *M. fortuitum*), and *M. kansasii* strains, as well as some Group III nonphotochromogens *(M. terrae, M. ulcerans, M. triviale)*, generally produce in excess of 45 mm of bubbles.[21] *M. tuberculosis, M. bovis, M. avium* complex, as well as clinically less significant strains of *M. kansasii*, will show less than 45 mm of foam; *M. marinum* and *M. xenopi* also produce less than 45 mm.

Tween 80 hydrolysis

Certain species of mycobacteria are capable of hydrolysing Tween 80, a derivative of sorbitan monoleate, with the production of oleic acid. In the presence of the pH indicator neutral red, this acid is indicated by a change in color from amber to pink.

The test is performed by the method of Wayne and co-workers[48] (see Chapter 43 for reagents). One 3-mm loopful of an actively growing egg slant culture is suspended in a tube of substrate, which is incubated at 35 C. A control tube consisting of a known positive culture *(M. kansasii)* and an uninoculated (negative control) tube also are incubated. The tubes are examined after 5 days' and 10 days' incubation; a **positive** test is indicated by a color change from amber to **pink** or **red.**

Generally, most strains of *M. kansasii* are positive within 5 days, whereas many strains of *M. tuberculosis* are positive in 10 to 20 days. Clinically significant Group II and Group III cultures usually remain negative for 3 weeks, whereas the clinically less significant strains are positive within 5 days.

Sodium chloride (NaCl) tolerance test[16]

This procedure is useful for the general separation of the rapid-growing mycobacteria (positive) from the slow-growing strains (negative), also to aid in the iden-

tification of *M. triviale* (positive) and *M. flavescens* (sometimes positive).

The medium can be prepared by adding NaCl in a final concentration of 5% to either L-J or American Trudeau Society medium before inspissation. A light suspension (barely turbid) of growth from an L-J slant is prepared, and 0.1 ml ,is inoculated to the surfaces of the NaCl medium* and a control slant containing no NaCl. The tests are incubated at 35 C and examined for growth at weekly intervals for 4 weeks.

Arylsulfatase test

The enzyme arylsulfatase is present in varying concentration in many mycobacterial species, and its concentration in a controlled 3-day test is particularly useful in differentiating the potential pathogens *M. fortuitum* and *M. chelonei* from other Group IV rapid growers.

This rapid test[44] is performed by growing the suspected organism in Tween–albumin broth (available commercially) for 7 days and inoculating 0.1 ml of this into a substrate containing 0.001 M tripotassium phenolphthalein disulfate,† along with known negative, weakly positive, and positive control cultures. These are incubated at 35 C for 3 days,‡ and then 6 drops of 1 molar sodium carbonate are added to each tube. If appreciable amounts of arylsulfatase have been produced, phenolphthalein will be split from the sulfate and detected by alkalinizing with sodium carbonate. The resulting color is then immediately compared with a set of standards (see Chapter 43 for preparation), since the color is not stable. A **positive** test shows a range from a faint pink (±) to a **light red** (3+); a negative test develops no color.

---

*Available from Difco Laboratories (No. 1423), Detroit.
†Nutritional Biochemical Corp., Cleveland; L. Light & Co., Colinbrook, Bucks, England.
‡A 2-week test using a 0.003 M substrate also may be used.

Growth on MacConkey agar

M. fortuitum and *M. chelonei* grow on MacConkey agar in 5 days, thereby differentiating them from other group IV rapid growers, which are inhibited.[13]

A Petri dish containing 15 ml of MacConkey agar* is inoculated on a turntable with a 3-mm loopful of a 7-day Tween–albumin broth culture of the organism in question by rotating the plate and slowly moving the loop along the surface of the medium. The plate is incubated a 35 C and observed at 5 days and 11 days for growth. A positive control of *M. fortuitum* and a negative control of *M. phlei* also should be included. Occasionally, other Group IV organisms will grow within 11 days.

Tellurite reduction

The tellurite reduction test[18] appears to be most valuable in the separation of the potentially pathogenic *M. avium* complex strains from the saprophytic nonphotochromogens. Middlebrook 7H9 broth with ACD enrichment and Tween 80 is the base medium, to which is added a solution of tellurite. (See Chapter 41 for preparation.)

The medium is inoculated with the mycobacterium and incubated at 35 C for 7 days. Two drops of a sterile 0.2% solution of potassium tellurite are then added to the medium, which is returned to the incubator and examined daily for reduction of the colorless tellurite salt to a black (or "dirty brown" in the case of highly pigmented strains) metallic tellurium.

Most *M. avium* complex strains and the majority of Group IV rapid growers will reduce tellurite in 3 to 4 days; other mycobacteria, such as *M. terrae* and *M. xenopi*, react more slowly.

---

*Plates, rather than slant cultures, should be used, since the latter medium will support growth of heavy inocula of *M. smegmatis*.

Other tests

Commercially available urease test disks[30] are convenient and reliable. Two additional tests recently introduced appear promising for identification of certain mycobacteria: the heat-stable acid phosphatase test[37] and the amidase test.[11] A recent paper (David, H. L., and Jahan, M. T.: J. Clin. Microbiol. **5**:383-384, 1977) indicates that a test for β-glucosidase activity may be useful in differentiating various mycobacteria.

Table 31-2 summarizes most of the tests described.

MYCOBACTERIUM LEPRAE (LEPROSY BACILLUS OR HANSEN'S BACILLUS)

The leprosy bacillus was discovered by Hansen in 1872 in lepra cells (mononuclear epithelioid cells) of patients with leprosy. The conditions concerning communicability of the disease are still unclear. Humans are the only host; there is no known animal or soil reservoir.

Typical cells of *Mycobacterium leprae* are found predominantly in smears and scrapings obtained from the skin (not in the epidermis, but the corium) and mucous membrane (particularly of the nasal septum) of patients with nodular leprosy. When stained by the Ziehl-Neelsen method and decolorized and counterstained with 0.2% methylene blue in 4% sulfuric acid, large numbers of acid-fast bacilli are seen packed in lepra cells in parallel bundles, suggesting packets of cigars (globi). The bacilli are also found within endothelial cells of blood vessels, although they are rare or absent in tuberculoid lesions and undifferentiated lesions.

The organism has not been cultured on artificial media or human tissue culture cells. In 1960, Shephard was successful in producing infection in mouse foot pads with material obtained from cases of human leprosy. Other mycobacteria, such as *M. marinum* and *M. ulcerans*, also will grow in foot pads of mice. Due to the prolonged generation time of these orga-

Table 31-2. Key differential features of mycobacteria recovered from humans*

Test or property	Tuberculosis complex		Group I			Group II				Group III						Group IV				
	M. tuberculosis	*M. bovis*	*M. kansasii*	*M. marinum*	*M. simiae*	*M. scrofulaceum*	*M. szulgai*	*M. gordonae*	*M. flavescens*	*M. xenopi*	*M. avium complex*	*M. ulcerans*	*M. gastri*	*M. terrae complex*	*M. triviale*	*M. fortuitum*	*M. chelonei*	*M. phlei*	*M. smegmatis*	*M. vaccae*
Growth rate	S	S	S	S	S	S	S	S	S	S	S	S	S	S	S	R	R	R	R	R
Growth temperature: 22 to 25 C	–	–	+	+	+	+	+	+	+	–	F	M	+	+	+	+	+	+	+	+
32 to 33 C	+	+	+	+	+	+	+	+	+	+	+	+	+	+	+	+	+	+	+	+
35 to 39 C	+	+	+	–	+	+	+	+	+	+	+	–	+	+	+	+	+	+	+	+
41 to 43 C	–	–	–	–	F	–	–	–	–	+	F	–	–	–	–	M	M	+	–	F
Niacin production	+	–	–	F	+	–	–	–	–	–	–	–	–	–	–	–	V	–	–	–
Nitrate reduction	+	–	+	–	–	–	+	–	+	–	–	–	–	+	+	+	–	+	+	+
Catalase (> 45 mm foam)	–	–	+	F	+	+	+	+	+	+	–	+	–	+	+	+	+	+	+	+
Catalase after 68 C, 20 min	–	–	+	F	+	+	+	+	+	+	+	+	–	+	+	+	+	+	M	+
Pigmentation in dark	–	–	–	–	–	3	2,3	3	+	3	–	–	–	–	–	–	–	+	–	+
Pigment photoactivated	–	–	+	+	+	–	F	–	–	–	–	–	–	–	–	–	–	–	–	–
Tween hydrolysis (5 or 10 days)	F	–	+	+	–	–	–	+	F	–	–	–	+	+	+	M	F	+	+	M
Tellurite reduction (3 days)	–	–	–	–	–	+	–	–	+	–	M	+	–	–	–	M	M	M	M	M
Resistant 10 μg TCH per ml	+	–	+	+	+	+	+	+	+	+	+	+	+	+	+	+	+	+	+	+
Growth in presence of 5% NaCl	–	–	–	–	–	–	–	–	+	–	–	–	–	+	+	+	M	+	+	+
Iron uptake (rusty colonies)	–	–	–	–	–	–	–	–	–	–	–	–	–	–	–	+	–	+	+	+
Arylsulfatase positive (3 days)	–	–	–	–	–	–	F	–	–	M	F	–	–	–	F	+	+	–	–	+
Growth on MacConkey agar	–	–	–	–	–	–	–	–	–	–	–	–	–	–	–	+	+	–	–	+
Urease activity	+	+	+	+	+	+	+	–	+	–	–	+	+	–	–	+	+	+	+	+
Clinical significance	+	+	+	+	+	+	+	–	–	+	+	+	–	–	†	+	+	–	–	–

S, slow; R, rapid; +, > 84% of strains +; M, 50% to 84% of strains +; F, 16% to 49% of strains +; V, variable; blank spaces, few or no data available; –, < 16% of strains +.

*Modified from Kubica et al.[26]

†One human infection (septic arthritis) reported.

1. Strains adapt to growth at 37 C; on primary isolation, usually grow only at 31 to 33 C.

2. *M. szulgai* is scotochromogenic when grown at 37 C but commonly photochromogenic at 25 C.

3. Pigment intensifies either with age or after prolonged (2 weeks) exposure to light.

nisms, especially *M. leprae,* this procedure is not considered practical as a diagnostic test. The armadillo is susceptible to leprosy and has been used experimentally. Humans are relatively resistant to infection.

There are no serologic tests of value for diagnosis; serologic tests for syphilis in lepers frequently yield biologically false-positive results. Lepromin, a sterile extract of leprous tissue containing numerous *M. leprae* organisms, is used to determine relative resistance or susceptibility to the disease.

Although they are primarily bacteriostatic in their action, the sulfone drugs are the current therapy of choice in the treatment of human leprosy.

REFERENCES

1. Adams, R. M., Remington, J. S., Steinberg, J., and Seibert, J. S.: Tropical fish aquariums; a source of *Mycobacterium marinum* infections resembling sporotrichosis, J.A.M.A. **211**:457-461,1970.
2. American Lung Association: Diagnostic standards and classifications of tuberculosis and other mycobacterial diseases, New York, 1974, the Association.
3. Bailey, W. C., Raleigh, J. W., and Turner, J. A. P.: Treatment of mycobacterial disease (official statement of American Thoracic Society), Am. Rev. Resp. Dis. **115**:185-187, 1977.
4. Chapman, J. S.: The atypical mycobacteria and human mycobacteriosis, New York, 1977, Plenum Medical Book Co.
5. Connor, D. H., and Lunn, H. F.: Buruli ulceration; a clinicopathologic study of 38 Ugandans with *Mycobacterium ulcerans* ulceration, Arch. Pathol. **81**:183-199, 1966.
6. Corper, H. J., and Uyei, N.: Oxalic acid as a reagent for isolating tubercle bacilli and a study of the growth of acid-fast nonpathogens on different mediums with their reaction to chemical reagents, J. Lab. Clin. Med. **15**:348-369, 1930.
7. Dizon, D., Mihailescu, C., and Bae, H. C.: Simple procedure for detection of *Mycobacterium gordonae* in water causing false-positive acid-fast smears, J. Clin. Microbiol. **3**:211, 1976.
8. Dross, I. C., Abbatiello, A. A., Jenney, F. S., and Cohen, A. C.: Pulmonary infections due to *M. fortuitum,* Am. Rev. Resp. Dis. **89**:923-925, 1964.
9. Gangadharam, P. R. J., Lockhart, J. A., Awe, R. J., and Jenkins, D. E.: Mycobacterial contamination through tap water, Am. Rev. Respir. Dis. **113**:894, 1976.
10. Gruft, H.: Isolation of acid-fast bacilli from contaminated specimens, Health Lab. Sci. **8**:79-82, 1971.
11. Helbecque, D. M., Handzel, V., and Eidus, L.: Simple amidase test for identification of mycobacteria, J. Clin. Microbiol. **1**:50-53, 1975.
12. Jones, W. D., Abbott, V. D., Vestal, A. L., and Kubica, G. P.: A hitherto undescribed group of nonphotochromogenic mycobacteria, Am. Rev. Resp. Dis. **94**:790-795, 1966.
13. Jones, W. D., and Kubica, G. P.: The use of MacConkey's agar for the differential typing of *M. fortuitum,* Am. J. Med. Technol. **30**:187-190, 1964.
14. Joseph, S. W., Vaichulis, E. M. K., and Houk, V. N.: Lack of auramine-rhodamine fluorescence of Runyon Group IV mycobacteria, Am. Rev. Resp. Dis. **95**:114-115, 1967.
15. Kenney, M., Loechel, A. B., and Lovelock, F. J.: Urine cultures in tuberculosis, Am. Rev. Resp. Dis. **82**:564-567, 1959.
16. Kestle, D. G., Abbott, V. D., and Kubica, G. P.: Differential identification of mycobacteria. II. Subgroups of Groups II and III (Runyon) with different clinical significance, Am. Rev. Resp. Dis. **95**:1041-1052, 1967.
17. Kestle, D. G., and Kubica, G. P.: Sputum collection for cultivation of mycobacteria; an early morning specimen or the 24- to 72-hour pool? Am. J. Clin. Pathol. **48**:347-349, 1967.
18. Kilburn, J. O., Silcox, A., and Kubica, G. P.: Differential identification of mycobacteria. V. The tellurite reduction test, Am. Rev. Resp. Dis. **99**:94-100, 1969.
19. Konno, K.: New chemical method to differentiate human-type tubercle bacilli from other mycobacteria, Science **124**:985, 1956.
20. Kubica, G. P.: Differential identification of mycobacteria, Am. Rev. Resp. Dis. **107**:9-21, 1973.
21. Kubica, G. P., and Dye, W. E.: Laboratory methods for clinical and public health mycobacteriology, Public Health Service Pub. No. 1547, Washington, D.C., 1967, U.S. Government Printing Office.
22. Kubica, G. P., Dye, W. E., Cohn, M. L., and Middlebrook, G.: Sputum digestion and decontamination with N-acetyl-L-cysteine-sodium hydroxide for culture of mycobacteria, Am. Rev. Resp. Dis. **87**:775-779, 1963.
23. Kubica, G. P., Jones, W. D., Jr., Abbott, V. D., Beam, R. E., Kilburn, J. O., and Cater, J. C., Jr.: Differential identification of mycobacteria. I. Tests on catalase activity, Am. Rev. Resp. Dis. **94**:400-405, 1966.
24. Kubica, G. P., and Pool, G. L.: Studies on the catalase activity of acid-fast bacilli. I. An attempt to subgroup these organisms on the basis of their catalase activities at different temperatures and pH, Am. Rev. Resp. Dis. **81**:387-391, 1960.

25. Kubica, G. P., and Vestal, A. L.: Tuberculosis —laboratory methods in diagnosis, Atlanta, 1959, National Communicable Disease Center.

26. Kubica, G. P., Gross, W. M., Hawkins, J. E., Sommers, H. M., Vestal, A. L., and Wayne, L. G.: Laboratory services for mycobacterial diseases, Am. Rev. Resp. Dis. **112:**773-787, 1975.

27. Listwan, W. J., Roth, D. A., Tsung, S. H., and Rose, H. D.: Disseminated *Mycobacterium kansasii* infection with pancytopenia and interstitial nephritis, Ann. Intern. Med. **83:**70-73, 1975.

28. Lorian, V.: Direct cord reading medium for isolation of mycobacteria, Appl. Microbiol. **14:**603-607, 1966.

29. Marks, J., and Schwabacher, H.: Infection due to *Mycobacterium xenopei*, Br. Med. J. **1:**32-33, 1965.

30. Murphy, D. B., and Hawkins, J. E.: Use of urease test disks in the identification of mycobacteria, J. Clin. Microbiol. **1:**465-468, 1975.

31. Oliver, J., and Reusser, T. R.: Rapid method for the concentration of tubercle bacilli, Am. Rev. Tuberc. **45:**450-452, 1945.

32. Petroff, S. A.: Some cultural studies on the tubercle bacillus, Bull. Johns Hopkins Hosp. **26:**276-279, 1915.

33. Runyon, E. H.: Anonymous mycobacteria in pulmonary disease, Med. Clin. North Am. **43:**273-290, 1959.

34. Runyon, E. H.: Identification of mycobacterial pathogens utilizing colony characteristics, Am. J. Clin. Pathol. **54:**578-586, 1970.

35. Runyon, E. H., Selin, M. J., and Harris, H. W.: Distinguishing mycobacteria by the niacin test —a modified procedure, Am. Rev. Tuberc. **79:**663-665, 1959.

36. Runyon, E. H., Karlson, A. G., Kubica, G. P., and Wayne, L. G.: *Mycobacterium.* In Lennette, E. H., Spaulding, E. H., and Truant, J. P., editors: Manual of clinical microbiology, ed. 2, Washington, D.C., American Society for Microbiology.

37. Saito, H., Yamaoka, K., Kiyotani, K., and Masai, H.: A new heat-stable acid phosphatase test for mycobacteria, Am. Rev. Resp. Dis. **114:**407-408, 1976.

38. Strumpf, I. J., Tsang, A. Y., Schork, M. A., and Weg, J. G.: The reliability of gastric smears by auramine-rhodamine staining technique for the diagnosis of tuberculosis, Am. Rev. Resp. Dis. **114:**971-976, 1976.

39. Truant, J. P., Brett, W. A., and Thomas, W., Jr.: Fluorescence microscopy of tubercle bacilli stained with auramine and rhodamine, Henry Ford Hosp. Med. Bull. **10:**287-296, 1962.

40. Vestal, A. L.: Procedures for the isolation and identification of mycobacteria, Publ. Health Serv. Publ. No. (CDC) 77-8230, Atlanta, 1977, Center for Disease Control.

41. Virtanen, S.: A study of nitrate reduction by mycobacteria, Acta Tuberc. Scand. (Suppl.) **48:** 1-119, 1960.

42. Washington, J. A. II, editor: Laboratory procedures in clinical microbiology, Boston, 1974, Little, Brown and Co., pp. 100-117.

43. Wayne, L. G.: The use of Millipore filters in clinical laboratories, Am. J. Clin. Pathol. **28:** 565-567, 1957.

44. Wayne, L. G.: Recognition of *Mycobacterium fortuitum* by means of a three-day phenolphthalein sulfatase test, Am. J. Clin. Pathol. **36:**185-187, 1961.

45. Wayne, L. G.: Two varieties of *M. kansasii* with different clinical significance, Am. Rev. Resp. Dis. **86:**651-656, 1962.

46. Wayne, L. G., and Doubek, J. R.: Diagnostic key to mycobacteria encountered in clinical laboratories, Appl. Microbiol. **16:**925-931, 1968.

47. Wayne, L. G., Doubek, J. R., and Diaz, G. A.: Classification and identification of mycobacteria. IV. Some important scotochromogens, Am. Rev. Resp. Dis. **96:**88-95, 1967.

48. Wayne, L. G., Doubek, J. R., and Russell, R. L.: Classification and identification of mycobacteria; tests employing Tween 80 as substrate, Am. Rev. Resp. Dis. **90:**588-597, 1964.

49. Wayne, L. G., Krasnow, I., and Kidd, G.: Finding the "hidden positive" in tuberculosis eradication programs; the role of the sensitive trisodium phosphate–benzalkonium chloride (Zephiran) culture technique, Am. Rev. Resp. Dis. **86:**537-541, 1962.

50. Wayne, L. G., Engbaek, H. C., Engel, H. W. B., et al.: Highly reproducible techniques for use in systematic bacteriology in the genus *Mycobacterium:* tests for pigment, urease, resistance to sodium chloride, hydrolysis of Tween 80, and β-galactosidase, Int. J. Syst. Bacteriol. **24:**412-419, 1974.

51. Wayne, L. G., Engel, H. W. B., Grassi, C., et al.: Highly reproducible techniques for use in systematic bacteriology in the genus *Mycobacterium:* tests for niacin and catalase and for resistance to isoniazid, thiophene 2-carboxylic acid hydrazide, hydroxylamine, and *p*-nitrobenzoate, Int. J. Syst. Bacteriol. **26:**311-318, 1976.

52. Wayne, L. G., David, H., Hawkins, J. E., Kubica, G. P., Sommers, H. M., and Wolinsky, E.: Referral without guilt or how far should a good lab go? ATS News **2:**8-12, 1976.

53. Whitcomb, F. C., Foster, M. C., and Dukes, C. D.: Increased carbon dioxide tension and the primary isolation of mycobacteria, Am. Rev. Resp. Dis. **86:**584-586, 1962.

54. Wolinsky, E.: Identification and classification of

mycobacteria other than *Mycobacterium tuberculosis*. In Hobby, G. L., editor: Handbook of tuberculosis laboratory methods, Washington, D.C., 1962, Veterans Administration, U.S. Government Printing Office.

55. Youmans, G. P.: The pathogenic "atypical" my-

cobacteria, Ann. Rev. Microbiol. **17**:473-494, 1963.

56. Zvetina, J. R., and Wichelhausen, R. H.: Pulmonary infection caused by niacin-positive *Mycobacterium avium*, Am. Rev. Resp. Dis. **113**:885-887, 1976.

CHAPTER 32

Miscellaneous and unclassified pathogens

Streptobacillus · *Actinobacillus* · *Calymmatobacterium* · *Mycoplasma* · *Cardiobacterium* · *Chromobacterium* · *Chlamydia* · *Agrobacterium* · Legionnaires' disease agent · Infections related to dog bites

STREPTOBACILLUS

Only one species, *Streptobacillus moniliformis*, comprises the genus *Streptobacillus*. This is an **extremely pleomorphic, facultatively anaerobic,** gram-negative organism, which forms irregular chains of fairly uniform, small, slender rods 2 to 4 μm in length, to long, interwoven, looped or curved filaments up to 150 μm in length. These may be interspersed with fusiform enlargements and large, round, *Candida*-like swellings along the length of the filament (L. *monile*, necklace), two to five times its width. These morphologic forms depend largely on the medium used, the conditions of incubation, and the age of the culture; the regular, rod-shaped cells predominate in clinical specimens and under favorable artificial conditions.

Streptobacillus moniliformis is a normal inhabitant of the throat and nasopharynx of wild and laboratory rats and mice. It is extremely virulent for mice. Humans usually acquire the infection by the **bite of a rat,** mouse, or other rodent, although ingestion of contaminated food, particularly milk, may also cause infection (Haverhill fever). The infection is characterized by fever, rash, and polyarthritis. The organism may be recovered from the blood, joint fluid, skin lesions, or pus.

Streptobacillus moniliformis requires the presence of natural body fluid, such as ascitic fluid, blood, or serum, for its growth on artificial media. The addition of 10% to 30% ascitic fluid to thioglycollate medium makes an excellent recovery medium. The streptobacillus grows in the form of "fluff balls" or "breadcrumbs" near the bottom of the tube or on the surface of the sedimented red cells if blood is cultured (see Chapter 7). These colonies may be removed with a sterile Pasteur pipet and transferred to solid media or to slides for microscopic examination. (Giemsa or Wayson stains are preferable to the Gram stain.)

On slightly soft media, such as ascitic fluid agar or serum agar, incubated aerobically (some strains require increased carbon dioxide, as in a candle jar) and with an excess of moisture, small discrete colonies develop in 2 to 3 days. These are smooth and glistening, irregularly round with sharp edges, and colorless or grayish. Beneath or adjacent to the colonies of *Streptobacillus*, L-form colonies may develop; these measure 100 to 200 μm in diameter and show a dark center and lacy periphery ("fried egg" appearance). This L form is a variant of *Streptobacillus moniliformis*, arising spontaneously. Morphologically, L-form colonies consist of tiny, bipolar-staining, coccoid or coccobacillary elements.

Agglutinins are formed in patients with *Streptobacillus* infections; titers of 1:80 or greater are considered indicative of infection. Penicillin is effective in therapy.

ACTINOBACILLUS

In *Bergey's Manual* (ed. 8), two species are recognized for the genus *Actinobacillus*: *A. lignieresii* and *A. equuli*, with *A. actinomycetemcomitans* considered a "species incertae sedis" (as *Bacterium actinomycetemcomitans*).

Organisms of this genus are primarily pathogenic in animals, with occasional infections occurring in humans. *Actinobacillus* can cause acute septicemia or localized granulomatous lesions or abscesses.[1] Under normal conditions, these organisms also are found in the mouth and gastrointestinal tract.

Gram stain of these bacteria reveals a gram-negative, nonmotile, non-spore-forming coccobacillus that can occur singly, in pairs, and in short chains. These bacteria also are pleomorphic, occurring as filaments, long rods, or coccal forms.

Considered primarily as a facultative organism, *Actinobacillus* grows best on blood agar or 10% serum agar at 35 C. Little or no growth occurs on MacConkey agar, and there is no growth on SS or eosin methylene blue agar.[1] Helpful biochemical tests for identification are: nitrates reduced to nitrites, urease positive, negative test results for indole, Voges-Proskauer, and sodium citrate. These bacteria utilize, with no gas formation, glucose, sucrose, and maltose and to a lesser extent lactose, xylose, fructose and galactose. Dulcitol, rhamnose, and inositol are not utilized. Because of the similarity of their reactions, the speciation of *A. lignieresii* and *A. equuli* is extremely difficult. Indeed, the valid existence of these two species has been seriously questioned.[18]

CALYMMATOBACTERIUM GRANULOMATIS

This organism is the etiologic agent of **granuloma inguinale.** This venereal disease is characterized by an initial single or multiple swellings or "bubos" in the groin area, followed by involvement of the genitalia and sometimes the buttocks and abdomen with a hypertrophic, sclerotic granulomatous lesion. There may be ulceration or bleeding.

Calymmatobacterium is a nonmotile, gram-negative, pleomorphic rod with rounded ends and may occur singly or in clusters. "Safety pin" forms may also occur as a result of bipolar condensation of chromatin. Wright's stain of the encapsulated forms reveals a blue rod surrounded by a large, well-defined, pink capsule.

In the cultivation of *C. granulomatis*, two factors are of importance: a low oxidation-reduction potential and growth factors found in egg yolk, phytone, and lactalbumin hydrolysate.[10] Inoculation of yolk sacs of 5-day-old chicken embryos and incubation for 72 hours at 35 C seems to be the method of choice.[10]

Therapy may be difficult. Active drugs include tetracycline, chloramphenicol, gentamicin, and streptomycin.

MYCOPLASMA (PPLO) SPECIES

Dienes and Edsall, in 1937, were the first to report pleuropneumonialike organisms (PPLO) from a pathologic process in humans. Since that time, the role of these small microorganisms, which can be grown on culture media, has been studied in infections of the urogenital and respiratory tracts, in wound infections, as tissue culture contaminants, and in animal hosts other than humans. Mycoplasmas are generally considered as physiologically intermediate between the bacteria and rickettsiae and may be either parasitic or saprophytic. Presently, seven identified species of *Mycoplasma* are indigenous to humans. One species, *Mycoplasma pneumoniae*, is a recognized human pathogen—the etiologic agent of **primary atypical pneumonia** and bronchitis. Other manifestations of *M. pneumoniae* infection include various rashes (including Stevens-Johnson syndrome), bullous hemorrhagic myringitis, arthritis, pericarditis, hemolytic anemia, meningoencephalitis, aseptic meningitis and

Guillain-Barré syndrome. *Ureaplasma urealyticum* may be responsible for some cases of nongonococcal urethritis (Bowie, W. R., et al.: J. Clin. Invest. **59:**735-742, 1977). *Mycoplasma hominis* rarely may cause significant infection in humans (Siber, G. R., et al.: J. Pediatr. **90:**625-627, 1977).

The parasitic mycoplasmas are best recognized by such characteristics as their growth requirements (enriched media with sterols, conditions of incubation, and so forth), colonial morphology, inhibition by specific antisera, and resistance to penicillin. Another clue to their identity is the anatomical site from which the specimen was obtained, for example, *M. pneumoniae* from lower respiratory tract secretions, T-strain mycoplasma from a genitourinary source, and so forth.

Growth of the parasitic mycoplasmas is best encouraged on an enriched medium containing heart infusion agar,* horse serum, yeast extract, and penicillin G (see Chapter 41 for preparation). Colonies are best observed microscopically under 45× magnification and appear large (250 to 750 μm in diameter) to small (1 to 10 μm), raised, pitted, and lacy to coarsely pebbled. The central portion of the colony grows into the agar medium,† appearing darker than the periphery when examined by transmitted light, thus giving it the characteristic "fried egg" appearance. Growth in enriched broth is not visible, due to the small size of the organisms— subculture to solid media is required— and organisms are not demonstrable by the usual staining methods. Stained preparations are best studied by the Dienes method, in which an agar block bearing the selected colony is cut out with a sterile scalpel and placed upright on a glass microscope slide. A coverslip previously coated with the Dienes stain* and dried is carefully lowered, stain side down, on the agar block, and the edges are sealed with Vaspar. The preparation may then be examined under oil immersion. Colonies of mycoplasma and bacteria (if selected in error) stain blue in a short time; after 15 minutes, however, the bacterial colonies will have lost their color due to the reduction of methylene blue by metabolizing organisms, while the mycoplasmas will retain their stain. The method of Clark and others[5] is also recommended for microscopic study of mycoplasmas.

Isolation from clinical material

Primary cultures of clinical specimens —usually sputum or swabs obtained from the throat, genital tract, and so forth—are placed in a tube of selective broth containing a bacterial inhibitor,† incubated for several days at 35 C, then subcultured to freshly prepared solid media and incubated as before. A transport medium consisting of 2 ml of trypticase soy broth with 0.5% bovine albumin has been used satisfactorily for swabs. *Mycoplasma pneumoniae* grows in air or 5% CO_2; *M. fermentans*, *M. orale*, and *M. salivarium* require an anaerobic atmosphere of 95% N_2 and 5% CO_2 in an evacuation-replacement jar. Plates are generally examined at 7, 14, and 21 days; those for *M. pneumoniae* should be held for 30 days before discarding. For primary cultivation of T-strain mycoplasmas, the use of a differential agar medium[13] containing urea and manganous sulfate is recommended. The T-strain mycoplasmas (now known as *Ureaplasma urealyticum*) have the unique property of hydrolysing urea and will appear on this medium as **dark, golden brown** colonies.

---

*Mycoplasma agar base, Baltimore Biological Laboratory, Cockeysville, Md.; PPLO agar, Difco Laboratories, Detroit. Broth media are also available.

†Colonies cannot be transferred with conventional needles or loops; an agar block cut from the plate is used.

---

*Dienes stain contains methylene blue, azure II, and other ingredients; see Chapter 41 for preparation.

†See Chapter 41 for preparation.

Identification by growth inhibition test

Early reports by various workers indicated that species-specific antisera prepared in rabbits were inhibitory to the growth of mycoplasmas and that this could be used in their identification.[6] Filter paper disks are saturated with individual growth-inhibiting antisera (made against prototype strains*) and pressed onto an agar plate that has been previously seeded with a pure culture of the unknown strain. After incubation at 35 C for 1 week (or until good growth occurs), the plate is examined microscopically (10 to 45× magnification). A **clear zone** around any disk greater than the clear zone around a control disk that contains normal rabbit serum identifies the species of mycoplasma being tested. Since inhibition zones have been shown to be a function of the size of the inoculum and may vary from 0.1 to 17 mm in diameter, the most frequent error of "no zones" around any disk indicates that too heavy an inoculum was used. This may be remedied by repeating the test with varying dilutions of the unknown strain or by cutting out an agar block of growth and pushing it across the surface of the test plate in **one direction only,** and placing two disks along the line of inoculation, 1 inch apart. Further details of the disk technique may be found in other sources.[14]

Other identification tests

Several other in vitro tests are available, directed primarily toward the identification of *M. pneumoniae,* and include the following:

1. *M. pneumoniae* colonies will produce beta hemolysis when coated with a layer of blood agar prepared with sheep or guinea pig erythrocytes; other species produce alpha, alpha prime, or nonhemolytic reactions,[16] but *Acholeplasma laidlawii,* which rarely may be isolated

---

*Hyperimmune rabbit antisera are available for the mycoplasma prototypes previously listed from Microbiological Associates, Inc., Bethesda, Md.

from the respiratory tract, also produces hemolysis. The unique cellular morphology of *M. pneumoniae* growing on glass has also been proposed as a rapid means of identification.[2]

2. Colonies of *M. pneumoniae* will reduce a tetrazolium salt incorporated in agar under aerobic conditions; the area around the colony becomes pink. *M. pneumoniae, M. fermentans,* and *A. laidlawii* will utilize dextrose with the production of acid but no gas, a useful presumptive test.[7]

3. Mycoplasma colonies stained with homologous FA conjugates also will show a characteristic yellow-green fluorescence when examined by incident light fluorescence microscopy.[8]

The **recommended** identification technique is use of specific antisera for detection of growth inhibition, as previously described.

Primary atypical pneumonia caused by *M. pneumoniae* has been most successfully treated with the tetracyclines and erythromycin; the penicillins have not proved effective, and clindamycin is less effective than tetracycline or erythromycin.

CARDIOBACTERIUM

Cardiobacterium hominis (formerly CDC Group II-D) is the sole species in this genus.[17] It is a **fermentative** gram-negative rod that is **oxidase positive, catalase negative** and **nonmotile.** The organisms are pleomorphic with bulbous ends on media without yeast extract; they tend to retain some crystal violet stain. Cells may occur in clusters resembling rosettes. Colonies on blood agar are tiny after 24 hours' incubation and 1 mm in diameter after 48 hours. They are convex, circular, entire, smooth, and soft; no hemolysis is seen on rabbit blood. Carbon dioxide and a moist atmosphere and yeast extract facilitate growth (Savage, D. D., et al.: J. Clin. Microbiol. **5:**75-80, 1977). Acid is produced throughout TSI (A/A), with no H_2S production, although lead acetate paper strips may give a positive reaction. The organism does not grow on enteric media. Urea is not hydrolyzed, and nitrate is not reduced. A small amount of indole is formed. Decarboxylase reactions are negative. Carbohydrate utilization may be determined in liquid peptone medium with added rabbit serum (2 drops). Acid is produced from glucose, mannitol, sucrose and maltose in 2 to 7 days; xylose and lactose are not fermented.

The organism is found in the human upper respiratory tract and in feces; it is not present in the genitourinary tract. Most isolates of this uncommon organism have been from cases of **endocarditis** (Slotnick, I. J., et al.: J. Infect. Dis. **114:** 503, 1964). Rarely, the organism has been recovered from sputum, pleural fluid, and spinal fluid.

C. hominis is quite susceptible to various penicillins, to cephalothin, tetracycline, chloramphenicol, aminoglycosides, and colistin (Savage, D. D., et al.: J. Clin. Microbiol. **5:**75-80, 1977).

CHROMOBACTERIUM

One species of this genus is encountered in humans, *C. violaceum.* It is found primarily in soil and water, but has been responsible for a small number of very serious infections in humans.

The genus is made up of gram-negative rods, sometimes slightly curved, motile by means of both one **polar flagellum** and one to four subpolar or lateral flagella. **Violet colonies** are produced, and in broth a **violet ring** is produced at the junction of the broth surface and the test tube wall. The pigment is soluble in ethanol but not in water or chloroform. The organism is usually **oxidase positive** (Kovacs method), but pigment may interfere with reading. It is **catalase positive** but **highly sensitive to hydrogen peroxide.** It is resistant to the vibriostatic agent 2,4-diamino-6,7-diiso-propyl pteridine.

C. violaceum forms a fragile pellicle in broth. Nonpigmented variants occur.

Gelatin is liquefied in 7 days. Carbohydrate attack is usually fermentative, sometimes oxidative. Acid is produced from glucose, fructose, and trehalose and often from maltose, mannose, sorbitol, and rhamnose. Rarely, strains will produce gas (Sivendra, R.: J. Clin. Microbiol. **3:**70-71, 1976). Chitin is usually digested. Casein is hydrolyzed. Nitrate is reduced, usually beyond nitrite but without gas. Arginine dihydrolase is positive. The organism produces HCN and cultures smell of ammonium cyanide. Turbidity is produced on egg yolk medium. The organism is facultatively anaerobic with best growth at 30 to 35 C. It will grow on MacConkey agar.

Infections in humans (only about 20 have been described) have usually involved abscess formation or bacteremia. Systemic infection with bacteremia has been uniformly fatal, at least partly due to delay in initiating appropriate therapy. One patient was cured recently with early institution of carbenicillin and genta-

Table 32-1. Laboratory methods of choice in etiologic diagnosis of chlamydial infections

Procedures	Psitta-cosis	LGV	TRIC
Direct microscopic examination (Giemsa, fluorescent antibody)	−	−	+
Isolation of agent			
Cell culture	+	±	−
Irradiated cell culture	+	+	+
Chick embryo yolk sac	+	+	+
Mice			
Intraperitoneal	+	−	−
Intracerebral	±	±	−
Serology (antibody titer rise)			
CF (group antigen)	+	+*	+
Fluorescent antibody	−	+	+
Skin test	−	±	−

From Hanna, L., Schacter, J., and Jawetz, E.: Chlamydiae (psittacosis–lymphogranuloma venereum–trachoma group). In Lennette, E. H., Spaulding, E. H., and Truant, J. P., editors: Manual of clinical microbiology, ed. 2, Washington, D.C., 1974, American Society for Microbiology.
*A single titer of 1:64 or higher supports the diagnosis of active LGV infection.

micin therapy (Victorica, B., et al.: J.A.M.A. **230**:578-580, 1974). Virtually all reported cases of human infection were from Southeast Asia and the southeastern United States. Bacteremia is usually accompanied by necrotizing metastatic lesions. While wounds not infrequently become infected with other soil and water organisms such as *Aeromonas* or *Edwardsiella*, this type of infection rarely occurs with *Chromobacterium*.

C. violaceum is resistant to penicillin and may produce a penicillinase at times. The strain studied by Victorica and coworkers was inhibited by 78 μg/ml of carbenicillin, an achievable level; by disk technique it was resistant to ampicillin, cephalothin, and colistin but sensitive to all aminoglycoside drugs, to chloramphenicol, and to tetracycline. The minimal inhibitory concentration (MIC) of gentamicin was 5 μg/ml.

CHLAMYDIA

Chlamydiae are nonmotile, gram-negative, obligately intracellular parasites that form characteristic intracellular microcolonies (inclusions). They differ from viruses in containing both RNA and DNA and in possessing bacterial-type cell walls, ribosomes, and some metabolically active enzymes. They multiply by binary fission and are susceptible to certain antimicrobial drugs. The organisms are large enough to be seen by light microscopy; they are best stained with the Giemsa, Macchiavello, or Giménez stains. Both group and specific antigens are found.

Table 32-1 summarizes laboratory methods of choice for diagnosis of various chlamydial infections.

Details regarding collection and storage of specimens and of techniques for direct examination, cultivation and identification of chlamydiae are given by Hanna and associates.[9]

These organisms are the causative agents of **psittacosis-ornithosis, lymphogranuloma venereum** (LGV), and **tracho-ma** and **inclusion conjunctivitis** (blennorrhea) (TRIC) in humans. Recently a distinctive pneumonia syndrome has been described in infants infected with *Chlamydia trachomatis* (Beem, M. O., and Saxony, E. M.: N. Engl. J. Med. **296**:306-310, 1977). As noted in Chapter 11, they are important causes of "nonspecific" urethritis. In the latter situation, the use of urethral swabs is superior to urine specimens for recovery of the organism.[15] These organisms also play a role in acute salpingitis (Mårdh, P-A, et al.: N. Engl. J. Med. **296**:1377-1379, 1977).

Tetracycline is active therapeutically in all chlamydial infections. Sulfonamides are also quite effective in lymphogranuloma venereum, trachoma, and inclusion blennorrhea. In psittacosis, in addition to tetracycline, chloramphenicol, penicillin in large doses, and erythromycin have been successful in some cases.

AGROBACTERIUM

The Center for Disease Control has studied 34 human isolates of a gram-negative bacillus previously called Vd-3 and concluded that they are probably identical with *Agrobacterium radiobacter*. Most isolates were from the respiratory tract, but three were from blood cultures (Riley, P. S., and Weaver, R. E.: J. Clin. Microbiol. **5**:172-177, 1977).

LEGIONNAIRES' DISEASE AGENT

The causative agent of Legionnaires' disease is an unusual and fastidious gram-negative bacillus. It may be difficult to stain by the Gram stain, but on electron microscopy the cell wall structure is typical of gram-negative bacilli. The organism in tissue is best stained by a silver impregnation technique, such as the Dieterle stain. The organisms are short pleomorphic rods, 1μm in diameter and 1 to 4 μm in length. They sometimes show bipolar staining. The cellular fatty acid composition is unique (Moss, C. W., et al.: J. Clin. Microbiol. **6**:140-143, 1977).

The organism is weakly catalase positive, oxidase positive (Kovac's reagent), gelatin positive (API), urease and ONPG negative, does not produce acid from carbohydrates, and produces a brown soluble pigment.

This agent causes a severe pneumonia, with mortality of 15% to 20%. It is susceptible to many drugs in vitro, but several of these are relatively ineffective in treating infected guinea pigs. Erythromycin is effective in the animal model and in humans. Tetracycline may also be effective. Cephalosporins are not very active.

Information on culture of this agent is found on p. 67.

INFECTIONS RELATED TO DOG BITES

A recent report describes significant illness in 17 patients due to a gram-negative bacillus that is distinct from all others previously described.[4] All 17 patients had positive blood cultures, seven had cellulitis, six pulmonary infiltrates, four purulent meningitis, and three endocarditis. Three patients died. Fifteen of the 17 were men, 14 had underlying diseases or conditions (including prior splenectomy in five), and 10 had had recent dog bites.

The organism is **oxidase and catalase positive,** is negative for nitrate reduction, indole production, and urease. It produces acid from glucose, lactose, and maltose, probably by fermentation. Growth is enhanced with most strains by incubation in candle jars and by addition of rabbit serum to heart infusion agar. The organism does not grow on MacConkey agar. It is nonmotile. Cells are thin and 1 to 3 μm in length, with some long rods and filaments, which are frequently curved. Two strains produced cigar-shaped rods and rods with spindle-shaped swellings. The organism is susceptible in vitro to penicillins, cephalothin, tetracycline, chloramphenicol, erythromycin, and clindamycin. It shows variable susceptibility to cotrimoxazole and is resistant to colistin and to aminoglycosides.

Aside from *Pasturella multocida* and the unusual organism discussed above, there are other organisms unique to the normal oral flora of dogs (Nyby, M. D., et al.: J. Clin. Microbiol. **6:**87-88, 1977; Wunder, J. A., et al.: J. Dent. Res. **55:** 1097-1102, 1976), which may be recovered from wound or other infection related to dog bites.

REFERENCES

1. Alexander, A. D.: *Actinobacillus.* In Lennette, E. H., Spaulding, E. H., and Truant, J. P., editors: Manual of clinical microbiology, ed. 2, Washington, D.C., 1974, American Society for Microbiology.
2. Bredt, W., Lam, W., and Berger, J.: Evaluation of a microscopy method for rapid detection and identification of *Mycoplasma pneumoniae*, J. Clin. Microbiol. **2:**541-545, 1975.
3. Buchanan, R. E., and Gibbons, N. E., editors: Bergey's manual of determinative bacteriology, ed. 8, Baltimore, 1974, The Williams and Wilkins Co.
4. Butler, T., Weaver, R. E., Venkata Ramani, T. K., et al.: Unidentified gram-negative rod infection. A new disease of man, Ann. Intern. Med. **86:**1-5, 1977.
5. Clark, H. W., Fowler, R. C., and Brown, T. M.: Preparation of pleuropneumonia-like organisms for microscopic study, J. Bacteriol. **81:** 500-502, 1961.
6. Clyde, W. A., Jr.: *Mycoplasma* species identification based upon growth inhibition by specific antisera, J. Immunol. **92:**958-965, 1964.
7. Crawford, Y. E.: A laboratory guide to the mycoplasmas of human origin, Naval Medical Research Unit No. 4, 1972, Great Lakes, Ill.
8. Del Guidice, R. A., Robillard, N. F., and Carski, T. R.: Immunofluorescence identification of *Mycoplasma* on agar by use of incident illumination. J. Bacteriol. **93:**1205-1209, 1967.
9. Hanna, L., Schachter, J., and Jawetz, E.: Chlamydiae (psittacosis–lymphogranuloma venereum–trachoma group). In Lennette, E. H., Spaulding, E. H., and Truant, J. P., editors: Manual of clinical microbiology, ed. 2, Washington, D.C., 1974, American Society for Microbiology.
10. Kellogg, D. S., Jr.: *Calymmatobacterium granulomatis.* In Lennette, E. H., Spaulding, E. H., and Truant, J. P., editors: Manual of clinical microbiology, ed. 2, Washington, D.C., 1974, American Society for Microbiology.
11. King, E. O., and Tatum, H. W.: *Actinobacillus actinomycetemcomitans* and *Hemophilus aphrophilus*, J. Infect. Dis. **111:**85-94, 1962.
12. Page, M. I., and King, E. O.: Infection due to

Actinobacillus actinomycetemcomitans and *Haemophilus aphrophilus,* N. Engl. J. Med. **275:**181-188, 1966.

13. Shepard, M. C., and Howard, D. R.: Identification of "T" mycoplasmas in primary agar cultures by means of a direct test for urease, Ann. N.Y. Acad. Sci. **174:**809-819, 1970.

14. Smith, T. F.: Isolation, identification, and serology of *Mycoplasma pneumoniae.* In Washington, J. A. II, editor: Laboratory procedures in clinical microbiology, Boston, 1974, Little, Brown and Co.

15. Smith, T. F., and Weed, L. A.: Comparison of urethral swabs, urine, and urinary sediment for the isolation of *Chlamydia,* J. Clin. Microbiol. **2:**134-135, 1975.

16. Somerson, N. L., Taylor-Robinson, D., and Chanock, R. M.: Hemolysin production as an aid in the identification and quantitation of Eaton agent *(Mycoplasma pneumoniae),* Am. J. Hyg. **77:**122-128, 1963.

17. Tatum, H. W., Ewing, W. H., and Weaver, R. E.: Miscellaneous gram-negative bacteria. In Lennette, E. H., Spaulding, E. H., and Truant, J. P., editors: Manual of clinical microbiology, ed. 2, Washington, D.C., 1974, American Society for Microbiology.

18. Wetmore, P. W., Thiel, J. F., Herman, Y. F., and Harr, J. R.: Comparison of selected *Actinobacillus* species with a hemolytic variety of *Actinobacillus* from irradiated swine, J. Infect. Dis. **113:**186-194, 1963.

CHAPTER 33

Laboratory diagnosis of viral and rickettsial diseases

In a text of this type it would be virtually impossible to provide adequate detail on the diagnosis of rickettsial and viral diseases. Therefore, we have attempted to provide only some guidelines and instructions on the collection and handling of appropriate specimens. For information on virus isolation and identification, the reader may refer to references 2, 3, 6-8, and 10.

Most diagnostic microbiology laboratories are not capable of performing isolation procedures for viruses and rickettsiae, due to lack of expertise and appropriate equipment. Serologic methods, however, can be carried out on a limited basis in many laboratories.

Smaller laboratories should probably not become involved in viral diagnosis, but all laboratories can become knowledgeable about collection, packaging, and transportation of specimens to virus reference laboratories, the services of which are now more available.

GENERAL CONSIDERATIONS

Viruses exist as obligate intracellular parasites and cannot be visualized by light microscopy. As an order they can be divided into two groups, depending on the type of nucleic acid they contain: ribonucleic acid (**RNA viruses**) or deoxyribonucleic acid (**DNA viruses**). On the basis of their host association, they may be placed taxonomically in suborders: plant, insect, bacterial, or animal viruses. The animal viruses are generally destroyed by temperatures sufficient to kill most pathogenic bacteria but are preserved by freezing at low temperatures (below −40 C).

Because of their unique metabolic and reproductive mechanisms, viruses and rickettsiae have not yet been propagated on cell-free media; they require **living cells** for replication. With advances in cell culture and the increasing availability of selectively sensitive cell lines, genetically sensitive strains of mice, embryonated eggs, and fluorescent microscopy, diagnosis of viral infection and the identification of viruses can be made with greater facility and speed in laboratories equipped for such procedures.

Antigens for serologic studies are available commercially,* thus permitting diagnostic tests to be performed routinely in the serology section of a hospital laboratory, especially those tests that involve complement fixation.

AVAILABLE TYPES OF LABORATORY EXAMINATION

Three types of laboratory examination are presently available for the diagnosis of viral and rickettsial infections:

1. Isolation and identification of the agent by inoculation of cell cultures, chick embryos, or susceptible animals
2. Detection and measurement of antibodies developing during the course of the disease (serologic tests)
3. Histologic examination of infected tissues (skin lesions, biopsies, postmortem specimens):

---

*Microbiological Associates, Inc., Bethesda, Md.; Flow Laboratories, Rockville, Md.; (Italdiagnostic) Telcolab Corporation, New York, N.Y.; Burroughs Wellcome Co., Research Triangle, N.C.; Beckman Instruments, Inc., Fullerton, Calif.

a. Examination by light microscopy for signs of infection
b. Detection of specific antigens, using fluorescent-antibody techniques
c. Detection of virus particles by electron microscopy

COLLECTION OF SPECIMENS FOR VIRAL AND RICKETTSIAL DISEASE DIAGNOSIS

Success of any laboratory diagnostic procedure depends on appropriate and timely specimen collection and the way in which specimens are handled after collection. So that a judicious choice of test systems can be made, the laboratory should be given information in the form of a brief clinical history. The following information is pertinent:

1. Date of onset of infection
2. Date specimen was collected
3. Clinical signs and symptoms
4. Suspected or differential diagnosis
5. Area of residence or travel
6. Similar cases in family or vicinity
7. Exposure to animals
8. Antimicrobial therapy

Finally, clearly specified test requisitions should accompany the specimen (cryptic requests for "viral studies" are useless).

The following is a brief description and outline for collecting specimens.[5]

Throat washings. The patient should gargle with approximately 10 ml of a saline solution and expectorate into a clean paper cup. The washing is then transferred to a screw-capped or rubber-stoppered glass or leak-proof plastic tube.

Nasal secretion. Nasal secretions should be collected with a cotton swab and placed in approximately 1 ml of sterile tryptose broth medium* with 0.5% to 1% gelatin.

Throat swabs. Best results are obtained by rubbing a dry cotton swab on the posterior pharynx. The swab tip should be broken into a tube containing approximately 1 ml of sterile broth medium.*

Autopsy and biopsy tissues. The tissues obtained depend on the nature of the disease. Lung and trachea specimens are most important in cases of respiratory illness. Brain, spinal cord, and colon tissue should be collected for central nervous system infections. Heart muscle and pericardial fluid should be collected in cases of myocarditis. Some infections are associated with widespread dissemination of the agent, and such tissues as liver, spleen, and kidney may be valuable sources of virus. Tissues submitted for virus isolation should **never** be placed in formalin, but should be placed in sterile jars or plastic freezer bags.

Vesicular fluid or skin scrapings. Vesicular lesions should be opened and the exudate absorbed on sterile dry cotton swabs. The swab should then be immersed in 1 ml sterile broth (see under throat swabs) if virus isolation is required. For histologic studies, cells should be scraped from the base of lesions, with as little blood as possible, and placed on clean microscope slides followed by immersion in ether-alcohol for a minimum of 5 minutes. Study by fluorescent microscopy requires an adequate specimen and fixation in cold acetone (-20 C).

Blood specimens. For serologic tests, at least 10 ml of blood should be collected aseptically, without preservative or anticoagulant. Do not freeze whole blood, but only separated serum. Smaller volumes may be acceptable from small children and infants. **Paired** sera, acute (onset) and convalescent (10 to 21 days later), are necessary for serodiagnosis.

---

*An excess volume of fluid dilutes the concentration of virus particles, making recovery difficult; it should be remembered that further dilution occurs when antibiotics are added. Commercially packaged sterile swabs with Stuart's transport medium have been found to be satisfactory.

---

*An excess volume of fluid dilutes the concentration of virus particles, making recovery difficult; it should be remembered that further dilution occurs when antibiotics are added. Commercially packaged sterile swabs with Stuart's transport medium have been found to be satisfactory.

Isolation of viruses from blood samples optimally requires coculture of blood leukocytes in tissue culture; therefore, a heparinized blood specimen is best for this purpose.

Stool specimens or rectal swabs. A fresh stool specimen (prune size) should be collected in a clean carton or jar. Rectal swabs may be used by passing a sterile swab, moistened with broth, into the anus so that the cotton tip is no longer visible. The swab is then placed in a tube with 1 ml broth.

Spinal fluid. Collect 3 to 5 ml of spinal fluid in a sterile screw-capped or rubber-stoppered glass or leak-proof plastic tube.

Urine. Collect voided urine in a sterile container.

Conjunctival swabs. Use sterile cotton swabs moistened with broth. Swab upper and lower palpebral conjunctivae. Submerge cotton tip in 1 ml of broth.

HOW TO SHIP SPECIMENS TO THE LABORATORY

The shorter the interval between collection of a specimen and its delivery to the laboratory, the greater the potential for isolating an agent. Storing specimens at temperatures above −60 C and freezing and thawing are **not** optimal proce-

dures. The nature of the agent suspected can influence handling of specimens, and, when doubt exists as to the most expedient method, a competent virologist should be consulted. In general, the following statements hold, but there may be a few exceptions:

1. Never leave a specimen at room or incubator temperature.
2. When it is impossible to deliver a specimen immediately, it should be **refrigerated** and packed in shaved ice for delivery to the laboratory within 12 hours of collection.
3. When the interval between collection and delivery is greater than 12 hours, **freeze** specimen below −40 C (preferably −70 C). If possible, split specimen and freeze one portion and pack the other in ice.

If carbon dioxide (dry) ice is needed for shipping frozen specimens, it can usually be obtained at an ice cream parlor or an ice plant. Specimens packed in dry ice must be in well-sealed containers to prevent contact with the carbon dioxide, which will lower pH and inactivate many viruses.

Specimens for isolation of cytomegalovirus and varicella virus should **not** be

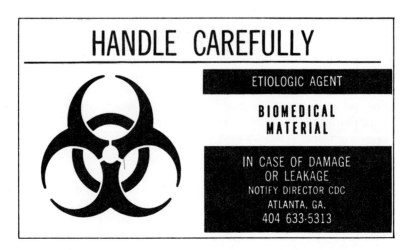

Fig. 33-1. Label for etiologic agents and biomedical material.

frozen. Respiratory syncytial virus is extremely labile and, if not processed immediately, will be lost unless the specimen is shell frozen. Arbovirus specimens should be frozen immediately (at least −40 C) if they must be held for delivery to the laboratory.

When diagnostic specimens are shipped by public carrier, packaging must conform with the Department of Transportation and Interstate Quarantine regulations (49 CFR, Section 173.386.388, and 42 CFR, Section 72.25, Etiologic Agents). These regulations can be obtained from the Biohazards Control Officer, Center for Disease Control, Atlanta, Georgia. In essence, regulations require that the specimen be wrapped in sufficient absorbent material to absorb the entire contents of the specimen in case of leakage or breakage. The wrapped specimen must then be enclosed in a durable water-tight container, which is then enclosed in an outer shipping container. When dry ice is used, it should be placed between the outer shipping container and the watertight container, which must be secured with shock-absorbent material or tape so that it does not become loose as the dry ice sublimates.

Labels. The label for Etiologic Agents/ Biomedical Material must appear on the outside of the shipping container (Fig. 33-1).

ISOLATION OF VIRUSES AND RICKETTSIAE FROM CLINICAL SPECIMENS

Table 33-1 provides a guideline for the types of specimens required for diagnosis of viral and rickettsial infections. Only generalizations relative to isolation and identification of viruses can be given. Additional information is provided in references cited at the end of this chapter.[1,7,8,10,11]

Successful isolation of a viral agent depends on collection of the appropriate material from the patient, careful preservation while the specimen is being transported to the laboratory, and elimination of viable bacteria and fungi before inoculating the material into indicator hosts. The available indicator hosts are cell cultures of various derivations, embryonated eggs, and small animals. The choice of one or more of these systems is influenced by the clinical history submitted with the specimen. Presence of virus in cell cultures is recognized by cellular changes (cytopathic effect, or **CPE**) observed microscopically.[4] Characteristic lesions or new antigens may develop in embryonated eggs, and animals must be observed for signs of infection ranging from ruffled fur to paralysis or death. Final identification of a virus requires immunologic procedures such as are outlined in a following section.

An attempt to isolate the virus is a preferred laboratory procedure under the following circumstances:

1. When death occurs early in the disease process; virus is present and antibody response inadequate
2. When the suspected virus is a member of a large virus group, the members of which have similar physical properties but share no common antigen
3. When the agent belongs to a virus group whose members provoke heterotypic response
4. When the agent has poor antigenic capacity
5. When the patient is a compromised host

Early diagnosis of Rocky Mountain spotted fever (as early as the fourth day of illness) may be made using a primary monocyte culture technique and demonstrating the organism in the monocytes by direct immunofluorescent or Giménez staining (De Shazo, R. D., et al., J.A.M.A. **235**:1353-1355, 1976).

SEROLOGIC DIAGNOSIS OF VIRAL AND RICKETTSIAL INFECTIONS

Serologic tests provide the most easily accessible diagnostic aid for viral and rickettsial infections. They may be the

Table 33-1. Specimens recommended for the isolation of viruses and rickettsiae

Type of infection	Stool/rectal	Throat swab/wash/ nasopharynx	Urine	Cerebrospinal fluid	Blood (leukocytes)	Vesicle fluid/ scrapings	Conjunctival swab	Acute and convalescent serum	Lung	Spleen, liver	Brain, spinal cord	Colon	Postmortem blood
Central nervous system													
Polioviruses	x	x						x		x	x		x
ECHO viruses	x	x		0				x (3)		x	x		x
Coxsackie viruses	x	x		0				x (3)		x	x		x
Mumps		x	x	x				x			x		x
Herpes simplex		x						x			x		x
Varicella-zoster				x	x	x		x			x		x
Rabies		x (4)						x			x		
Arboviruses													
Western, eastern, equine, Venezuelan				0	x			x			x		x
Japanese B, St. Louis, California								x			x		
Respiratory													
Influenza		x						x	x				x
Parainfluenza		x						x	x				
Respiratory syncytial		x (1)						x	x				
Adenovirus		x	0					x	x				
Psittacosis								x	x	x			
Q fever								x	x	x			x
Enteroviruses	x	x						x (3)	x			x	x
Morbilliform eruptions													
Measles		x			x			x	x	x			x
ECHO viruses	x	x						x (3)					
Coxsackie viruses	x	x						x (3)					
Typhus					x			x					
Rocky Mountain spotted fever					x			x					
Vesicular eruptions													
Herpes simplex		x				x		x	x	x	x		
Smallpox						x		x	x	x	x		
Vaccinia						x							
Rubella		x (2)						x					
Varicella-zoster						x		x	x	x	x		
Enteroviruses	x	x				x		x (3)	x	x	x		
Eye infections													
Adenoviruses							x	x					
Herpes simplex							x						
Miscellaneous diseases													
Herpangina	x	x						x (3)					
Pleurodynia	x	x						x					
Mumps		x	x					x					
Dengue					x			x					
Lymphogranuloma venereum								(B) x					
Cytomegalovirus		x	x		x			x					
Gastroenteritis	x	x											

0, Not usually required; may be submitted optionally. 1, Very labile; inoculate immediately or shell freeze. 2, Swab from nose optimal. 3, Serology not routinely available; useful if agent is isolated. 4, Saliva. (B), Buboes and lymph nodes.

only reliable diagnostic tests available and provide presumptive assessment of a disease etiology in the absence of an isolate. However, there are limitations to the usefulness of serologic tests: (1) sudden death, (2) the compromised host, (3) heterotypic serologic responses, (4) multiplicity of antigens in a virus group, or (5) when a virus is a poor antigen. Antibody titer in a single specimen can rarely be considered significant; the titer merely indicates that infection has occurred at some nonspecified time. To be diagnostically significant, the acute and convalescent specimens must be titrated at the same time and **at least a fourfold rise** in antibody level should be demonstrated. Unfortunately, this provides the diagnosis rather late after onset of illness. It may be advantageous to study spinal fluid as well as blood.

Serologic tests in present use include: complement fixation, neutralization, hemagglutination inhibition, passive hemagglutination, fluorescent antibody, immunoelectron microscopy, and nonspecific agglutination tests. The choice of test depends on the nature of the infecting agent and the type of information required.[8] Immune electron microscopy, when it is feasible, may provide early diagnosis, since it does not depend on developing antibody.

Each one of the above tests can be adapted to demonstrate antibody response and to identify a particular agent.

Complement fixation is a highly satisfactory serologic test. Antibodies measured by this system generally develop slightly later in the course of an illness than those measured by other techniques; this lag provides a greater opportunity to demonstrate titer differences between acute and convalescent sera. Complement-fixing antibody response occurs within the first week of illness and plateaus within the next 2 weeks. Titers decay and disappear within 1 to 2 years. Members of some virus groups, such as the adenoviruses, influenza A viruses,

and influenza B viruses, possess common antigens demonstrable by complement fixation. Thus, antibody response to infection by any member of the particular group can be observed without resorting to a multiplicity of tests with individual antigens. A negative complement fixation titer does not necessarily indicate susceptibility to infection.

The **neutralization test** is essentially a protection test. When a virus is incubated with homologous type-specific antibody, the virus is rendered incapable of producing infection in an indicator host system. The test is technically more exacting than other serologic tests and provides the principal method used for identifying virus isolates. A neutralizing-antibody response is virus type specific and develops with the onset of symptoms. Titers peak rapidly to a plateau, persist for long intervals, and measurable titers may be maintained indefinitely.

The **hemagglutination test** can be performed with a variety of viruses that have the capacity to agglutinate selectively red blood cells of various animal species (chicken, guinea pig, human O group, and others). The hemagglutinating capacity of a virus is inhibited by specific immune or convalescent serum. Hemagglutination-inhibiting antibody develops rapidly after the onset of symptoms, plateaus rapidly, declines slowly, and may last indefinitely at low levels.

In the **passive hemagglutination** procedure, certain viruses can be chemically coupled to the surface of erythrocytes, which then serve as indicators of the presence of homologous antibody in serum by the hemagglutination reaction.

For the **indirect fluorescent antibody test,** virus-infected cells are placed in prepared wells on microscope slides, then fixed in cold acetone and dried. Serum antibody is applied and, following incubation for antigen-antibody coupling, antihuman globulin-fluorescein conjugate is added to delineate, by fluorescence, the sites of antigen-antibody reaction. It is possible to identify specifically a virus in any adequately prepared specimen containing virus-infected cells, but current lack of high-titer-specific antisera limits the usefulness of this method.

Immunoelectron microscopy has been used to visualize viruses and in very limited situations may provide the method of choice. Immune electron microscopy has not proved useful for the diagnosis of respiratory infections but has been applied to visualization of rotaviruses and the Norwalk agent associated with diarrhea of infancy and to detection of hepatitis A virus in feces. Conventional electron microscopy has been used to identify smallpox virus.

The problem of obtaining sufficiently pure antigen precludes the use of specific virus agglutination tests. However, **nonspecific agglutination** tests are used for the diagnosis of rickettsial infections (Weil-Felix), as is the heterophile agglutination test for diagnosis of infectious mononucleosis (see Chapter 38).

Other studies, such as gel diffusion, hemolysis in gel, reversed passive hemagglutination, radioimmunoassay, and enzyme-linked immunoassay, have been used to test for Australia antigen, for antibodies against hepatitis A virus, and for antibodies against various other viruses and rickettsiae.

WHERE TO SEND SPECIMENS

If there is no local laboratory facility that can handle the desired viral or rickettsial studies, the following sources can be investigated:
1. Municipal, county or state health department laboratories
2. The Center for Disease Control (**only** through the local public health or state laboratory)
3. Local medical or research centers
4. Private laboratories engaged in virology
5. Military and Veterans Administration laboratories

In cases of suspected viral disease of obscure etiology or in instances of major outbreaks, **prompt consultation with a virologist should be the rule rather than the exception.**

REFERENCES

1. Acton, I. D., Kucera, L. S., Myrvik, Q. M., and Weiser, R. S.: Fundamentals of medical virology, Philadelphia, 1974, Lea & Febiger.
2. Fenner, F., and White, D.: Medical virology, New York, 1970, Academic Press, Inc.
3. Horsfall, F. L., and Tamm, I., Editors: Viral and rickettsial diseases of man, ed. 4, Philadelphia, 1965, J. B. Lippincott Co.
4. Hsiung, G. D., Diagnostic virology, New Haven, Conn. 1973, Yale University Press.
5. Laboratory diagnosis of viral diseases, Course No. 8241-C, U.S. Department of Health, Education, and Welfare, Public Health Service, Atlanta, 1976, Center for Disease Control.
6. Lennette, E. H.: Laboratory diagnosis of virus infections; general principles, Am. J. Clin. Pathol. **57:**737-750, 1972.
7. Lennette, E. H., Spaulding, E. H., and Truant, J. P., editors: Manual of clinical microbiology, ed. 2, Washington, D.C., 1974, American Society for Microbiology.
8. Lennette, E. H., and Schmidt, N. J.: Diagnostic procedures for viral and rickettsial diseases, ed. 3, New York, 1964, American Public Health Association, Inc.
9. Pumper, R. W., and Yamashiroya, H. M.: Essentials of medical virology, Philadelphia, 1975, W. B. Saunders Co.
10. Smith, T. F. In Washington, J. A. II, editor: Laboratory procedures in clinical microbiology, Boston, 1974, Little, Brown & Co., pp . 217-277.
11. Swain, R. H. A., and Dodds, T. T.: Clinical virology, Edinburgh, 1967, E. & S. Livingston, Ltd.

CHAPTER 34

Laboratory diagnosis of mycotic infections

Although the pathogenicity of certain fungi has been recognized since the first half of the nineteenth century, laboratory expertise in the handling of clinical specimens and in the subsequent isolation and identification of the causative agents of fungal disease has been developed only in comparatively recent years. It is recognized now that the **mycoses,** those diseases of fungal etiology, are more common than before; their incidence has increased through the widespread use of antibacterial agents and immunosuppressive drugs. It is incumbent on microbiologists, therefore, to become more knowledgeable about the pathogenic fungi and their identifying characteristics.

From basic microbiology, one is reminded that fungi are multicellular heterotrophic members of the plant kingdom that lack roots and stems and are referred to as **thallophytes.** They are larger than the bacteria and more complex in their morphology. In addition, they are devoid of chlorophyll and fundamentally consist of a basic, branching, intertwining structure called a **mycelium,** composed of tubular filaments known as **hyphae** (sing. **hypha**). The latter may possess cross walls, or **septa,** in which case the mycelium is referred to as being **septate;** in the absence of septa, where the filaments are continuous, the mycelium is said to be **aseptate.** The mycelial cottony mass constitutes the **colony** of a **mold.** Colonies of **yeasts,** the other major form of fungi, are more like bacterial colonies. Certain fungi are **dimorphic;** these have both a mold phase and a yeast phase.

In septate hyphae, only one nucleus is found per segment or cell, whereas aseptate hyphae exhibit a multinucleated condition in the continuous filament. The mycelium also has two parts: the **vegetative** part that grows in or on the substrate, absorbing nutrients, and the **reproductive** or aerial part that projects above the substrate, producing fruiting bodies bearing characteristic spores. Disseminated mature spores, when arriving on a suitable substrate, germinate by producing a **germ tube** that finally leads to a new mature organism.

Fungi may reproduce sexually or asexually or by both means. **Sexual** reproduction is associated with the formation of specialized structures that facilitate fertilization and nuclear fusion, resulting in the production of specialized spores called **oospores, ascospores,** and **zygospores.** Fungi that exhibit a sexual phase are known as **perfect** fungi. **Imperfect** fungi are those in which no sexual phase has been demonstrated; the spores are produced directly by or from the mycelium. Most of the fungi of medical importance belong to the imperfect group, although the possibility of future recognition of a perfect phase must not be excluded.*

Since the form of sporulation and the type of spore are important criteria in the identification of the various fungi, the following morphologic data are provided to aid in the characterization of these microorganisms.

The simplest type of sporulation is the

---

*The perfect state of some dermatophytes has been described.[4,13,15]

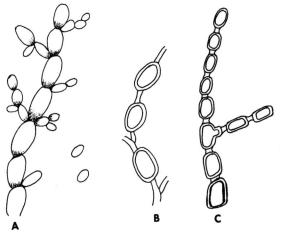

Fig. 34-1. Asexual spores produced from hyphae. **A**, Blastospores. **B**, Chlamydospores. **C**, Arthrospores.

development of the spore directly from the vegetative mycelium; the spore is known as a **thallospore.** Three types of thallospores are recognized: **blastospores** (Fig. 34-1, *A*), simple budding forms in which the daughter cell is abstricted from a single mother cell or elongated budding cells that have not detached, called **pseudohyphae.** The resulting mycelium is called a pseudomycelium, as in *Candida* species; **chlamydospores** (Fig. 34-1, *B*), thick-walled, resistant, resting spores produced by the rounding up and enlargement of the terminal cells of the hyphae, as in *Candida albicans;* and **arthrospores** (Fig. 34-1, *C*), resulting from simple fragmentation of the mycelium into cylindrical or cask-shaped, thickwalled spores, as in *Geotrichum candidum.*

Conidia are asexual spores produced singly or in groups by specialized vegetative hyphal stalks called **conidiophores** (Fig. 34-2, *B* and *C*). The conidia are freed from the point of attachment by pinching off, or abstriction. Some conidiophores become swollen at the end, and over their swollen surfaces are formed numerous small flask-shaped stalks, from which conidia in chains (**catenate**) are pushed

out. The swollen portion of the conidiophore is called a **vesicle;** the flask-shaped structures are **sterigmata** (Fig. 34-2, *C*). Many fungi produce conidia of two sizes: **microconidia** (Fig. 34-2, *D* and *E*) are small, unicellular, round, elliptical or pear shaped (pyriform); **macroconidia** (Fig. 34-2, *F* to *H*) are large, usually septate, club shaped (clavate) or spindle shaped (fusiform). The microconidia may be borne directly on the hyphae (Fig. 34-2, *D*) and are said to be **sessile,** or they may develop directly from the end of a short conidiophore (Fig. 34-2, *E*) and are called **pedunculate.** If conidia have a rough or spiny surface, they are called **echinulate.**

Sporangiospores (Fig. 34-2, *A*) are asexual spores contained in **sporangia** produced terminally on **sporangiophores** (aseptate stalks). Sporulation takes place by a process called progressive cleavage. During maturation within the sporangium, the protospores divide into definitive uninucleate sporangiospores, which are released by irregular rupture of the sporangial wall (Fig. 34-2, *A*). This form of sporulation occurs in the Phycomycetes, which exhibit an aseptate mycelium.

Other types of spores, produced sexually by the perfect fungi, have been referred to above. Additional information may be gained from texts on general mycology.[7,10]

Since the techniques commonly employed in medical bacteriology are not always practical for the identification of fungi, these microorganisms must often be recognized primarily by their **gross** and **microscopic** characteristics. This brings into focus certain questions. Is the colony rapid growing (2 to 5 days) or slow growing (2 to 3 weeks)? Is it flat, heaped up, or regularly or irregularly folded? Is its texture creamy and yeastlike, or is it smooth and skinlike (glabrous)? Does it produce a mycelium that is powdery, granular, velvety, or cottony? Is a distinct surface pigment observed, and is the pigment similar on the reverse side? These

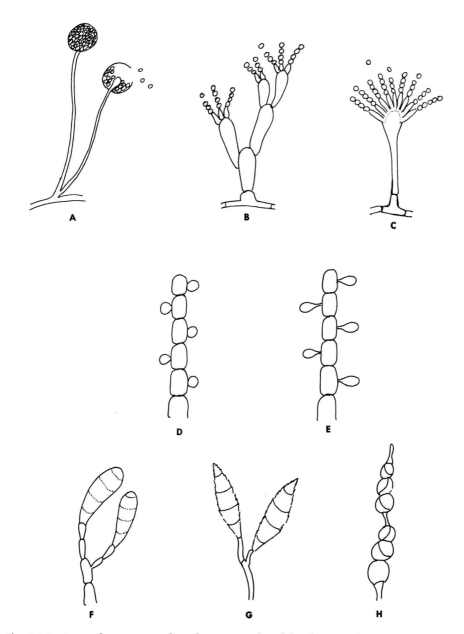

Fig. 34-2. Asexual spores produced on specialized hyphae. **A,** Sporangiospores in a sporangium. **B,** Conidia on branched conidiophore. **C,** Conidia on unbranched conidiophore. **D,** Sessile microconidia. **E,** Pedunculate microconidia. **F** to **H,** Various macroconidia.

are the criteria that are important in the **gross** characterization of an unknown fungus.

A small portion of the colony and some of the medium may be removed with a 22-gauge nichrome needle or loop, and the mycelium teased apart in a drop of lactophenol cotton blue mounting fluid on a glass slide This is covered with a thin coverglass heated *gently* over the pilot flame of a Bunsen burner, and examined **microscopically,** using both low- and high-power objectives. The size, shape, septation, and color of spores, if present, may be observed, and the morphology of the specialized structures bearing the spores is noted.

Frequently, it may be possible to identify a culture solely by this direct examination. Often, however, more sophisticated procedures may be required, and the preparation of a **slide culture** or hanging drop for the demonstration of the undisturbed relation of spores to specialized hyphae, such as conidiophores, is recommended. This technique is described in a later section.

Special culture media designed to suppress vegetative growth and stimulate **sporulation** of a fungus or to produce a special type of growth may also be required. Such media include potato dextrose agar, wort agar, brain-heart infusion agar, rice grain medium, chlamydospore agar, and others; their preparation and use is discussed in the section on culture media.

To identify a fungal culture properly, it may be necessary on occasion to carry out various **biochemical** studies, such as sugar fermentations, nitrate assimilation tests, carbon and nitrogen utilization studies, and specific vitamin requirements. When indicated, these are described.

Finally, it may be necessary in certain instances to carry out **animal pathogenicity tests** to identify some species. These are discussed under the specific diseases.

Saprophytic fungi are frequently found as **contaminants** of clinical specimens as well as in the laboratory, where they readily contaminate cultures. It is essential, therefore, that the medical mycologist become well versed in differentiating these forms from recognized pathogens. Furthermore, some of these common saprophytes are important **opportunistic invaders** in debilitated patients who may have been treated with immunosuppressive drugs and the multiplicity of antimicrobial agents presently available to the clinician. Their recognition, therefore, is becoming important. The laboratorian should become familiar with about 12 genera commonly considered as contaminants. Much help will be obtained from the excellent texts referred to at the end of this chapter.[13,15,19,38] Excellent training courses also are offered by the Mycology Branch of the Center for Disease Control (CDC).*

In familiarizing oneself with the characteristics of the saprophytic genera, including *Penicillium, Aspergillus, Paecilomyces, Fusarium, Scopulariopsis, Rhizopus, Mucor,* and others, not only does the student acquire experience in mycologic techniques but gains an opportunity to recognize fungal morphology and its relation to the taxonomy of the group.

GENERAL LABORATORY METHODS

Mycologic examination of all clinical material should include a direct microscopic examination, culturing on appropriate media, and if indicated, inoculation of susceptible laboratory animals. Although it is true that mycelial fragments, spores, and spore structures can sometimes be demonstrated in unstained preparations, it is essential that **all specimens be cultured** in an attempt to isolate the causative agent.

---

*Contact the Office of Training Activities, Bureau of Laboratories, Center for Disease Control, Atlanta, Ga. 30333.

Collection of specimens

A prime requisite to good medical mycology is a **properly collected and properly handled specimen.** Procedural details are given under specific diseases; therefore, the following are general instructions only.

To obtain specimens of **skin** or **nails,** the affected site is carefully washed with 70% isopropanol and, after drying, the lesion is scraped with a sterile scalpel and the material obtained is placed in a sterile Petri dish or on a piece of white paper carefully folded in a packet to prevent loss of the specimen. **Hairs** from infected areas are clipped or plucked and sent to the laboratory in a similar manner.

In the **subcutaneous mycoses** (see later section in this chapter), a variety of materials may be submitted, including pus or exudate from draining lesions, material aspirated with syringe and needle from unopened abscesses or sinus tracts, or biopsied tissue.* These should be placed in sterile tubes or Petri dishes and submitted directly to the laboratory. If mailing of the specimen is necessary, the material first should be inoculated to a suitable culture medium. **Under no circumstances** should glass or plastic Petri dishes containing clinical specimens or fungus cultures be sent through the mails; they invariably break in transit. Neither should inoculated cotton swabs be mailed, as they are usually dried out on arrival. For best results, **only pure cultures** on agar slants should be mailed. Scrapings of skin or nails and hair can be mailed in appropriate containers.

Material from suspected cases of **systemic mycoses** includes such varied specimens as blood, cerebrospinal fluid, sputum, bronchial secretions, gastric washings, pus and exudates from ab-

scesses and draining sinuses, bone marrow, and tissue. These specimens should be placed in sterile tubes or bottles and submitted **promptly** to the laboratory.

Sputum treated with cetylpyridinium chloride for shipment to a mycobacteriology laboratory is **not** satisfactory for culture for fungi, but the morphology and stainability of fungi is not affected.[28] To aid the laboratory in the proper examination of the specimen, some indication of the **suspected disease** should be noted on the laboratory request slip, along with the specimen source. This is necessary to guide laboratory personnel in the selection of the proper media and methods of incubation.

Direct microscopic examination

The following clinical specimens are preferably examined in the **unstained state:** sputum and bronchoscopic secretions, gastric washings, pus and exudates, sediments of cerebrospinal fluid, pleural effusions, and urine. Several loopfuls of the material are placed on a clean glass slide, covered with a thin coverglass, and examined under both the low- and high-power objectives, using reduced light. If the material is opaque, 10% sodium hydroxide may be added and gentle heat applied. These preparations should be carefully searched for the following: broad mycelial fragments with septa (*Aspergillus* and *Penicillium* species) or without septa (*Mucor* species), arthrospores (*Geotrichum* species), and budding cells (*Candida, Cryptococcus,* and *Blastomyces* species).

Dried and fixed films may be stained by the Gram method (mycelium and spores are gram positive), periodic acid–Schiff (PAS), and Wright's or Giemsa stain to reveal the presence of *Histoplasma capsulatum* in the macrophages or phagocytes of blood or bone marrow.* An India

---

*Tissue for fungus culturing should **not** be ground in a tissue grinder; it fragments and may kill the larger fungal elements. Rather, the specimen should be minced with a sterile scalpel blade and pieces embedded directly into the culture medium.

---

*Special histology stains for fungi are also extremely useful, including the Gridley and Gomori stains.[21]

ink preparation of cerebrospinal fluid may reveal encapsulated forms of *Cryptococcus neoformans*.

Cultural procedures

Although the basic principles of microbiologic technique apply to the mycologist as well as the bacteriologist, certain differences should be noted. A 22-gauge nichrome needle, flattened at the end to a spade shape, is used to transfer mycelial growth. When the needle contains infectious material, care should be exercised to prevent spattering when flaming it; the needle should be gradually heated in the less intense part of the flame. Two stiff, sharp-pointed, teasing needles in holders are useful in tearing apart the mycelial mat when immersed in mounting fluid on a glass slide.

Large (18 by 150 mm) borosilicate test tubes without lips are recommended for solid culture media rather than Petri dishes, to minimize the hazard of spore dissemination. Small, strong bottles with flat sides are also suitable for culture of fungi, such as *Coccidioides immitis*, to minimize the hazard to the laboratory worker. Screw-capped test tubes used for cultures are **not** recommended; they promote anaerobiosis and retain moisture, both of which prevent maximum sporulation. The large cotton-plugged test tubes afford ease of storage and handling, are less easily broken, and allow for a **thick butt** of agar that will withstand drying during extended incubation. Pigment production is enhanced when there is free circulation of air, and a dry surface encourages the development of an aerial mycelium and spores.

Sabouraud dextrose agar (SAB agar) at pH 5.6, brain-heart infusion (BHI) agar, with or without added blood, and a combination of both (SABHI agar) are the most useful media for primary isolation of most pathogens. (The preparation of these media is discussed in Chapter 41.) The addition of 0.5 mg per milliliter of cycloheximide and 0.05 mg per milliliter

of chloramphenicol to these media will effectively inhibit the growth of contaminating saprophytic molds and bacteria, especially when material likely to contain these contaminants in large numbers is cultured, such as skin and nail scrapings, sputum, pus, or autopsy material. One study recommends gentamicin in lieu of chloramphenicol[26]; this would be better for prevention of growth of gram-negative rods, but would be less effective against gram-positive cocci. On BHI agar with added antibiotics, pathogenic fungi will develop their typical colonial morphology, color, and microscopic appearance and can generally be identified without further subculturing. It is well to note, however, that certain pathogens are partially or completely inhibited by these antibiotics. Included here are *Cryptococcus neoformans*, *Candida* species (including *C. parapsilosis* and *C. krusei*), and *Trichosporon beigelii (cutaneum)*. The yeast phases of *Histoplasma capsulatum* and *Blastomyces dermatitidis* are susceptible to cycloheximide when incubated at 35 C but not at 25 C; *Petriellidium boydii* and *Aspergillus fumigatus* are partially sensitive, but cycloheximide may inhibit sporulation. Since the concentration procedure for mycobacteria is very detrimental to fungi, one must not rely on mycobacterial cultures to recover fungal pathogens.[6,30]

Some cultures neither sporulate nor produce pigment satisfactorily on SAB. To induce these, special media, including potato-dextrose agar, potato-carrot agar, cornmeal agar, rice grain agar, or SAB agar with added thiamine and inositol, have proved useful and can be recommended.

All isolation media should be held for a minimum of 4 weeks before discarding.

Slide culture

Microscopic observation of fungi in the natural state is often necessary for identification. The following method (Fig. 34-3)

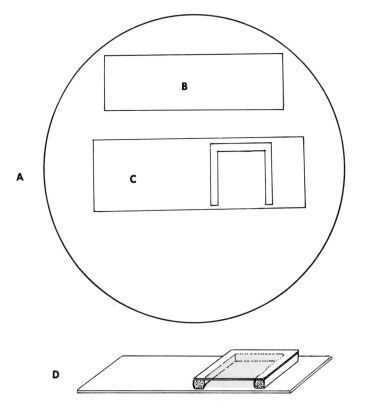

Fig. 34-3. Slide culture. **A,** Petri dish. **B,** Strip of filter paper. **C,** Slide with cement walls. **D,** Slide with agar medium and coverglass in place.

has proved successful in our hands for the culture of fungi on a glass slide:

1. With forceps dip a clean slide in alcohol, flame, and burn off. Repeat the procedure. Place in a sterile Petri dish to cool.
2. Carefully apply aseptically quick-drying, tubed, plastic cement to the slide to form three sides of a square, equal in area to that of a coverglass or slightly less. The cement walls should be 1 to 2 mm high.
3. With a sterile dropping pipet, place sufficient melted SAB agar at 45 C in the square to fill the space to the top of the walls. Place the slide in the Petri dish and allow the agar to set firmly.
4. Apply a minimal amount of inoculum (spores) to the center of the agar surface and then cover squarely with a sterile coverglass (alcohol flamed and cooled).
5. Place a wide strip of filter paper in the Petri dish and moisten with a few drops of water. Cover with lid, which assures a moist atmosphere.
6. Incubate at the desired temperature and examine after 48 hours or until sporulation occurs. This is determined by quick microscopic examination under the low-power objective.
7. At the desired stage, remove the damp filter paper and add a dry strip. Add several drops of formalin to the strip and cover the dish.
8. Allow to remain for 30 minutes. Remove the slide, blot away excess moisture, and examine under high-dry objective, with appropriate light adjustment.

9. Make a permanent mount by applying nail polish or asphalt tar varnish to all four sides of the square.

Hanging-drop culture

A hanging-drop culture preparation is recommended also for examining fungi in the natural state. Introduce a minute inoculum of the culture into a drop of Sabouraud broth on a coverglass, which is inverted over a glass microslide ring* coated at both ends with petroleum jelly. Place this assembly on a 3- by 1-inch glass slide (with several drops of water for moisture) and incubate at 25 C for 2 to 3 days until sporulation occurs.

Do not make slide cultures of *Histoplasma, Blastomyces,* or *Coccidioides.* The procedure is too hazardous.

STUDY OF THE MYCOSES

Of the more than 50,000 valid species of fungi, only about 50 are generally recognized as being pathogenic for humans. These organisms normally live a saprophytic existence in soil enriched by decaying nitrogenous matter, where they are capable of maintaining a separate existence, with a parasitic cycle in humans or animals. The systemic mycoses are not communicable in the usual sense of person-to-person or animal-to-person transfer; humans become **accidental hosts** by the inhalation of spores or by their introduction into tissues through trauma. Unusual circumstances, reflecting an **altered susceptibility** of the host, may also lead to infection by fungi that are normally considered saprophytes. Such conditions may occur in patients with debilitating diseases or diabetes mellitus or those suffering from impaired immunologic mechanisms resulting from steroid or antimetabolite therapy. Prolonged administration of antibiotic agents may also upset the host's normal microbiota, resulting in a **superinfection** by one of these fungi. The neophyte mycologist is cau-

---

*A. H. Thomas Co., Philadelphia.

tioned, therefore, not to discard cultures as "contaminants" without first checking the clinical history of the patient and discussing findings with the patient's physician.

The **mycoses** may be conveniently considered in three groups, based on the tissues involved[5]:

1. The **dermatophytoses,** including the superficial and cutaneous mycoses.
2. The **subcutaneous** mycoses, which involve the subcutaneous tissues and muscles.
3. The **systemic** mycoses, which involve the deep tissues and organs. These are the most serious of the groups.

DERMATOPHYTOSES

As the name suggests, these fungal diseases include those infections that involve the superficial areas of the body, namely the **skin, hair,** and **nails.** The etiology for these diseases for the most part relates to the genera *Microsporum, Trichophyton,* and *Epidermophyton.* Such cutaneous mycoses are probably the most common fungal infections of humans and are usually referred to as **tinea** (Latin for gnawing worm or ringworm). The gross appearance is that of an outer ring of active, progressing infection with healing centrally within the ring. They may be characterized or qualified by another Latin noun in the genitive form to designate the area involved, for example, tinea corporis (body), tinea cruris (groin), tinea capitis (scalp and hair), tinea barbae (beard), tinea unguium (nail), and others. These fungi break down and use keratin (keratinolytic) as a source of nitrogen, but are incapable of penetrating the subcutaneous layers.

Tinea versicolor (pityriasis), a disease of skin characterized by superficial brownish scaly areas on the trunk, arms, shoulders, and face, is widely distributed throughout the world. It is caused by *Pityrosporum (Malassezia) furfur,* whose mycelial fragments and clusters of thick-walled, yeast-like spores may be observed microscopically in skin scrapings.

The reader is referred to some excellent texts for a further description of the less common dermatophytoses.[13,15,38]

Direct examination

The presence of the fungi can be readily demonstrated in direct slide preparations of digested skin scales, nail scrapings, or hair. The skin of the involved area is cleansed with 70% alcohol, and some epidermal scales at the active edge of the lesion are stripped off and placed on a microscope slide containing a drop of 10% sodium hydroxide. A coverglass is added, and the slide is gently warmed over a small flame to just short of boiling. The slide is then examined under low- and high-power magnification, using much-reduced light, for the presence of long branching threads of young hyphae or of older septate hyphae and barrel-shaped arthrospores. Occasionally, budding yeasts may be seen. Fungal elements must be differentiated from fibers of cotton, wool, and other fabrics as well as from a mosaic of cholesterol crystals and other artifacts. Phase contrast microscopy can be useful in this regard.

Specimens of nail scrapings must be secured from the deeper layers of the infected nail; they are handled as previously described for skin scrapings.

Infected hairs are selected either by their characteristic appearance (broken-off hairs and twisted grayish stubs) or by their bright yellow-green fluorescence when examined under a Wood's lamp, using filtered ultraviolet light. Invasion of the inside of the hair (**endothrix**) or outside the hair shaft (**ectothrix**), as determined microscopically, can be helpful in identifying the fungus involved.

Some workers find a stained preparation easier to examine than an unstained sodium hydroxide mount. The most convenient stain digestant is prepared from equal parts of Parker 51 blue-black ink and 10% sodium hydroxide and is used as previously described. One part of ink plus four parts of hydroxide will give a lighter stain and is preferred by some.

Cultural procedures

Specimens of skin and nail scrapings are obtained as described previously; hair stubs or scrapings of areas showing loss of hair (alopecia) are obtained without attempting to cleanse the scalp; however, a gentle wiping with 70% alcohol will result in less contamination.

The specimen may be either inoculated directly to media in the clinic or submitted to the laboratory in sterile disposable Petri dishes or clean paper envelopes. The upper portion of the hairs should be clipped off with alcohol-flamed scissors; only the lower ends should be inoculated to media.

Primary isolation of the dermatophytes is readily accomplished by inoculating the hairs or scrapings on the surfaces of SAB agar slants; the specimen should be partially embedded in the agar. It is advisable to inoculate **duplicate** sets of media, one of plain medium and one containing cycloheximide and chloramphenicol to inhibit the common bacterial and mold contaminants. These agents do not alter the cultural characteristics of dermatophytes and facilitate their isolation.

Another medium, dermatophyte test medium (DTM), is valuable when trained personnel are not immediately available to identify cultures. This medium inhibits bacteria and saprophytic molds and becomes alkaline (red) when a dermatophyte has grown on it. The culture is then purified and maintained on media free of antibiotics. Various commercial DTM media have proved satisfactory (Sinski, J. T., et al.: J. Clin. Microbiol. **5**:34-38, 1977).

All media should be incubated at room temperature (not over 30 C) and examined at 5-day intervals for at least a month before discarding. Sporulation of the dermatophytes generally occurs within 5 to 10 days of inoculation, and cultures should be examined during this period because characteristic colonial appearance and microscopic morphology are more easily recognized before the cultures age.

Common species

Species of *Microsporum* attack the hair and skin (and nails very rarely) and include *M. audouinii, M. canis,* and *M. gypseum. Trichophyton* species are responsible for infection of the hair, skin, and nails and include principally *T. mentagrophytes, T. rubrum, T. tonsurans, T. schoenleinii, T. violaceum,* and *T. verrucosum. Epidermophyton* causes infection of the skin and nails but not the hair and includes a single species, *E. floccosum.*

Since these dermatophytes generally present an identical appearance on microscopic examination of infected skin or nails,[8] final identification can only be made by culture.

Descriptions of the 10 principal species of fungi involved in dermatophytoses in the United States[3] follow; other geographically limited species are described in the references cited.

Genus Microsporum

The genus *Microsporum* is immediately identified by the presence of large (8 to 15 μm by 35 to 150 μm), spindle-shaped, rough or spiny macroconidia with thick (up to 4 μm) walls and containing 4 to 15 septa (Fig. 34-4). The microconidia are small (3 to 7 μm), club shaped, and borne on the hyphae, either sessile or on short sterigmata. Cultures of *Microsporum* develop slowly or rapidly and produce an aerial mycelium that may be velvety, powdery, glabrous, or cottony, varying in color from whitish, buff, and bright yellow to a deep cinnamon brown, with varying shades on the reverse side of the colony.

M. audouinii is the most important cause of epidemic tinea capitis among school children in the United States, but

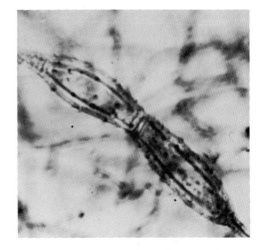

Fig. 34-4. *Microsporum audouinii,* showing macroconidium. (1800×.)

it rarely infects adults. The fungus is known as an **anthropophilic,** or "man-loving," fungus and is spread directly by means of infected hairs on headwear, upholstery, combs, or barbers' clippers. The majority of infections are chronic in nature; some heal spontaneously, whereas others may persist for several years. Infected hair shafts fluoresce yellow-green.

On SAB agar, *M. audouinii* grows slowly, producing a flat gray to tan colony with short aerial hyphae and a radially folded surface. The reverse of the colony is generally reddish brown in color. *M. audouinii* sporulates poorly on SAB agar, and the characteristic macroconidia may be lacking in some cultures. The addition of yeast extract will stimulate growth and production of both macroconidia and small club-shaped microconidia borne laterally along the hyphae. Abortive and bizarrely shaped macroconidia, hyphal cells with swollen ends (racquet hyphae), abortive branches (pectinate bodies), and chlamydospores are commonly observed.

*M. canis** is primarily a pathogen of animals **(zoophilic);** it is the most common cause of ringworm in dogs and cats in the United States. Children and adults ac-

---

*Ascomycetous state: *Nannizzia otae.*

Fig. 34-5. *Microsporum canis,* showing several spindle-shaped, thick-walled, multicelled macroconidia. (500×.)

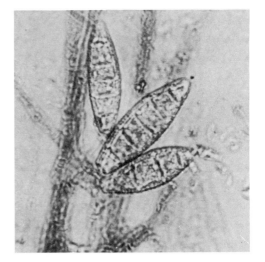

Fig. 34-6. *Microsporum gypseum,* showing ellipsoidal multicelled macroconidia. (750×.)

quire the disease through contact with infected animals, particularly puppies and kittens, although human-to-human transfer has been reported. Hairs infected with *M. canis* **fluoresce** a bright yellow-green under a Wood's lamp, which is a useful tool for screening pets as possible sources of human outbreaks. On direct examination in 10% sodium hydroxide, small spores (2 to 3 μm) are found outside the hair (ectothrix), although cultural procedures must be carried out for specific identification.

On SAB agar, *M. canis* grows rapidly as a flat, disklike colony with a bright yellow periphery and possesses a short aerial mycelium. On aging (2 to 4 weeks), the mycelium becomes dense and cottony, a deeper brownish yellow or orange, and frequently shows an area of heavy growth in the center. The reverse side of the colony is **bright yellow,** becoming orange or reddish brown with age. Rarely, strains are isolated that show no reverse-side pigment. Microscopically, *M. canis* shows an abundance of large (15 μm by 60 to 125 μm), spindle-shaped, multicelled (4 to 8) macroconidia (Fig. 34-5) with knoblike ends. These are thick walled and bear warty (echinulate) pro-

jections on their surfaces. Microconidia, pectinate hyphae, racquet hyphae, and chlamydospores are found.

*M. gypseum** is a free-living saprophyte of the soil (**geophilic**) that only rarely causes human or animal infection. Infected hairs generally do not fluoresce under a Wood's lamp. However, microscopic examination of infected hairs shows them to be irregularly covered with clusters of spores (5 to 8 μm), some in chains. These arthrospores of the ectothrix type are considerably larger than those of other *Microsporum* species.

On SAB agar *M. gypseum* grows rapidly as a flat, irregularly fringed colony with a coarse powdery surface with a fawn to buff or **cinnamon brown** color. The underside of the colony is a conspicuous orange to brownish color. Macroconidia are seen in large numbers and are characteristically large, ellipsoidal, and multicelled (3 to 9) with echinulate surfaces (Fig. 34-6). Although spindle shaped, these macroconidia are not as pointed at the distal end as are those of *M. canis.* Microconidia are rare.

_____

*Ascomycetous state: *Nannizzia gypsea.*

Genus Trichophyton

Species of this genus are widely distributed and are the most important causes of **ringworm** of the feet and nails; they are occasionally responsible for tinea corporis, tinea capitis, and tinea barbae. They are most commonly seen in adult infections and vary considerably in their clinical manifestations. Most cosmopolitan species are anthropophilic; a few are zoophilic.

Generally, trichophyton-infected hairs **do not fluoresce** under a Wood's lamp; the demonstration of fungal elements either inside the hair shaft, surrounding and penetrating the hair shaft, or within skin scrapings is needed to make a diagnosis of ringworm. Isolation and identification of the fungus are necessary for confirmation.

Microscopically, *Trichophyton* is characterized by club-shaped, **smooth**, thin-walled macroconidia with 8 to 10 septa ranging in size from 8 by 4 μm to 15 by 8 μm. The macroconidia are borne singly at the terminal ends of hyphae or on short branches; the microconidia are usually spherical or clavate and 2 to 4 μm in size. Although a large number of *Trichophyton* species have been described, many have proved to be colonial variants. Only the common species will be described.

*T. mentagrophytes** occurs in **two** distinct colonial forms: the so-called "**downy**" variety commonly isolated from human tinea pedis, and the "**granular**" variety isolated from ringworm acquired from animals (zoophilic). It is possible to convert the downy form to the granular form by animal passage; the reverse may occur spontaneously in laboratory cultures.

Growth of *T. mentagrophytes* is rapid and abundant on SAB agar, appearing as white, cottony, or downy colonies to flat, cream-colored, or peach-colored colonies that are coarsely granular to powdery. The reverse side of the colony is rose

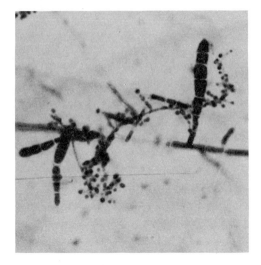

Fig. 34-7. *Trichophyton mentagrophytes*, showing numerous microconidia in grapelike clusters. Also shown are several thin-walled macroconidia. (500×.)

brown, occasionally orange to deep red in color. The white downy colonies produce only a few clavate microconidia; the granular colony sporulates freely with numerous small, globose to club-shaped microconidia and thin-walled, slightly clavate, spindle- or pencil-shaped macroconidia measuring 6 by 20 μm to 8 by 50 μm in size, with 2 to 5 septa (Fig. 34-7). Spiral hyphae and nodular bodies may be present. Macroconidia are demonstrated best in 5- to 10-day-old cultures.

T. rubrum is a slow-growing species, producing a flat or heaped-up colony with a white to reddish, cottony or velvety surface. This characteristic **cherry red** color is best observed on the reverse side of the colony, commencing at the margin or spreading concentrically, but it may disappear on subculture. Occasional strains may lack the deep red pigmentation on first isolation.* Microconidia are rare in most of the fluffy strains and more common in the velvety or granular strains, oc-

---

*Ascomycetous state: *Arthroderma benhamiae*.

---

*A useful method, utilizing in vitro hair cultures to differentiate aberrant forms of *T. mentagrophytes* from *T. rubrum*, has been described by Ajello and Georg.[4]

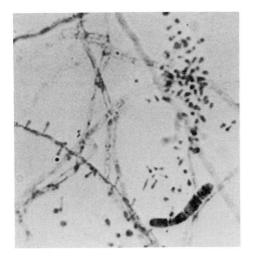

Fig. 34-8. *Trichophyton rubrum,* showing a sausage-shaped macroconidium and numerous pyriform microconidia borne singly on hyphae. (750×.)

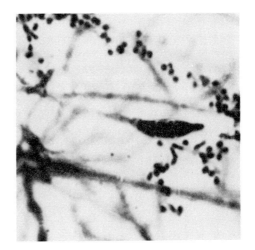

Fig. 34-9. *Trichophyton tonsurans,* showing numerous microconidia borne singly or in clusters. A single macroconidium (rare) is also present. (600×.)

curring as globose to clavate spores, 2 to 5 µm in size, and growing in clusters on the lateral sides of the mycelium. Macroconidia are rarely seen, although they are more common in the granular strains where they appear as thin-walled, sausagelike cells with blunt ends, containing 3 to 8 septa (Fig. 34-8).

T. tonsurans, along with *M. audouinii,* is responsible for an **epidemic** form of tinea capitis occurring most commonly in children but occasionally in adults. The fungus causes a low-grade superficial lesion of varying chronicity and produces circular, scaly patches of alopecia. The stubs of hair remain in the epidermis of the scalp after the brittle hairs have broken off and give the typical "black dot" ringworm appearance. Since the infected hairs do not fluoresce under a Wood's lamp, a careful search for the embedded stub should be carried out in a bright light, using a magnifying head loop.

The direct microscopic examination of infected stubs mounted in 10% sodium hydroxide reveals the hair shaft to be filled with masses of large (4 to 7 µm) arthrospores in chains, an endothrix type of invasion. Cultures of *T. tonsurans* de-velop slowly on SAB agar as flat, white, powdery colonies, later becoming velvety and varying in color from a gray through a sulfur yellow to tan. The colony surface shows radial folds, often developing a **craterlike** depression in the center with deep fissures. The reverse side of the colony is a yellowish to reddish brown color. Microscopically, numerous microconidia are observed, borne laterally on undifferentiated hyphae or in clusters. These vary greatly in size, from 2 by 3 µm to 5 by 7 µm. Macroconidia, although rarely encountered, are clavate or irregular in shape with thin walls (Fig. 34-9). Chlamydospores are abundant in old cultures; swollen and fragmented hyphal cells resembling arthrospores are also seen. The addition of thiamine to the isolation medium will enhance the growth of *T. tonsurans.*

Favus is a severe type of ringworm of the scalp caused by *T. schoenleinii.* The infection is characterized by the formation of yellowish cup-shaped crusts, or scutulae, resulting in considerable scarring of the scalp and sometimes permanent baldness. A distinctive invasion of the infected hair, the favic type, is demonstrated by the presence of large inverted cones of hyphae and arthrospores at the mouths of the hair follicles along with a branching mycelium throughout the length of the hair (endothrix). Longitudinal tunnels or empty spaces appear in the hair shaft where the hyphae have disintegrated, which, in sodium hydroxide preparations, readily fill with fluid; air bubbles also can be seen in these tunnels.

T. schoenleinii grows slowly as a gray, glabrous, and waxy colony on SAB agar, somewhat hemispherical at first, but later spreading to resemble a sponge placed on the medium. The irregular border consists mostly of submerged mycelium, which tends to crack the agar. The surface of the colony is yellow to tan, furrowed, and irregularly folded. Old atypical strains show a powdery or downy surface with short aerial hyphae. The reverse side of the colony is usually tan in color or nonpigmented.

Microscopically, one sees only a few microconidia, which vary greatly in size and shape. Macroconidia are not pro-

duced; the mycelium is highly irregular. The hyphae tend to become knobby and club shaped at the terminal ends (pin heads), with the production of many short lateral and terminal branches (favic chandeliers) (Fig. 34-10). Chlamydospores are generally numerous. All strains of *T. schoenleinii* may be cultivated in a vitamin-free medium and grow equally well at both room temperature and 35 C.

T. violaceum causes ringworm of the scalp and body, mainly in the Mediterranean region, the Middle and Far East, and occasionally in the United States. Hair invasion is of the endothrix type; clinically, the typical black-dot ringworm is observed. Microscopically, direct examination of sodium hydroxide mounts of the short nonfluorescing hair stubs shows dark thick hairs filled with masses of arthrospores arranged in chains, similar to the appearance in *T. tonsurans* infections. On SAB agar, the fungus is very slow growing, beginning as a cream-colored, glabrous, cone-shaped colony, later becoming heaped up, verrucose (warty), and lavender to deep purple. The reverse side of the colony is purple or nonpigmented. Older cultures may develop a velvety aerial mycelium and sometimes lose their purple pigment. Microscopically, microconidia or macroconidia are generally absent; only sterile, thin, and irregular hyphae and chlamydospores are found. *T. violaceum* requires thiamine-enriched media to produce conidia.

T. verrucosum causes a variety of ringworm lesions in cattle (zoophilic) and in humans; it is seen most often in farmers who are infected from cattle. The lesions are found chiefly on the beard, neck, wrist, and back of the hand; they are deep, boggy, and suppurating with sinus tracts. On pressure, short stubs of hair can be recovered from the purulent lesion. On direct examination, the outside of the hair shaft reveals sheaths of isolated chains of large (5 to 10 μm) spores and mycelium within the hair (ectothrix and endothrix type). Masses of these spores

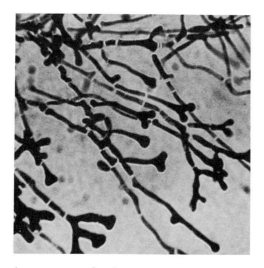

Fig. 34-10. *Trichophyton schoenleinii*, showing swollen hyphal tips, resembling antlers, with lateral and terminal branching (favic chandeliers). Microconidia and macroconidia are absent. (500×.)

are also seen in the pus and germinate to form long thin filaments.

T. verrucosum grows very slowly (10 to 14 days) and poorly on SAB agar at room temperature but better at 35 C. Maximal growth is obtained on media enriched with thiamine and inositol or yeast extract; no growth occurs on vitamin-free media.[5] The colony on SAB agar is small, heaped, and folded, occasionally flat and disk shaped. At first glabrous and waxy, the colony sometimes develops a short aerial mycelium on enriched media; colonies vary from a gray, waxlike color to a bright ochre. The reverse of the colony is yellow but may be nonpigmented.

On SAB agar a thin, irregular mycelium is produced, with many terminal and intercalary (between two hyphal segments) chlamydospores; sometimes favic chandeliers are formed. Chlamydospores are more numerous on SAB agar incubated at 35 C. On enriched media, *T. verrucosum* forms more regular mycelia and numerous small microconidia, borne singly along the hyphae. Macroconidia (Fig. 34-11) are rarely formed and vary considerably in size and shape.

Genus Epidermophyton

Epidermophytosis is caused by *Epidermophyton floccosum*, but a similar picture can be caused also by species of *Trichophyton* and *Microsporum*. The skin and nails are usually attacked, but not the hair. In the direct examination of scrapings mounted in sodium hydroxide, the fungus is seen as fine branching filaments in young lesions that form chains of arthrospores in older lesions.

E. floccosum grows slowly on SAB agar; primary growth appears as yellowish white spots, developing into powdery or velvety colonies with centrally radiating furrows of a distinctive greenish yellow color. The reverse side of the colony is a yellowish tan color. After several weeks, the colony develops a white aerial mycelium (pleomorphic), which completely overgrows the colony.

Microscopically, numerous smooth, thin-walled, multiseptate (2 to 4) macroconidia are seen, rounded at the tip and borne singly or in groups of two or three on the hyphae (Fig. 34-12). Microconidia are absent, spiral hyphae are rare, and chlamydospores are usually numerous.

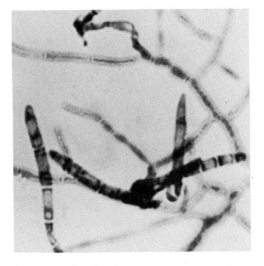

Fig. 34-11. *Trichophyton verrucosum*, showing multicelled, smooth, thin-walled macroconidia, which are rarely seen. (500×.)

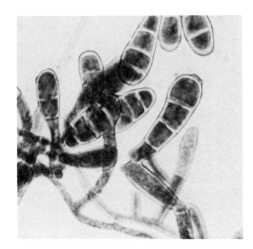

Fig. 34-12. *Epidermophyton floccosum*, showing numerous smooth, multiseptate, thin-walled macroconidia with rounded ends (microconidia absent). (1000×.)

SUBCUTANEOUS MYCOSES

Subcutaneous mycoses are fungal infections that involve the skin and subcutaneous tissue, generally without dissemination to the internal organs of the body. This classification is artificial; for example, sporotrichosis on occasion involves the lungs, other viscera, or joints or may disseminate. The other agents causing subcutaneous infection may also produce visceral infection occasionally. The agents are found in several unrelated fungal genera, all of which probably exist as saprophytes in nature. Humans and animals serve as **accidental hosts** through inoculation of the fungal spores into cutaneous and subcutaneous tissue after trauma. Three subcutaneous mycoses are considered here: **sporotrichosis, chromomycosis,** and **maduromycosis.**

Sporotrichosis

Sporotrichosis is a chronic infection of worldwide distribution caused by the **dimorphic** fungus *Sporothrix schenckii,* whose natural habitat is in the soil and on living or dead vegetation. Humans acquire the infection through an accidental wound (thorn, splinter) of the hand, arm, or leg. The infection is characterized by the development of a nodular lesion of the skin or subcutaneous tissue at the point of contact and later involves the lymphatic vessels and nodes draining the area. The lesion then breaks down to form an indolent ulcer that later becomes chronic. Only rarely is the disease disseminated. The infection is an occupational hazard for farmers, nurserymen, gardeners, florists, miners, and others.*

The **tissue forms** of *S. schenckii* appear as small, oval, budding, yeastlike cells, which are not usually demonstrable in unstained or stained smears of material

---

*An outbreak of sporotrichosis associated with sphagnum moss as the source of infection has been reported.[14] A report, in lighter vein, of an outbreak involving beer cans, bricks, and medical students is described in Arch. Intern. Med. **127:**482-483, 1971.

from suspected lesions, except by immunofluorescence procedures (see Chapter 39). However, the tissue form may be produced readily by inoculating mice or rats (see p. 295). The organism can be demonstrated in tissue sections stained by the Gomori or Gridley procedure.

Pus from unopened subcutaneous nodules or from open draining lesions is inoculated to BHI agar incubated at 35 C and on SAB agar at room temperature. Chloramphenicol and cycloheximide should be added to the medium if contamination is anticipated. *S. schenckii* is not inhibited by these agents.

The tissue (yeast or parasitic) phase develops at 35 C, appearing in 3 to 5 days as smooth, tan, yeastlike colonies. Microscopically, such colonies show cigar-shaped (fusiform) cells, measuring 1 to 4 μm by 1 μm or less, and round or oval budding cells 2 to 3 μm in diameter (Fig. 34-13, *A*). Occasionally, a few large, pyriform cells, 3 to 5 μm in size, may be produced.

On SAB agar at room temperature, growth appears in 3 to 5 days as small, moist, white to cream-colored colonies. On further incubation these become membranous, wrinkled, and coarsely tufted, the color becoming irregularly dark **brown** or **black.** Microscopically, the mycelium is made up of delicate (1 to 2 μm thick), branching, septate hyphae that bear pyriform or ovoid to spherical microconidia 2 to 5 μm in diameter. These are borne, bouquetlike, in clusters from the tips of the conidiophores or directly on the sides of hyphae as dense sleeves of conidia (Fig. 34-13, *B*). Older cultures may produce larger, thick-walled chlamydospores.

Because of their morphology, saprophytic species of the genus may be confused with *S. schenckii,* and it is necessary to distinguish between them by in vitro and in vivo culture. For the former, moist slants of BHI agar containing 5% blood are inoculated and incubated at 35 C. It may require several transfers before

Chromomycosis

Chromomycosis is a chronic noncontagious skin disease characterized by the development of a papule at the site of infection that spreads to form warty or tumorlike lesions, later resembling a cauliflower in appearance. There may be secondary infection and ulceration. The lesions are usually confined to the feet and legs but may involve the head, face, neck, and other body surfaces. Brain abscess may also be caused by species in the genera *Phialophora* and *Cladosporium*.

The disease is widely distributed, but most cases occur in tropical and subtropical areas. Occasional cases are reported from temperate zones, including the United States. The infection is seen most often in areas where barefoot workers suffer thorn or splinter puncture wounds, through which the spores enter from the soil.

The etiologic agents of chromomycosis comprise a group of closely related fungi that produce a slow-growing, heaped-up, and slightly folded **dematiaceous** (dark-colored) colony with a grayish velvety mycelium. The reverse side of the colony is jet black. The different species are distinguished by the type of conidiophores they produce and include:

1. Cladosporium type *(Cladosporium carrionii)*. Conidia in branched chains are produced by conidiophores of various lengths.
2. Phialophora type *(Phialophora verrucosa)*. Conidia are produced endogenously in flasklike conidiophores or phialides (Fig. 34-14).
3. Acrotheca type *(Phialophora [Hormodendrum] pedrosoi, P. compacta, P. dermatitidis)*. Conidia are formed along the sides of irregular club-shaped conidiophores (Fig. 34-15).

A laboratory diagnosis of chromomycosis is essential and is easily made. Scrapings or scales from encrusted areas mounted in 10% sodium hydroxide, xylol, or balsam show the presence of long, dark

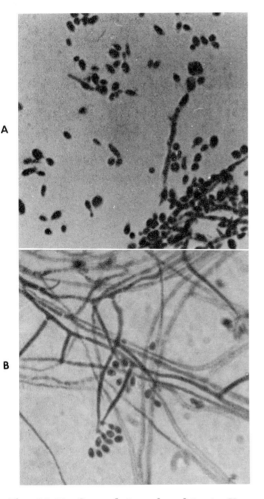

Fig. 34-13. *Sporothrix schenckii.* **A,** Yeast phase, showing cigar-shaped and oval-budding cells. (500×.) **B,** Mycelial phase, showing pyriform to ovoid microconidia borne bouquetlike at the tip of the conidiophore. (750×.)

the characteristic yeast form (tissue phase) develops. Animal inoculation may be employed if laboratory culture is nonproductive; to this end, white rats are injected intratesticularly with pus, yeast cells, or mycelial fragments, using approximately 0.2 ml. In 3 weeks the animals are killed and examined for a purulent orchitis. Gram-stained pus will reveal gram-positive cigar-shaped or oval budding forms of S. *schenckii.*

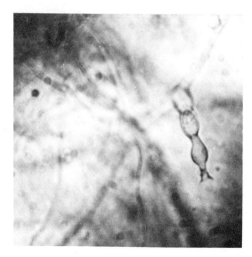

Fig. 34-14. *Phialophora verrucosa,* showing a single flasklike conidiophore. (1000×.)

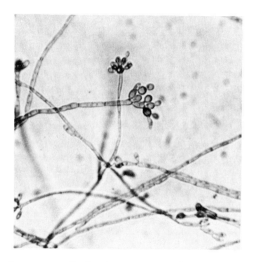

Fig. 34-15. *Phialophora (Hormodendrum) pedrosoi,* showing conidia produced terminally and in clusters on club-shaped conidiophores (cladosporium type). (400×.)

brown, thick-walled, branching septate hyphae 2 to 5 μm in width. In pus, tissue, or biopsy specimens, thick-walled, rounded brown cells 4 to 12 μm in diameter may be observed. All of the fungi causing chromomycosis have the same appearance.

Crusts, pus, and biopsy tissue are cultured on SAB agar with antibiotics (see

Chapter 41) and incubated at room temperature. Identification of the dematiaceous isolates is based on the type of sporulation observed. *C. carrionii* exhibits only the cladosporium type of sporulation, and the conidial chains are quite long. *P. compacta* and *P. pedrosoi* may exhibit all three types of sporulation concurrently, although the cladosporium type predominates, with short chains of conidia. *P. verrucosa* exhibits only the Phialophora type of sporulation. A *Cladosporium* species, considered to be a saprophyte, also produces a cladosporium type of sporulation but, unlike *C. carrionii,* will liquefy gelatin and hydrolyze a Loeffler serum slant.

Mycetoma (maduromycosis)*

Mycetoma is a chronic granulomatous infection that usually involves the lower extremities but may occur on any part of the body. It was first described by Gill in 1842 while he was working in a dispensary near Madura, India. The term Madura foot probably originated from the natives in describing the deformed foot seen in infected patients. The disease is characterized by swelling, purplish discoloration and tumorlike deformities of the subcutaneous tissue, and multiple sinus tracts that drain pus containing yellow, red, or black granules. The infection gradually progresses to involve bone, muscle, or other contiguous tissue, ultimately requiring amputation. Occasionally there may be more significant invasion, with involvement of the brain or other internal organs.

Maduromycosis is common among the natives of the tropical and subtropical regions, whose outdoor occupations and shoeless habits often predispose them to trauma. These are significant factors in exposure to the fungus. More than 60 cases of mycetoma have been reported in the United States.[18]

*The interested reader is referred to Vanbreuseghem's excellent monograph on mycetoma.[37]

There are two types of mycetoma: so-called actinomycotic mycetoma, or nocardiomycosis, caused by at least six species of *Nocardia* and *Streptomyces*, and fungal mycetoma, or maduromycosis, caused by a heterogeneous group of 16 species of septate true fungi with broad hyphae. The most common cause of maduromycosis in the United States is the perfect fungus, *Petriellidium (Allescheria) boydii*,* in the class Ascomycetes, since it produces sexual ascospores. The fungus is a common saprophyte in soil and sewage, and humans acquire the infection after such injuries as a scratch, bruise, or penetrating wound or by contaminating an open wound.

P. boydii is a **hyaline** (glassy, transparent) organism, producing white or yellow granules in pus. These are composed of tightly meshed, wide, septate mycelia and numerous large, hyaline chlamydospores. On SAB agar without antibiotics,

---

*Imperfect state: *Monosporium apiospermum.*

P. boydii grows rapidly at room temperature as a white fluffy colony that changes in several weeks to a brownish gray mycelium. The reverse of the colony is gray black. Microscopic examination shows large, septate, hyaline hyphae and many conidia borne singly on conidiophores. The conidia are pyriform to oval in shape, are unicellular, and measure approximately 6 by 9 μm. Clusters of conidiophores (coremia) with conidia at the ends sometimes occur; these resemble ripened grains on sheaves of wheat.

Some strains of *P. boydii* produce **perithecia,** closed structures containing asci with ascospores. When the latter are fully developed, the large (50 to 200 μm), thin-walled perithecia rupture, liberating the asci and spores (Fig. 34-16). The ascospores are yellow, oval, and delicately pointed at each end and are somewhat smaller than the conidia.

It should be noted that media containing antibiotics are **not to be used alone** in culturing clinical specimens from myce-

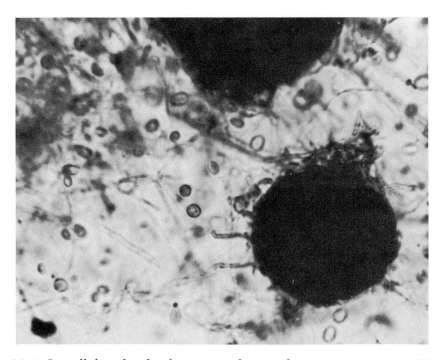

Fig. 34-16. *Petriellidium boydii,* showing perithecia and numerous ascospores. (750×.)

tomas or draining sinuses, since some of the agents of maduromycosis may be inhibited in their growth. This is particularly true with species of *Nocardia, Aspergillus,* and *P. boydii.* Since "actinomycotic" lesions may respond to specific therapy, whereas maduromycoses may not, this etiologic differentiation must be made.

For information on other fungi involved in maduromycosis, please refer to sources listed in the references.

EYE INFECTIONS

Keratomycosis (fungus infection of the cornea) is an uncommon but important fungal infection. Failure to recognize and treat it early and failure to avoid therapeutic use of corticosteroids may lead to deeper penetration of the infection and loss of the involved eye. Corneal involvement is typically in the form of an elevated ulcer with surrounding infiltrate and satellite lesions.

Except for *Candida albicans* infection, keratomycosis is **exogenous** in origin. Outdoor material, especially vegetable matter, is the primary source of the organism. Some 80 species of fungi in 35 genera may be causative agents. Most are saprophytes or plant pathogens; only a few are known as causes of other types of fungal infection. Accordingly, identification may often require the aid of mycologists with specialized knowledge. The most common causes of keratomycosis, in order of frequency, are *Fusarium solani, Candida albicans, Aspergillus fumigatus, Curvularia* species and other dematiaceous hyphomycetes, *Acremonium* species and related genera, *Aspergillus flavus* and other *Aspergillus* species, *Penicillium, Fusarium episphaeria* and other *Fusarium* species, *Cylindrocarpon, Volutella* species, *Petriellidium boydii,* and *Lasiodiplodia theobromae.*[29]

Other eye infections involving fungi include canaliculitis, dacryocystitis, orbital cellulitis, endophthalmitis following surgery or trauma, and the extension of cutaneous or systemic mycotic infection to the eye. Recently there was an outbreak of endophthalmitis associated with implantation of lenses; this was due to contamination of the lens prostheses with *Paecilomyces lilacinus.*[12]

YEASTLIKE FUNGI

This group of imperfect fungi resembles the true yeasts both morphologically and culturally. They produce yeastlike, creamy colonies on solid media and are generally unicellular, although some produce a pseudomycelium or true mycelium. The genera described here include *Candida, Geotrichum,* and *Torulopsis.*

Cryptococcus is also a yeast; this organism is discussed under **systemic mycoses.** Again, as with the subcutaneous mycoses, the division is arbitrary. The three yeasts discussed here can all cause systemic disease.

Candidiasis

Candidiasis is an acute or subacute infection caused by members of the genus *Candida,* chiefly *C. albicans,* although all species may be pathogenic.* The fungus may be isolated from the stools, genitourinary tract, throat, and skin of normal persons (**endogenous**), and it may produce lesions in the mouth, esophagus, genitourinary tract, skin, nails, bronchi, lungs, and other organs in patients whose normal defense mechanisms may have been altered by underlying disease, antimicrobial therapy, or immunosuppressive agents.[36] Bloodstream infection, endocarditis (primarily in drug addicts), and meningitis caused by *Candida* species have also been reported.

The isolation of *Candida* species from clinical materials is difficult to evaluate, since positive cultures may be obtained

---

Candida tropicalis and *C. parapsilosis,* although less commonly isolated from human infections, have been increasingly implicated in endocarditis and fungemia.

from various anatomic sites of a large percentage of normal persons. The organisms must be recovered repeatedly in significant numbers from **fresh** specimens and to the exclusion of other known etiologic agents in a patient with an appropriate clinical picture before a diagnosis of candidiasis can be entertained. Indeed, a Mayo Clinic group[27] concluded recently that *Candida* and other yeasts in respiratory secretions (aside from *Cryptococcus neoformans*) probably represent normal flora and that their routine identification is not warranted.

Skin and nail scrapings should be examined directly; they are mounted in 10% sodium or potassium hydroxide with a coverglass and heated gently. Other material, such as exudate from the oropharynx or vagina or material from the intestinal tract, should be pressed under a coverglass and examined fresh, either unstained or Gram stained. *Candida* appears as small (2 to 4 μm), oval or budding, yeastlike cells, along with mycelial-like fragments of varying thickness and length **(tissue phase)**. The yeastlike cells and pseudomycelial elements are strongly gram positive. It is well to report the approximate number of such forms seen, since the presence of large numbers in a fresh specimen may have diagnostic significance.

Since saprophytic yeasts are similar microscopically to the pathogenic species, all infected material should be **cultured** on duplicate sets of SAB agar with and without cycloheximide* and incubated at both room temperature and 35 C. Colonies of *Candida* species (and saprophytic yeasts) appear in 3 to 4 days as medium-sized, cream-colored, smooth, and pasty, with a characteristic yeastlike appearance. Most strains grow well at either temperature. On microscopic examina-

tion, a slide mount will show budding cells along with elongated unattached cells (pseudomycelia) with clusters of blastospores at constrictions (Fig. 34-17).

If the unknown culture is suspected of belonging to the genus *Candida,* subsequent procedures must be carried out, including the demonstration of chlamydospore production, germ tube production, and sugar fermentation and assimilation tests. Although other species of *Candida* may be encountered in candidiasis, *C. albicans* is the most frequently isolated species and is the usual etiologic agent in oral or vaginal thrush, intertriginous or cutaneous monilial infection, paronychial infection, or bronchopulmonary candidiasis.

A simple test for production of germ tubes recommended by Ahearn[1] is as follows:

1. Cells from a young (not more than 96 hours) colony are transferred by means of the tip of a plastic straw* into about 0.3 ml pooled **human** serum contained in a clean 12- × 75-mm test tube, leaving the straw immersed in the serum.
2. The tube is incubated for 3 hours at 35 C; a drop of the suspension is placed on a glass slide, using the straw for transfer, and a coverslip is applied.
3. Microscopic examination of typical *C. albicans* reveals thin **germ tubes** 3 to 4 μm in diameter and up to 20 μm in length; unlike pseudohyphae, they are **not constricted** at their point of origin.
4. Arthrospores of *Geotrichum* or *Trichosporon*† species and structures produced by other organisms may be mistaken for germ tubes by the

---

*A number of *Candida* species are inhibited by 0.5 mg/ml of cycloheximide; these include *C. parapsilosis, C. krusei,* and strains of *C. tropicalis;* most strains of *C. albicans* are resistant.

*Commercial cocktail straws cut into approximately 100-mm lengths, clean but not sterile.

†*Trichosporon* species produce pseudomycelia, true mycelia, blastospores, and arthrospores. This rapidly growing yeast may be part of the normal skin flora; it has also been isolated from infected fingernails.

inexperienced. For this reason, a **known isolate** of *C. albicans* and *C. tropicalis* should be included as controls.

Another method of identification of *Candida* relates to morphology when grown on corn meal agar containing 1% Tween 80 at room temperature in a Petri dish for 24 to 48 hours. The dish is inverted and examined microscopically under 100× magnification. All *Cryptococcus* and *Torulopsis* strains fail to produce hyphae or pseudohyphae, whereas *Candida,** *Trichosporon,* and *Geotrichum* strains do show such structures.

Identification of Candida by fermentation tests†

1. Obtain a pure culture by inoculating a tube of SAB dextrose broth and incubating overnight at 35 C.
2. Shake the tube and inoculate a loopful to a blood agar plate and incubate overnight at 35 C.
3. Examine the plate and pick single colonies to SAB agar slants; incubate overnight at 35 C.
4. Transfer to sugar-free beef extract agar slants for three successive transfers, incubating each transfer overnight at 35 C.
5. Inoculate growth from the third transfer to sugar media in the following manner:
   a. Make a suspension of the growth in 2 ml of sterile saline.
   b. Pipet 0.2 ml to each of five tubes containing 9.5 ml of beef extract broth with 0.04% bromthymol blue indicator.
   c. To each of these tubes add, respectively, 0.5 ml of a filter-sterilized 20% stock solution (Millipore or Seitz) of glucose,

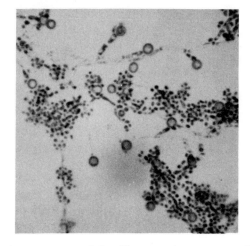

Fig. 34-17. *Candida albicans,* showing round, thick-walled chlamydospores, pseudomycelia, and numerous blastospores. (750×.)

maltose, sucrose, lactose, and galactose.
   d. Overlay each tube with sterile melted petrolatum or paraffin and petrolatum, to form a plug about 1 cm thick.
   e. Hold five uninoculated tubes containing the sugars as sterility controls.
   f. Incubate all tubes at 35 C for 10 days and record the presence of acid or acid and gas.

Refer to Table 34-1 for test results.

Identification of Candida by assimilation tests*

1. Prepare a sterile solution of yeast-nitrogen base† by weighing out 6.7 g of the dehydrated medium. To this add 5.0 g of the appropriate carbo-

---

*Occasional strains of *C. guilliermondii* and *C. parapsilosis* do not produce obvious hyphal structures.

†A simple, rapid (24 hours) technique using tablets (available from Key Scientific Products, Los Angeles) and a Vaspar seal is described by Huppert et al.[22]

---

*A rapid technique using carbohydrate-impregnated disks was proposed by Huppert et al.[22] and supported by Segal and Ajello.[33] Correlation with reference techniques is 90% after incubation for 1 day, 97% after 2 days, and 98% after 3 days. Certain of the carbohydrates for assimilation studies, as well as certain other tests important in identification of yeasts, are available in commercial kits. These provide results rapidly and have received favorable evaluations.[11,31]

†Difco Laboratories, Detroit.

hydrate* and dissolve in 100 ml distilled water.
2. Sterilize by membrane filtration and add to an agar solution (20 g in 900 ml distilled water) that has been autoclaved and cooled to approximately 50 C.
3. Dispense into sterile, cotton-stoppered test tubes; solidify in the slanting position. The use of agar slants, rather than plates, facilitates handling and storage.
4. The inoculum is prepared from 24- to 36-hour cultures of the isolate in yeast-nitrogen broth plus 1 mg per liter of glucose, and 1 drop (approximately 0.01 ml) is added to each carbohydrate slant. The tests are read after 96 hours' incubation, and evidence of growth is noted on each of the carbohydrate slants when compared with a control slant of the basal medium.

Refer to Table 34-1 for test results.

Geotrichosis

Geotrichosis is a rather rare infection caused by the yeastlike fungus *Geotri-*

---

*Recommended: lactose, inositol, melibiose, cellobiose, erythritol, xylose, trehalose, and raffinose.

chum candidum (not a true yeast), which reproduces by fragmentation of the hyphae into rectangular arthrospores. It may produce lesions in the mouth, bronchi, or lungs. Since *Geotrichum* has been isolated from the mouths and intestinal tracts of normal persons, it must be recovered repeatedly and in large numbers from freshly obtained clinical specimens in patients with an appropriate clinical picture and no other likely pathogens present to be considered of etiologic significance.

Sputum or pus is pressed in a thin layer on a a slide under a coverglass and examined directly. *Geotrichum* appears as rectangular (4 by 8 μm) or large spherical (4 to 10 μm) arthrospores that stain heavily gram-positive; no budding forms are seen.

Since these cells may be confused with those of the saprophytic *Oospora*, which frequently occurs as a contaminant, or with the filamentous *Coccidioides immitis*, the etiologic agent of coccidioidomycosis, or with *Blastomyces dermatitidis*, the etiologic agent of North American blastomycosis, all infected material should be **cultured.** The specimen is inoculated on duplicate sets of SAB agar and BHI blood agar slants with and without chloramphenicol and cycloheximide;

Table 34-1. Fermentation and assimilation patterns of *Candida* species

Candida species	Glucose	Maltose	Sucrose	Lactose*	Inositol*	Melibiose*	Cellobiose*	Erythritol*	Xylose*	Trehalose*	Raffinose*	Urease	Growth at 35 C
C. albicans	+	+	0	0	0	0	0	0	+	+	0	0	+
C. stellatoidea	+	+	0	0	0	0	0		+	+/0	0	0	+
C. parapsilosis	+	0	0	0	0	0	0	0	+	+	0	0	+
C. tropicalis	+	+	+	0	0	0	+/0	0	+	+	0	0	+
C. pseudotropicalis	+	0	+	+	0	0	+		+/0	0	+	0	+
C. krusei	+	0	0	0	0	0	0	0	0	0	0	+/0	+
C. guilliermondii	+	0	+	0	0	+	+	0	+	+	+	0	+
C. rugosa	0	0	0	0	0	+	0	0	+	0	0	0	0

Abbreviations: +, positive; 0, no reaction.
*Used in assimilation tests only.

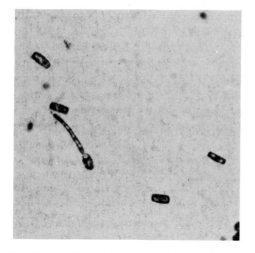

Fig. 34-18. *Geotrichum candidum,* showing barrel- to rectangular-shaped arthrospores, some with developing germ tubes. (500×.)

one set is incubated at room temperature and one at 35 C for at least 3 weeks. At room temperature the fungus develops rather rapidly as a moist, creamy colony, or as a colony with a dry, mealy surface and radial furrows, or as one with a fluffy aerial mycelium. At 35 C the slowly growing fungus develops only a small waxlike surface growth with a distinct zone of mycelium penetrating the subsurface of the medium. Microscopically, the septate, branching hyphae are fragmented into chains of rectangular, barrel-shaped, or spherical arthrospores that break apart readily. The rectangular cells frequently germinate by germ tubes from **one corner,** which are at first rounded and later elongated, a characteristic of *Geotrichum* (Fig. 34-18). Blastospores are **not** produced. *G. candidum* does not ferment carbohydrates but assimilates glucose and xylose.

Animal inoculation or serologic testing procedures are of no value in diagnosis.

Torulopsis glabrata

This organism, closely related to *Cryptococcus* and *Candida* species, was once considered a nonpathogenic saprophyte from the soil, being widely distributed in nature. However, it is now clear that it is a potentially important opportunistic pathogen, particularly in the compromised host.[24]

On sheep blood agar, *T. glabrata* appears as **tiny,** white, raised, nonhemolytic colonies after 1 to 3 days' incubation at 35 C. Gram stain of these colonies reveals round to oval budding yeasts, 2 to 4 μm in diameter; no hyphae or capsules are seen. Cultures on cornmeal-Tween agar are negative for mycelia or pseudohyphae; the germ tube test is also negative. *T. glabrata* ferments glucose and trehalose only and does not assimilate carbohydrates.[25]

SYSTEMIC MYCOSES

Systemic mycotic infections may involve any of the internal organs of the body, as well as lymph nodes, bone, subcutaneous tissue, and skin. **Asymptomatic** infections may go unrecognized clinically and may be detected only through skin sensitivity tests or serologic procedures; in some cases, x-ray examination may reveal healed lesions. **Symptomatic** infection may present signs of only a mild or more severe but self-limited disease, with positive supportive evidence by culture or immunologic findings. **Disseminated** or **progressive** infection may reveal severe symptoms, with spread of the initial disease to visceral organs as well as the bone, skin, and subcutaneous tissues. This type of disease is frequently fatal. Some cases of disseminated infection may exhibit little in the way of signs or symptoms of the disease for long periods, only to exacerbate later.

Collection of specimens

The obtaining of a proper specimen for the laboratory diagnosis of systemic mycoses is of prime importance; success or failure in isolating the etiologic agent may well depend on it.

The most satisfactory **sputum** specimen is a single, early morning, coughed specimen, before eating and after vigorous rinsing of the mouth with water after brushing the teeth. Twenty-four-hour specimens or those containing excessive amounts of postnasal discharge are **not** satisfactory. Sputum raised after inhalation of a heated 5% saline aerosol (prepared fresh and sterilized) and material obtained by bronchial aspiration are also satisfactory for mycologic examination. In all instances of suspected pulmonary infection, Georg[17] recommended that **at least six** sputum specimens be obtained on successive days. It should also be emphasized that all of these types of specimens should be delivered promptly to the mycology laboratory and **cultured promptly,** since *Histoplasma capsulatum* dies rapidly in specimens held at room temperature; furthermore, saprophytic fungi, *C. albicans,* and commensal bacteria may multiply rapidly and prevent the isolation of significant pathogens.

Biopsy specimens, such as scalene nodes, direct lung biopsies, and so forth, are excellent specimens for the recovery of fungal pathogens; they should be submitted in sterile tubes or Petri dishes that are slightly moistened with sterile salt solution. Empyema fluid should be anticoagulated with heparin during aspiration to prevent clotting; gastric aspirates, cerebrospinal or synovial fluid, blood and bone marrow, urine, prostatic secretions (particularly in blastomycosis), and lesions of skin and mucous membranes afford significant sources for recovery of systemic fungal pathogens.

Introduction to systemic mycoses

The systemic mycoses to be considered in this section include **cryptococcosis, coccidioidomycosis, histoplasmosis, blastomycosis,** and **paracoccidioidomycosis.** The fungi responsible for these infections, although unrelated generically and dissimilar morphologically and culturally, have (except for *Cryptococcus neoformans*) one characteristic in common —that of **dimorphism.** The dimorphic or-

ganisms involved exist in nature as the **saprophytic** form, sometimes called the **mycelial phase,** which is quite distinct from the **parasitic,** or tissue-invading form, sometimes called the **tissue phase.** The reader will note that reference has been made previously to this diphasic phenomenon in the fungal diseases candidiasis, sporotrichosis, and chromomycosis, where distinct morphologic differences may be observed both in vivo and in vitro. Temperature (35 C), certain nutritional factors, and stimulation of growth in tissue independent of temperature have been among the factors considered necessary to effect the transformation of mycelial forms to the parasitic phase.

Cryptococcosis (torulosis)

Cryptococcosis is a subacute or chronic mycotic infection involving primarily the brain and meninges and the lungs; at times the skin or other parts of the body may be involved. It is caused by a single species of yeastlike organism, *Cryptococcus neoformans.* The organism was first isolated by Sanfelice in 1894 from peach juice. Subsequently, the source of infection in humans and animals was erroneously assumed to be endogenous until Emmons, in 1950, reported the isolation of virulent strains of *C. neoformans* from barnyard soil. In 1955 he further reported a frequent association of virulent strains of *C. neoformans* with the excreta of pigeons and indicated that, until other sources are discovered, exposure to pigeon excreta was the most significant and important source of infection in humans and animals. This hypothesis has been substantiated by numerous reports that pigeon habitats serve as reservoirs for human infection; the pigeon manure apparently serves as an enrichment for *C. neoformans* due to its chemical makeup.[2] The organism is apparently the only pathogenic yeast not found in the normal human flora.

There is a strong association of cryptococcal infection with such debilitating diseases as leukemia and malignant lymphoma and the immunosuppressive therapy that may be required for these and other underlying diseases. The infection is probably more frequent than is commonly supposed; it is estimated that there are 2,000 undiagnosed cases for every proved case of infection due to *C. neoformans.*

All clinical material, especially the cerebrospinal fluid,* should be mixed with a drop of **India ink** (a cool loop must be used, since heat will precipitate the ink particles) on a slide and examined under a coverglass using the oil-immersion objective with reduced light. The India ink serves to delineate the large capsule, since the ink particles cannot penetrate the capsular material. *C. neoformans* appears as an oval to spherical, single-budding, thick-walled yeastlike organism 5 to 15 μm in diameter, surrounded by a wide, **refractile, gelatinous capsule** (Fig. 34-19). This characteristic

---

*The demonstration of encapsulated forms in the urine may precede their presence in cerebrospinal fluid; urine may also be a good source for the isolation of *C. neoformans.*

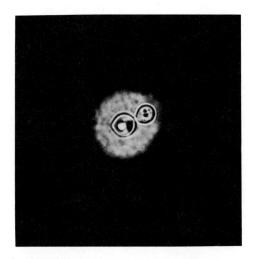

Fig. 34-19. *Cryptococcus neoformans* in spinal fluid, showing a large, encapsulated, single-budding, yeastlike cell. (India ink; 1000×.)

morphology occurs in India ink preparations of cerebrospinal fluid, sputum, pus, urine, infected tissue, or gelatinous exudates. Frequently these capsules are more than twice the width of the individual cells. In cerebrospinal fluid, *C. neoformans* may be mistaken for a lymphocyte and is often observed first in the spinal fluid counting chamber.

Dried, heat-fixed, or stained preparations are not generally recommended; distortion of the cryptococci may render them unrecognizable.

The infected material should be promptly cultured on infusion blood agar **without cycloheximide** (*C. neoformans* is inhibited) at 35 C and on SAB agar without cycloheximide at room temperature. In culturing cerebrospinal fluid, Utz[36] recommends that the **uncentrifuged** fluid should be inoculated in generous amounts (15 to 20 ml) to a series of culture tubes, since the cryptococci may be present in very small numbers and the centrifugation may destroy the more fragile cells. After several days of incubation at either temperature, the organism produces a wrinkled, whitish colony, which on microscopic examination may show only budding cells without capsules. On further incubation the typical slimy (*Klebsiella*-like), mucoid, cream- to brown-colored colony develops. This colony has no mycelium and flows down to the bottom of the slant. At this time budding cells with large capsules can be readily demonstrated in India ink wet mounts, although some strains do not form large capsules (unless they are transferred several times) and generally produce a shiny, dry colony. Incubation in a candle jar may stimulate capsule production.

Of the yeastlike fungi, only members of the genus *Cryptococcus* (both saprophytic strains and *C. neoformans*) consistently produce **urease.** This can be detected by inoculating a urea agar slant (Christensen) with the suspected cryptococcus. If it is a *Cryptococcus* species, it will produce a positive reaction (red color) in the medium after 1 to 2 days of incubation at room temperature.* *C. neoformans* characteristically does **not** assimilate nitrate or lactose but can assimilate glucose, maltose, and sucrose as carbon sources.

The incorporation of cycloheximide and chloramphenicol in SAB agar suffices in most instances for the isolation of pathogenic fungi from heavily contaminated material, **except** for *C. neoformans*, which is inhibited by cycloheximide. Staib[34] introduced a medium ("birdseed agar") containing creatinine and an extract of *Guizotia abyssinica* (a canary seed constituent) as a color marker for the selective isolation of *C. neoformans*, the growth of which produces a **brown** color. This medium, however, was rapidly overgrown with saprophytic fungi when inoculated with material from pigeon nests. The addition of diphenyl and chloramphenicol apparently increased the efficiency of the original preparation.† Caffeic acid, a constituent of *Guizotia* seeds, together with ferric citrate, has been used in a 6-hour paper disk test for brown pigment formation.[20] Other methods for rapid identification of *C. neoformans*, as well as other yeasts, are described by Huppert and associates.[22]

Most saprophytic strains of cryptococci will **not** grow at 35 C.‡ The pathogenicity of suspected strains of *C. neoformans*, especially from the sputum or skin, should be demonstrated in mice. This is carried out by injecting two to four white mice with either 1 ml intraperitoneally or 0.02 ml intracerebrally (under light ether anesthesia) of a heavy saline suspension of a 4- or 5-day-old culture.

Mice injected **intraperitoneally** will develop lesions in the brain in about **3 weeks,** whereas mice injected **intracerebrally** will generally develop them in **less than 1 week.** Mice are killed at the end of 3 weeks if death has not occurred. At autopsy the animals will show gelatinous masses in the abdominal viscera, lungs, and brain. India ink preparations will reveal the typical budding, encapsulated *C. neoformans*, and the fungus may be cultivated from these lesions.

Coccidioidomycosis

Coccidioidomycosis is an infectious disease caused by a single species of fungus, *Coccidioides immitis*. Generally an acute, benign, and self-limiting respiratory tract infection, the disease less frequently becomes disseminated, with extension to other visceral organs, bone, lymphatic tissue, skin, and subcutaneous tissue.

Coccidioides immitis spores are found in the semiarid regions of the southwestern part of the United States and northern Mexico as well as in Central and South America; the disease, however, may be seen anywhere in the United States and can be traced to previous travel or residence in an endemic area.

Humans acquire the infection by inhaling **arthrospores** from contaminated soil, particularly during the dry and dusty season. Less than 0.5% of persons who acquire the infection ever become seriously ill; dissemination does, however, occur most frequently in dark-skinned races.

In **direct** microscopic examination (using wet, unstained preparations) of sputum, sediment from gastric washings or cerebrospinal fluid, exudates, or pus, *C. immitis* appears as a nonbudding, thick-walled (up to 2 μm) **spherule** or sporangium 20 to 200 μm in diameter, containing either granular material or

---

*Some strains of *Rhodotorula, Candida,* and *Trichosporon* occasionally hydrolyze urea.

†See Shields, A. B., and Ajello, L.: Medium for selective isolation of *C. neoformans*, Science **151:** 208-209, 1966.

‡*C. laurentii, C. albidus,* and *C. luteolus* sometimes grow at 35 C and may show a mild degree of mouse virulence. Therefore, when reporting to the clinician, microbiologists must make certain whether the cryptococcus isolated is or is not *C. neoformans*.

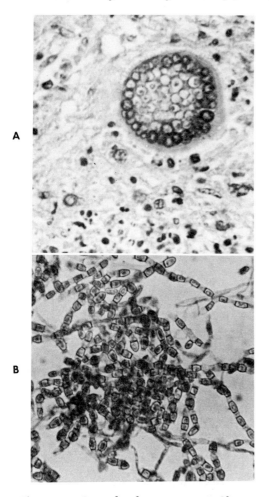

Fig. 34-20. *Coccidioides immitis.* **A,** Showing spherule containing many spherical endospores. (1000×.) **B,** Showing thick-walled, rectangular- or barrel-shaped arthrospores in mycelial phase. (500×.)

numerous small (2 to 5 μm in diameter) **endospores** (Fig. 34-20, *A*). These endospores are freed by rupture of the cell wall, and empty and collapsed "ghost" spherules may be present. Small, immature spherules measuring 10 to 20 μm may be confused with nonbudding forms of *Blastomyces dermatitidis,* since they are thick walled and endospores are not yet apparent. To check such structures, seal the edges of the coverglass with petrolatum and incubate overnight. If spherules are present, mycelial filaments

will have developed from the endospores.

The tissue phase cannot always be demonstrated by direct microscopic examination. Consequently, **all material** from suspected cases should be cultured on **duplicate** sets of SAB agar and BHI agar with and without chloramphenicol and cycloheximide,* one set incubated at room temperature and one set at 35 C. Animal inoculation is also indicated on occasion.

Growth appears in 3 to 5 days at room temperature (more positive isolations at this temperature) as a moist, membranous colony growing close to the surface of the medium. This soon develops a white, cottony mycelium, turning from buff to brown with age. Frequently, the central area of the colony will remain moist and glabrous. The slant should be **wetted down** before removing any of the mycelium (see following list of precautions). Microscopically, these cultures show a branching, septate mycelium, forming chains of thick-walled, rectangular or barrel-shaped **arthrospores,** 2 by 3 μm to 3 by 5 μm. In lactophenol cotton blue mounts, these chains show only **alternate** deeply stained arthrospores, with dried-out transparent cell tags on either side (Fig. 34-20, *B*). If such structures are observed, the identification should be confirmed by animal inoculation (see below).

At 35 C only the saprophytic or mycelial phase develops, since spherule production generally cannot be induced on the usual artificial media. The tissue phase can be obtained, however, in embryonated eggs or by injecting ground-up mycelia intratesticularly into guinea

---

*The mycelial growth of *C. immitis* is not appreciably affected by cycloheximide, whereas the white cottony growth of most saprophytes is inhibited. However, polymyxin B, which is incorporated in some media for fungi, is inhibitory to *C. immitis* (Collins, M. S.: J. Clin. Microbiol. **1:**335-336, 1975).

pigs.* Mycelial suspension, 0.1 ml, is injected, and if orchitis does not develop (generally within 1 week), the animal is killed in 2 to 4 weeks. In either case, the testicular tissue is examined for the presence of typical spherules, which verifies the identification of *C. immitis*. It has been demonstrated recently (Sun, S. H., et al.: J. Clin. Microbiol. **3**:186-190, 1976) that it is possible to get conversion to the spherule phase rapidly and consistently by means of slide culture on modified Converse liquid medium at 40 C in a candle jar. Slide culture, however, presents **significant hazards** to laboratorians working with *Coccidioides immitis* unless specialized safety facilities are available.

Old (more than 10 days) cultures in the arthrospore stage are the **most dangerous** phase of the fungus, and dissemination of the highly infectious arthrospores in the air can lead to infection of laboratory personnel. Therefore, the following precautions must be taken in handling such cultures:

1. Never use Petri dishes—always employ cotton-plugged test tubes or bottles.
2. To prevent the escape of arthrospores, **as soon as a cottony mold grows out** (usually within 3 days), **wet down the slant with sterile saline before introducing an inoculating needle.** Carry out all procedures in a biological safety hood.
3. Make mounts for microscopic examination in lactophenol cotton blue, which kills the spores; subculture to SAB slants if indicated.
4. Sterilize all contaminated equipment by autoclaving promptly.

---

*If guinea pigs are not available, white mice may be injected intraperitoneally with 1 ml of the spore suspension. After about 1 week they will develop lesions containing the mature spherules of *C. immitis*. These methods may also be used with sputum or gastric washings by adding 0.05 mg per milliliter of chloramphenicol and incubating 1 hour with frequent shaking prior to injection.

In culture, *C. immitis* must be differentiated from *Geotrichum* and *Oospora* (and other saprophytes), which produce arthrospores by mycelial fragmentation. The following features may be noted:

1. *Geotrichum* remains yeastlike on SAB agar.
2. *Oospora* does not produce alternately stained arthrospores and is not virulent for animals.
3. *Coccidioides immitis*, on animal injection, produces the characteristic endospore-filled spherules.

Histoplasmosis

Histoplasmosis is a mycotic infection of the reticuloendothelial system that may involve the lymphatic tissue, lungs, liver, spleen, kidneys, skin, central nervous system, and other organs. It is caused by the **dimorphic** (saprophytic and tissue forms) fungus *Histoplasma capsulatum*,* which exists as a saprophyte in the soil. Humans and animals acquire the infection by the inhalation of spores from the environment; the severity of the disease is generally related directly to the intensity of the exposure. The growth of *H. capsulatum* in nature appears to be associated with decaying or composted manure of chickens, birds (especially starlings), and bats ("cave disease"). A typical human case may result from the cleaning of a chicken house or silo that has not been disturbed for a long period or from working in soil under trees that have served as roosting places for starlings, grackles, or other birds. Although histoplasmosis may affect dogs and cats, there is no evidence of contagion between animal and human or between persons. The domestic animals, as well as several species of wild animals, appear to be accidental hosts and play no role in distributing or encouraging growth of *H. capsulatum* in the soil.

---

*Perfect form: *Emmonsiella capsulata* (Kwon-Chung, K. J.: Science **177**:368-369, 1972).

Histoplasmosis, once considered a rare and generally fatal illness, is now recognized as a common and benign disease in endemic areas in the eastern and central United States, where it is estimated that 500,000 persons are infected annually. Further studies and proper utilization of laboratory facilities will probably reveal the disease to be global in distribution.

The most frequent site of **primary** infection in humans is the respiratory tract, usually resulting in a mild or asymptomatic pulmonary infection with cough, fever, and malaise. In some areas a positive histoplasmin skin test, indicating exposure to *H. capsulatum,* is elicited in 60% to 90% of the inhabitants, most of whom give no history of an unusual respiratory illness at all.

A chronic **cavitary** form of histoplasmosis also occurs in humans, with a productive cough, low-grade fever, and an x-ray picture of pulmonary cavitation that resembles tuberculosis. Undoubtedly, many such cases have been misdiagnosed and patients have been hospitalized and given treatment for pulmonary tuberculosis.

In less than 1% of cases of histoplasmosis, a severe, **disseminated** form of the disease develops, with involvement of the reticuloendothelial system and many other sites.

Since *H. capsulatum* is primarily a parasite of the reticuloendothelial system, it is **rarely found extracellularly** in tissue. Therefore, direct and stained smears of clinical material are generally inadequate to demonstrate the fungus. Films of the buffy coat of the blood (white cell layer following sedimentation), bone marrow, cut surfaces of lymph nodes, splenic and liver punch biopsies, sputum, and scrapings should be stained with the Giemsa or Wright's stains and carefully examined with the **oil-immersion objective.** *H. capsulatum* occurs **intracellularly** as small, round or oval yeastlike cells, 2 by 3 μm to 3 by 4 μm in size, with a large vacuole and a crescent or half-moon-shaped mass of red-stained protoplasm at the larger end of the cell. These may be found within the cytoplasm of macrophages and occasionally in the polymorphonuclear leukocytes or free in the tissue.

The following methods for the isolation of *H. capsulatum* from clinical material are those used by the Mycology Branch of the CDC[5] and are highly recommended. **Sputum** specimens should be requested in all cases in which pulmonary or disseminated disease is suspected. A series of **six** early-morning specimens should be collected in sterile bottles; 1- to 10-ml quantities are adequate. If the specimen cannot be inoculated promptly to culture media (**immediate inoculation is recommended**), add 1 ml of a stock solution of chloramphenicol* to 1 to 10 ml of the specimen. It is advisable, however, to inoculate the specimen directly to duplicate sets of SAB agar and BHI blood agar, with and without antibiotics; with antibiotics, pretreatment with chloramphenicol is unnecessary. **Never hold specimens at room temperature;** *Histoplasma* **will not survive.**

Gastric washings should be requested when sputum is unobtainable; a series of three to six specimens is adequate. These are centrifuged and the sediments inoculated to the media previously described. Induced sputum may also be useful. Cerebrospinal fluid is submitted only when cerebral or meningeal involvement is evident. It is also centrifuged, and the sediment is inoculated as described. Citrated blood and bone marrow are of value only in acute disseminated cases. The blood is centrifuged and the buffy coat used for inoculation of media or laboratory animals. Bone marrow is handled identically, without centrifugation.

---

*To prepare a stock solution, suspend 20 mg of chloramphenicol in 10 ml of 95% ethanol and add 90 ml of distilled water. Heat gently to dissolve. The solution is stable. The concentration is approximately 0.2 mg per milliliter of sputum.

Incubate the BHI blood agar without antibiotics at 35 C, and incubate the other media at room temperature. The yeast phase of *H. capsulatum* and other dimorphic fungi does not develop at 35 C on media containing the antibiotics.

The use of **mouse inoculation** is often helpful in the isolation of *H. capsulatum* from clinical specimens. Sputum and gastric washings are liquefied by agitation with glass beads and an equal part of physiologic saline; tissues are ground with alundum and physiologic saline in a tissue grinder. Chloramphenicol is added (0.05 mg per milliliter) for decontamination, and the specimen incubated at 35 C for 1 hour. This is not necessary for buffy coat of blood or cerebrospinal fluid. Two to four mice are inoculated intraperitoneally with 1-ml aliquots of the material. They are killed in 4 weeks, and cultures are made of portions of liver and spleen on cycloheximide media at room temperature and on BHI agar without antibiotics at 35 C. These are examined at intervals for development of colonies of *H. capsulatum*.

On SAB agar at **room temperature,** *H. capsulatum* grows slowly (10 to 14 days) as a raised, **white, fluffy mold,** becoming tan to brown with age.* Microscopically, these cultures show septate, branching hyphae bearing delicate, round to pyriform, smooth **microconidia** (2 to 4 μm in diameter), either on short lateral branches or attached directly by the base (sessile). Although at this stage the culture may be mistaken for *Blastomyces dermatitidis*, further incubation will usually reveal the diagnostic **tuberculate macroconidia** (chlamydospores) (Fig. 34-21, A). These spores are large (7 to 25 μm in diameter), round, thick walled, and covered with knoblike or spikelike projections (tuberculate) that are some-

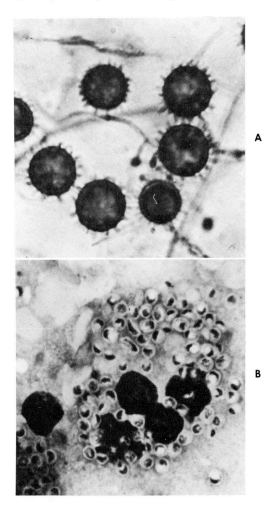

Fig. 34-21. *Histoplasma capsulatum.* **A,** Mycelial phase, showing characteristic tuberculate macroconidia. (1000×.) **B,** Blood smear, showing intracellular oval- to pear-shaped yeastlike cells, deeply stained. (2000×.)

times difficult to see when focusing in only one plane.

On **moist** BHI blood agar incubated at 35 C,* *H. capsulatum* grows slowly as a white to brown, membranous, convoluted (cerebriform), yeastlike colony, resembling that of *Staphylococcus aureus,*

---

*A white (albino) colony, showing only rare, smooth chlamydospores, may overgrow the typical colonies; the latter must be subcultured early to separate the two forms.

***Note:** The yeast phase of the dimorphic fungi is suppressed by cycloheximide; media containing this agent must be incubated at **room** temperature.

or a very mucoid colony. A mycelial or mixed type of growth also may be produced. Although these colonies do not produce typical spores, a higher percentage of isolations results on this medium.* Transfer of these colonies to SAB agar held at room temperature with ready access to air results in development of the typical macroconidia previously described, and this is recommended as a confirmatory procedure.

To convert the mycelial to the typical **yeast phase,** inoculate slants of moist BHI blood agar, seal with Parafilm, and incubate at 35 C for several weeks. Growth appears as dull, white, yeastlike colonies, the contents of which appear microscopically as small (1 to 3 μm), oval, **budding cells** similar to those seen in tissues. Strains in the mycelial phase that cannot readily be converted to the yeast phase culturally may be converted by animal inoculation.

Confirmation of the identification of *H. capsulatum* may also be made by **animal injection.** Inoculate several white mice intraperitoneally with a suspension of yeast phase cells harvested from several tubes of BHI blood agar in 5% hog gastric mucin† or a suspension of a 4- to 6-week mycelial growth ground in saline. One mouse is killed after 2 weeks and the others at weekly intervals thereafter; impression smears of the involved organs (generally liver and spleen) are made and stained with Giemsa (**not** hematoxylin-eosin) stain. Microscopic demonstration of the typical intracellular organisms confirms the identification (Fig. 34-21, *B*). Yeast phase cultures also may be obtained from these tissues by culturing on enriched media incubated at 35 C.

Sepedonium, a saprophytic fungus found on mushrooms, may be confused with *H. capsulatum,* since it produces tuberculate chlamydospores. However, it will not form a yeast phase and is not virulent for animals.

North American blastomycosis

North American blastomycosis is a chronic granulomatous and suppurative disease caused by the **dimorphic** fungus *Blastomyces dermatitidis.** The disease is limited to the continent of North America, extending southward from Canada to the Mississippi Valley, Mexico, and Central America. Some isolated cases also have been reported from Africa.[2] The largest number of cases occurs in the Mississippi Valley region.

Blastomycosis, first described by Gilchrist in 1894, originates as a respiratory infection. Humans probably acquire the infection through inhalation of the spores from the dust of their environment. Recently, a case of blastomycosis was traced to a bag of pigeon manure used for fertilizer.[32] The infection may spread and involve the lungs, bone, and soft tissue. It is not spread from person to person and generally occurs as a sporadic case. Small outbreaks appear to have been related to a common exposure; although blastomycosis is more common in the male, there is no apparent association with occupational exposure.

Material from cutaneous lesions is collected by scraping bits of tissue or taking swabs of pus from the edge of the lesion. Pus from unopened subcutaneous abscesses should be aspirated with a sterile syringe and needle. Sputum, urine, and cerebrospinal fluid also should be examined in suspected systemic blastomycosis.

Such material is prepared for direct microscopic examination by placing it on a slide and pressing it into a thin layer with a coverglass. If the material is opaque, it

---

*Incubation under an increased CO_2 tension appears to improve the growth of *H. capsulatum.*
†Intravenous injection of 0.2 ml of **chilled** (must be kept at 8 to 10 C until the moment of injection) cell suspension (without mucin) into the tail vein of white mice frequently gives better results.[5]

*The perfect stage, *Ajellomyces dermatitidis,* has been described (McDonough, E. S., and Lewis, A. L.: Mycologia **60:**76, 1968).

may be cleared in 10% sodium hydroxide with gentle heating. The specimen is examined under high power, using subdued light. *B. dermatitidis* appears as a large, spherical, **thick-walled** cell, 8 to 20 μm in diameter, usually with a single bud that is connected to the mother cell by a **wide base.** Some walls may be sufficiently thick to give a double-contoured effect.

The infected material should be cultured on BHI blood agar incubated at 35 C and on SAB agar incubated at room temperature. Material likely to be contaminated with bacteria should be inoculated on the aforementioned media with chloramphenicol and cycloheximide and incubated at room temperature. Growth of the yeast phase of *B. dermatitidis* may be suppressed when grown on these media incubated at 35 C. In suspected pulmonary disease, fresh morning sputum specimens should be examined and cultured as previously described.

On SAB agar at **room temperature,** *B. dermatitidis* generally forms a slowly growing, moist, grayish, mealy or prickly colony, which soon develops a **white, cottony,** aerial mycelium, becoming tan (rarely, dark brown to black) with age. Microscopically, these filamentous colonies are made up of septate hyphae bearing small, oval (2 to 3 μm) or pyriform (4 to 5 μm) **conidia** (aleurospores) laterally, near the point of septation (Fig. 34-22, *A*). Older cultures develop conidia 7 to 15 μm in diameter with thickened outer walls that suggest the appearance of chlamydospores.

At 35 C incubation on either SAB agar or BHI blood agar, *B. dermatitidis* grows as a **yeastlike** organism. The fungus develops slowly (1 week) as a creamy, wrinkled, waxy colony (similar to that of *Mycobacterium tuberculosis*) with a verrucose (warty) surface texture, cream to tan in color. Microscopic examination reveals thick-walled, budding, **yeastlike cells** 8 to 20 μm in diameter, resembling those seen in tissues or exudates (Fig. 34-22, *B*).

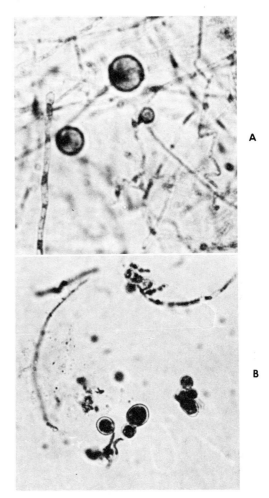

A

B

Fig. 34-22. *Blastomyces dermatitidis.* **A,** Mycelial phase, showing oval microconidia borne laterally on branching hyphae. (1000×.) **B,** Yeast phase, showing thick-walled, oval to round, single-budding, yeastlike cells. (500×.)

To identify an organism as *B. dermatitidis*, it is necessary to **convert** the mycelial phase at 25 C to the tissue phase at 35 C. This is done by subculturing to **fresh** media (BHI blood agar) and incubating at 35 C. Animal inoculation usually is unnecessary if the budding cells are demonstrated in direct smears of the infected material and the dimorphism is shown as described here.

If, however, the two phases are poorly defined, **animal injection** is required. A heavy suspension of either the mycelial or yeast phase is prepared, using physiologic saline, and 1 ml of this suspension is then injected intraperitoneally into several mice. These are killed in 3 weeks; microscopic examination of caseous material from the lesions or peritoneal fluid will show the thick-walled, budding, yeastlike tissue forms of *B. dermatitidis*. If these are not demonstrated, portions of the organs are cultured on BHI blood agar and incubated at 35 C. Yeast phase cultures are generally obtained.

Paracoccidioidomycosis (South American blastomycosis)

Paracoccidioidomycosis is a chronic progressive infection of the mucous membranes of the mouth (portal of entry) and nose and the lymph nodes of the neck, which may give rise to metastatic lesions of the internal organs. It is caused by the **dimorphic** fungus *Paracoccidioides brasiliensis*. The disease is most common in Brazil, although it is seen in many areas, including Mexico, Central America, and Africa.

Materials for **direct examination*** are secured and prepared as described for North American blastomycosis. *P. brasiliensis* appears as large, round to oval, budding cells 8 to 40 μm in diameter, thick walled, refractile, and with characteristic **multiple** buds (Fig. 34-23). Cells with single buds are indistinguishable from those of *B. dermatitidis;* therefore, a search should be made for the diagnostic multiple budding cells. The daughter buds, 1 to 2 μm in diameter, usually are attached to the thick-walled mother cell by narrow connections, giving the whole a **steering wheel** appearance.

Infected material should be cultured on BHI blood agar at 35 C and on SAB agar at room temperature, as in the study of North American blastomycosis. At room temperature the fungus develops as

*Recommended as the most practical diagnostic method.[2]

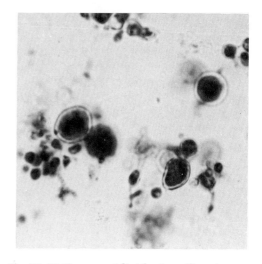

Fig. 34-23. *Paracoccidioides brasiliensis*, yeast phase, showing multiple budding. (400×.)

a very **slowly growing** (2 to 3 weeks), heaped-up folded colony with a short nap of white, velvety mycelium. Microscopically, small, delicate (3 to 4 μm), round or oval conidia may be seen, sessile or on very short sterigmata on septate hyphae. Usually, however, only a fine septate mycelium and chlamydospores are seen.

At 35 C, *P. brasiliensis* grows slowly as a smooth, soft, yeastlike colony, cream to tan in color. These yeastlike colonies, which may appear either verrucous or smooth and shiny, are composed of single cells and multiple budding forms, identical with those seen in tissue and exudates. All cultures should be held at least 4 weeks before being discarded as negative.

As with *B. dermatitidis*, **conversion** of the mycelial to the yeast phase must be demonstrated by subculture and incubation at 35 C. If animal inoculation is required for further confirmation, guinea pigs injected intratesticularly (1 ml of a heavy saline suspension of yeast phase organisms) may be used. The guinea pigs are killed in 8 to 12 days and examined for the presence of the diagnostic multiple budding tissue forms of the fungus in pus from the draining sinuses.

LESS COMMON MYCOSES
Aspergillosis

Aspergilli are among the most common and troublesome contaminants in the laboratory; several are pathogenic and may produce either inflammatory or chronic granulomatous lesions in the bronchi or lungs, often with hematogenous spread to other organs. The external ear, cornea of the eye, nasal sinuses, and other tissues of humans or animals are frequently infected. *Aspergillus fumigatus* is the species most frequently associated with pathologic processes. This may include pulmonary aspergillosis of a severe and invasive type or generalized infection, which is being observed with increasing frequency in debilitated patients receiving antibiotic or corticosteroid therapy and immunosuppressive or antimetabolite drugs.

Since aspergilli are found frequently in cultures of sputum, skin scrapings, and other specimens, it is essential that the fungus be **repeatedly demonstrated** in large numbers in direct smears of the fresh material and **repeatedly isolated** on culture in a patient with an appropriate clinical picture to be considered etiologically significant. Direct microscopic examination of sputum or other infected material may reveal fragments of branched, septate mycelium (3 to 6 μm in width) and, sometimes, conidial heads.

Suspected material should be inoculated on SAB agar without cycloheximide* and incubated at room temperature. *A. fumigatus* grows rapidly (2 to 5 days) and appears first as a flat, white filamentous growth, which rapidly becomes **blue green and powdery** as a result of the production of spores.

Microscopically, *A. fumigatus*† is characterized by branching, septate hyphae,

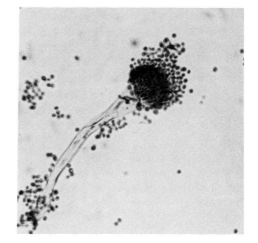

Fig. 34-24. *Aspergillus fumigatus,* showing conidia in chains, arising from a single row of sterigmata on the upper portion of the vesicle. (500×.)

some of which terminally bear a conidiophore that expands into a large, inverted, flask-shaped vesicle (sac) covered with small sterigmata (Fig. 34-24). These sterigmata occur only in a **single row** and around the **upper half** of the vesicle; from their tips are extruded parallel chains of small rough-surfaced, green conidia, giving the whole structure a flaglike appearance. A number of other species may cause serious infection, particularly in the compromised host.*

Animal inoculation is not necessary for the identification of *A. fumigatus.*

Mucormycosis

Mucormycosis is a rare but often fatal disease caused by fungi that produce aseptate mycelia and are ordinarily considered nonpathogenic laboratory contaminants. The genera involved include *Mucor, Rhizopus, Absidia,* and others.

Aspergillus fumigatus may be inhibited by cycloheximide.
†Although *A. fumigatus* is the most common agent in pulmonary aspergillosis, other species, such as *A. flavus, A. niger,* and others, have been incriminated.

*For further aid in identifying other species of aspergilli, the reader is referred to Thom, C. T. and Raper, K. B.: The aspergilli, Baltimore, 1920, The Williams & Wilkins Co., and to Austwick, P. K. C. In Lennette, E. H., Spaulding, E. H., and Truant, J. P., editors: Manual of clinical microbiology, ed. 2, Washington, D.C., 1974, American Society for Microbiology.

The fungus most often enters the nose of susceptible patients, particularly uncontrolled diabetics and patients receiving prolonged antibiotic, corticosteroid, or cytotoxic therapy, and penetrates the arteries, producing thrombosis and death of the segment of tissue normally receiving its blood supply from the affected vessel. Later it invades the veins and lymphatics. The disease assumes cerebral and pulmonary forms and, rarely, intestinal, ocular, and disseminated forms. It is usually fatal.

The diagnosis of mucormycosis is usually made by examination of tissue specimens taken at biopsy or autopsy, in sections of which can be demonstrated broad (4 to 200 μm thick), branching, predominantly **nonseptate hyphae.** The culture of sputum, cerebrospinal fluid, or exudate should be attempted in suspected cases.

On SAB agar incubated at room temperature, *Rhizopus* species produce a rapidly growing (2 to 4 days), coarse, wooly colony, which soon fills the test tube with a loose, grayish mycelium dotted with brown or black sporangia. The fungus is characterized microscopically by a large, broad, **nonseptate,** hyaline mycelium that produces horizontal runners (stolons), which attach at contact points (medium or glass) by rootlike structures called **rhizoids** (Fig. 34-25). From these contact points arise clusters of long stalks, known as **sporangiophores,** the ends of which are terminated in large, round, dark-walled **sporangia** (spore sacs). When mature, these sporangia are filled with spherical hyaline spores. Since *Rhizopus* species are common contaminants, the recovery of this organism in culture is not in itself diagnostic.

On SAB agar at room temperature, *Mucor* species produce a rapidly growing colony that fills the test tube with a white fluffy mycelium, becoming gray to brown

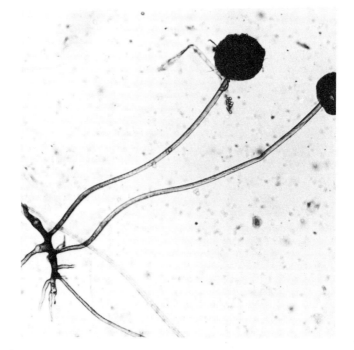

Fig. 34-25. *Rhizopus* species, showing sporangium on long sporangiophore arising from a nonseptate mycelium. Characteristic rhizoids are seen at the base of the sporangiophore. (250×.)

with age. Microscopically, the fungus is characterized by a nonseptate, colorless mycelium **without** rhizoids, the sporangiophores arise singly from stolons and branch profusely, with sporangia containing many spores arising from the apex of each branch. The columellae (the persistent dome-shaped apices of the sporangiophore) are usually well developed and of various shapes but **never hemispherical** as in *Rhizopus*. Empty sporangial sacs, still attached to the conidiophores after release of the spores, may be observed in both genera.

Organisms of the genus *Absidia* are similar to *Rhizopus* except that the sporangiophores arise on stolons at a **point between two nodes** from which rhizoids are formed. The sporangia are pear shaped (10 to 70 μm in diameter) rather than round.

Miscellaneous systemic mycoses

Occasionally, other fungi, such as *Petriellidium boydii* (Rippon, J. W., and Carmichael, J. W.: Mycopathologia **58:** 117-124, 1976) and *Alternaria* (Garau, J., et al.: Ann. Intern. Med. **86:**747-748, 1977) may cause systemic infection.

TREATMENT OF SYSTEMIC MYCOSES

The available agents for treatment of systemic fungal infections include the polyene antibiotic amphotericin B, the synthetic drug flucytosine, iodides, and others. Amphotericin B appears to be the drug of choice in blastomycosis, cryptococcal meningitis, systemic candidiasis, disseminated histoplasmosis, and coccidioidomycosis and is given parenterally. Flucytosine, given by mouth, has proved effective in candidemia and candidal urinary tract infections and may prove useful as an adjunct to amphotericin in some infections, such as cryptococcosis. Iodides may be useful in selected cases of sporotrichosis.

The interested reader is advised to consult the publication by Bennett[9] for further information on therapy.

USE AND INTERPRETATION OF HYPERSENSITIVITY TESTS AND SEROLOGIC REACTIONS IN MYCOTIC INFECTIONS

Conclusive evidence of the presence of a fungal infection is best offered by the demonstration of the fungi, either by direct examination or by cultural procedures, in the exudate or in diseased tissue. However, **indirect** evidence, by the demonstration of a hypersensitive state or an increasing titer of specific antibody, may prove exceedingly helpful.

Intradermal skin tests have been widely used to aid in the diagnosis, treatment, and epidemiologic survey of the systemic mycoses, particularly in blastomycosis, histoplasmosis, and coccidioidomycosis. In the hypersensitive patient, the injection of mycelial or whole-cell extracts of these fungal agents (blastomycin, histoplasmin, coccidioidin, and spherulin*) will elicit a **tuberculinlike response**— redness and induration—within 24 to 48 hours. The significance of this positive reaction is simply that a mycotic infection has occurred at some time during the patient's life. Since cross-reactions occur among these three antigens, some workers have applied all three **simultaneously,** using a separate needle and syringe uncontaminated by previous use with other skin test antigens. Skin testing is of questionable diagnostic value, however, in blastomycosis. Testing with histoplasmin causes an antibody rise, which may prove confusing; immunodiffusion studies may avert the problem. In general, skin testing with histoplasmin also is of little diagnostic value. Results should be interpreted with caution; when negative, they **may be** valuable in ruling out the diagnosis of a systemic infection, particularly in histoplasmosis (patients who are very ill or who have serious underlying illness may be anergic). Skin tests

---

*Standardized skin test antigens may be obtained from Parke-Davis & Co., Eli Lilly & Co., Cutter Laboratories, and Michigan State Health Laboratories.

have been used to demonstrate hypersensitivity to *Paracoccidioides brasiliensis, Sporothrix schenckii, Aspergillus fumigatus,* and other fungi, although standardized reagents are not yet commercially available.

Serum from patients infected with mycotic agents may contain **specific antibodies.** These may be demonstrated in the laboratory by utilizing complement fixation, precipitin, latex or colloidin particle agglutination, hemagglutination, agar gel immunodiffusion, or indirect immunofluorescence procedures.* The need for standardized reference antigen-antibody systems has been stressed.[21] These serologic examinations are useful adjuncts in the diagnosis and prognosis of a mycotic infection. For example, in a case of a pulmonary infection in which no sputum can be obtained, or in a case where it is difficult or impossible to secure infected tissue or exudate, the diagnosis may well depend on the application of serologic tests and the interpretation of their results.

Precipitins and agglutinins appear early in mycotic diseases, and their presence may indicate subsequent cultural examinations. Complement fixation tests using antigens prepared either from the yeast or mycelial form, or both, of a dimorphic fungus are useful in the diagnosis of histoplasmosis and especially in generalized coccidioidomycosis (coccidioidin is a more specific antigen than spherulin, according to Huppert et al.: J. Clin. Microbiol. **6:**33-41, 1977); it is less useful in blastomycosis. A recent cooperative study (Merz, W. G., et al.: J. Clin. Microbiol. **5:**596-603, 1977) confirms the value of serologic tests for the diagnosis of systemic candidiasis, provided that reliable standardized reagents are used. As is true of all other serologic examinations, a dependable diagnostic interpretation ordinarily **cannot be made on a single**

serum specimen. A **sharp rise** (fourfold or greater) or subsequent fall in titer usually corroborates a clinical diagnosis. Because of cross-reactions, all three antigens should be used in the serologic tests performed.

The latex slide agglutination test for detection of **cryptococcal antigen** in spinal fluid and in serum is extremely useful diagnostically and prognostically. Antibody to *C. neoformans* may also be detected by a variety of tests. Serologic tests are very useful in South American blastomycosis and may be helpful in sporotrichosis as well.

Prognostic interpretation of the immunologic and serologic tests is based on the general observation that a patient reacting positively to an intradermal test and not showing a significant serologic titer has a better prognosis than the patient with a negative skin test and a high titer of complement-fixing antibodies. This is particularly true in coccidioidomycosis. It is well to note that serum for histoplasmosis serologic testing should be drawn **before,** or not later than a few days after, an intradermal test is carried out, because the latter may induce the production of antibodies, as noted earlier.

Immunologic procedures may also be used for identification of fungi. Antisera to the H and M precipitinogens of *Histoplasma* have been used successfully for rapid identification of mycelial forms of this organism.[35] A microimmunodiffusion precipitin technique has provided rapid, specific identification of *Coccidioides immitis* cultures (Standard, P. G., and Kaufman, L.: J. Clin. Microbiol. **5:**149-153, 1977). Antibody-coated bacteria in urine have been thought to represent upper urinary tract infection (although some evidence to the contrary has come forth recently). In the case of yeast in urine, however, immunofluorescence is common but is not indicative of upper urinary tract infection, nor is it an indication for therapy.[16]

---

*The reader is referred to an excellent summary of the serology of the systemic mycoses by Kaufman.[23]

REFERENCES

1. Ahearn, D. G.: Identification and ecology of yeasts of medical importance. In Prier, J. E., and Friedman, H., editors: Opportunistic pathogens, Baltimore, 1973, University Park Press.
2. Ajello, L.: Comparative ecology of respiratory mycotic disease agents, Bacteriol. Rev. **31**:6-24, 1967.
3. Ajello, L.: A taxonomic review of the dermatophytes and related species, Sabouraudia **6**: 147-159, 1968.
4. Ajello, L., and Georg, L. K.: In vitro hair cultures for differentiating between atypical isolates of *Trichophyton mentagrophytes* and *Trichophyton rubrum*, Mycopathologia **8**:3-17, 1957.
5. Ajello, L., Georg, L. K., Kaplan, W., and Kaufman, L.: CDC manual for medical mycology, Public Health Service Publication No. 994, Washington, D.C., 1963, U. S. Government Printing Office.
6. Ajello, L., Grant, V.Q., and Gutzke, M. A.: The effect of tubercle bacillus concentration procedures on fungi causing pulmonary mycoses, J. Lab. Clin. Med. **38**:486-491, 1951.
7. Alexopoulous, C. J.: Introductory mycology, ed. 2, New York, 1962, John Wiley & Sons, Inc.
8. Beneke, J. E. In Thomas, B. A., editor: Scope monograph on human mycoses, Kalamazoo, Mich., 1972, The Upjohn Co.
9. Bennett, J. E.: Chemotherapy of systemic mycoses, N. Engl. J. Med. **290**:30-32, 320-323, 1974.
10. Bessey, E. A.: Morphology and taxonomy of fungi, New York, 1950, Blakiston Division, McGraw-Hill Book Co.
11. Bowman, P. I., and Ahearn, D. G.: Evaluation of the Uni-Yeast-Tek kit for the identification of medically important yeasts, J. Clin. Microbiol. **2**:354-358, 1975.
12. Center for Disease Control: Endophthalmitis associated with implantation of intraocular lens prosthesis—United States, Morbid. and Mortal. Weekly Report **25**:369, 1976.
13. Conant, N. F., Smith, D. T., Baker, R. D., and Callaway, J. L.: Manual of clinical mycology, ed. 3, Philadelphia, 1971, W. B. Saunders Co.
14. D'Alessio, D. J., Leavens, L. J., Strumpf, G. B., and Smith, C. D.: An outbreak of sporotrichosis in Vermont associated with sphagnum moss as the source of infection, N. Engl. J. Med. **272**:1054-1058, 1965.
15. Emmons, C. W., Binford, C. H., Utz, J. P. and Kwon-Chung, K. J.: Medical mycology, ed. 3, Philadelphia, 1977, Lea & Febiger.
16. Everett, E. D., Eickhoff, T. C., and Ehret, J. M.: Immunofluorescence of yeast in urine, J. Clin. Microbiol. **2**:142-143, 1975.
17. Georg, L. K.: Personal communication (EGS), 1972.
18. Green, W. O., Jr., and Adams, J. E.: Mycetoma in the United States, Am. J. Clin. Pathol. **42**:75-91, 1964.
19. Haley, L. D.: Diagnostic medical mycology, New York, 1964, Appleton-Century-Crofts.
20. Hopfer, R. L., and Gröschel, D.: Six-hour pigmentation test for the identification of *Cryptococcus neoformans*, J. Clin. Microbiol. **2**:96-98, 1975.
21. Huppert, M., Sun, S. H., and Vukovich, K. R.: Standardization of mycological reagents, Proc. Int. Conf. on Standardization of Diagnostic Materials, Atlanta, 1974, Center for Disease Control, p. 187-194.
22. Huppert, M., Harper, G., Sun, S. H., and Delanerolle, V.: Rapid methods for identification of yeasts, J. Clin. Microbiol. **2**:21-34, 1975.
23. Kaufman, L.: Serodiagnosis of fungal disease. In Lennette, E. H., Spaulding, E. H., and Truant, J. P., editors: Manual of clinical microbiology, ed. 2, Washington, D.C., 1974, American Society for Microbiology.
24. Marks, M. I., Langston, C., and Eickhoff, T. C.: *Torulopsis glabrata*—an opportunistic pathogen in man, N. Engl. J. Med. **283**:1131-1135, 1970.
25. Marks, M. I., and O'Toole, E.: Laboratory identification of *Torulopsis glabrata;* typical appearance on routine bacteriological media, Appl. Microbiol. **19**:184-185, 1970.
26. Merz, W. G., Sandford, G., and Evans, G. L.: Clinical evaluation of the addition of gentamicin to commercially prepared mycological media, J. Clin. Microbiol, **3**:496-500, 1976.
27. Murray, P. R., Van Scoy, R. E., and Roberts, G. D.: Should yeasts in respiratory secretions be identified? Mayo Clin. Proc. **52**:42-45, 1977.
28. Phillips, B. J., and Kaplan, W.: Effect of cetylpyridinium chloride on pathogenic fungi and *Nocardia asteroides* in sputum, J. Clin. Microbiol. **3**:272-276, 1976.
29. Rebell, G. C., and Forster, R. K.: Fungi of keratomycosis. In Lennette, E. H., Spaulding, E. H., and Truant, J. P. editors: Manual of clinical microbiology, ed. 2, Washington, D.C., 1974, American Society for Microbiology.
30. Roberts, G. D., Karlson, A. G., and DeYoung, D. R.: Recovery of pathogenic fungi from clinical specimens submitted for mycobacteriological culture, J. Clin. Microbiol. **3**:47-48, 1976.
31. Roberts, G. D., Wang, H. S., and Hollick, G. E.: Evaluation of the API 20 C microtube system for the identification of clinically important yeasts, J. Clin. Microbiol. **3**:302-305, 1976.
32. Sarosi, G. A., and Serstock, D. S.: Isolation of *Blastomyces dermatitidis* from pigeon manure, Am. Rev. Respir. Dis. **114**:1179-1183, 1976.
33. Segal, E., and Ajello, L.: Evaluation of a new system for the rapid identification of clinically

important yeasts, J. Clin. Microbiol. **4:**157-159, 1976.

34. Staib, F.: Membranfiltration und Negersaat (Guizotia abyssinica)—Nährboden für den *Cryptococcus neoformans*—Nachweis (Braunfarbeffekt), Z. Hyg. Infektionskr. **149:**329-336, 1963.

35. Standard, P. G., and Kaufman, L.: Specific immunological test for the rapid identification of members of the genus *Histoplasma*, J. Clin. Microbiol. **3:**191-199, 1976.

36. Utz, J. P.: Recognition and current management of the systemic mycoses, Med. Clin. North Am. **51:**519-527, 1967.

37. Vanbreuseghem, R.: Early diagnosis, treatment and epidemiology of mycetoma, Rev. Med. Vet. Mycol. **6:**49-60, 1967.

38. Wilson, J. W., and Plunkett, D. A.: The fungous diseases of man, Berkeley, Calif., 1965, University of California Press.

CHAPTER 35

Laboratory diagnosis of parasitic infections

LYNNE SHORE GARCIA

The field of parasitology is often associated with tropical areas; however, many parasitic organisms that infect humans are worldwide in their distribution and occur with some frequency in the temperate zones. Many organisms endemic elsewhere are seen in the United States in persons who have lived or traveled in those areas. Consequently, laboratory personnel should be aware of the possibility that these organisms may be present and should be trained in the performance of appropriate procedures for their recovery and identification.

The identification of parasitic organisms is dependent on morphologic criteria; these criteria are in turn dependent on correct specimen collection and adequate fixation. Improperly submitted specimens may result in failure to find the organisms or in their misidentification. The information presented here should provide the reader with appropriate laboratory techniques and examples of morphologic criteria to permit the correct identification of the more common parasitic organisms.

FECAL SPECIMENS
Collection

The ability to detect and identify intestinal parasites (particularly protozoa) is directly related to the quality of the specimen submitted to the laboratory. Certain guidelines are recommended to ensure proper collection and accurate examination of specimens.[42]

Collection of fecal specimens for intestinal parasites should always be performed **prior to** radiologic studies involving barium sulfate. Due to the excess crystalline material in the stool specimen, the intestinal protozoa may be impossible to detect for at least 1 week after the use of barium. Certain medications may also prevent the detection of intestinal protozoa; these include mineral oil, bismuth, nonabsorbable antidiarrheal preparations, antimalarials, and some antibiotics (tetracyclines, for example). The organisms may be difficult to detect for several weeks after the medication is discontinued.

Fecal specimens should be collected in clean, wide-mouthed containers; most laboratories use a waxed, cardboard half-pint container with a tight-fitting lid. The specimen should not be contaminated with water that may contain free-living organisms. Contamination with urine should also be avoided to prevent destruction of motile organisms in the specimen. All specimens should be identified with the patient's name, physician's name, hospital number if applicable, and the time and date collected. Every fecal specimen represents a potential source of infectious material (e.g., bacteria, viruses, parasites) and should be handled accordingly.

The number of specimens required to demonstrate intestinal parasites will vary depending on the quality of the specimen submitted, the accuracy of the examination performed, and the severity of the infection. For a routine examination for

319

parasites prior to treatment, **a minimum of three fecal specimens** is recommended, two specimens collected from normal movements and one specimen collected after a cathartic, such as magnesium sulfate or Fleet Phospho-Soda. A cathartic with an oil base should **not** be used, and all laxatives are contraindicated if the patient has diarrhea or significant abdominal pain. Stool softeners are inadequate for producing a purged specimen. The examination of at least six specimens will ensure detection of 90% of infections[52]; six are usually recommended when amebiasis is suspected.

The number of specimens to be examined after therapy will vary depending on the diagnosis; however, a series of three specimens collected as previously outlined is usually recommended. A patient who has received treatment for a protozoan infection should be checked 3 to 4 weeks after therapy. Patients treated for helminth infections may be checked 1 to 2 weeks after therapy, and those treated for *Taenia* infections, 5 to 6 weeks after therapy.

Many organisms do not appear in fecal specimens in consistent numbers on a daily basis[36]; thus, collection of specimens on alternate days will tend to yield a higher percentage of positive findings. The series of three specimens should be collected within no more than 10 days, and a series of six within no more than 14 days.

Since the age of the specimen will directly influence the recovery of protozoan organisms, the time the specimen was **collected** should be recorded on the laboratory request form. Freshly passed specimens are mandatory for the detection of trophic amebae or flagellates. Liquid specimens should be examined **within 30 minutes of passage** (not 30 minutes from the time they reach the laboratory), or the specimen should be placed in polyvinyl alcohol fixative (PVA) or another suitable preservative (see following section on preservation of

specimens). Semiformed or soft specimens should be examined within 1 hour **of passage;** if this is not possible, the stool material should be preserved. Although the time limits are not as critical for the examination of a formed specimen, it is recommended that the material be examined on the day of passage. If these time limits cannot be met, portions of the sample should be preserved. Stool specimens should not be held at room temperature, but refrigerated at 3 to 5 C and stored in closed containers to prevent dessication. At this temperature eggs, larvae, and protozoan cysts will remain viable for several days. Fecal specimens should never be incubated or frozen prior to examination. When the proper criteria for collection of fecal specimens are not met, the laboratory should request additional samples.

Collection kit for clinic use

A collection kit that can be used for outpatient laboratory services contains the following items:

1. One 5- to 7-dram brown glass, screw-capped vial containing approximately 10 ml of PVA. Some laboratories also request some fecal material be placed in a separate vial containing 10 ml of 10% formalin, while others may also request a vial containing a portion of the unpreserved sample. Some workers recommend using a vial of Schaudinn's fixative in place of PVA.[53]
2. One half-pint cardboard carton with a tight-fitting lid.
3. Four 6-inch applicator sticks.
4. Instruction sheet containing information on proper collection procedures.
5. Small paper bag in which all supplies can be returned to the laboratory after specimen collection.

Since the examination of three specimens is usually recommended, the kit can be prepared with three vials of preservatives, three cartons, and 10 or 12 applicator sticks. If the patient's clinical status permits, all preserved specimens in the requested series may be collected prior to delivery.

Caution: PVA solution contains a large amount of mercury; for safety reasons and protection of the patient, each vial containing PVA or any type of preservative should have a child-proof cap and be marked POISON. In some areas of the country it may be helpful to label the vials in more than one language, depending on the population using the medical facility.

Collection kit delivered by regular mail service

Specimens may be submitted to a referral laboratory through the regular postal service; however, the following United States postal regulations must be followed: the final kit size may vary according to individual needs; there must be two separate containers—one screw-capped metal container, which is placed inside a screw-capped cardboard container.

The inner mailing tube should contain one vial of PVA (individual laboratories may wish to include a vial of 10% formalin, one of Schaudinn's fixative, or an empty vial) and several applicator sticks. The instruction sheet may be placed around the inner tube, which is then placed in the cardboard mailing container. This kit would be adequate for a single specimen; if three examinations were requested, the patient should submit three separate mailing kits.

The PVA-preserved portion of the specimen may be used for the complete examination,[20] although some laboratories may prefer to use the sample in 10% formalin for the concentration procedure. Other laboratories may prefer to use Schaudinn's fixative for collection. The unpreserved portion of the specimen (many laboratories do not request this sample unless an occult blood procedure is requested) may be examined to determine the specimen type (e.g., liquid, soft, formed).

Preservation

Often, depending on specimen-to-laboratory transportation time, the laboratory work load, and the availability of trained personnel, it may be impossible to examine the specimen within specified time limits. To maintain protozoan morphology and prevent further development of certain helminth eggs and larvae, the fecal specimen should be placed in an appropriate preservative for examination at a later time. A number of preservatives are available; three of these methods—PVA, formalin, and MIF—will be discussed. When selecting an appropriate fixative, it is important to realize the limitations of each. PVA is the only fixative included here from which a permanent stained smear can be easily prepared. The stained smear is extremely important in providing a complete and accurate examination for intestinal protozoa. The other fixatives mentioned will permit the examination of the specimen as a wet mount only, a technique much less accurate than the stained smear for the identification of protozoa.

PVA fixative. Polyvinyl alcohol (PVA) fixative solution is highly recommended as a means of preserving protozoan cysts and trophozoites for examination at a later time. The use of PVA also permits specimens to be shipped by regular mail service to a laboratory for subsequent examination. PVA, which is a combination of modified Schaudinn's fixative and a water-soluble resin, should be used in the ratio of 3 parts PVA to 1 part fecal material. Perhaps the greatest advantage in the use of PVA is that permanent stained slides can be prepared from PVA-preserved material. This is not the case with many other preservatives that permit the specimen to be examined as a wet preparation only, a technique that may not be adequate for the correct identification of protozoan organisms. PVA can be prepared in the laboratory[20] or purchased commercially.* This fixative remains stable for long periods (months to years) when kept in **sealed** containers at room temperature. However, when dispensed in small vials, PVA may become viscous and turn cloudy or white after 2 or 3 months; these vials should be discarded.

The following procedure for preparation of PVA fixative is modified from Brooke and Goldman.[7]

FORMULA

PVA, Elvanol 71-24	10.0 g
95% ethyl alcohol	62.5 ml
Mercuric chloride, saturated aqueous	125.0 ml
Glacial acetic acid	10.0 ml
Glycerin	3.0 ml

PREPARATION

1. Mix liquid ingredients in a 500-ml beaker.
2. Add PVA powder (stirring not recommended).
3. Cover beaker with either a large Petri dish, heavy waxed paper, or foil and allow to soak overnight.
4. Heat solution slowly to 75 C. When this temperature is reached, remove beaker and swirl mixture until a homogeneous, slightly milky solution is obtained (30 seconds).

Formalin preservation (10% formalin)

FORMULA

Formaldehyde (USP)	100 ml
0.85% saline solution	900 ml

---

*Elvanol, grades 71-24, 71-30, and 90-25, can be obtained from E. I. du Pont de Nemours and Company, Electrochemical Department, Niagara Falls, N.Y. (or their local representatives), in a minimum of 50-pound bags. Small quantities may be obtained from Delkote, Inc., 76 S. Virginia Ave., Penns Grove, N.J. 08069, or one can check with a local chemical supply house. When ordering PVA powder, be sure to specify the pretested powder for use in PVA fixative. The PVA powder should be water soluble and of medium viscosity. Prepared liquid PVA (ready for use) can be obtained from Delkote, Inc., and from Medical Chemical Corporation, P.O. Box 445, Santa Monica, Calif. 90404.

PREPARATION. Dilute 100 ml of formaldehyde with 900 ml of 0.85% saline solution (distilled water may be used instead of saline).

Protozoan cysts, helminth eggs, and larvae are well preserved for long periods in 10% formalin. It is recommended that hot formalin (60 C) be used for helminth eggs to prevent further development of the eggs to the infective stage. Formalin should be used in the ratio of at least 3 parts formalin to 1 part fecal material; thorough mixing of the fresh specimen and fixative is necessary to ensure good preservation.

MIF solution. The Merthiolate (thimerosal)-iodine-formalin solution of Sapèro and Lawless[51] can be used as a stain-preservative for most kinds and stages of intestinal parasites and may be helpful in field surveys. Helminth eggs and larvae and certain protozoa can be identified without further staining in wet mounts, which can be prepared immediately after fixation or several weeks later. This type of wet preparation may not be adequate for the diagnosis of all intestinal protozoa, and another fixative preparation may be necessary to provide a permanent stained smear. There are certain disadvantages with the MIF method, which include the instability of the iodine component of the fixative. For a more thorough discussion of this technique, refer to Dunn.[14] This publication also contains a discussion of the concentration procedure that uses MIF-preserved material; this technique is referred to as the MIFC or TIFC method.

Macroscopic examination

The consistency of the stool (formed, semiformed, soft, liquid) may give an indication of the protozoan stages present. When the moisture content of the fecal material is decreased during **normal** passage through the intestinal tract, the trophozoite stages of the protozoa will encyst to survive. **Trophozoites** (motile forms) of the intestinal protozoa are usually found in soft or liquid specimens and occasionally in a semiformed specimen; the **cyst stages** are normally found in formed or semiformed specimens, rarely in liquid stools.

Helminth eggs or larvae may be found in any type of specimen, although the chances of finding any parasitic organism in a liquid specimen will be reduced due to the dilution factor.

Occasionally, adult helminths, such as *Ascaris lumbricoides* or *Enterobius vermicularis* (pinworm), may be seen in or on the surface of the stool. Tapeworm proglottids may also be seen on the surface, or they may actually crawl under the specimen and be found on the bottom of the container. Other adult helminths, such as *Trichuris trichiura* (whipworm), hookworms, or perhaps *Hymenolepis nana* (dwarf tapeworm), may be found in the stool, but usually this occurs only after medication.

The presence of blood in the specimen may indicate a number of things and should always be reported. Dark stools may indicate bleeding high in the gastrointestinal tract, while fresh (bright red) blood most often is the result of bleeding at a lower level. In certain parasitic infections blood and mucus may be present; a soft or liquid stool may be highly suggestive of an amebic infection. These areas of blood and mucus should be carefully examined for the presence of trophic amebae. Occult blood in the stool may or may not be related to a parasitic infection and can result from a number of different conditions. Ingestion of various compounds may give a distinctive color to the stool (iron, black; barium, light to tan to white).

Microscopic examination

The identification of intestinal protozoa and helminth eggs is based on recognition of specific morphologic characteristics; these studies require a good binocular microscope, good light source, and the use of a calibrated ocular micrometer.

The microscope should have 5× and 10× oculars (widefield oculars are often recommended) and three objectives: low power (10×), high-dry (40× to 44×), and oil immersion (97× to 99×). The microscope should be kept covered when not in use, and all lenses should be carefully cleaned with lens paper. The light source should provide light of variable intensity in the blue-white range.

Calibration of microscope

Parasite identification depends on several parameters, one of which is size; any laboratory doing diagnostic work in parasitology should have a calibrated microscope available for precise measurements.

Measurements are made by means of a micrometer disk placed in the ocular of the microscope; the disk is usually calibrated as a line divided in 50 units. Since the divisions in the disk will represent different measurements, depending on the objective magnification used, the ocular disk divisions must be compared with a known calibrated scale, usually a stage micrometer with a scale of 0.1- and 0.01-mm divisions. Specific directions may be found in Garcia and Ash.[20]

Note: After *each* objective power has been calibrated on the microscope, the oculars containing the disk or these objectives **cannot** be interchanged with corresponding objectives or oculars on another microscope. **Each** microscope that will be used to measure organisms must be calibrated as a unit; the original oculars and objectives that were used to calibrate the microscope **must** also be used when an organism is measured.

Direct smears

Normal mixing in the intestinal tract will usually ensure even distribution of helminth eggs or larvae and protozoa. However, examination of the fecal material as a direct smear may or may not reveal organisms, depending on the parasite density. The direct smear is prepared by mixing a small amount of fecal material (approximately 2 mg) with a drop of physiologic saline; this mixture will provide a uniform suspension under a 22- × 22-mm coverslip. Some workers prefer a 1½- × 3-inch glass slide for the wet preparations, rather than the standard 1- × 3-inch slide most often used for the permanent stained smear. A 2-mg sample of fecal material forms a low cone on the end of a wooden applicator stick. If more material is used for the direct mount, the suspension is usually too thick for an accurate examination; any less than 2 mg will result in the examination of too thin a suspension, thus decreasing the chances of finding any organisms. If present, blood and mucus should always be examined as a direct mount. The entire 22- × 22-mm coverslip should be systematically examined using the low-power objective (10×) and low light intensity; any suspect objects may then be examined on high-dry power (43×). The use of the oil immersion objective on mounts of this kind is not recommended unless the coverslip (No. 1 thickness coverslip is recommended when the oil immersion objective is used) is sealed to the slide with a cotton-tipped applicator stick dipped in equal parts· of heated paraffin and petroleum jelly. Many workers feel the use of oil immersion on this type of preparation is impractical, especially since morphologic detail is most easily seen and the diagnosis confirmed with oil immersion examination of the permanent stained smear.

The **direct wet mount** is used primarily to detect motile trophozoite stages of the protozoa. These organisms are very pale and transparent, two characteristics that require the use of low light intensity. Protozoan organisms in a saline preparation will usually appear as refractile objects. If suspect objects are seen on high-dry power, allow at least 15 seconds to detect motility of slow-moving protozoa. Heat applied by placing a hot penny on the edge of a slide may enhance the motility of trophic protozoa.

Note: With few exceptions, protozoan organisms should never be identified on the basis of a wet mount alone. Permanent stained smears should be examined to confirm the specific identification of suspected organisms.

Helminth eggs or larvae and protozoan cysts may also be seen on the wet film, although these forms are more often detected after fecal concentration procedures.

After the wet preparation has been thoroughly checked for trophic amebae, a drop of iodine may be placed at the edge of the coverslip or a new wet mount can be prepared with iodine alone. A weak iodine solution is recommended; too strong a solution may obscure the organisms. Several types of iodine are available: Dobell and O'Connor's, Lugol's, and D'Antoni's (which is included here). Gram's iodine used in bacteriologic work is **not** recommended for staining parasitic organisms.

Modified D'Antoni's iodine[40]
FORMULA

Distilled water	100.0 ml
Potassium iodide (KI)	1.0 g
Powdered iodine crystals	1.5 g

PREPARATION. The potassium iodide solution should be saturated with iodine, with some excess remaining in the bottle. Store in brown, glass-stoppered bottles in the dark. The solution is ready for use immediately and should be decanted into a brown-glass dropping bottle; when the solution lightens, discard and replace with fresh stock. The stock solution remains good as long as an excess of iodine remains on the bottom of the bottle. The iodine solutions will eventually lighten in color and lose their staining strength; fresh solutions should be prepared every 2 to 3 weeks.

Protozoan cysts correctly stained with iodine will contain yellow-gold cytoplasm, brown glycogen material, and paler refractile nuclei. The chromatoidal bodies may not be as clearly visible as they were in the saline mount.

Several staining solutions are available that may be used to reveal nuclear detail in the trophozoite stages. Nair's[45] buffered methylene blue stain is effective in showing nuclear detail when used at a low pH; a pH range of 3.6 to 4.8 will allow more active penetration of the organism with the biologic dye. After 5 to 10 minutes Nair's buffered methylene blue stain will stain the cytoplasm a pale blue and the nuclei a darker blue. Methylene blue (0.06%) in an acetate buffer at pH 3.6 usually gives satisfactory results; the mount should be examined within 30 minutes.

ACETATE BUFFER SOLUTION

Stock solution A: 0.2 M solution of acetic acid (11.55 ml in 1,000 ml of distilled water).

Stock solution B: 0.2 M solution of sodium acetate (16.4 g of $C_2H_3O_2Na$, or 27.2 g of $C_2H_3O_2Na \cdot 3H_2O$ in 1,000 ml of distilled water).

Proportions of A and B for specific pH: mix indicated quantity of stock solutions A and B and dilute with distilled water to a total of 100 ml.

Desired pH	Stock solution A (ml)	Stock solution B (ml)
3.6	46.3	3.7
3.8	44.0	6.0
4.0	41.0	9.0
4.2	36.8	13.2
4.4	30.5	19.5
4.6	25.5	24.5

Concentration procedures. Often a direct mount of fecal material will fail to reveal the presence of parasitic organisms in the gastrointestinal tract. Fecal concentration procedures should be included for a complete examination for parasites; these procedures allow the detection of small numbers of organisms that may be missed using only the direct mount.

A number of concentration procedures are available, which are either **flotation** or **sedimentation** techniques designed to separate the parasitic components from excess fecal debris through differences in specific gravity.[15] A flotation procedure permits the separation of protozoan cysts and certain helminth eggs through the use of a liquid with a high specific gravity. The parasitic elements are recovered in the surface film, while the debris will be found in the bottom of the tube. This technique yields a cleaner preparation than does the sedimentation procedure; however, some helminth eggs (operculated eggs or very dense eggs, such as unfertilized *Ascaris* eggs) and some protozoa do not concentrate well with the flotation method. The specific gravity may be increased, although this may produce more distortion in the eggs and protozoa. Any laboratory that uses a flotation procedure only may fail to recover all the parasites present; to ensure detection of all organisms in the sample, both the surface film and the sediment should be carefully examined.

Note: Directions for any flotation technique must be followed exactly to produce reliable results.

Sedimentation procedures (using gravity or centrifugation) will allow the recovery of all protozoa, eggs, and larvae present; however, the sediment preparation will contain more fecal debris. If a single technique is selected for routine use, the sedimentation procedure is **recommended** as the easiest to perform and least subject to technical error.

FORMALIN-ETHER[20,49] SEDIMENTATION TECHNIQUE

1. Transfer ¼ to ½ teaspoon of fresh stool into 10 ml of 10% formalin in a 15-ml shell vial, unwaxed paper cup, or 16- × 125-mm tube (container may vary depending on individual preferences) and comminute thoroughly. Let stand 30 minutes for adequate fixation.
2. Filter this material (funnel or pointed paper cup with end cut off) through two layers of gauze into a 15-ml centrifuge tube.
3. Add physiologic saline to within ½ inch of the top and centrifuge for 2

minutes at 1,500 rpm (or 1 minute at 2,000 to 2,500 rpm).

4. Decant, resuspend sediment (should have 0.5 to 1 ml sediment) in saline to within ½ inch of the top, and centrifuge again for 2 minutes at 1,500 rpm (or 1 minute at 2,000 to 2,500 rpm). This second wash may be eliminated if the supernatant fluid after the first wash is light tan or clear.

5. Decant and resuspend sediment in 10% formalin (fill tube half full only). If the amount of sediment left in the bottom of the tube is very small, do **not** add ether in step 6; merely add the formalin, then spin, decant, and examine the remaining sediment.

6. Add approximately 3 ml of ether (**do not use near open flames**), stopper, and shake vigorously for 30 seconds. Hold the tube so that the stopper is directed away from your face; remove stopper carefully to prevent spraying of material due to pressure within the tube.

7. Centrifuge for 2 to 3 minutes at 1,500 rpm. Four layers should result: a small amount of sediment in the bottom of the tube, containing the parasites; a layer of formalin; a plug of fecal debris on top of the formalin layer; and a layer of ether at the top.

8. Free the plug of debris by ringing with an applicator stick and decant all the fluid. After proper decanting, a drop or two of fluid remaining on the side of the tube will drain down to the sediment. Mix the fluid with the sediment and prepare a wet mount for examination.

The formalin-ether sedimentation procedure may be used on PVA-preserved material.[20] Steps 1 and 2 will differ as follows:

1. Fixation time with PVA should be at least 30 minutes. Mix contents of PVA bottle (stool-PVA mixture: 1 part stool to 2 or 3 parts PVA) with applicator sticks. Immediately after mixing, pour approximately 2 to 5 ml (amount will vary depending on the viscosity and density of the mixture) of the stool-PVA mixture into a 15-ml shell vial, 16- × 125-mm tube, or such, and add approximately 10 ml physiologic saline.

2. Filter this material (funnel or paper cup with pointed end cut off) through two layers of gauze into a 15-ml centrifuge tube.

Steps 3 through 8 will be the same for both fresh and PVA-preserved material.

Note: Tap water may be substituted for physiologic saline throughout this procedure; however, saline is recommended. Some workers prefer to use 10% formalin for all the rinses (steps 3 and 4).

When examining the sediment in the bottom of the tube:

1. Prepare a saline mount (1 drop sediment and 1 drop saline solution mixed together) and scan the whole 22- × 22-mm coverslip under low power for helminth eggs or larvae.

2. Iodine may then be added to aid in the detection of protozoan cysts and should be examined under high-dry power. If iodine is added prior to low-power scanning, be certain that the iodine is not too strong; otherwise, some of the helminth eggs will stain so darkly that they will be mistaken for debris.

3. Occasionally a precipitate is formed when iodine is added to the sediment obtained from a concentration procedure using PVA-preserved material. The precipitate is formed from the reaction between the iodine and excess mercuric chloride that has not been thoroughly rinsed from the PVA-preserved material. The sediment can be rinsed again to remove any remaining mercuric chloride, or the sediment can be examined as a saline mount without the addition of iodine.

ZINC SULFATE FLOTATION PROCEDURE (modified).[16] Protozoan cysts may be more distorted using this method, and the technique is unsuitable for stool specimens containing large amounts of fatty material. The specific gravity of the zinc sulfate should be 1.18 and should be checked frequently with a hydrometer. If this technique is used to concentrate formalin-preserved material, the specific gravity should be increased to 1.20.

Zinc sulfate solution ($ZnSO_4$, specific gravity 1.18; approximately 330 g of dry crystals in 670 ml of distilled water). A 33% solution will usually approximate the correct specific gravity but may be adjusted to 1.18 by the addition of zinc sulfate or distilled water.

1. Prepare a fecal suspension of ¼ to ½ teaspoon of feces (more if specimen is diarrheal) in 10 to 15 ml of tap water.
2. Filter this material through two layers of gauze into a small tube (Wassermann tube). Fill the tube with tap water to within 2 to 3 mm of the top and centrifuge for 1 minute at 2,300 rpm.
3. Decant supernatant fluid, fill the tube with water, and resuspend the sediment by stirring with an applicator stick. Centrifuge for 1 minute at 2,300 rpm.
4. Decant water, add 2 to 3 ml zinc sulfate solution, resuspend sediment, and fill the tube with zinc sulfate solution to within 0.5 cm from top.
5. Centrifuge for 1 to 2 minutes at 2,500 rpm. Do not "brake" the centrifuge; allow the tubes to come to a stop without interference or vibration.
6. Without removing the tubes from the centrifuge, touch the surface film of the suspension with a wire loop (diameter 5- to 7-mm; loop should be parallel with the surface of the fluid). **Do not go below the surface of the film with the loop.** Add the material in the loop to a slide containing a drop of dilute iodine or saline.

Permanent stained smears

The detection and correct identification of **intestinal protozoa** are frequently dependent on the examination of the permanent stained smear. These slides not only provide the microscopist with a permanent record of the protozoan organisms identified but also may be used for consultations with specialists when unusual morphologic characteristics are found. In view of the number of morphologic variations possible, organisms may be found that are very difficult to identify and do not fit the pattern for any one species.

Many times the smaller protozoan organisms will be seen on the stained smear and missed using only the direct smear and concentration methods. Although an experienced microscopist can occasionally identify certain organisms on a wet preparation, most identifications should be considered tentative until confirmed by the permanent stained slide. For these reasons, **the permanent stain is recommended for every stool sample submitted for a routine examination for parasites.**

A number of staining techniques are available; individual selection of a particular method may depend on the degree of difficulty and amount of time necessary for staining. The older classical method is the long Heidenhain's iron-hematoxylin method; however, for routine diagnostic work most laboratories select one of the shorter procedures, such as the **trichrome** method or one of several methods using iron-hematoxylin. Other procedures are available[21]; however, those included here generally tend to give the best and most reliable results with both fresh and PVA-preserved specimens.

Preparation of fresh material. When the specimen arrives, use an applicator stick or brush to smear a small amount of stool on two clean slides and **immediately** immerse them in Schaudinn's fixative. If the slides are prepared correctly, one should be able to read newsprint through the fecal smear. The smears should fix for a minimum of 30 minutes; fixation time may be decreased to 5 minutes if the Schaudinn's solution is heated to 60 C.

If a liquid specimen is received, mix 3 or 4 drops of PVA with 1 to 2 drops of fecal material on a slide, spread the mixture, and allow the slides to dry for several hours at 35 C or overnight at room temperature. The following fixative solutions may be used.

SATURATED MERCURIC CHLORIDE

Mercuric chloride (HgCl$_2$)	110 g
Distilled water	1,000 ml

Use a beaker as a water bath; boil (use a hood if available) until the HgCl$_2$ is dissolved; let stand until crystals form.

SCHAUDINN'S FIXATIVE (STOCK SOLUTION)

Saturated aqueous solution of HgCl$_2$	600 ml
95% ethyl alcohol	300 ml

Immediately prior to use add 5 ml glacial acetic acid per 100 ml stock solution.

Preparation of PVA-preserved material. Stool specimens preserved in PVA should be allowed to fix at least 30 minutes. After fixation the sample should be thoroughly mixed and a small amount of the material poured onto a paper towel to absorb the excess PVA. This is an important step in the procedure; allow the PVA to soak into the paper towel for 2 to 3 minutes before preparing the slides. With an applicator stick apply some of the stool material from the paper towel to two slides and let them dry for several hours at 37 C or overnight at room temperature. The PVA-stool mixture should be spread to the edges of the glass slide; this will cause the film to adhere to the slide during staining. It is also important to thoroughly dry the slides to prevent the material from washing off during staining.

Trichrome stain. This stain was originally developed by Gomori[22] for tissue differentiation and was adapted by Wheatley[69] for intestinal protozoa. It is an uncomplicated procedure that will pro-

duce well-stained smears from both fresh and PVA-preserved material.

FORMULA

Chromotrope 2R	0.6 g
Light green SF	0.3 g
Phosphotungstic acid	0.7 g
Acetic acid (glacial)	1.0 ml
Distilled water	100.0 ml

PREPARATION. The stain is prepared by adding 1.0 ml glacial acetic acid to the dry components. Allow the mixture to stand for 15 to 30 minutes to "ripen," then add 100 ml of distilled water. This preparation gives a highly uniform and reproducible stain; the stain should be purple.

PROCEDURE

1. Prepare fresh fecal smears or PVA smears as described.
2. Place in 70% ethyl alcohol for 5 minutes.* (Step 2 may be eliminated for PVA smears.)
3. Place in 70% ethyl alcohol plus D'Antoni's iodine (dark reddish brown) for 2 to 5 minutes.
4. Place in two changes of 70% ethyl alcohol: one for 5 minutes* and one for 2 to 5 minutes.*
5. Place in trichrome stain solution for 10 minutes.
6. Place in 90% ethyl alcohol, acidified (1% acetic acid) for up to 3 seconds (**do not leave the slides in this solution too long**).
7. Dip once in 100% ethyl alcohol.
8. Place in two changes of 100% ethyl alcohol for 2 to 5 minutes each.*
9. Place in two changes of xylene or toluene for 2 to 5 minutes each.*
10. Mount in Permount or some other mounting medium; use a No. 1 thickness coverglass.

The trichrome stain can be used repeatedly, and stock solution may be added to the dish when the volume is decreased. Periodically, the staining strength can be restored by removing the lid and allowing the 70% alcohol carried over from the preceding dish to evapo-

---

*Slides can be held several hours or overnight.

rate. Each lot number or batch of stain (either purchased commercially or prepared in the laboratory) should be checked to determine the optimum staining time, which is usually a few minutes longer for PVA-preserved material.

The 90% acidified alcohol is used as a destaining agent that will provide good differentiation; however, prolonged destaining (more than 3 seconds) may result in a poor stain. To prevent continued destaining, the slides should be quickly rinsed in 100% alcohol and then dehydrated through two additional changes of 100% alcohol.

INTERPRETATION OF STAINED SMEARS. Many problems in interpretation may arise when poorly stained smears are examined; these smears are usually the result of specimen collection and submission requirements not being followed or inadequate fixation. An old specimen or inadequate fixation may result in organisms that fail to stain or which appear as pale pink or red objects with very little internal definition. This type of staining reaction may occur with *Entamoeba coli* cysts, which require a longer fixation time; mature cysts in general (need additional fixation time) may not be as well stained as immature cysts. Degenerate forms or those that have been understained or destained too much may stain pale green.

When the smear is well fixed and correctly stained, the background debris will be green and the protozoa will have a blue-green to purple cytoplasm with red or purple-red nuclei and inclusions. The differences in colors between the background and organisms provide more contrast than in hematoxylin-stained smears.

Helminth eggs and larvae usually stain dark red or purple; they are often distorted and difficult to identify. White blood cells, macrophages, tissue cells, yeast cells, and other artifacts still present diagnostic problems, since their color range on the stained smear will approximate that of the parasitic organisms.

Iron-hematoxylin stain. Although the original method produces excellent results,[37] most laboratories that use an iron-hematoxylin stain select one of the shorter methods. A number of procedures are available; both of those presented here can be used with either fresh or PVA-preserved material. Both background debris and the organisms will stain gray-blue to black, with the cellular inclusions and nuclei appearing darker than the cytoplasm.

The method described by Spencer and Monroe[58] is a bit longer than the trichrome procedure. Although the slides do not require destaining, decolorizing in 0.5% hydrochloric acid after a longer initial staining time may provide better differentiation.

SOLUTION I. Hematoxylin, 10 g in 1,000 ml of absolute ethyl alcohol. Keep stain in stoppered flask and allow to ripen in sunlight for at least a week.

SOLUTION II

Ferrous ammonium sulfate	10 g
Ferric ammonium sulfate	10 g
HCl, concentrated	10 ml
Distilled water	1000 ml

WORKING SOLUTION. Mix equal parts of solutions I and II; this working solution will last for about 7 days.

PROCEDURE

1. Prepare fresh fecal smears or PVA-preserved smears as previously described.
2. Place in 70% ethyl alcohol for 5 minutes.*
3. Place in 70% ethyl alcohol plus D'Antoni's iodine (dark reddish brown) for 2 to 5 minutes.
4. Place in 70% ethyl alcohol for 5 minutes.*
5. Wash in running tap water for 10 minutes.
6. Place in working solution of iron-

---

*Slides can be held several hours or overnight.

hematoxylin staining solution for 4 to 5 minutes.

7. Wash in running tap water for 10 minutes.
8. Place in 70% ethyl alcohol for 5 minutes.*
9. Place in 95% ethyl alcohol for 5 minutes.
10. Place in two changes of 100% ethyl alcohol for 5 minutes each.*
11. Place in two changes of xylene or toluene for 5 minutes each.*
12. Mount in Permount or some other mounting medium; use a No. 1 thickness coverglass.

Another iron-hematoxylin method, described by Tompkins and Miller,[65] includes the use of phosphotungstic acid as a destaining agent. This procedure also gives good, reproducible results.

PROCEDURE

1. Prepare fresh fecal smears or PVA-preserved smears as previously described.
2. Place in 70% ethyl alcohol D'-Antoni's iodine (dark reddish brown) for 2 to 5 minutes.
3. Place in 50% ethyl alcohol for 3 minutes.
4. Wash in running tap water for 3 minutes.
5. Place in 4% ferric ammonium sulfate mordant for 5 minutes.
6. Wash in tap water for 1 minute.
7. Place in 0.5% aqueous hematoxylin for 2 minutes.
8. Wash in tap water for 1 minute.
9. Place in 2% aqueous phosphotungstic acid for 2 to 5 minutes.
10. Wash in running tap water for 10 minutes.
11. Place in 70% ethyl alcohol (plus a few drops of saturated aqueous lithium carbonate) for 3 minutes.
12. Place in 95% ethyl alcohol for 5 minutes.
13. Place in two changes of 100% ethyl alcohol for 5 minutes each.*

14. Place in xylene or toluene for 5 minutes.*
15. Mount in Permount or some other mounting medium; use a No. 1 thickness coverglass.

General information

The most important step in preparing a well-stained fecal smear is adequate fixation of a specimen that has been submitted within specified time limits. To ensure best results, the acetic acid component of Schaudinn's fixative should be added just prior to use; fixation time (room temperature) may be extended overnight with no adverse effects on the smears.

After fixation it is very important to completely remove the mercuric chloride residue from the smears. The 70% alcohol-iodine mixture removes the mercury complex; the iodine solution should be changed often enough (at least once a week) to maintain a dark reddish brown color. If the mercuric chloride is not completely removed, the stained smear may contain varying amounts of highly refractive granules, which may prevent finding or identifying any organisms present.

Good results in the final stages of dehydration (100% alcohol) and clearing (xylene) depend on the use of **fresh** reagents. It is recommended that solutions be changed at least weekly and more often if large numbers of slides (10 to 50 per day) are being stained. Stock containers and staining dishes should have well-fitting lids to prevent evaporation and absorption of moisture from the air. If the clearing agent turns cloudy on addition of the slides from 100% alcohol, there is water in the solution. When clouding occurs, **immediately** return the slides to 100% alcohol, replace all dehydrating and clearing agents with fresh stock, and continue with the dehydration process.

---

*Slides can be held several hours or overnight.

*Slides can be held several hours or overnight.

ADDITIONAL PROCEDURES
Sigmoidoscopy material

When repeated fecal examinations fail to reveal the presence of *Entamoeba histolytica,* material obtained from sigmoidoscopy may be valuable in the diagnosis of amebiasis. However, this procedure does **not** take the place of routine fecal examinations; a series of at least three (six is preferable) fecal specimens should be submitted for each patient having a sigmoidoscopy examination. Material from the mucosal surface should be obtained by **aspiration or scraping,** not with a cotton-tipped swab. If swabs must be used, most of the cotton should be removed (leave just enough to safely cover the end) and should be tightly wound to prevent absorption of the material to be examined.

The specimen should be processed and examined **immediately;** the number of techniques used will depend on the amount of material obtained. If the specimen is sufficient for both wet preparations and permanent stained smears, proceed as follows. The direct mount should be examined immediately for the presence of moving trophozoites; it may take time for the organisms to become acclimated to this type of preparation, thus motility may not be obvious for several minutes. Care should be taken not to confuse protozoan organisms with macrophages or other tissue cells; any suspect cells should be confirmed with the use of the permanent stained slide. The smears for permanent staining should be prepared at the same time the direct mount is made by gently smearing some of the material onto several slides and **immediately** placing them into Schaudinn's fixative. The slides can then be stained by any of the techniques mentioned for routine fecal smears. If the material is bloody, contains a lot of mucus, or is a "wet" specimen, one or two drops of the sample can be mixed with three or four drops of PVA right on the slide. Allow the smears to dry (overnight if possible) prior to staining.

Duodenal contents

In some instances, repeated fecal examinations may fail to confirm a diagnosis of *Giardia lamblia* and *Strongyloides stercoralis* infections. Since these two parasites are normally found in the duodenum, the physician may submit duodenal drainage fluid to the laboratory for examination. The specimen should be submitted without preservatives and should be received and examined within 1 hour after being taken. The amount of fluid may vary. It should be centrifuged and the sediment examined as wet mounts for the detection of motile organisms. Several mounts should be prepared and examined; due to the dilution factor, the organisms may be difficult to recover using this technique.

Another convenient method of sampling duodenal contents, which eliminates the necessity for intubation, is the use of the Entero-Test.[3] This device consists of a gelatin capsule containing a weighted, coiled length of nylon yarn. The end of the line protrudes through the top of the capsule and is taped to the side of the patient's face. The capsule is then swallowed, the gelatin dissolves, and the weighted string is carried by peristalsis into the duodenum. After approximately 4 hours, the string is recovered and the bile-stained mucus attached to the string is examined as a wet mount for the presence of organisms. This type of specimen should also be examined immediately after the string is recovered.

Cellophane-covered thick smear

This procedure, which is commonly referred to as the Kato thick-smear technique, was originally developed in Japan by Kato and Miura.[30] It can be used for examination for **helminth eggs** but is not suitable for larvae or protozoa; it is not recommended for examination of stool containing large amounts of fiber or gas.

PROCEDURE

1. Cut wettable cellophane of medium thickness (30 to 50 μm) in 22- $\times$ 30-mm strips and soak for 24 hours or longer in a mixture of 100 parts glycerine, 100 parts water, and 1 part 3% aqueous malachite green.
2. Place 50 to 60 mg of feces on a clean slide and cover with a strip of cellophane prepared as above.
3. Turn the slide upside down on paper towels and press to spread the fecal material to the edges.
4. Reverse the slide and allow to dry at 40 C for 30 minutes or at room temperature for 1 hour.
5. Examine the slide under low power; higher magnification can be used if necessary.

As the film dries, the fecal debris will clear more rapidly than the helminth eggs; however, with time the eggs will also clear, making accurate identification impossible. Thus the optimum drying time must be determined; overdrying will cause distortion in many of the delicate eggs.

Martin and Beaver[41] modified the technique and concluded that egg counts made using their technique provided reliable results for the quantitative diagnosis of helminth infections.

Estimation of worm burdens

Circumstances may arise when it is helpful to know the degree of infection in a patient or perhaps to follow the effectiveness of therapy. In certain helminth infections that have little clinical significance, the patient may not be given treatment if the numbers of parasites are small. The **parasite burden** may be estimated by counting the number of eggs passed in the stool. In addition to the procedures presented here, the direct smear method of Beaver[4,5] has proved to be very helpful in estimating the parasite burden.

Stoll's dilution egg-count technique. This technique developed by Stoll and Hausheer[60] has been widely used to estimate the number of adult worms present in several helminth infections, specifically, hookworm, *Ascaris,* and *Trichuris.* The value of this type of procedure is based on repeated egg counts to detect changes in the numbers present.

PROCEDURE

1. Save entire 24-hour stool specimen and determine weight in grams.
2. Weigh out accurately 4 g of feces.
3. Place feces in calibrated bottle or large test tube and add sufficient 0.1N sodium hydroxide to bring volume to 60 ml.
4. Add a few glass beads and shake mixture vigorously to make a uniform suspension. If specimen is hard, the mixture may be placed in a refrigerator overnight, before shaking, to aid in its comminution.
5. With a pipet, quickly remove 0.15 ml of the suspension and drain onto a slide.
6. Do not use coverglass; place slide on mechanical stage and count all the eggs.
7. Multiply egg count by 100 to obtain the number of eggs per gram of feces and by weight of 24-hour specimen to get total number of eggs per 24 hours.
8. The estimate (eggs per gram) obtained will vary depending on the consistency of the feces. The following correction factors should be used to convert the estimate to a formed-stool basis: mushy formed, $\times 1.5$; mushy, $\times 2$; mushy diarrheal, $\times 3$; diarrheal, $\times 4$; watery, $\times 5$.

The following figures indicate the correlation between egg counts and need for therapy. *Trichuris trichiura* and hookworm are generally the only helminth infections where the egg count may determine whether the patient will receive therapy; low egg counts usually correlate with a lack of clinical symptoms in patients infected with these parasites. The effectiveness of therapy for any helminth infection may be checked by doing repeated egg counts.

Note: The presence of even one *Ascaris* is potentially dangerous; when irritated, the parasite tends to migrate while in the gastrointestinal tract, and these migratory habits may cause severe clinical symptoms in the patient.

With *Trichuris trichiura*, approximately 30,000 eggs per gram indicate the presence of several hundred worms. This type of worm burden will usually cause definite symptoms.

With hookworm, approximately 2,500 to 5,000 eggs per gram usually indicate a clinically significant infection.

Recovery of larval-stage nematodes

Nematode infections that give rise to larval stages, which hatch either in the soil or in tissues, may be diagnosed by using specific culture techniques designed to concentrate the larvae. These procedures are used in hookworm, *Strongyloides*, and trichostrongyle infections. Some of these techniques, which permit the recovery of infective-stage larvae, may be helpful, since the eggs of many species are identical and specific identifications are based on larval morphology.

Harada-Mori filter paper strip culture. As a means of detecting light infections and providing specific identifications, the Harada-Mori filter paper strip culture technique is very useful. The method was originally introduced by Harada and Mori[24] in 1955 and has been modified by several workers.[25] Fecal specimens should not be refrigerated prior to culture; some of the nematodes are susceptible to cold and will not undergo further development. Since infective-stage larvae may be recovered from the culture system, it is recommended that gloves be worn to handle the filter paper strip and other equipment.

PROCEDURE

1. To each 15-ml centrifuge tube, add approximately 3 to 4 ml distilled water.
2. In the center of each filter paper strip (⅜ inch × 5 inches), smear, in a relatively thin film, approximately 0.5 to 1 g of feces.
3. The identification of the specimen can be written in pencil on the piece of filter paper between the waterline and the fecal material.
4. Insert the strip in the tube so that the end of the filter paper strip, usually cut so that it is slightly tapered, is near the bottom of the tube. Caps are not required for the tubes.
5. Maintain tube in rack at 24 to 28 C and add water to original level as it is needed. Usually there is rapid evaporation over the first day or two, and then the culture becomes stabilized.
6. The capillary flow of water up the filter paper strip keeps the feces moist. Soluble elements in the feces will be carried out of the fecal mass and accumulate as a dark area at the top of the paper.
7. Tubes should be kept for approximately 10 days, but infective larvae may be found any time after the fifth day.
8. Using a glass pipet, a small amount of fluid from the bottom of the tube may be withdrawn; larvae will generally be alive and very active. They may be heat killed within the tube or after removal to the slide; iodine may also be used to kill larvae.
9. Examination of the larvae for typical morphologic features will reveal either hookworm, *Strongyloides*, or *Trichostrongylus*.

Baermann technique. When the stools are repeatedly negative in a patient suspected of having strongyloidiasis, the Baermann technique may be helpful in recovering larvae. The apparatus is designed to allow the larvae to migrate from the fecal material through several layers

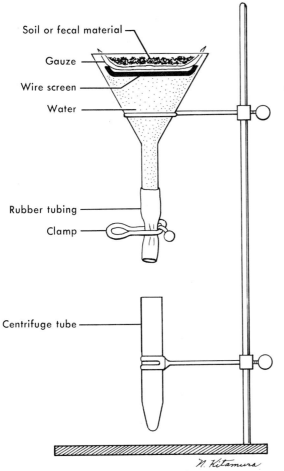

Soil or fecal material

Gauze

Wire screen

Water

Rubber tubing

Clamp

Centrifuge tube

N. Kitamura

Fig. 35-1. Baermann apparatus.

of damp gauze into water, which is centrifuged, thus concentrating the larvae in the bottom of the tube (Fig. 35-1). Specimens for this technique should be collected after a mild saline cathartic, not a stool softener.

PROCEDURE
1. Attach rubber tubing with pinch clamp to bottom of 6-inch funnel. Fill funnel with water. Place wire gauze, with one or two layers of gauze padding, in the funnel.
2. Place a large amount of fecal material on the gauze so that it is covered with water. If fecal matter is too firm, it should be broken up slightly.
3. Allow apparatus to stand for at least

2 hours. Draw off 10 ml of fluid by releasing the pinch clamp, spin down in a centrifuge, and examine the sediment for larvae.

Hatching procedure for schistosome eggs. When schistosome eggs are recovered from either urine or stool, they should be carefully examined to determine viability. The presence of living miracidia within the eggs indicates an active infection, which may require therapy. The viability of the miracidium larvae can be determined in two ways: (1) the cilia on the flame cells (primitive excretory cells) may be seen on high-dry power and are usually actively moving, and (2) the larvae may be released from the eggs with the use of a hatching procedure. The eggs will usually hatch within several hours when placed in 10 volumes of dechlorinated or spring water. The eggs, which are recovered in the urine, are easily obtained from the sediment and can be examined under the microscope to determine viability.

PROCEDURE
1. Thoroughly mix a stool specimen in saline and strain through two layers of gauze.
2. Allow the material to settle, pour off the supernatant fluid, and repeat the process.
3. Decant the saline, add spring water, and pour the solution into a 500- or 1,000-ml Erlenmeyer flask or side-arm flask. Add enough fluid that the level rises into the neck of the flask.
4. Cover the flask with foil or black paper, leaving 1 to 2 ml of fluid in the neck of the flask exposed to light.
5. Leave the flask at room temperature in subdued light for 2 to 3 hours.
6. Place a bright light at the side of the flask directly opposite and close to the surface of exposed water.
7. The miracidia will come to the illuminated portion of the fluid and can be identified with a hand lens.

Cellulose tape preparations[8,23] *Enterobius vermicularis* is a roundworm that is

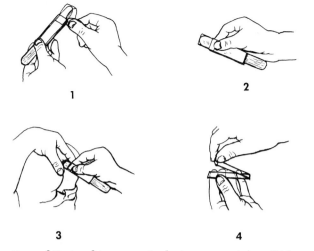

Fig. 35-2. Collection of *Enterobius vermicularis* eggs using cellulose tape method. (Illustration by Nobuko Kitamura.)

worldwide in distribution. It is very common in children and is known as the **pinworm** or the seatworm. The adult female migrates from the anus during the night and deposits her eggs on the perianal area. Since the eggs are deposited outside the gastrointestinal tract, examination of a stool specimen may produce negative results. Although some laboratories use the anal swab technique, pinworm infections are most frequently diagnosed by using the cellulose tape method for egg recovery (Fig. 35-2). Occasionally the adult female may be found on the surface of a formed stool or on the cellulose tape. Specimens should be taken in the morning **before bathing** or going to the bathroom. A series of at least four consecutive negative tapes should be obtained before ruling out infection with pinworms.

PROCEDURE

1. Place a strip of cellulose tape on a microscope slide, starting ½ inch from one end and, running toward the same end, continuing around this end across the slide; tear off the strip even with the other end. Place a strip of paper, ½ inch × 1 inch, between the slide and the tape at the end where the tape is torn flush.

2. To obtain the sample from the perianal area, peel back the tape by gripping the label, and with the tape looped adhesive side outward over a wooden tongue depressor held against the slide and extended about 1 inch beyond it, press the tape firmly against the right and left perianal folds.

3. Spread the tape back on the slide, adhesive side down.

4. Place name and date on the label. **Note:** Do not use Magic transparent tape, but regular clear cellulose tape.

5. Lift one side of the tape and apply 1 **small** drop of toluene or xylene; press the tape down onto the glass slide.

6. The tape is now cleared; examine under low power and low illumination. The eggs should be visible if present; they are described as football shaped with one slightly flattened side.

Identification of adult worms[20]

Most adult worms or portions of worms that are submitted to the laboratory for identification are either *Ascaris lumbricoides*, *Enterobius vermicularis*, or segments of tapeworms. The adult worms present no particular problems in identification; however, identification of the *Taenia* species tapeworms is dependent on the gravid proglottids, which contain the fully developed uterine branches. Identification as to species is based on the number of lateral uterine branches that arise from the main uterine stem in the gravid proglottids. Often the uterine branches are not clearly visible; one technique that can be used is the injection of the branches with India ink, which will allow them to be easily seen and counted. With a 1-ml syringe and 25- to 26-gauge needle, India ink can be injected into the central stem or into the uterine pore, filling the uterine branches with ink. The proglottid can then be pressed between two slides, held up to the light, and the branches counted.

Note: Caution should be used in handling proglottids of *Taenia* species, since the eggs of *T. solium* are infective for humans.

UROGENITAL SPECIMENS

The identification of *Trichomonas vaginalis* is usually based on the examination of wet preparations of vaginal and urethral discharges and prostatic secretions. These specimens are diluted with a drop of saline and examined under low power with reduced illumination for the

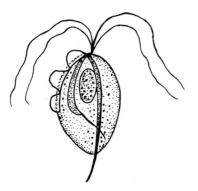

Fig. 35-3. *Trichomonas vaginalis* trophozoite. (Illustration by Nobuko Kitamura.)

presence of actively motile organisms; urine sediment can be examined in the same way. As the jerky motility of the organisms begins to diminish, it may be possible to observe the undulating membrane, particularly under high-dry power (Fig. 35-3). Stained smears are usually not necessary for the identification of this organism; often the number of false-positive and false-negative results reported on the basis of stained smears would strongly suggest the value of confirmation (i.e., observation of the motile organisms).[46]

SPUTUM

When sputum is submitted for examination it should be "deep sputum" from the lower respiratory passages, not a specimen that is mainly saliva. The specimen should be collected early in the morning, before eating or brushing teeth, and immediately delivered to the laboratory. Sputum is usually examined as a saline or iodine wet mount under low and high-dry microscope power. If the quantity is sufficient, the formalin-ether sedimentation technique can be used. A very mucoid or thick sputum can be centrifuged after the addition of an equal volume of 3% sodium hydroxide. With any technique, the sediment should be carefully examined for the presence of brownish spots or "iron filings," which may be *Paragonimus* eggs.

Note: Care should be taken not to confuse *Entamoeba gingivalis*, which may be found in the mouth and might be seen in the sputum, with *E. histolytica* from a pulmonary abscess. *E. gingivalis* will contain ingested PMNs; *E. histolytica* will not.

ASPIRATES

The diagnosis of certain parasitic infections may be based on procedures using aspirated material. These techniques include microscopic examination, animal inoculation, or culture.

Examination of aspirated material from

lung or liver abscesses may reveal the presence of *Entamoeba histolytica;* however, the demonstration of these parasites is often extremely difficult for several reasons. Hepatic abscess material taken from the peripheral area, rather than the necrotic center, may reveal organisms, although they may be trapped in the thick pus and not exhibit any motility. The Amoebiasis Research Unit, Durban, South Africa, has recommended using proteolytic enzymes to free the organisms from the aspirate material.[35]

PROCEDURE

1. A minimum of two separate portions of exudate should be removed. The first portion, usually yellowish white, seldom contains organisms. Later portions, which are reddish, are more likely to contain amebae. The best material to examine is the final portion from the wall of the abscess.

2. Ten units of the enzyme streptodornase are added to each 1 ml of thick pus; this mixture is incubated

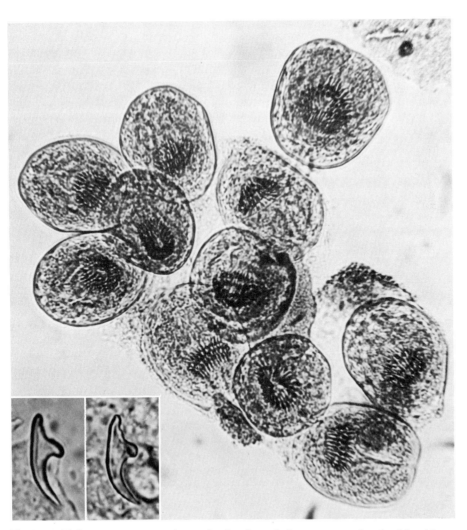

Fig. 35-4. *Echinococcus granulosus,* hydatid sand. *Inset,* two individual hooklets.

for 30 minutes at 37 C, with repeated shaking.

3. Centrifuge the mixture at 1,000 rpm for 5 minutes. The sediment may be examined microscopically as wet mounts or used to inoculate culture media. Some of the aspirate can be mixed directly with PVA on a slide and examined as a permanent stained smear.

Aspiration of cyst material (usually liver or lung) for the diagnosis of hydatid disease is usually performed when open surgical techniques are used for cyst removal. The aspirated fluid is submitted to the laboratory and examined for the presence of hydatid sand (scolices) or hooklets; the absence of this material does not rule out the possibility of hydatid disease, since some cysts are sterile (Fig. 35-4).

Material from lymph nodes, spleen, liver, bone marrow, or spinal fluid may be examined for the presence of trypanosomes or leishmanial forms. Part of the specimen should be examined as a wet preparation to demonstrate motile organisms. Impression smears can also be prepared and stained with Giemsa stain (see Detection of Blood Parasites). This type of material can also be cultured; specific details are presented in the section Culture Techniques.

Specimens obtained from cutaneous ulcers should be aspirated from below the ulcer bed rather than the surface; this type of sample will be more likely to contain the intracellular leishmanial organisms and will be free of bacterial contamination. A few drops of sterile saline may be introduced under the ulcer bed (through uninvolved tissue) through needle (25 gauge) and syringe (1 or 2 ml). The aspirated fluid should be examined as stained smears and should be inoculated into appropriate media (see Culture Techniques).

SPINAL FLUID

Cases of primary meningoencephalitis are infrequently seen, but the examina-

tion of spinal fluid may reveal the causative agent, *Naegleria fowleri,* if present. The spinal fluid may range from cloudy to purulent (with or without red blood cells). The cell count ranges from a few hundred to over 20,000 white blood cells per milliliter, primarily neutrophils; the failure to find bacteria in this type of spinal fluid should alert one to the possibility of primary meningoencephalitis. Motile amebae may be found in unstained spinal fluid; however, one should be very careful not to confuse organisms with various blood and tissue cells that may also be motile. Isolation of these organisms from tissues or soil samples may be accomplished with the use of the *Acanthamoeba* medium developed by Culbertson, and co-workers.[10]

BIOPSY MATERIAL

In some cases biopsy material may be used to confirm the diagnosis of certain parasitic infections. Most of these specimens will be sent for routine tissue processing (fixation, embedding, sectioning, and staining).[54] However, fresh material may be sent directly to the laboratory for examination; it is imperative that these specimens be received immediately to prevent deterioration of any organisms present.

Pneumocystis carinii is usually classified with the sporozoa and is recognized as an important cause of pulmonary infection in patients who are immunosuppressed as a result of therapy or in patients with congenital or acquired immunologic disorders. The organisms can be demonstrated in stained impression smears of lung material obtained by open or brush biopsy. *Pneumocystis* can also be seen in stained smears of tracheobronchial aspirates, although preparations of lung tissue are more likely to reveal the organisms. Sputum specimens are generally considered unacceptable for the recovery of *Pneumocystis.* The recommended stain is Gomori's silver methenamine,[37] which clearly outlines *Pneumocystis* organisms (or various fungi) in dark

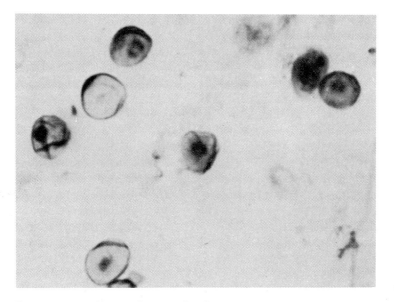

Fig. 35-5. *Pneumocystis carinii* from tracheobronchial aspirate; stained with methenamine silver. (From Markell, E. K., and Voge, M.: Medical parasitology, ed. 4, Philadelphia, 1976, W. B. Saunders Co.)

brown or black (Fig. 35-5). The staining procedure is relatively complex, and positive control slides should be included with each patient specimen to ensure accurate interpretation of the smears.

Skin biopsies for the diagnosis of cutaneous amebiasis or cutaneous leishmaniasis should be submitted for tissue processing. A portion of the tissue for the diagnosis of leishmaniasis can be teased apart with sterile needles and inoculated into appropriate culture media (see Culture Techniques).

The diagnosis of onchocerciasis (*Onchocerca volvulus*) may be confirmed by the examination of "skin snips," very thin slices of skin, which are teased apart in saline to release the microfilariae.

Biopsy specimens taken from lymph nodes are submitted for routine tissue processing; impression smears can also be prepared and stained with Giemsa stain (see Detection of Blood Parasites).

The diagnosis of trichinosis is usually based on clinical findings; however, confirmation may be obtained by the examination of a muscle biopsy. The encapsulated làrvae can be seen in small pieces of fresh tissue, which are pressed between two slides and examined under low power of the microscope. At necropsy the larvae are most abundant in the diaphragm, masseter muscle, or tongue. Larvae can also be recovered from tissue that has undergone digestion in artificial digestive fluid at 37 C.[20]

Tapeworm larvae may occasionally be recovered from a muscle specimen and should be carefully dissected from the capsules. They should then be pressed between two slides and examined under low power for the presence of a scolex with four suckers and a circle of hooks. If no hooks are present, it may be a species other than *Taenia solium.*

In some cases of schistosomiasis the eggs may not be recovered in the stool or urine; however, examination of the rectal or bladder mucosa may reveal eggs of the appropriate species. The mucosal tissue should be compressed between two slides and examined under low power and decreased illumination. The eggs should be carefully examined to determine viability (see p. 334). Small pieces of tissue may also be digested with 4% sodium hydroxide for 2 to 3 hours at 60 to 80 C. The eggs, which are recovered by sedimentation or centrifugation, can be examined under the microscope.

CULTURE TECHNIQUES

Most clinical laboratories do not provide culture techniques for the diagnosis of parasitic organisms; however, the lack of culture procedures should not prevent the correct identification of the majority of parasites. Isolation of intestinal amebae will yield a higher number of positive results, providing fresh specimens are received by the laboratory within specified time limits.[35] Most of the intestinal amebae do not culture as well as *Entamoeba histolytica;* however, once the organisms are established in culture they must be speciated on the basis of morphology. Accurate identification of the organisms can be determined by the examination of permanent stained smears of culture sediment material.

Many different media have been developed for the culture of protozoan organisms (some of which are available commercially), and specific directions for their preparation are available in the literature. Types of media that have been widely used include: amebae—Balamuth's aqueous egg yolk infusion[40] and Boeck and Drbohlav's Locke-egg-serum medium[40]; *Trichomonas vaginalis*—Lash's casein hydrolysate serum medium[20] and Feinberg's medium[17]; leishmaniae and trypanosomes—Novy-MacNeal-Nicolle medium[20] and diphasic blood agar medium (NIH method).[35] Techniques for the culture of other organisms (*Giardia lamblia, Plasmodium* spp., some of the helminths) are more difficult and are generally reserved for research purposes.

ANIMAL INOCULATION

Most laboratories have neither the time nor facilities for animal care to provide animal inoculation procedures for the diagnosis of parasitic infections. Host specificity for many parasites will also limit the kinds of laboratory animals available for these procedures. Occasionally, animal studies may be requested; included here are several procedures[20] that can be used.

The hamster is the animal of choice for inoculation procedures designed to recover leishmanial organisms. After intraperitoneal or intratesticular inoculation, the infection may develop very slowly over a period of several months; in some cases a generalized infection develops more quickly, and the animal may die in several days. Splenic and testicular aspirates should be examined for the presence of intracellular organisms; stained smears should be prepared and carefully examined with the oil immersion lens.

Mice are generally used for the isolation of *Toxoplasma gondii,* although most cases are diagnosed on clinical and serologic findings. Mice that are inoculated through the peritoneum develop a fulminating infection that leads to death within a few days. Organisms can be easily recovered from the ascitic fluid and should be examined as stained smears. Giemsa stain is recommended for both types of inoculation studies listed above; specific staining techniques are found in the section Detection of Blood Parasites.

SERODIAGNOSIS

Although serologic procedures for the diagnosis of parasitic diseases have been available for many years, they are generally not performed by most clinical laboratories. The procedures vary both in sensitivity and specificity and at times may be difficult to interpret. The Center for Disease Control (CDC), Atlanta, offers a number of serologic procedures for diagnostic purposes, some of which are still in the experimental stages and not available elsewhere. For a detailed discussion of specific procedures, consult Kwapinski.[33] The procedures mentioned

Table 35-1. Immunodiagnostic tests for parasitic diseases

Parasitic disease	Intradermal	Com-plement fixation	Bentonite floccu-lation	Indirect hemag-glutination	Latex	Indirect fluorescent-antibody	Precipitin
Amebiasis	▲	●	▲	●	●	▲	○
Chagas' disease	○	●		●	○	●	○
African trypanosomiasis		▲		○		▲	●
Leishmaniasis	●	●		●	○	▲	●
Malaria		▲		▲	○	●	○
Pneumocystis		▲				▲	
Toxoplasmosis	●	●		●	○	●	
Ancylostomiasis	▲	○		▲	▲	○	
Ascariasis	▲	○	●	●		▲	▲
Filariasis	▲	●	●	●	○	▲	
Toxocariasis	▲	●	●	●		▲	○
Trichinellosis	●	●	●	●	●	●	●
Clonorchiasis	●	●		●			○
Fascioliasis	▲	●		●		▲	○
Paragonimiasis	●	●	○	○			
Schistosomiasis	●	●	▲	●	○	●	○
Cysticercosis	○	●		●	○	▲	○
Echinococcosis	●	●	●	●	●	●	▲

From Kagan, I. G.: Serodiagnosis of parasitic diseases. In Lennette, E. H., Spaulding, E. H., and Truant, J. P., editors: Manual of clinical microbiology, ed. 2, Washington, D.C., 1974, American Society for Microbiology.

Symbols: ●, evaluated; ▲, experimental test; ○, reported in the literature.

here have been found to give fairly reproducible results at CDC. Although commercial antigens and diagnostic kits are available, Kagan[35] emphasizes the variability of the results from different reagents. Diagnostic tests for 18 parasitic infections are presented in Table 35-1.

At the present time immunodiagnostic procedures are most widely used for amebiasis, toxoplasmosis, leishmaniasis, Chagas' disease, trichinosis, schistosomiasis, cysticercosis, and hydatid disease.

Amebiasis

The sensitivity of the procedures for amebiasis depend on the type of disease present; a very low degree of sensitivity is found with sera from asymptomatic carriers, increased sensitivity from patients with amebic dysentery, and the greatest sensitivity with sera from those patients with extraintestinal disease. The complement fixation (CF) procedure has generally been replaced by the indirect hemagglutination (IHA), gel diffusion, and indirect fluorescent-antibody (IF) procedures; these three have approxi-

mately the same degree of sensitivity, although reports on the specificity of the IF procedure vary.

Toxoplasmosis

The methylene blue dye test (MBD; Sabin-Feldman dye test) has been used for many years for the serologic diagnosis of toxoplasmosis.[35] However, this procedure is being replaced by the IHA and IF procedures, which are both technically simple to peform and utilize a killed antigen rather than live organisms (used in the MBD test). Although all three procedures are approximately the same in terms of specificity and sensitivity, the IF procedure can be performed using specific conjugates of the immunoglobulin M type (IgM). Congenital infections are indicated when sera from newborns are positive with IgM conjugates.

Since two different types of antigens are used, the combination of the IF and IHA procedures will allow more accurate interpretation, particularly if the titers are borderline in terms of clinical significance.

Leishmaniasis

The serologic procedures (IHA, IF, CF) available for visceral leishmaniasis are quite helpful in making the diagnosis; however, the diagnosis of cutaneous leishmaniasis is less satisfactory, usually because of problems encountered in antigen production and subsequent poor specificity and sensitivity. Many of these problems should be resolved as additional research is directed toward preparation and purification of the antigens.

Chagas' disease

Although both the IHA and CF procedures are used for the diagnosis of Chagas' disease, the CF is more sensitive. A direct agglutination technique introduced by Vattuone and Yanovsky[66] in 1971 is very sensitive when used with sera from patients with acute disease; the procedure also provides good specificity in terms of cross-reactivity with leishmaniasis.

Trichinosis

A number of procedures are available for the diagnosis of trichinosis. The bentonite flocculation (BF) test has a high degree of specificity and is the standard procedure used at CDC; it is used to measure an increase or decrease in serum titers during the acute phase of the infection, yet is not too sensitive and does not react with residual antibodies from past exposure to the parasite. It is important diagnostically when a series of serum samples from one patient show a rise in titers. When low titers are obtained, it is recommended that additional serologic tests (different types) may be valuable in confirming the diagnosis. The IF procedure is the most sensitive[29,34,61] and can be used to detect antibodies in pigs infected with less than one larva per gram of diaphragm muscle tissue.[29]

Schistosomiasis

Several different tests (cholesterol-lecithin flocculation, BF, CF, IF) are used for the diagnosis of schistosomiasis;

however, they all share problems associated with both specificity and sensitivity. Although results with the CF procedure correlate closely with active clinical infections, Buck and Anderson[35] reported poor sensitivity with sera from infected children and from people with chronic schistosomiasis infections. There has been increased use of the IF procedure, which uses sections of adult worms for the antigen and which has proved to be the most sensitive technique (less cross-reactivity with trichina sera).

Cysticercosis

Extensive evaluation of the diagnostic tests for cysticercosis has been difficult to achieve due to the low number of sera available from proved human infections in the United States. Workers[47,48] in South Africa reported the IHA procedure as providing 85% positive results with sera from proved cases of human infection. Although the IHA was 100% reactive with sera from heavily infected animals, only 26% of those that were lightly infected were reactive. Biagi and associates[6] in Mexico reported the best results with the IHA. There are still difficulties with both specificity and sensitivity, and a higher positive titer with human sera may have to be selected as the standard to rule out false-positive reactions. The double-diffusion procedure, when evaluated by CDC, was found to be insensitive using both animal and human sera.[35]

Echinococcosis

The IHA, IF, and immunoelectrophoresis (IE) procedures are considered to be the tests of choice for the diagnosis of echinococcosis. The IHA and BF procedures are routinely used at CDC, the IHA being the more sensitive test. High IHA titers usually indicate the presence of hydatid disease; however, low titers are difficult to interpret and have been found in sera from patients with collagen diseases or liver cirrhosis.[27] Hydatid cysts in the lung (sera, 33% to 50% sensitivity) are less likely to be diagnosed by serologic means than those found in the liver (sera, 82% to 86% sensitivity).[28] Additional studies with the IF and IE techniques may produce a procedure that will provide greater specificity than is now available.

DETECTION OF BLOOD PARASITES
Malaria

Malaria is caused by four species of the protozoan genus *Plasmodium: P. vivax, P. falciparum, P. malariae,* and *P. ovale.* Humans become infected when the sporozoites are introduced into the blood from the salivary secretion of the infected mosquito when the mosquito vector takes a blood meal. These sporozoites then leave the blood and enter the parenchymal cells of the liver, where they undergo asexual multiplication. This development in the liver prior to red cell invasion is called the preerythrocytic cycle; if further liver development takes place after red cell invasion, it is called the exoerythrocytic cycle. The length of time for the preerythrocytic cycle and the number of asexual generations will vary depending on the species; however, the schizonts will eventually rupture, releasing thousands of merozoites into the bloodstream, where they will invade the erythrocytes.

The early forms in the red cells are called ring forms or young trophozoites. As the parasites continue to grow and feed, they become actively ameboid within the red cell. They feed on hemoglobin, which is incompletely metabolized; the residue left is called malarial pigment and is a compound of hematin and protein (hemozoin).

During the next phase of the cycle the chromatin (nuclear material) becomes fragmented throughout the organism and the cytoplasm begins to divide, each portion being arranged with a fragment of nuclear material. These forms are called mature schizonts and are composed of individual merozoites. The infected red cell then ruptures, releasing the merozoites and also metabolic products into the bloodstream. If large numbers of red cells rupture simultaneously, a malarial paroxysm may result from the amount of toxic materials released into the bloodstream. In the early stages of infection or in a mixed infection with two species, rupture of the red cells is usually not synchronous; consequently, the fever may be continuous or daily rather than intermittent. After several days a 48- or 72-hour periodicity is usually established.

After several generations of erythrocytic schizogony, the production of gametocytes begins. These forms are derived from merozoites, which do not undergo schizogony but continue to grow and form the male and female gametocytes, which circulate in the bloodstream. When the mature gametocytes are ingested by the appropriate mosquito vector, the sexual cycle is initiated within the mosquito, with the eventual production of the sporozoites, the infective stage for humans.

The asexual and sexual forms just described circulate in the human bloodstream in three species of *Plasmodium.* However, in *P. falciparum* infections, as the parasite continues to grow, the red cell membrane becomes sticky and the cells tend to adhere to the endothelial lining of the capillaries of the internal organs; thus, only the ring forms and crescent-shaped gametocytes occur in the peripheral blood. Interference with normal blood flow in these vessels gives rise to additional problems, which are responsible for the different clinical manifestations of this type of malaria.

Laboratory diagnosis of malaria

The definitive diagnosis of malaria is based on the demonstration of the parasites in the blood. Two types of blood films are used. The **thick film** allows the examination of a larger amount of blood and is used as a screening procedure[18];

Table 35-2. Microscopic identification of plasmodia of humans in Giemsa-stained thin blood smears[19,68,70]

	Plasmodium vivax	Plasmodium malariae	Plasmodium falciparum	Plasmodium ovale
Appearance of parasitized red blood cells: size and shape	1½ to 2 times larger than normal; oval to round	Normal shape; size may be normal or slightly smaller	Both normal	60% of cells larger than normal and oval; 20% have irregular, frayed edges
Schuffner's dots (eosinophilic stippling)	Usually present in all cells except early ring forms	None	None; occasionally comma-like red dots are present (Maurer's dots)	Present in all stages including early ring forms, dots may be larger and darker than in P. vivax
Color of cytoplasm	Decolorized, pale	Normal	Normal, bluish tinge at times	Decolorized, pale
Multiple infections	Occasional	Rare	Common	Occasional
All developmental stages present in peripheral blood	All stages present	Ring forms few, as ring stage brief; mostly growing and mature trophozoites and schizonts	Young ring forms and no older stages; few gametocytes	All stages present
Appearance of parasite: young trophozoite (early ring form)	Ring is ⅓ diameter of cell; cytoplasmic circle around vacuole; heavy chromatin dot	Ring often smaller than in P. vivax, occupying ⅙ of cell; heavy chromatin dot; vacuole at times "filled in"; pigment forms early	Delicate, small ring with small chromatin dot (frequently 2); scanty cytoplasm around small vacuoles; sometimes at edge of red cell (appliqué form) or filamentous slender form; may have multiple rings per cell.	Ring is larger and more ameboid than in P. vivax, otherwise similar to P. vivax
Growing trophozoite	Multishaped irregular ameboid parasite; streamers of cytoplasm close to large chromatin dot; vacuole retained until close to maturity; increasing amounts of brown pigment	Nonameboid rounded or bandshaped solid forms; chromatin may be hidden by coarse dark brown pigment	Heavy ring forms fine pigment grains	Ring shape maintained until late in development
Mature trophozoite	Irregular ameboid mass; 1 or more small vacuoles retained until schizont stage; fills almost entire cell; fine brown pigment	Vacuoles disappear early; cytoplasm compact, oval, band shaped, or nearly round, almost filling cell; chromatin may be hidden by peripheral coarse dark	Not seen in peripheral blood (except in severe infections); development of all phases following ring form occurs in capillaries of viscera	Compact; vacuoles disappear; pigment dark brown, less than in P. malariae

Schizont (presegmenter)	Progressive chromatin division; cytoplasmic bands containing clumps of brown pigment	Similar to *P. vivax* except smaller, darker, larger pigment granules peripheral or central	Not seen in peripheral blood (see above)	Smaller and more compact than *P. vivax*
Mature schizont	16 (12 to 24) merozoites, each with chromatin and cytoplasm, filling entire red cell, which can hardly be seen	8 (6 to 12) merozoites in rosettes or irregular clusters filling normal-sized cells, which can hardly be seen; central arrangement of brown-green pigment	Not seen in peripheral blood	¾ of cells occupied by 8 (8 to 12) merozoites in rosettes or irregular clusters
Macrogametocyte	Rounded or oval homogeneous cytoplasm; diffuse delicate light brown pigment throughout parasite; eccentric compact chromatin	Similar to *P. vivax*, but fewer in number, pigment darker and more coarse	Sex differentiation difficult; "crescent" or "sausage" shapes characteristic; may appear in "showers"; black pigment near chromatin dot, which is often central	Smaller than *P. vivax*
Microgametocyte	Large pink to purple chromatin mass surrounded by pale or colorless halo; evenly distributed pigment	Similar to *P. vivax* but fewer in number, pigment darker and more coarse	See above	Smaller than *P. vivax*
Main criteria	Large, pale red cell; trophozoite irregular; pigment usually present; Schuffner's dots not always present; several phases of growth seen in one smear; gametocytes appear early	Red cell normal in size and color; trophozoites compact, stain usually intense, band forms not always seen; coarse pigment; no stippling of red cells; gametocytes appear late	Development following ring stage takes place in blood vessels of internal organs; delicate ring forms and crescent-shaped gametocytes are only forms normally seen in peripheral blood	Red cell enlarged, oval, with fimbriated edges; Schuffner's dots seen in all stages

the **thin film** allows speciation of the parasite.

Blood films are usually prepared when the patient is admitted; samples should be taken at intervals of 6 to 18 hours for at least 3 successive days. One-hundred microscopic fields should be examined before a film is signed out as negative. If possible, the smears should be prepared from blood obtained from the finger or earlobe; the blood should flow freely. If patient contact is not possible and the quality of the submitted slides may be poor, request a tube of fresh blood (EDTA anticoagulant is recommended) and prepare smears immediately after the blood is received.

To prepare the thick film place two or three small drops of fresh blood (no anticoagulant) on an alcohol-cleaned slide. With the corner of another slide, and using a circular motion, mix the drops and spread the blood over an area about 2 cm in diameter. Continue stirring for about 30 seconds to prevent formation of fibrin strands, which may obscure the parasites after staining. If the blood is too thick or any grease remains on the slide, the blood will flake off during staining. Allow the film to air dry (room temperature) in a dust-free area. Never apply heat to a thick film, since heat will fix the blood, causing the red blood cells to remain intact during staining; the result is stain retention and subsequent inability to identify any parasites present.

The thin blood film is used primarily for specific parasite identification, although the number of organisms per field is much reduced compared with the thick film (see Table 35-2 and color plates 1 to 7). The thin film is prepared exactly as one used for the differential blood count. After the film has air dried (do not apply heat), it may be stained. The necessity for fixation prior to staining will depend on the stain selected.

Staining blood films.[55,56] For accurate identification of blood parasites, it is very important that a laboratory develop proficiency in the use of at least one good staining method. As a general rule, blood films should be stained as soon as possible, since prolonged storage results in stain retention.

The stains that are generally used are of two types. One has the fixative in combination with the staining solution, so that both fixation and staining occur at the same time. Wright's stain is an example of this type of staining solution. Giemsa stain represents the other type of staining solution, in which the fixative and stain are separate; thus, the thin film must be fixed prior to staining.

When slides are removed from either type of staining solution, they should be dried in a vertical position. After being air dried, they may be examined under oil immersion by placing the oil directly on the uncovered blood film.

Giemsa stain. Giemsa stain is available

Plate 1. *Plasmodium vivax.* **1,** Normal-sized red cell with marginal ring form trophozoite. **2,** Young signet ring form trophozoite in macrocyte. **3,** Slightly older ring form trophozoite in red cell showing basophilic stippling. **4,** Polychromatophilic red cell containing young tertian parasite with pseudopodia. **5,** Ring form trophozoite showing pigment in cytoplasm, in enlarged cell containing Schüffner's stippling (dots). (Schüffner's stippling does not appear in all cells containing growing and older forms of *P. vivax,* as would be indicated by these pictures, but it can be found with any stage from fairly young ring form onward.) **6, 7,** Very tenuous medium trophozoite forms. **8,** Three ameboid trophozoites with fused cytoplasm. **9, 11-13,** Older ameboid trophozoites in process of development. **10,** Two ameboid trophozoites in one cell. **14,** Mature trophozoite. **15,** Mature trophozoite with chromatin apparently in process of division. **16-19,** Schizonts showing progressive steps in division (presegmenting schizonts). **20,** Mature schizont. **21, 22,** Developing gametocytes. **23,** Mature microgametocyte. **24,** Mature macrogametocyte. (From Wilcox, A.: Manual for the microscopical diagnosis of malaria in man, Washington, D.C., 1960, Department of Health, Education, and Welfare, Public Health Service, U.S. Government Printing Office.)

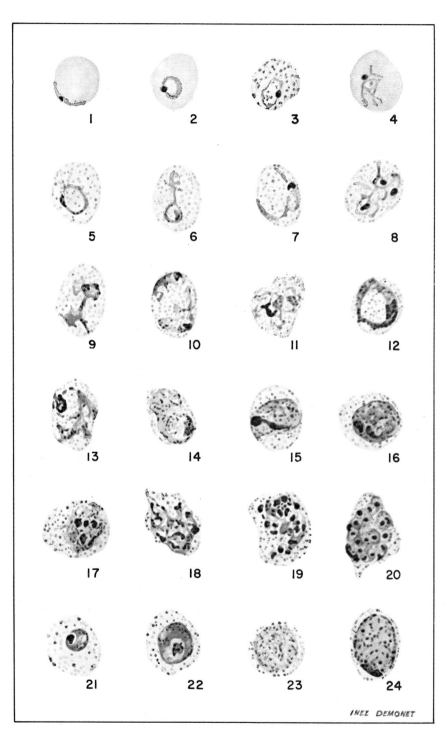

Plate 1. For legend see opposite page.

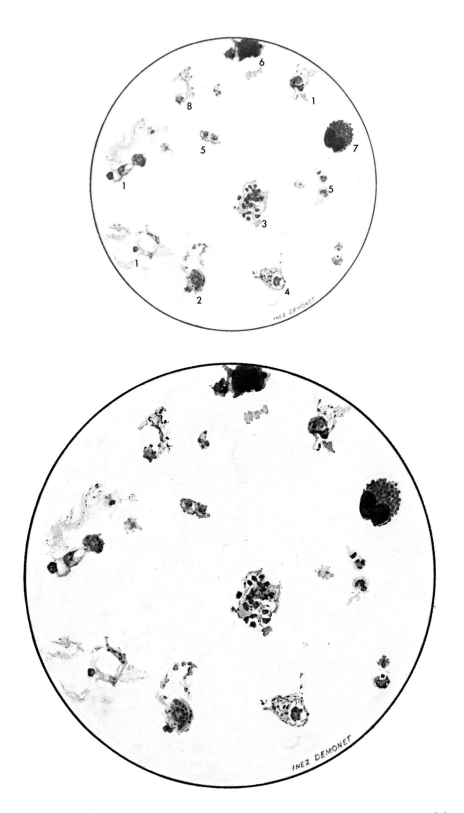

Plate 2. *Plasmodium vivax* in thick smear. **1,** Ameboid trophozoites. **2,** Schizont, two divisions of chromatin. **3,** Mature schizont. **4,** Microgametocyte. **5,** Blood platelets. **6,** Nucleus of neutrophil. **7,** Eosinophil. **8,** Blood platelet associated with cellular remains of young erythrocytes. (From Wilcox, A.: Manual for the microscopical diagnosis of malaria in man, Washington, D.C., 1960, Department of Health, Education, and Welfare, Public Health Service, U.S. Government Printing Office.)

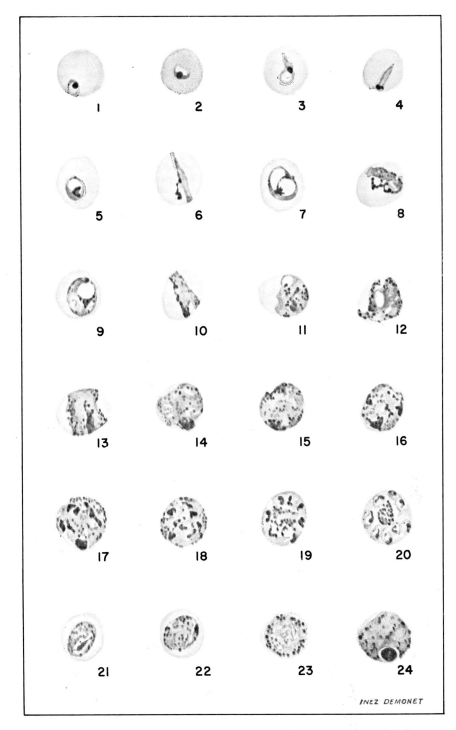

Plate 3. *Plasmodium malariae.* **1,** Young ring form trophozoite of quartan malaria. **2-4,** Young trophozoite forms of parasite showing gradual increase of chromatin and cytoplasm. **5,** Developing ring form trophozoite showing pigment granule. **6,** Early band form trophozoite, elongated chromatin, some pigment apparent. **7-12,** Some forms that developing trophozoite of quartan may take. **13, 14,** Mature trophozoites, one a band form. **15-19,** Phases in development of schizont (presegmenting schizonts). **20,** Mature schizont. **21,** Immature microgametocyte. **22,** Immature macrogametocyte. **23,** Mature microgametocyte. **24,** Mature macrogametocyte. (From Wilcox, A.: Manual for the microscopical diagnosis of malaria in man, Washington, D.C., 1960, Department of Health, Education, and Welfare, Public Health Service, U.S. Government Printing Office.)

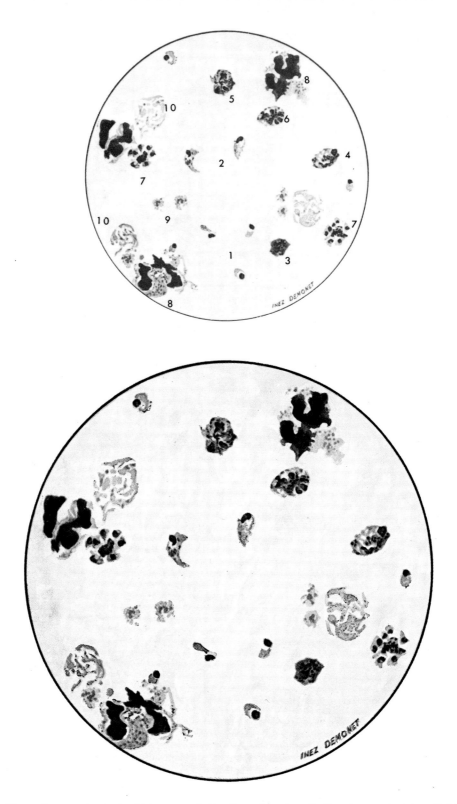

Plate 4. *Plasmodium malariae* in thick smear. **1,** Small trophozoites. **2,** Growing trophozoites. **3,** Mature trophozoites. **4-6,** Schizonts (presegmenting) with varying numbers of divisions of chromatin. **7,** Mature schizonts. **8,** Nucleus of leukocyte. **9,** Blood platelets. **10,** Cellular remains of young erythrocytes. (From Wilcox, A.: Manual for the microscopical diagnosis of malaria in man, Washington, D.C., 1960, Department of Health, Education, and Welfare, Public Health Service, U.S. Government Printing Office.)

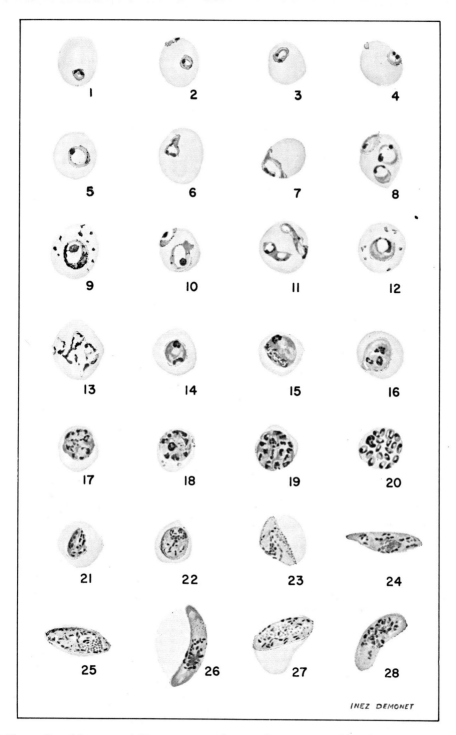

INEZ DEMONET

Plate 5. *Plasmodium falciparum.* **1,** Very young ring form trophozoite. **2,** Double infection of single cell with young trophozoites, one a marginal form, the other signet ring form. **3, 4,** Young trophozoites showing double chromatin dots. **5-7,** Developing trophozoite forms. **8,** Three medium trophozoites in one cell. **9,** Trophozoite showing pigment in cell containing Maurer's dots. **10, 11,** Two trophozoites in each of two cells, showing variations of forms that parasites may assume. **12,** Almost mature trophozoite showing haze of pigment throughout cytoplasm. Maurer's dots in cell. **13,** Estivoautumnal "slender forms." **14,** Mature trophozoite showing clumped pigment. **15,** Parasite in process of initial chromatin division. **16-19,** Various phases of development of schizont (presegmenting schizonts). **20,** Mature schizont. **21-24,** Successive forms in development of gametocyte, usually not found in peripheral circulation. **25,** Immature macrogametocyte. **26,** Mature macrogametocyte. **27,** Immature microgametocyte. **28,** Mature microgametocyte. (From Wilcox, A.: Manual for the microscopical diagnosis of malaria in man, Washington, D.C., 1960, Department of Health, Education, and Welfare, Public Health Service, U.S. Government Printing Office.)

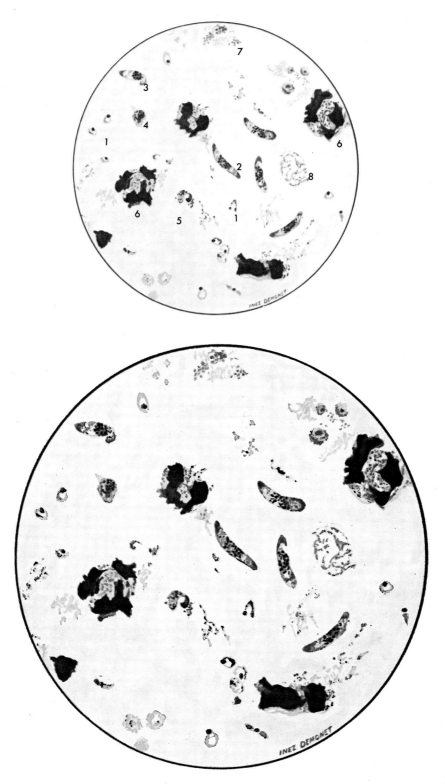

Plate 6. *Plasmodium falciparum* in thick film. **1,** Small trophozoites. **2,** Normal gametocytes. **3,** Slightly distorted gametocyte. **4,** "Rounded-up" gametocyte. **5,** Disintegrated gametocyte. **6,** Nucleus of leukocyte. **7,** Blood platelets. **8,** Cellular remains of young erythrocyte. (From Wilcox, A.: Manual for the microscopical diagnosis of malaria in man, Washington, D.C., 1960, Department of Health, Education, and Welfare, Public Health Service, U.S. Government Printing Office.)

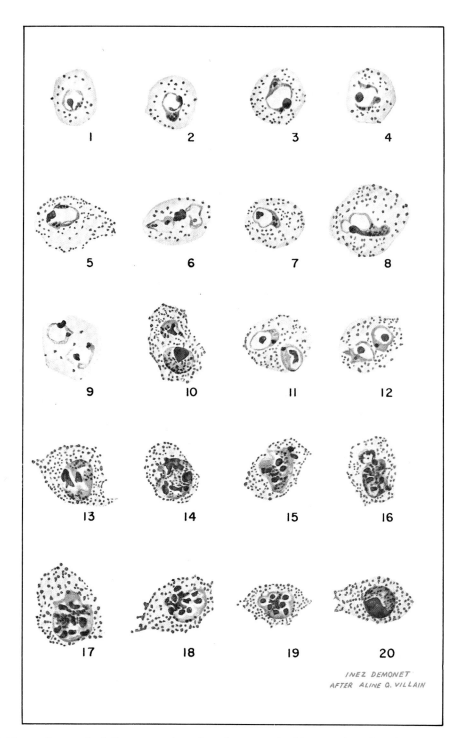

Plate 7. *Plasmodium ovale.* **1,** Young ring-shaped trophozoite. **2-5,** Older ring-shaped trophozoites. **6-8,** Older ameboid trophozoites. **9, 11, 12,** Doubly infected cells, trophozoites. **10,** Doubly infected cell, young gametocytes. **13,** First stage of the schizont. **14-19,** Schizonts, progressive stages. **20,** Mature gametocyte. (From Markell, E. K., and Voge, M.: Medical parasitology, ed. 4, Philadelphia, 1976, W. B. Saunders Co.)

commercially as a concentrated stock solution or as a powder for those who wish to make their own stain; there seems to be very little difference between the two preparations.

REAGENTS

1. Stock Giemsa stain:

Giemsa stain powder, certified	600 mg
Methanol, absolute and certified neutral, acetone free	50 ml
Glycerine, neutral, certified	50 ml

Grind well small portions of stain and glycerine in mortar and collect mixtures in a 500- or 1,000-ml flask until all measured material is mixed. Stopper flask with cotton plug, cover with heavy paper, place in 55 to 60 C water bath for 2 hours, making sure that the water reaches the level of the stain. Shake gently at ½-hour intervals. Allow to cool; add alcohol. Use a portion of the measured alcohol to wash out the mortar and add to the flask. Store in brown bottle. Allow to stand for 2 to 3 weeks. Filter before use.

2. 10% stock solution of Triton X-100:

Triton X-100	10 ml
Distilled water	100 ml

3. Stock buffers:
   a. Disodium phosphate buffer:

Na_2HPO_4 anhydrous	9.5 g
Distilled water	1,000 ml

   b. Sodium acid phosphate buffer:

$NaH_2PO_4H_2O$	9.2 g
Distilled water	1,000 ml

4. Buffered water: pH range 7.0 to 7.2; check with pH meter before use.

Disodium phosphate buffer	61 ml
Sodium acid phosphate buffer	39 ml
Distilled water	900 ml

5. Triton–buffered water solutions:
   a. 0.01% Triton–buffered water:

Stock 10% aqueous Triton X-100	1 ml
Buffered water	1,000 ml

Use for thin blood films or combination of thin and thick blood films.

   b. 0.1% Triton–buffered water:

Stock 10% aqueous Triton X-100	10 ml
Buffered water	1,000 ml

Use for thick blood films.

PROCEDURE FOR STAINING THIN FILMS

1. Fix blood films in absolute methyl alcohol (acetone free) for 30 seconds.
2. Allow slides to air dry.
3. Immerse slides in a solution of 1 part Giemsa stock (commercial liquid stain or stock prepared from powder) to 10 to 50 parts of Triton–buffered water (pH 7.0 to 7.2). Stain 10 to 60 minutes (see note below). Fresh working stain should be prepared from stock solution each day.
4. Dip slides briefly in Triton–buffered water.
5. Drain thoroughly in vertical position and allow to air dry.

Note: A good general rule for stain dilution versus staining time is that if dilution is 1:20, stain for 20 minutes; if 1:30, stain for 30 minutes; and so forth. However, a series of stain dilutions and staining times should be tried to determine the best dilution/time for each batch of stock stain.

PROCEDURE FOR STAINING THICK FILMS. The procedure to be followed for thick films is the same as for thin films, except that the first two steps are omitted. If the slide has a thick film at one end and a thin film at the other, fix only the thin portion, and then stain both parts of the film simultaneously.

RESULTS. Giemsa stain colors the components of blood as follows: erythrocytes, pale gray-blue; nuclei of white blood cells, purple with pale purple cytoplasm; eosinophilic granules, bright purple-red; neutrophilic granules, deep pink-purple.

There have been many studies using immunodiagnostic procedures for the diagnosis of malaria[26,32,39,57,64,67]; however, these procedures are not routinely performed in most laboratories. Sulzer and associates[62] have reported the use of thick-smear antigens prepared from washed parasitized blood cells. This type of antigen is used in the IF procedure, which has a 95% sensitivity and a false-positive rate of 1% at a titer of 1:16.[63] The IHA has also been used and evaluated by a number of workers.[13,43,50,59]

In some areas of the world where *P. falciparum* is endemic, there is also a high incidence of hemoglobin S (HbS), thalassemia, and glucose-6-phosphate dehydrogenase (G-6-PD) deficiency carriers. The young heterozygous carrier of HbS gains some protection against *P. falciparum*.[1] Apparently, G-6-PD deficiency[2] and thalassemia[9] are also associated with increased resistance.

Miller and co-workers[44] reported that Duffy-positive human erythrocytes are easily infected with *P. knowlesi;* however, Duffy-negative human erythrocytes are resistant to infection. They suggest that the resistance of many West Africans and approximately 70% of American blacks to *P. vivax* may be related to the high incidence of Duffy-negative erythrocytes in these groups. In East Africa, where there is a higher incidence of Duffy-positive red cells, *P. vivax* is more common.

Babesiosis

Babesia are tick-borne sporozoan parasites, which have generally been considered parasites of animals (Texas cattle fever) rather than humans. However, there are now a number of documented human cases, some infections occurring in splenectomized patients and others in patients with intact spleens.[40] *Babesia* organisms infect the red blood cells and appear as pleomorphic ringlike structures when stained with any of the recommended stains used for blood films (Fig. 35-6). They may be confused with the ring forms in *Plasmodium* infections; however, in a *Babesia* infection there are often many rings (four or five) per red cell, and the individual rings are quite

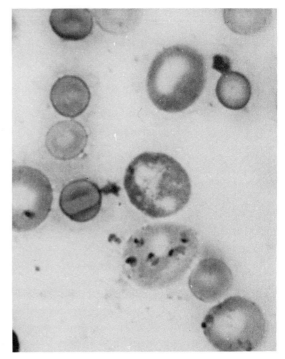

Fig. 35-6. *Babesia* in red blood cells. (Photomicrograph by Zane Price.) (From Markell, E. K., and Voge, M.: Medical parasitology, ed. 4, Philadelphia, 1976, W. B. Saunders Co.)

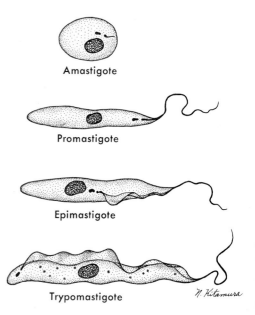

Amastigote

Promastigote

Epimastigote

Trypomastigote

Fig. 35-7. Characteristic stages of species of *Leishmania* and *Trypanosoma* in humans and in insect host. (Illustration by Nobuko Kitamura.)

small compared with those found in malaria infections.

Hemoflagellates

Hemoflagellates are blood and tissue flagellates, two genera of which are medically important for humans: *Leishmania* and *Trypanosoma*. Some species may circulate in the bloodstream or at times may be present in lymph nodes or muscle. Other species tend to parasitize the reticuloendothelial cells of the hemopoietic organs. The hemoflagellates of human beings have four morphologic types (Fig. 35-7): amastigote (leishmanial form, or Leishman-Donovan body), promastigote (leptomonal form), epimastigote (crithidial form), and trypomastigote (trypanosomal form).

The amastigote form is an intracellular parasite in the cells of the reticuloendothelial system and is oval, measuring approximately 1½ to 5 μm, and contains a nucleus and kinetoplast. Species of the genus *Leishmania* usually exist as the amastigote form in humans and in the promastigote form in the insect host. The life cycle is essentially the same for all three species, and the clinical manifestations will vary depending on the species involved. As the vector takes a blood meal, the promastigote form is

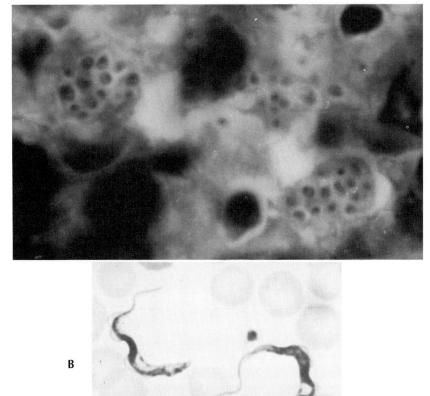

Fig. 35-8. A, *Leishmania donovani* parasites in Küpffer cells of liver. **B,** *Trypanosoma gambiense* in blood film.

introduced into a human, thus initiating the infection. Depending on the species, the parasites then move from the site of the bite to the organs of the reticuloendothelial system (liver, spleen, bone marrow) or to the macrophages of the skin.

The three species found in humans are morphologically the same; however, there are differences in serologies and in growth requirements for culture. There is a great deal of biologic variation among the many strains that comprise the three species: *Leishmania tropica* causes oriental sore or cutaneous leishmaniasis of the Old World; *L. braziliensis* causes mucocutaneous leishmaniasis of the New World; and *L. donovani* causes visceral leishmaniasis (Dumdum fever, or kala azar) (Fig. 35-8, *A*).

In tissue impression smears or sections, *Histoplasma capsulatum* must be differentiated from the Leishman-Donovan (L-D) bodies. *H. capsulatum* does not have a kinetoplast and stains with both PAS and Gomori methenamine silver stain, neither of which stains L-D bodies.

Diagnosis of leishmanial organisms is based on the demonstration of the L-D

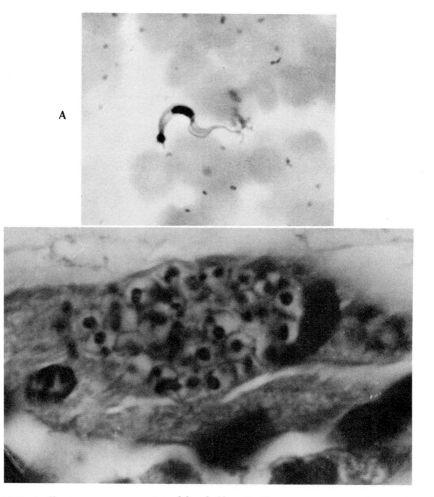

Fig. 35-9. A, *Trypanosoma cruzi* in blood film. **B,** *Trypanosoma cruzi* parasites in cardiac muscle. (From Markell, E. K., and Voge, M.: Medical parasitology, ed. 4, Philadelphia, 1976, W. B. Saunders Co.)

bodies or the recovery of the promastigote culture stages.

Three species of trypanosomes are pathogenic for humans: *Trypanosoma gambiense* causes West African sleeping sickness; *T. rhodesiense* causes East African sleeping sickness; and *T. cruzi* causes South American trypanosomiasis, or Chagas' disease. The first two species are morphologically similar and produce African sleeping sickness (Fig. 35-8, *B*),

an illness characterized by both acute and chronic stages. In the acute stage of the disease the organisms can usually be found in the peripheral blood or lymph node aspirates. As the disease progresses to the chronic stage, the organisms can be found in the cerebrospinal fluid (comatose stage: "sleeping sickness"). *T. rhodesiense* produces a more severe infection, usually resulting in death within a year.

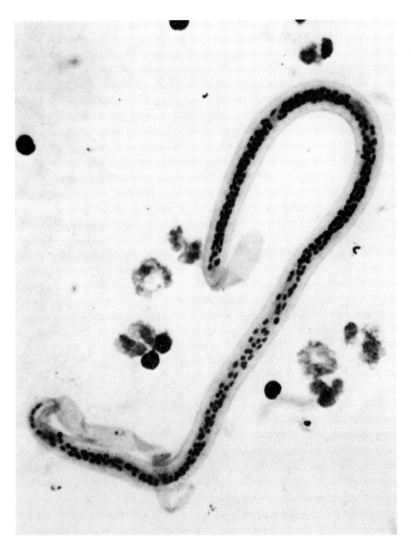

Fig. 35-10. Microfilaria of *Wuchereria bancrofti* in thick blood film. (From Markell, E. K., and Voge, M.: Medical parasitology, ed. 4, Philadelphia, 1976, W. B. Saunders Co.)

In the early stages of infection with *T. cruzi*, the trypomastigote forms appear in the blood but do not multiply (Fig. 35-9, *A*). They then invade the endothelial or tissue cells and begin to divide, producing many L-D bodies, which are most often found in cardiac muscle (Fig. 35-9, *B*). When these forms are liberated into the blood, they transform into the trypomastigote forms, which are then carried to other sites, where tissue invasion again occurs.

Diagnosis of the organisms is based on demonstration of the parasites, most often on wet unstained or stained blood films. Both thick and thin films should be examined; these can be prepared from peripheral blood or buffy coat. The sediment recovered from cerebrospinal fluid can also be examined for the presence of trypomastigotes. Specific techniques for culture, animal inoculation, handling of aspirate and biopsy material, and serologic procedures are presented in earlier sections of this chapter.

Another technique often used in endemic areas for the diagnosis of Chagas' disease is xenodiagnosis. Triatomids, the insect vector, are raised in the laboratory and are free from infection with *T. cruzi*. These insects are allowed to feed on the blood of an individual suspected of having Chagas' disease, and after 2 weeks the intestinal contents are checked for the presence of the epimastigote forms. For additional information consult Maekelt.[38]

Filariae

The filarial worms are long, thin nematodes, which inhabit parts of the lymphatic system and the subcutaneous and deep connective tissues. Most species produce microfilariae, which can be found in the peripheral blood; two species, *Onchocerca volvulus* and *Dipetalonema streptocerca,* produce microfilariae found in the subcutaneous tissues and dermis.

Diagnosis of filarial infections is often based on clinical grounds, but demonstration of the parasite is the only accurate means of confirming the diagnosis (Fig. 35-10). Fresh blood films may be prepared; actively moving microfilariae will be observed in a preparation of this type. If the patient has a light infection, thick blood films can be prepared and stained. The Knott concentration procedure[31] and the membrane filtration technique[11,12] may also be hepful in recovering the organisms. Microfilariae of some strains tend to exhibit nocturnal periodicity; thus, the time the blood is drawn may be critical in demonstrating the parasite. The microfilariae of *Onchocerca volvulus* and *Dipetalonema streptocerca* are found in "skin snips," very thin slices of skin, which are teased apart in normal saline to release the organisms. Differentiation of the species is dependent on (1) the presence or absence of the sheath and (2) the distribution of nuclei in the tail region of the microfilaria (Fig. 35-11).

IDENTIFICATION OF ANIMAL PARASITES

The animal parasites of humans are found in five major groups of phyla: the Protozoa; the Platyhelminthes, or flatworms; the Nematoda, or roundworms; the Acanthocephala, or thorny-headed worms; and the Arthropoda, which include ticks, mites, spiders, insects, and various other groups. The Protozoa, Platyhelminthes, and Nematoda phyla are discussed here, since they contain the majority of organisms parasitic for humans. Within each phylum are various classes, which can be subdivided into orders and further subdivided into families with their own genera and species. These subdivisions are based on morphologic criteria; thus, identification of the organisms requires a certain knowledge of basic structures and their definitions.

Intestinal protozoa

The protozoa are unicellular organisms, most of which are microscopic. They possess a number of specialized organelles, which are responsible for

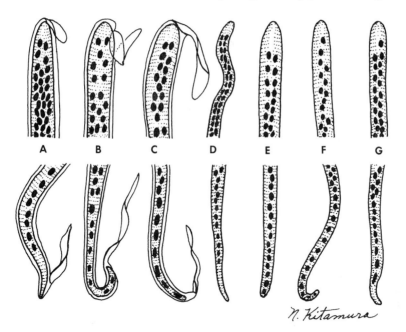

Fig. 35-11. Anterior and posterior ends of microfilariae found in humans. **A,** *Wuchereria bancrofti.* **B,** *Brugia malayi.* **C,** *Loa loa.* **D,** *Onchocerca volvulus.* **E,** *Dipetalonema perstans.* **F,** *Dipetalonema streptocerca.* **G,** *Mansonella ozzardi.*

life functions and which allow further division of the group into classes.

The class Sarcodina contains those organisms that move by means of cytoplasmic protrusions called pseudopodia. Included in this group are free-living organisms, as well as nonpathogenic and pathogenic organisms found in the intestinal tract and other areas of the body.

The Mastigophora, or flagellates, contain specialized locomotor organelles called flagella: long, thin cytoplasmic extensions, which may vary in number and position depending on the species. Different genera may live in the intestinal tract, the bloodstream, or various tissues. The blood- and tissue-dwelling flagellates are discussed in the section Detection of Blood Parasites.

The class Ciliata contains a number of species that move by means of cilia, short extensions of cytoplasm that cover the surface of the organism. This group contains only one organism that infects humans: *Balantidium coli* infects the intestinal tract and may produce severe symptoms.

Members of the class Sporozoa are found in the blood and other tissues and have a complex life cycle that involves both sexual and asexual generations. The four species of *Plasmodium,* the cause of malaria, are found in this group and are discussed in the section Detection of Blood Parasites. Members of the genus *Isospora* are found in the intestinal mucosa.

The majority of the intestinal protozoa live in the colon, with the exception of the flagellate *Giardia lamblia* and the coccidian parasite *Isospora belli,* which are found in the small intestine. *I. belli* is the only protozoan that is an obligate tissue parasite in the intestinal tract and is passed in the stool as oocysts; the other members of the group exist in the intestinal tract in the trophozoite or cyst stages. The important characteristics of the intestinal protozoa are found in Tables 35-3 to 35-7. The clinically important protozoa are generally con-

Text continued on p. 359.

Table 35-3. Morphologic criteria used to identify intestinal protozoa[8,20,40]. Amebae, trophozoites

	Entamoeba histolytica	*Entamoeba hartmanni*	*Entamoeba coli*	*Endolimax nana*	*Iodamoeba bütschlii*
Size (diameter or length)	10 to 60 μm; usual range, 15 to 20 μm; invasive forms may be over 20 μm	5 to 12 μm; usual range, 8 to 10 μm	15 to 50 μm; usual range, 20 to 25 μm	6 to 12 μm; usual range, 8 to 10 μm	8 to 20 μm; usual range, 12 to 15 μm
Motility	Progressive with hyaline, fingerlike pseudopods; motility may be rapid	Usually nonprogressive	Sluggish, nondirectional, with blunt pseudopods	Sluggish, usually nonprogressive	Sluggish, usually nonprogressive
Number of nuclei	Difficult to see in unstained preparations; usually not seen; 1	Usually not seen in unstained preparations; 1	Often visible in unstained preparations; 1	Occasionally visible in unstained preparations; 1	Usually not visible in unstained preparations; 1
Nucleus Peripheral chromatin (stained)	Fine granules, uniform in size, usually evenly distributed; may have beaded appearance	Nucleus may stain more darkly than *E. histolytica*, although morphology is similar; chromatin may appear as solid ring rather than beaded	May be clumped and unevenly arranged on membrane; may also appear as solid, dark ring with no beads or clumps	No peripheral chromatin	No peripheral chromatin
Karyosome (stained)	Small, usually compact; centrally located, but may also be eccentric	Usually small and compact; may be centrally located or eccentric	Large, not compact; may or may not be eccentric; may be diffuse and darkly stained	Large, irregularly shaped; may appear "blot-like"; many nuclear variations common	Large, may be surrounded by refractile granules that are difficult to see
Cytoplasm Appearance (stained)	Finely granular, "ground glass" appearance; clear differentiation of ectoplasm and endoplasm; if present, vacuoles usually small	Finely granular	Granular with little differentiation into ectoplasm and endoplasm; usually vacuolated	Granular, vacuolated	Coarsely granular; may be highly vacuolated
Inclusions (stained)	Noninvasive organism may contain bacteria; presence of red blood cells	May contain bacteria; *no red blood cells*	Bacteria, yeast, other debris	Bacteria	Bacteria, yeast, other debris

Table 35-4. Morphologic criteria used to identify intestinal protozoa[8,20,40]: Amebae, cysts

	Entamoeba histolytica	*Entamoeba hartmanni*	*Entamoeba coli*	*Endolimax nana*	*Iodamoeba bütschlii*
Size	10 to 20 μm; usual range, 12 to 15 μm	5 to 10 μm; usual range, 6 to 8 μm	10 to 35 μm; usual range, 15 to 25 μm	5 to 10 μm; usual range, 6 to 8 μm	5 to 20 μm; usual range, 10 to 12 μm
Shape	Usually spherical	Usually spherical	Usually spherical; occasionally oval, triangular, or other shapes; may be distorted on stained slide if fixation poor	Spherical, ovoidal, or ellipsoidal	Ovoidal, ellipsoidal, or other shapes
Number of nuclei	Mature cyst, 4; immature, 1 or 2 nuclei may be seen; nuclear characteristics difficult to see on wet preparation	Mature cyst, 4; immature, 1 or 2 nuclei may be seen; 2 nucleated cysts very common	Mature cyst, 8; occasionally 16 or more nuclei may be seen; immature cysts with 2 or more nuclei occasionally seen	Mature cyst, 4; immature cysts, 2 (very rarely seen and may resemble cysts of *Enteromonas hominis*)	Mature cyst, 1
Nucleus					
Peripheral chromatin (stained)	Peripheral chromatin present; fine, uniform granules, evenly distributed; nuclear characteristics may not be as clearly visible as in trophozoite	Fine granules evenly distributed on membrane; nuclear characteristics may be difficult to see	Coarsely granular and may be clumped and unevenly arranged on membrane; nuclear characteristics not as clearly defined as in trophozoite; may resemble *E. histolytica*	No peripheral chromatin	No peripheral chromatin
Karyosome (stained)	Small, compact, usually centrally located	Small, compact, usually centrally located	Large, may or may not be compact or eccentric; occasionally appears to be centrally located	Smaller than karyosome seen in trophozoite, but generally larger than those of genus *Entamoeba*	Large, usually eccentric refractile granules may be on one side of karyosome ("basket nucleus")

Continued.

Table 35-4. Morphologic criteria used to identify intestinal protozoa[8,20,40]: Amebae, cysts—cont'd

	Entamoeba histolytica	Entamoeba hartmanni	Entamoeba coli	Endolimax nana	Iodamoeba bütschlii
Cytoplasm					
Chromatoidal bodies (stained)	May be present; bodies usually elongate with blunt, rounded, smooth edges	Often present; bodies elongate with blunt, rounded, smooth edges	May be present (less frequently than E. histolytica); splinter shaped with rough, pointed ends	No chromatoidal bodies present; occasionally small granules or inclusions seen; also, *fine* linear structures may be faintly visible on well-stained smears	No chromatoidal bodies present; occasionally, small granules may be present
Glycogen (stained)	May be diffuse or absent in mature cyst; clumped chromatin mass may be present in early cysts (will stain reddish brown with iodine)	May or may not be present, as in E. histolytica	May be diffuse or absent in mature cysts; clumped mass occasionally seen in immature cysts (will stain reddish brown in iodine)	Usually diffuse if present (will stain reddish brown in iodine)	Large, compact, well-defined mass (will stain reddish brown in iodine)

Table 35-5. Morphologic criteria used to identify intestinal protozoa [8,20,40]: Flagellates, trophozoites

	Dientamoeba fragilis	*Trichomonas hominis*	*Giardia lamblia*	*Chilomastix mesnili*	*Enteromonas hominis*	*Retortamonas intestinalis*
Shape and size	Shaped like amebae; 5 to 15 μm; usual range, 9 to 12 μm	Pear shaped; 8 to 20 μm; usual range, 11 to 12 μm	Pear shaped; 10 to 20 μm; usual range, 12 to 15 μm	Pear shaped; 6 to 24 μm; usual range, 10 to 15 μm	Oval; 4 to 10 μm; usual range, 8 to 9 μm	Pear shaped or oval; 4 to 9 μm; usual range, 6 to 7 μm
Motility	Usually nonprogressive; pseudopodia angular, serrated, or broad lobed and almost transparent	Jerky and rapid	"Falling leaf"	Stiff, rotary	Jerky	Jerky
Number of nuclei	Percentage may vary, but approximately 40% of organisms will have 1 nucleus, and 60% 2 nuclei; not visible in unstained preparations; no peripheral chromatin, karyosome composed of cluster of 4 to 8 granules	Not visible in unstained mounts; 1	Not visible in unstained mounts; 2	Not visible in unstained mounts; 1	Not visible in unstained mounts; 1	Not visible in unstained mounts; 1
Number of flagella (usually difficult to see)	No visible flagella	3 to 5 anterior, 1 posterior	4 lateral, 2 ventral, 2 caudal	3 anterior, 1 in cytostome	3 anterior, 1 posterior	1 anterior, 1 posterior
Other features	Cytoplasm finely granular and may be vacuolated with ingested bacteria, yeasts, and other debris	Axostyle (slender rod) protrudes beyond posterior end and may be visible; undulating membrane extends length of body	Sucking disk occupying ⅓ to ½ of ventral surface; pear-shaped front view, spoon-shaped side view	Prominent cytostome extending ⅓ to ½ length of body; spiral groove across ventral surface	One side of body flattened; posterior flagellum extends free posteriorly or laterally	Prominent cytostome extending approximately ½ length of body

Table 35-6. Morphologic criteria used to identify intestinal protozoa[8,20,40]: Flagellates, cysts

	Dientamoeba fragilis	Trichomonas hominis	Giardia lamblia	Chilomastix mesnili	Enteromonas hominis	Retortamonas intestinalis
Shape	No cyst	No cyst	Oval, ellipsoidal, or may appear round	Lemon shaped with anterior hyaline knob	Elongate or oval	Pear shaped or slightly lemon shaped
Size			8 to 9 μm; usual range, 11 to 12 μm	6 to 10 μm; usual range, 8 to 9 μm	6 to 8 μm; usual range, 4 to 10 μm	4 to 7 μm; usual range 4 to 9 μm
Number of nuclei			Not distinct in unstained preparations; usually located at one end; 4	Not visible in unstained preparations; 1	Usually 2 lying at opposite ends of cyst; not visible in unstained mounts; 1 to 4	Not visible in unstained mounts; 1
Other features			Longitudinal fibers in cyst may be visible in unstained preparations; deep-staining fibers usually lie across longitudinal fibers; there is often shrinkage, and cytoplasm pulls away from cyst wall; may also be "halo" effect around outside of cyst wall	Cytostome with supporting fibrils, usually visible in stained preparation; curved fibril along side of cytostome usually referred to as "shepherd's crook"	Resembles *E. nana* cyst; fibrils or flagella usually not seen	Resembles *Chilomastix* cyst; shadow outline of cytostome with supporting fibrils extending above nucleus

Table 35-7. Morphologic criteria used to identify intestinal protozoa[8,20,40]: Ciliates and coccidia

	Balantidium coli		Isospora belli
	Trophozoite	**Cyst**	
Shape and size	Ovoid with tapering anterior end; 50 to 100 μm in length, 40 to 70 μm wide, usual range, 40 to 50 μm	Spherical or oval; 50 to 70 μm; usual range, 50 to 55 μm	Ellipsoidal oocyst; usual range, 30 μm long, 12 μm wide; sporocysts rarely seen broken out of oocysts, but measure 9 × 11 μm
Motility	Rotary, boring, may be rapid		Nonmotile
Number of nuclei	1 large kidney-shaped macronucleus; 1 small round micronucleus, which is difficult to see even in stained smear; macronucleus may be visible in unstained preparation	1 large macronucleus visible in unstained preparation	
Other features	Body covered with cilia, which tend to be longer near cytosome; cytoplasm may be vacuolated	Macronucleus and contractile vacuole visible in young cysts; in older cysts, internal structure appears granular	Mature oocyst contains 2 sporocysts with 4 sporozoites each; usual diagnostic stage is immature oocyst containing spherical mass of protoplasm

sidered to be *Entamoeba histolytica, E. hartmanni* (pathogenicity still open to question), *Dientamoeba fragilis, Giardia lamblia, Isospora belli,* and *Balantidium coli. E. histolytica* is the most important species and may invade other tissues of the body, resulting in severe symptoms and possible death. *D. fragilis* has been associated with diarrhea, nausea, vomiting, and other nonspecific abdominal compliants. *G. lamblia* is probably the most common protozoan organism found in persons in this country and is known to cause symptoms ranging from mild diarrhea, flatulence, and vague abdominal pains to steatorrhea and a typical malabsorption syndrome.

The identification of intestinal protozoan parasites is difficult, at best, and the importance of the permanent stained slide should be reemphasized. It is important to remember that many artifacts (vegetable material, debris, cells of human origin) may mimic protozoan organisms on a wet mount (Fig. 35-12). The important diagnostic characteristics will be visible on the stained smear, and the final identification of protozoan parasites should be confirmed with the permanent stain.

Amebae

Occasionally when fresh stool material is examined as a direct wet mount, motile trophozoites may be observed. *E. histolytica* is described as having directional and progressive motility, while the other amebae tend to move more slowly and at random. The cytoplasm will usually appear finely granular, less frequently coarsely granular, or vacuolated. Bacteria, yeast cells, or debris may be present in the cytoplasm. The presence of red blood cells in the cytoplasm is usually considered to be diagnostic for *E. histolytica.*

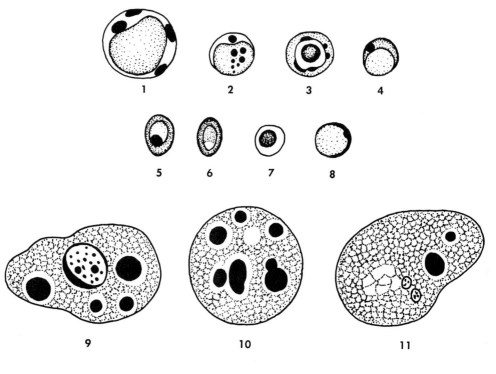

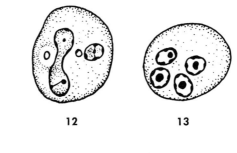

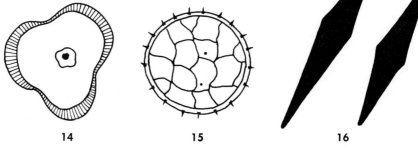

Fig. 35-12. Various structures that may be seen in stool preparations. **1, 2, 4,** *Blastocystis hominis* yeast cells; **3, 5-8,** various yeast cells; **9,** macrophage with nucleus; **10, 11,** deteriorated macrophage without nucleus; **12, 13,** polymorphonuclear leukocytes; **14, 15,** pollen grains; **16,** Charcot-Leyden crystals. (Adapted from Markell, E. K., and Voge, M.: Medical parasitology, ed. 4, Philadelphia, 1976, W. B. Saunders Co.) (Illustration by Nobuko Kitamura.)

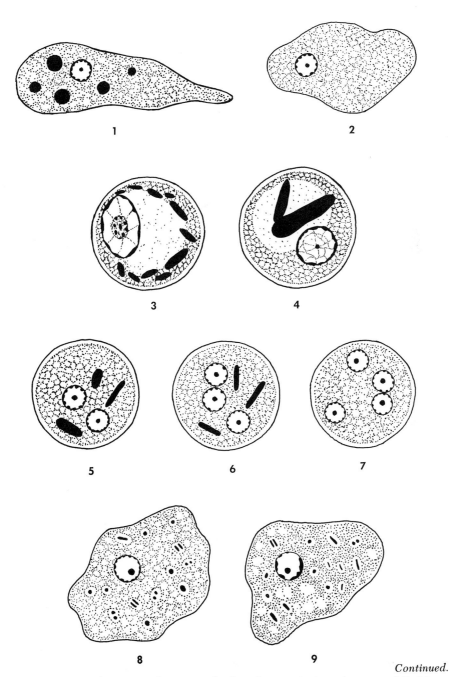

Continued.

Fig. 35-13. **1, 2,** Trophozoites of *Entamoeba histolytica;* **3, 4,** early cysts of *E. histolytica;* **5-7,** cysts of *E. histolytica.* **8, 9,** Trophozoites of *Entamoeba coli;* **10, 11,** early cysts of *E. coli;* **12-14,** cysts of *E. coli.* **15, 16,** Trophozoites of *Entamoeba hartmanni;* **17, 18,** cysts of *E. hartmanni.* (From Garcia, L. S., and Ash, L. R.: Diagnostic parasitology: clinical laboratory manual, St. Louis, 1975, The C. V. Mosby Co.) (Illustrations **4** and **11** by Nobuko Kitamura.)

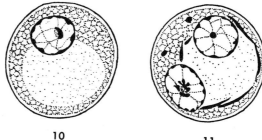

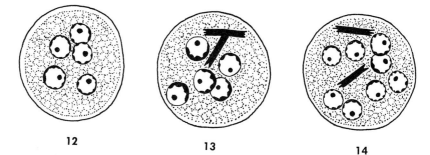

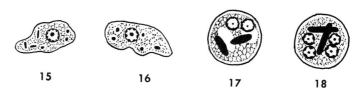

Fig. 35-13, cont'd. For legend see p. 361.

Nuclear morphology is one of the most important criteria used for identification; nuclei of the genus *Entamoeba* contain a relatively small karyosome and have chromatin material arranged on the nuclear membrane (Figs. 35-13 to 35-15). The nuclei of the other two genera, *Endolimax* and *Iodamoeba*, tend to have very large karyosomes with no peripheral chromatin on the nuclear membrane (Figs. 35-16 to 35-18).

The trophozoite stages may often be pleomorphic and asymmetrical, while the cysts are usually less variable in shape with more rigid cyst walls. The number of nuclei in the cysts may vary, but their general morphology is similar to that found in the trophozoite stage. There are various inclusions in the cysts, such as chromatoidal bars or glycogen material, which may be helpful in identification.

Flagellates

Four common species of flagellates are found in the intestinal tract: *Giardia lamblia, Chilomastix mesnili, Trichomonas hominis,* and *Dientamoeba fragilis.* Several other smaller flagellates, such as *Enteromonas hominis* and *Retortamonas intestinalis* (Fig. 35-19), are rarely seen, and none of the flagellates in the intestinal tract, with the exception of *G. lamblia* and *D. fragilis,* are considered pathogenic. *T. vaginalis* is pathogenic, but occurs in the urogenital tract. *T. tenax* is occasionally found in the mouth and may be associated with poor oral hygiene.

Text continued on p. 368.

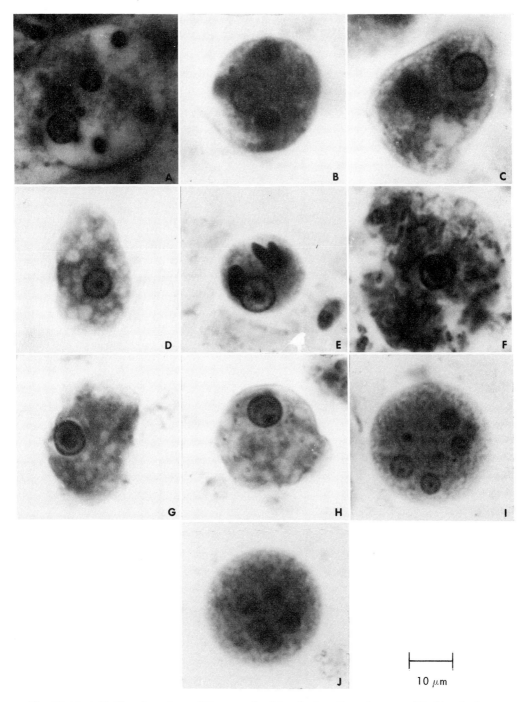

Fig. 35-14. A-D, Trophozoites of *Entamoeba histolytica;* **E,** early cyst of *E. histolytica.*
F-H, Trophozoites of *Entamoeba coli;* **I, J,** cysts of *E. coli.*

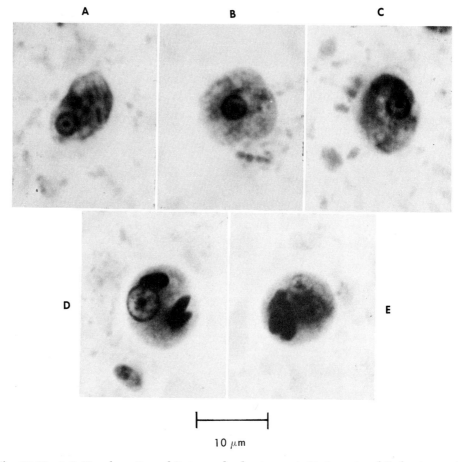

Fig. 35-15. A-C, Trophozoites of *Entamoeba hartmanni;* **D, E,** cysts of *E. hartmanni.*

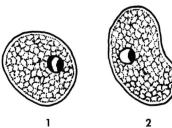

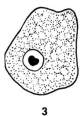

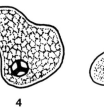

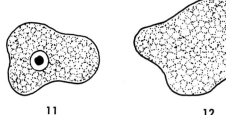

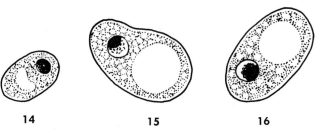

Fig. 35-16. 1-5, Trophozoites of *Endolimax nana;* **6-10,** cysts of *E. nana.* **11-13,**
Trophozoites of *Iodamoeba bütschlii;* **14-16,** cysts of *I. bütschlii.* (From Garcia, L. S.,
and Ash, L. R.: Diagnostic parasitology: clinical laboratory manual, St. Louis, 1975, The
C. V. Mosby Co.)

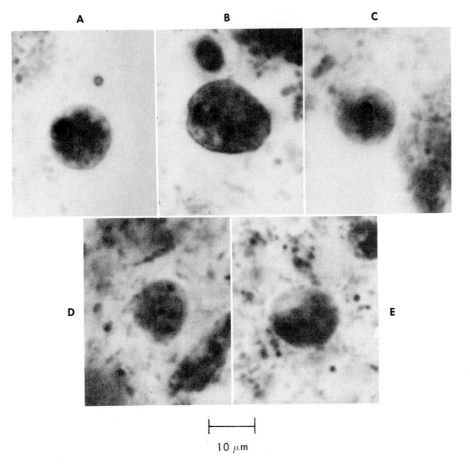

10 μm

Fig. 35-17. A-C, Trophozoites of *Endolimax nana;* **D, E,** cysts of *E. nana.*

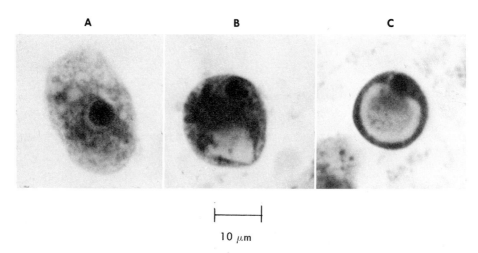

10 μm

Fig. 35-18. A, Trophozoite of *Iodamoeba bütschlii;* **B, C,** cysts of *I. bütschlii.*

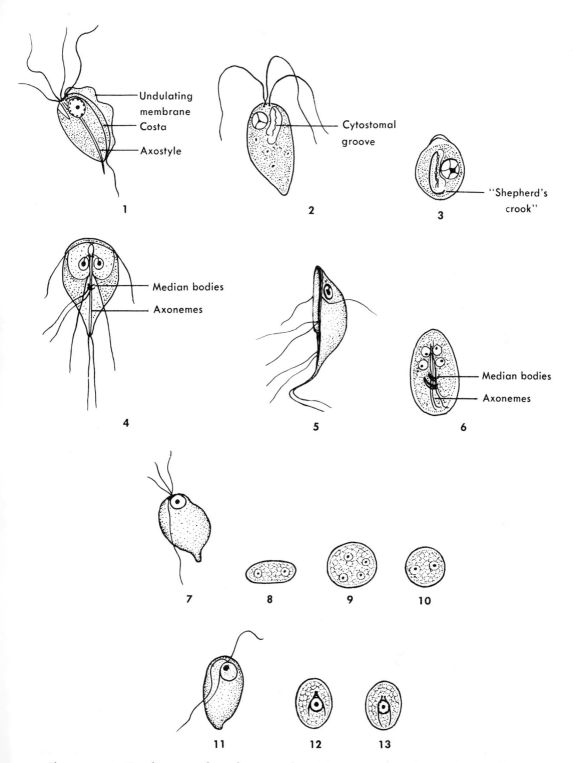

Fig. 35-19. 1, Trophozoite of *Trichomonas hominis.* **2,** Trophozoite of *Chilomastix mesnili;* **3,** cyst of *C. mesnili.* **4,** Trophozoite of *Giardia lamblia* (front view); **5,** Trophozoite of *G. lamblia* (side view); **6,** cyst of *G. lamblia.* **7,** Trophozoite of *Enteromonas hominis;* **8-10,** cysts of *E. hominis.* **11,** Trophozoite of *Retortamonas intestinalis;* **12, 13,** cysts of *R. intestinalis.* (From Garcia, L. S., and Ash, L. R.: Diagnostic parasitology: clinical laboratory manual, St. Louis, 1975, The C. V. Mosby Co.) (Illustration **5** by Nobuko Kitamura; illustrations **7** to **13** adapted from Markell, E. K., and Voge, M.: Medical parasitology, ed. 4, Philadelphia, 1976, W. B. Saunders Co.)

With the exception of *Dientamoeba*, the flagellates are easily recognized by their characteristic rapid motility, which has been described as a "falling leaf" motion for *Giardia* and a jerky motion for the other species. Most of the flagellates have a characteristic pear shape and possess different numbers and arrangements of flagella, depending on the species. The sucking disk and axonemes of *Giardia*, the cytostome and spiral groove of *Chilomastix*, and the undulating membrane of *Trichomonas* are all distinctive criteria for identification (Figs. 35-19 to 35-21).

Until recently, *Dientamoeba* was grouped with the amebae; however, electron microscopy studies have confirmed its correct classification with the flagellates, specifically the trichomonads.[40] *Dientamoeba* has no known cyst stage and is characterized by having one or two nuclei, which have no peripheral chromatin and which have four to eight chromatin granules in a central mass. This organism is quite variable in size and shape and may contain large numbers of ingested bacteria and other debris. *Dientamoeba* is inconspicuous in the wet mount and will be consistently overlooked without the use of the stained smear (Figs. 35-22 and 35-23).

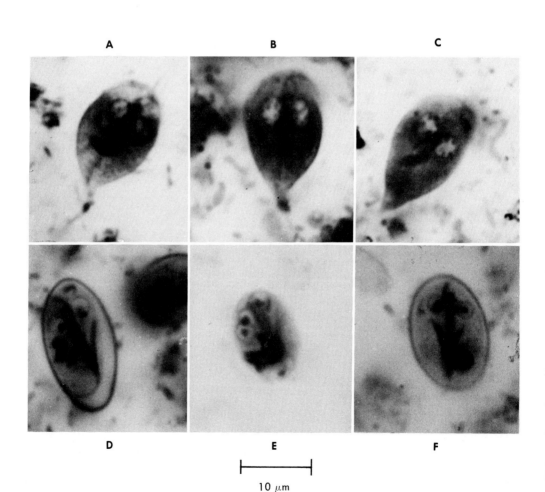

A   B   C

D   E   F

10 μm

Fig. 35-20. A-C, Trophozoites of *Giardia lamblia;* **D-F,** cysts of *G. lamblia.*

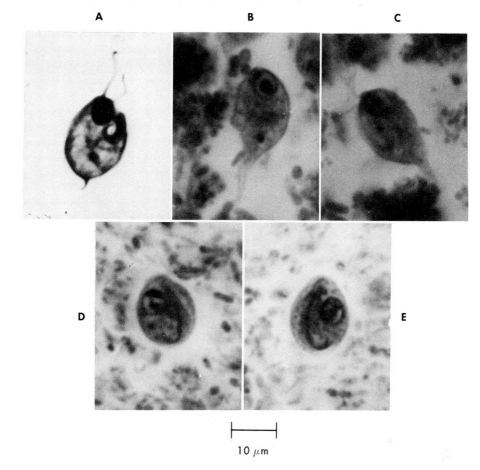

Fig. 35-21. A-C, Trophozoites of *Chilomastix mesnili* (**A,** silver stain); **D, E,** cysts of *C. mesnili*).

Ciliates

Balantidium coli is the largest protozoan and the only ciliate that infects humans. The living trophozoites have a rotatory, boring motion, which is usually rapid. The surface of the organism is covered by cilia, and the cytoplasm contains both a kidney-shaped macronucleus and a smaller, round micronucleus that is often difficult to see. The number of nuclei in the cyst remain the same as that in the trophozoite. *B. coli* infections are rarely seen in this country; however, the organisms are quite easily recognized (Fig. 35-24).

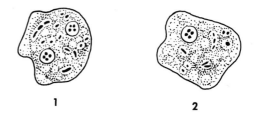

Fig. 35-22. 1, 2, Trophozoites of *Dientamoeba fragilis*. (From Garcia, L. S., and Ash, L. R.: Diagnostic parasitology: clinical laboratory manual, St. Louis, 1975, The C. V. Mosby Co.)

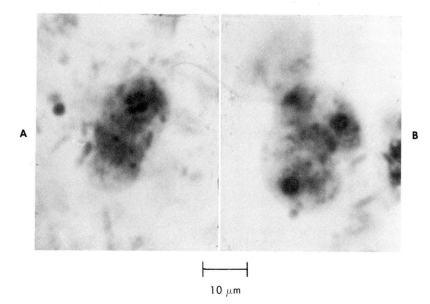

10 μm

Fig. 35-23. A, B, Trophozoites of *Dientamoeba fragilis.*

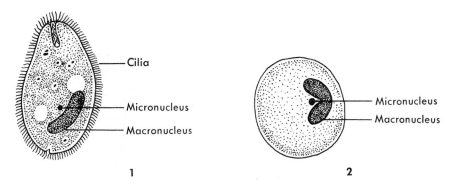

Cilia

Micronucleus

Macronucleus

Micronucleus

Macronucleus

1   2

Fig. 35-24. 1, Trophozoite of *Balantidium coli;* **2,** cyst of *B. coli.* (From Garcia, L. S., and Ash, L. R.: Diagnostic parasitology: clinical laboratory manual, St. Louis, 1975, The C. V. Mosby Co.)

Sporozoa

Isospora belli is now considered to be the only valid species of the genus *Isospora* that infects humans. This organism is released from the intestinal wall as immature oocysts, so all stages from the immature oocyst containing a mass of undifferentiated protoplasm to those containing fully developed sporocysts and sporozoites will be found in the stool (Fig. 35-25). If passed in the immature condi-tion, they will mature within 4 or 5 days to form sporozoites. This organism is rarely seen; however, it is not easily recognized in the stool and may be more common than statistics would indicate.

Intestinal helminths

The intestinal helminths that infect humans belong to two phyla: the Nematoda, or roundworms; and the Platyhelminthes, or flatworms. The Platyhelminthes, most

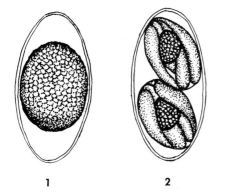

1   **2**

Fig. 35-25. 1, Immature oocyst of *Isospora belli;* **2,** mature oocyst of *I. belli.* (Illustration by Nobuko Kitamura.)

of which are hermaphroditic, have a flat bilaterally symmetrical body. The two classes, Trematoda and Cestoda, contain organisms that are parasitic for human beings.

The trematodes (flukes) are leaf-shaped or elongate and slender organisms (blood flukes: *Schistosoma* sp.) that possess hooks or suckers for attachment. Members of this group, which parasitize humans, are found in the intestinal tract, the liver, the blood vessels, and lungs.

The cestodes (tapeworms) typically have a long, segmented, ribbonlike body, which has a special attachment portion, or scolex, at the anterior end. Adult forms inhabit the small intestine; however, human beings may be host to either the adult or larval forms, depending on the species. The cestodes, as well as the trematodes, require, with few exceptions, one or more intermediate hosts for the completion of the life cycle.

The phylum Nematoda, or roundworms, are elongate, cylindrical worms containing a well-developed digestive tract. The sexes are separate, the male usually being smaller than the female. Intermediate hosts are required for larval development in certain species; a large number of species parasitize the intestinal tract and certain tissues of humans.

Diagnosis of most intestinal helminth

Key to helminth eggs*

a. Egg nonoperculated, spherical or subspherical, containing a six-hooked embryo   b
    Egg other than above   e
b. Eggs separate   c
    Eggs in packets of twelve or more
                   Dipylidium caninum
c. Outer surface of egg consists of a thick, radially striated capsule or embryophore   *Taenia* sp.
    Outer surface of egg consists of very thin shell, separated from inner embryophore by gelatinous matrix   d
d. Filamentous strands occupy space between embryophore and outer shell
                  Hymenolepis nana
    No filamentous strands between embryophore and outer shell   *H. diminuta*
e. Egg operculated   f
    Egg nonoperculated   j
f. Egg less than 35 μm long   *Clonorchis (Opisthorchis)* sp. or *Heterophyes heterophyes* or *Metagonimus yokogawai*
    Egg 38 μm or over   g
g. Egg 38 to 45 μm in length   *Dicrocoelium dendriticum*
    Egg over 60 μm in length   h
h. Egg with shoulders into which operculum fits
                  Paragonimus westermani
    Egg without opercular shoulders   i
i. Egg more than 85 μm long   *Fasciolopsis buski* or *Fasciola hepatica* or *Echinostoma* sp.
    Egg less than 75 μm long   *Diphyllobothrium latum*
j. Egg 75 μm or more in length, spined   k
    Egg less than 75 μm long, not spined   m
k. Spine terminal   *Schistosoma haematobium*
    Spine lateral   l
l. Lateral spine inconspicuous (perhaps absent)
                  S. japonicum
    Lateral spine prominent   *S. mansoni*
m. Egg with thick tuberculated capsule
                  Ascaris lumbricoides
    Egg without thick tuberculated capsule   n
n. Egg barrel shaped, with polar plugs   o
    Egg not barrel shaped, without polar plugs   p
o. Shell nonstriated   *Trichuris trichiura*
    Shell often striated   *Capillaria* spp.
p. Egg flattened on one side
                  Enterobius vermicularis
    Egg symmetrical   q
q. Egg with large blue-green globules at poles
                  Heterodera marioni
    Egg without polar globules   r
r. Egg bluntly rounded at ends, 56 to 76 μm long
                  hookworm
    Egg pointed at one or both ends, 73 to 95 μm long   *Trichostrongylus* sp.

*From Markell, E. K., and Voge, M.: Medical parasitology, ed. 4, Philadelphia, 1976, W. B. Saunders Co.

A B C

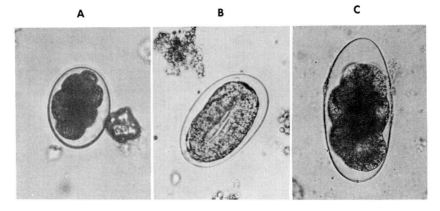

Fig. 35-26. A, Immature hookworm egg; **B,** embryonated hookworm egg. **C,** *Trichostrongylus orientalis,* immature egg. **D,** *Strongyloides stercoralis,* rhabditiform larva (200 μm). **E,** *Enterobius vermicularis* egg. **F,** *Trichuris trichiura* egg. **G,** *Ascaris lumbricoides,* fertilized egg; **H,** *A. lumbricoides,* fertilized egg, decorticate; **I,** *A. lumbricoides,* unfertilized egg; **J,** *A. lumbricoides,* unfertilized egg, decorticate.

infections is based on the detection of the characteristic eggs and larvae in the stool; occasionally, adult worms or portions of worms may also be found. No permanent stains are required, and most diagnostic features can easily be seen on direct wet mounts or in mounts of the concentrated stool material.

Nematodes

The majority of nematodes are diagnosed by finding the characteristic eggs in the stool (Fig. 35-26). The eggs of *Ancylostoma duodenale* and *Necator americanus* are essentially identical, so an infection with either species is reported as "Hookworm eggs present." *Trichostrongylus* eggs may easily be mistaken for those of hookworms; however, the eggs of *Trichostrongylus* are somewhat larger and one end tends to be more pointed.

Strongyloides stercoralis is passed in the feces as the noninfective rhabditiform larva. Although hookworm eggs are normally passed in the stool, these eggs may continue to develop and hatch if the stool is left at room temperature for several days. These larvae may be mistaken for those of *Strongyloides.* Fig. 35-27 shows

the morphologic differences between the rhabditiform larvae of hookworm and *Strongyloides.* Recovery of *Strongyloides* larvae in duodenal contents is mentioned in the section Duodenal Contents.

The appropriate techniques for recovery of *Enterobius vermicularis* (pinworm) eggs are given on pp. 334-335. Eggs of the other nematodes are fairly easy to find and to differentiate from one another.

Cestodes

With the exception of *Diphyllobothrium latum,* tapeworm eggs are embryonated and contain a six-hooked oncosphere (Fig. 35-28 and Table 35-8).

Taenia saginata and *T. solium* cannot be speciated on the basis of egg morphology; gravid proglottids or the scolices must be examined (Fig. 35-29). *T. saginata* (beef tapeworm) proglottids have approximately 15 to 30 main lateral branches, and the scolex has no hooks; the proglottids of *T. solium* have 7 to 12 main lateral branches, and the scolex has a circle of hooks.

The eggs of *Hymenolepis nana* (more common) and *H. diminuta* are very similar; however, *H. nana* eggs are

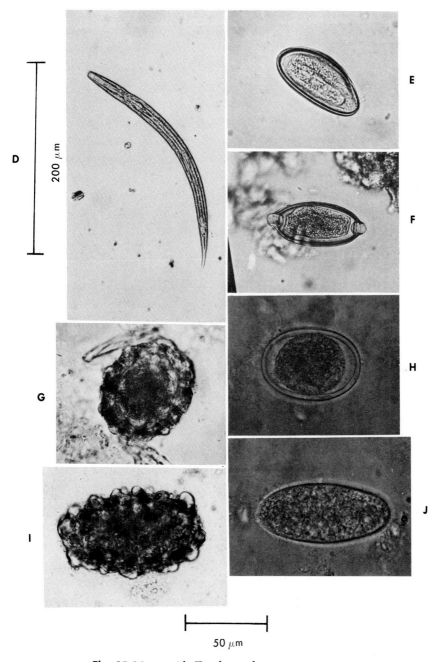

Fig. 35-26, cont'd. For legend see opposite page.

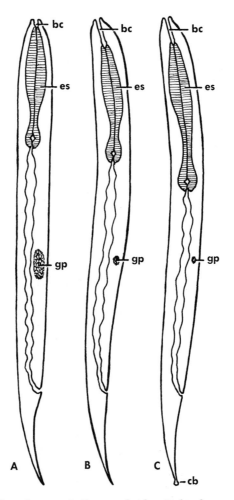

Fig. 35-27. Rhabditiform larvae: **A,** *Strongyloides;* **B,** hookworm; **C,** *Trichostrongylus.* *bc,* Buccal cavity; *es,* esophagus; *gp,* genital primordia; *cb,* beadlike swelling of caudal tip. (Illustration by Nobuko Kitamura.)

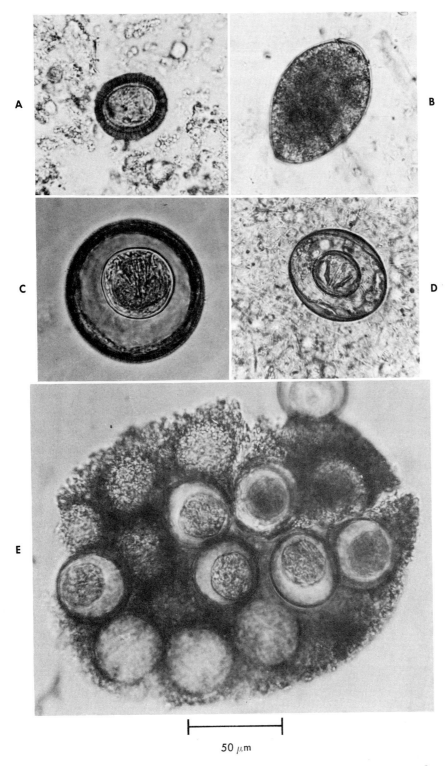

Fig. 35-28. A, *Taenia* spp. egg. **B,** *Diphyllobothrium latum* egg. **C,** *Hymenolepis diminuta* egg. **D,** *Hymenolepsis nana* egg. **E,** *Dipylidium caninum* egg packet.

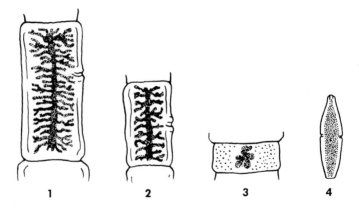

Fig. 35-29. Gravid proglottids: **1,** *Taenia saginata;* **2,** *Taenia solium;* **3,** *Diphyllobothrium latum;* **4,** *Hymenolepis nana.* (From Garcia, L. S., and Ash, L. R.: Diagnostic parasitology: clinical laboratory manual, St. Louis, 1975, The C. V. Mosby Co.)

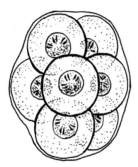

Fig. 35-30. *Dipylidium caninum* egg packet. (Illustration by Nobuko Kitamura.)

Key to gravid proglottids of the major human tapeworms*

a. Uterus forms rosette in center of proglottid
　　　　　　　　　Diphyllobothrium latum
　 Uterus otherwise disposed　　　　　　b
b. Uterus with central stem running length of
　　 proglottid　　　　　　　　　　　c
　 Uterus without central stem　　　　　d
c. Central stem with 7 to 13 main lateral branches
　　　　　　　　　　　　　Taenia solium
　 Central stem with 15 to 20 main lateral branches
　　　　　　　　　　　　　T. saginata
d. Proglottid wider than long　*Hymenolepis* sp.
　 Proglottid longer than wide
　　　　　　　　　　　　Dipylidium caninum

*From Markell, E. K., and Voge, M.: Medical parasitology, ed. 4, Philadelphia, 1976, W. B. Saunders Co.

smaller and have polar filaments, which are present in the space between the oncosphere and the egg shell.

Eggs of *Dipylidium caninum* are occasionally found in humans (particularly children) and are passed in the feces in packets of 5 to 15 eggs each (Fig. 35-30). The proglottids may also be found; they may resemble cucumber seeds or, when dry, may look like rice grains (white).

The fish tapeworm, *Diphyllobothrium latum,* does not have embryonated eggs; these eggs have a somewhat thicker shell and are operculated, much like the trematode eggs.

Trematodes

Most of the trematodes have operculated eggs, which are best recovered by the sedimentation concentration technique rather than the flotation method. Many of these eggs are very similar, both in size and morphology, and often careful measurements must be taken to speciate the eggs (Fig. 35-31). The eggs of *Clonorchis (Opisthorchis), Heterophyes,* and *Metagonimus* are very similar and quite small; they are easily missed if the concentration sediment is examined with

Table 35-8. Differential characteristics of some important tapeworms of humans[8,20,40]

	Taenia saginata	*Taenia solium*	*Hymenolepis nana*	*Diphyllobothrium latum*
Length	4 to 8 m	3 to 5 m	2.5 to 4 cm	4 to 10 m
Scolex				
Shape	Quadrilateral	Globular	Usually not seen	Almondlike
Size	1 × 1.5 mm	1 × 1 mm		3 × 1 mm
Rostellum and hooklets	No	Yes		No
Suckers	4	4		2 (grooves)
Terminal proglottids (gravid)				
Size	19 × 7 mm, longer than wide	11 × 5 mm	Usually not seen	3 × 11 mm, wider than long
Primary lateral uterine branches	15 to 30 on each side	6 to 12 on each side		Rosette shaped
Color	Milky white	Milky white		Ivory
Appearance in feces	Usually appear singly	5 or 6 segments		Varies from a few inches to a few feet in length
Ova				
Shape	Spheroid	Spheroid	Broadly oval	Oval
Size	35 µm	35 µm	30 × 47 µm	70 × 45 µm
Color	Rusty brown	Rusty brown	Pale	Yellow-brown
Embryo with hooklets	Yes	Yes	Yes	No
Operculum	No	No	No	Yes (difficult to see)

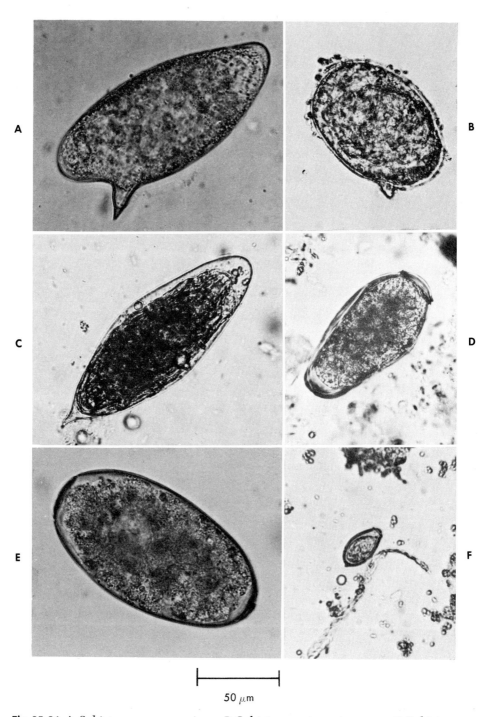

50 μm

Fig. 35-31. A, *Schistosoma mansoni* egg. **B,** *Schistosoma japonicum* egg. **C,** *Schistosoma haematobium* egg. **D,** *Paragonimus westermani* egg. **E,** *Fasciola hepatica* egg. **F,** *Clonorchis (Opisthorchis) sinensis* egg.

the 10× objective only. The eggs of *Fasciola hepatica* and *Fasciolopsis buski* are also very similar but much larger than those just mentioned.

Paragonimus westermani eggs are not only found in the stool but may be found in sputum. These eggs are very similar in size and shape to the egg of the fish tapeworm, *D. latum.*

Probably the easiest trematode eggs to identify are those of the schistosomes: *Schistosoma mansoni* eggs are characterized by having a very prominent lateral spine, *S. haematobium* a terminal spine, and *S. japonicum* a small lateral spine that may be difficult to see (Fig. 35-31). These eggs are nonoperculated. Specific procedures for their recovery and identification are found on p. 334.

REFERENCES

1. Allison, A. C.: Genetic factors in resistance to malaria, Ann. N.Y. Acad. Sci. **91:**710, 1961.
2. Allison, A. C., and Clyde, D. F.: Malaria in African children with deficient erythrocyte glucose-6-phosphate dehydrogenase, Br. Med. J. **1:** 1346, 1961.
3. Beal, C. B., Viens, P., Grant, R. G. L., and Hughes, J. M.: A new technique for sampling duodenal contents; demonstration of upper small-bowel pathogens, Am. J. Trop. Med. Hyg. **19:**349-352, 1970.
4. Beaver, P. C.: A nephelometric method of calibrating the photoelectric meter for making egg-counts by direct fecal smear, J. Parasitol. **35**(sec. 2): 13, 1949.
5. Beaver, P. C.: The standardization of fecal smears for estimating egg production and worm burden, J. Parasitol. **36:**451-456, 1950.
6. Biagi, F., Navarrete, F., Piña, A., Santiago, A. M., and Tapia, L: Estudio de tres reacciones serologicas en el diagnostico de la cisticercosis, Rev. Med. Hosp. Gen. Mexico City **24:**501-508, 1961.
7. Brooke, M. M., and Goldman, M.: Polyvinyl alcohol-fixative as a preservative and adhesive for Protozoa in dysenteric stools and other liquid material, J. Lab. Clin. Med. **34:**1554-1560, 1949.
8. Brooke, M. M., and Melvin, D.: Morphology of diagnostic stages of intestinal parasites of man, DHEW Publ. No. (HSM) 72-8116, Atlanta, 1969, U.S. Government Printing Office.
9. Chatterjea, J. B., Saha, T. K., Ray, R. N., and Chaudluri, R. N.: Response of tropical splenomegaly and thalassemia to induce malaria, Bull. Calcutta School Trop. Med. **4:**105-106, 1956.
10. Culbertson, C. G., Ensminger, P. W., and Overton, W. M.: The isolation of additional strains of pathogenic *Hartmanella* sp. (Acanthamoeba). Proposed culture method for application to biological material, Am. J. Clin. Pathol. **43:**383-387, 1965.
11. Dennis, D. T., and Kean, B. H.: Isolation of microfilariae: report of a new method, J. Parasitol. **57:**1146-1147, 1971.
12. Desowitz, R. S., and Hitchcock, J. C.: Hyperendemic brancroftian filariasis in the kingdom of Tonga: the application of the membrane filter concentration technique to an age-stratified blood survey, Am. J. Trop. Med. Hyg. **23:**877-879, 1974.
13. Desowitz, R. G., Saave, J. J., and Stein, B.: Application of the indirect hemagglutination test in recent studies on the immunoepidemiology of human malaria and the immune response in experimental malaria, Mil. Med. (Suppl.) **131:** 1157-1166, 1966.
14. Dunn, F. L.: The TIF direct smear as an epidemiological tool, Bull. WHO **39:**439-449, 1968.
15. Faust, E. C., D'Antoni, J. S., Odom, V., Miller, M. F., Peres, C., Sawitz, W., Thomen, L. F., Tobie, J., and Walker, J. H.: A critical study of clinical laboratory technics for the diagnosis of protozoan cysts and helminth eggs in feces, Am. J. Trop. Med. **18:**169-183, 1938.
16. Faust, E. C., Russell, P. F., and Jung, R. C.: Craig and Faust's clinical parasitology, ed. 8, Philadelphia, 1970, Lea & Febiger.
17. Feinberg, J. G., and Whittington, M. J.: A culture medium for *Trichomonas vaginalis* Donne and species of Candida, J. Clin. Pathol. **10:**327-329, 1957.
18. Field, J. W., Sandosham, A. A., and Fong, Y. L.: The microscopical diagnosis of human malaria. I. A morphological study of the erythrocytic parasites in thick blood films, ed. 2, Malaya, 1963, Institute of Medical Research.
19. Field, J. W., and Shute, P. G.: The microscopic diagnosis of human malaria. II. A morphological study of the erythrocytic parasites, Malaya, 1956, Institute of Medical Research.
20. Garcia, L. S., and Ash, L. R.: Diagnostic parasitology: clinical laboratory manual, St. Louis, 1975, The C. V. Mosby Co.
21. Gleason, N. N., and Healy, G. R.: Modification and evaluation of Kohn's one-step staining technic for intestinal protozoa in feces or tissue, Am. J. Clin. Pathol. **43:**494-496, 1965.
22. Gomori, G.: A rapid one-step trichrome stain, Am. J. Clin. Pathol. **20:**661-663, 1950.
23. Graham, C. F.: A device for the diagnosis of *Enterobius* infection, Am. J. Trop. Med. **21:**159-161, 1941.
24. Harada, Y., and Mori, O.: A new method for culturing hookworm, Yonago Acta Medica **1:**177-179, 1955.

25. Hsieh, H. C.: A test-tube filter-paper method for the diagnosis of *Ancylostoma duodenale*, *Necator americanus* and *Strongyloides stercoralis*, WHO Tech. Rept. Ser. **255**:27-30, 1962.

26. Ingram, R. L., Otken L. B., Jr., and Jumper, J. R.: Staining of malarial parasites by the fluorescent antibody technic, Proc. Soc. Exp. Biol. Med. **106**:52-54, 1961.

27. Kagan, I. G., Norman, L., Allain, D. S., and Goodchild, C. G.: Studies on echinococcosis: nonspecific serologic reactions of hydatid-fluid antigen with serum of patients ill with diseases other than echinococcosis, J. Immunol. **84**:635-640, 1960.

28. Kagan, I. G., Osimani, J. J., Varela, J. C., and Allain, D. S.: Evaluation of intradermal and serologic tests for the diagnosis of hydatid disease, Am. J. Trop. Med. Hyg. **15**:172-179, 1966.

29. Kagan, I. G., and Quist, K. D.: An evaluation of five serologic tests for the diagnosis of trichinosis in lightly infected swine. In Singh, K. S., Tandan, B. K., editors: H. D. Srivastava Commemoration Volume, Iznatnagar, U. P. India, 1970, India Veterinary Research Institute.

30. Kato, K., and Miura, M.: Comparative examinations (Japanese text), Jap. J. Parasitol. **3**:35, 1954.

31. Knott, J. I.: A method for making microfilarial surveys on day blood, Trans. R. Soc. Trop. Med. Hyg. **33**:191-196, 1939.

32. Kuvin, S. F., Tobie, J. E., Evans, C. B., Coatney, G. R., and Contacos, P. G.: Fluorescent antibody studies on the course of antibody production and serum gamma globulin levels in normal volunteers infected with human and simian malaria, Am. J. Trop. Med. Hyg. **11**:429-436, 1962.

33. Kwapinski, J. B.: Methods of serological research, New York, 1965, John Wiley & Sons, Inc.

34. Labzoffsky, N. A., Baratawidjaja, R. K., Kuitunen, E., Lewis, F. N., Kavelman, D. A., and Morrissey, L. P.: Immunofluorescence as an aid in the early diagnosis of trichinosis, Can. Med. Assoc. J. **90**:920-921, 1964.

35. Lennette, E. H., Spaulding, E. H., and Truant, J. P., editors: Manual of clinical microbiology, ed. 2, Washington, D. C., 1974, American Society for Microbiology.

36. Lincicome, D. R.: Fluctuation in numbers of cysts of *Endamoeba histolytica* and *Endamoeba coli* in the stools of rhesus monkeys, Am. J. Hyg. **36**:321-337, 1942.

37. Luna, L. G.: Manual of histologic staining methods of the Armed Forces Institute of Pathology, ed. 3, New York, 1968, McGraw-Hill Book Co.

38. Maekelt, G. A.: A modified procedure of xenodiagnosis for Chagas' disease, Am. J. Trop. Med. Hyg. **13**:11-15, 1964.

39. Mahoney, D. F., Redington, B. C., and Schoenbechler, M. J.: Preparation and serologic activity of plasmodial fractions, Mil. Med. (Suppl.) **131**:1141-1151, 1966.

40. Markell, E. K., and Voge, M.: Medical parasitology, ed. 4, Philadelphia, 1976, W. B. Saunders Co.

41. Martin, L. K., and Beaver, P. C.: Evaluation of Kato thick-smear technique for quantitative diagnosis of helminth infections, Am. J. Trop. Med. Hyg. **17**:382-391, 1968.

42. Melvin, D. M., and Brooke, M. M.: Laboratory procedures for the diagnosis of intestinal parasites, DHEW Publ. No. (CDC) 75-8282, Washington, D.C., 1974, U.S. Government Printing Office.

43. Meuwissen, J. H. E. T., and Leeuwenberg, A. D. E. M.: Indirect haemagglutination test for malaria with lyophilized cells, Trans. R. Soc. Trop. Med. Hyg. **66**:666-667, 1972.

44. Miller, L. H., Mason, S. J., Dvorak, J. A., McGinniss, M. H., and Rothman, I. K.: Erythrocyte receptors for *(Plasmodium knowlesi)* malaria: Duffy blood group determinants, Science **189**:561-562, 1975.

45. Nair, C. P.: Rapid staining of intestinal amoebae on wet mounts, Nature **172**:1051, 1953.

46. Perl, G.: Errors in the diagnosis of *Trichomonas vaginalis* infection, Obstet. Gynecol. **39**:7-9, 1972.

47. Proctor, E. M., and Elsdon-Dew, R.: Serological tests in porcine cysticerosis, S. Afr. J. Sci. **62**:264-267, 1966.

48. Proctor, E. M., Powell, S. J., and Elsdon-Dew, R.: The serological diagnosis of cysticercosis, Ann. Trop. Med. Parasitol. **60**:146-151, 1966.

49. Ritchie, L. S.: An ether sedimentation technique for routine stool examinations, Bull. U.S. Army Med. Dept. **8**:326, 1948.

50. Rogers, W. A., Jr., Fried, J. A., and Kagan, I. G.: A modified indirect microhemagglutination test for malaria, Am. J. Trop. Med. Hyg. **17**:804-809, 1968.

51. Sapero, J. J., and Lawless, D. K.: The MIF stain-preservation technique for the identification of intestinal protozoa, Am. J. Trop. Med. Hyg. **2**:613-619, 1953.

52. Sawitz, W. G., and Faust, E. C.: The probability of detecting intestinal protozoa by successive stool examinations, Am. J. Trop. Med. **22**:131-136, 1942.

53. Scholten, T. H., and Yang, J.: Evaluation of unpreserved and preserved stools for the detection and identification of intestinal parasites, Am. J. Clin. Pathol. **62**:563-567, 1974.

54. Sheehen, D. C., and Hrapchak, B. B.: Theory and practice of histotechnology, St. Louis, 1973, The C. V. Mosby Co.

55. Shute, P. G.: The staining of malaria parasites,

Trans. R. Soc. Trop. Med. Hyg. **60**:412-416, 1966.

56. Shute, P. G., and Maryon, M. E.: Laboratory technique for the study of malaria, ed. 2, London, 1966, J. & A. Churchill, Ltd.

57. Sodeman, W. A., Jr., and Jeffery, G. M.: Indirect fluorescent antibody test for malaria antibody, Pub. Hlth. Rep. **81**:1037-1041, 1966.

58. Spencer, F. M., and Monroe, L. S.: The color atlas of intestinal parasites, ed. 2, Springfield, Ill., 1976, Charles C Thomas, Publisher.

59. Stein, B., and Desowitz, R. S.: The measurement of antibody in human malaria by a formalized sheep cell hemagglutination test, Bull. WHO **30**:45-49, 1964.

60. Stoll, N. R., and Hausheer, W. C.: Concerning two options in dilution egg counting: small drop and displacement, Am. J. Hyg. **6**:134-145, 1926.

61. Sulzer, A. J., and Chisholm, E. S.: Comparison of the IFA and other tests for *Trichinella spiralis* antibodies, Pub. Hlth. Rep. **81**:729-734, 1966.

62. Sulzer, A. J., and Wilson, M.: The use of thick-smear antigen slides in the malaria indirect fluorescent antibody test, J. Parasitol. **53**:1110-1111, 1967.

63. Sulzer, A. J., Wilson, M., and Hall, E. C.: Indirect fluorescent antibody tests for parasitic diseases. V. An evaluation of a thick-smear antigen in the IFA test for malaria antibodies, Am. J. Trop. Med. Hyg. **18**:199-205, 1969.

64. Tobie, J. E., and Coatney, G. R.: Fluorescent antibody staining of human malaria parasites, Exp. Parasitol. **11**:128-132, 1961.

65. Tompkins, V. N., and Miller, J. K.: Staining intestinal protozoa with iron-hematoxylin-phosphotungstic acid, Am. J. Clin. Pathol. **17**:755-758, 1947.

66. Vattuone, N. H., and Yanovsky, J. F.: *Trypanosoma cruzi:* agglutination activity of enzyme-treated epimastigotes, Exp. Parasitol. **30**:349-355, 1971.

67. Voller, A., and Bray, R. S.: Fluorescent antibody staining as a measure of malarial antibody, Proc. Soc. Exp. Biol. Med. **110**:907-910, 1962.

68. Walker, A. J.: Manual for the microscopic diagnosis of malaria, ed. 3 (Sci. Publ. No. 161), Washington, D.C., 1968, Pan American Health Organization.

69. Wheatley, W. B.: A rapid staining procedure for intestinal amoebae and flagellates, Am. J. Clin. Pathol. **21**:990-992, 1951.

70. Wilcox, A: Manual for the microscopical diagnosis of malaria in man, Washington, D.C., 1960, U.S. Public Health Service.

PART V

Antimicrobial susceptibility tests and assays

Determination of susceptibility of bacteria to antimicrobial agents; assay of antimicrobial agents

According to Isenberg,[12] the value of the clinical laboratory can be measured only by the significance of the guidance it gives the practicing physician in the treatment of his patients. In no other area of clinical microbiology does this statement become more pertinent than in the testing of clinical isolates for their susceptibility to antimicrobial agents. With the increasing number of these agents at the physician's disposal and the changing pattern of resistance and susceptibility among bacteria—particularly the gram-negative enteric bacilli—the clinician must rely more and more on susceptibility testing to guide the selection of appropriate drugs or the altering of an already imposed regimen. Therefore, to a large extent, a laboratory report showing susceptibility or resistance to a particular antimicrobial agent becomes an endorsement of its usefulness or withdrawal.

Since the clinical microbiologist is in a position to advise the physician regarding proper antimicrobial therapy, it follows that he or she must maintain (1) a high level of accuracy in testing procedures and (2) a high degree of reproducibility for the results.[12] **Only through close cooperation and exchange of information between the laboratory staff and the clinician can the best possible management of an infectious process be achieved.**

The principle methods used by the laboratory to determine susceptibility of a microorganism to an antimicrobial agent include the **dilution tests,** such as the tube and agar plate dilution procedures, and the **disk agar diffusion test,** utilizing antibiotic-impregnated disks. Each method has its advantages and its limitations, and these must be fully understood and appreciated to obtain maximum usefulness from the results.[22] Since all of these methods have a place in the clinical laboratory, the procedures and directions for the use of each will be described in detail.

In the interpretation of any in vitro susceptibility tests, it is well to remember that they are essentially **artificial measurements;** the data obtained from these tests give only the approximate range of effective inhibitory action against the microorganisms. The only absolute criterion of the efficacy of an antibiotic is the **clinical response** of the patient after an adequate dose of the appropriate drug is administered.

A recent general reference of interest is Thornsberry, C., Gavan, T. L., and Gerlach, E. H.: New developments in antimicrobial agent susceptibility testing. In Sherris, J. C., editor: Cumitech 6, Washington, D.C., 1977, American Society for Microbiology.

TUBE DILUTION METHOD FOR DETERMINING SUSCEPTIBILITY TO ANTIBIOTICS

In the tube dilution method for determining the susceptibility of an organism to antimicrobial agents, specific amounts of the antibiotic, prepared in decreasing concentration in broth by serial dilution technique, are inoculated with a broth culture of the bacterium to be tested. The susceptibility of any organism is determined, after a suitable period of incubation, by macroscopic observation of the presence or absence of growth in the varying concentrations of the antimicrobial agent. This end point is a measure of the bacteriostatic effect of the agent on the bacterium and is commonly referred to as the **minimal inhibitory concentration** (MIC). The amount of drug in the tube containing the least amount of antibiotic with no observable growth is the MIC. With minor additions, the technique can be adapted to the determination of bactericidal levels of an antibiotic, or the **minimal bactericidal concentration** (MBC); this is discussed in another section.

A number of factors must be considered in establishing the procedures and in evaluating the results of these tests.[8] They include (1) the medium in which the tests are performed, (2) the stability of the drug, (3) the size of the inoculum, (4) the rate of growth of the organism, and (5) the period of incubation of the tests. Any variation in one or more of these factors may influence the test results, and those obtained by one procedure may not agree with those arrived at by a slightly different method.[5] However, if a **standard procedure using only pure cultures** is adopted and strictly adhered to, reproducible results can usually be obtained and the reports from a given laboratory can be interpreted readily by the clinical staff.

The tube dilution method is considered the most accurate for determination of susceptibility to measured amounts (either units or micrograms) of an antimicrobial agent. It is a time-consuming and expensive procedure, however, especially when the clinician wants to know the susceptibility of an organism to a number of drugs. For this reason its use may well be restricted to special cases when quantitative results may be of value. In any event, it is strongly recommended that all clinical laboratories should be prepared to provide this service to the clinician, either directly or through a referral laboratory.

The tube dilution method may be recommended for determining the susceptibility of organisms isolated in the following instances: (1) from blood cultures, (2) from patients who fail to respond to apparently adequate therapy, and (3) from patients who relapse while undergoing such therapy. The study of organisms isolated from patients who relapse while on therapy usually involves determination of any increase in resistance on subsequent isolations and may require special methods.

Routine procedure for tube dilution tests
Preparation of stock solution of antimicrobial agents

Stock solutions of drugs are prepared from concentrated, dehydrated sterile material of known potency that may be obtained directly from the pharmaceutical manufacturer. Generally, they are prepared in concentrations of 1,000 μg/ml, using phosphate buffer or Mueller-Hinton broth as the diluent, and are tubed in 1-ml amounts in screw-capped vials.

When stored in the frozen state at -20 C, most antibiotics will remain stable for at least 8 weeks; when refrigerated at 5 C, most show no appreciable loss of potency in 1 week. However, ampicillin requires storage at -60 to -70 C to prevent loss of potency. Any unused thawed solutions of drugs should be discarded; each aliquot should be sufficient for 1 day's use only and should not be refrozen.

Table 36-1 is provided as a guide to the preparation of stock solutions of the most frequently used antimicrobial agents.

Selection of media

The broth media in which the tube dilution sensitivity tests are carried out must be of a kind that will support optimal rapid growth of the test organism in pure culture. A medium that will support

Table 36-1. Procedures for preparing stock solutions

Antimicrobial agent	Manufacturer	Method of preparation
Ampicillin	Bristol-Myers Co.	Weigh out material and multiply by "activity standard" provided by manufacturer.* Add 0.1 ml of pH 8 phosphate buffer to dissolve; dilute with pH 6 phosphate buffer.
Penicillin G	Bristol-Myers Co.	Weigh out material and multiply by "activity standard" provided by manufacturer* (1 unit = 0.62 μg). Dilute with sterile deionized water.
Methicillin	Bristol-Myers Co.	Weigh out material and multiply by "activity standard" provided by manufacturer.* Dilute with pH 6 phosphate buffer.
Oxacillin	Bristol-Myers Co.	Weigh out material and multiply by "activity standard" provided by manufacturer.* Dilute with sterile deionized water.
Cephalothin	Eli Lilly & Co.	Weigh out material and multiply by the "activity standard" provided by manufacturer.* Dilute with pH 6 phosphate buffer.
Cephaloridine	Eli Lilly & Co.	Weigh out exactly 30 mg; add 30 ml of pH 6 phosphate buffer to give 1,000 μg/ml.
Carbenicillin	Pfizer Inc.	Weigh out material and multiply by "activity standard" provided by manufacturer.* Dilute in sterile deionized water. Refrigerate for 2 to 3 hours to dissolve completely.
Tetracycline	Bristol-Myers Co.	Weigh out material and multiply by "activity standard" provided by manufacturer.* Dilute in sterile deionized water; use acid-washed glassware.
Chloramphenicol	Parke, Davis & Co.	Weigh out exactly 50 mg; add 1 ml of ethyl alcohol to dissolve drug and sufficient water to give 50 ml. This gives a concentration of 1,000 μg/ml.
Erythromycin	Eli Lilly & Co.	Weigh out material and multiply by "activity standard" provided by manufacturer.* Dilute in 70% ethanol. Freeze at −60 C.
Lincomycin	The Upjohn Co.	Add 20 ml of water to a vial to make 1,000 μg/ml.
Clindamycin	The Upjohn Co.	Add 15 ml of water to a vial containing 150 mg of drug. Dilute 1:10 to give 1,000 μg/ml.
Kanamycin	Bristol-Myers Co.	Weigh out material and multiply by "activity standard" provided by manufacturer.* Dilute in 0.1 M phosphate buffer, pH 8.
Gentamicin	Schering Corp.	Weigh out material and multiply by "activity standard" provided by manufacturer.* Dilute in 0.1 M phosphate buffer, pH 8.
Tobramycin	Eli Lilly & Co.	Obtain from manufacturer in 2-ml aliquots of 1,000 μg/ml solution.
Amikacin	Bristol-Myers Co.	Weigh out material and multiply by "activity standard" provided by manufacturer.* Dilute in sterile pH 8 phosphate buffer.
Polymyxin B	Burroughs Wellcome & Co.	Add 5 ml of water to a vial containing 50 mg to give 10,000 μg/ml. Dilute to desired concentration.
Colistin (polymyxin E)	Warner-Chilcott Lab.	Weigh out material and multiply by "activity standard" provided by manufacturer.* Dilute with water to desired concentration.
Bacitracin	The Upjohn Co.	Add 10 ml of water to a vial containing 10,000 units to give 1,000 units/ml.

Courtesy John A. Washington II, Head, Section of Clinical Microbiology, Mayo Clinic.
*Example: 1 mg = 825 μg ("activity standard" per microgram); 50 mg = 50 × 825 = 41,250 μg; therefore, add 41.2 ml of diluent to give 1,000 μg/ml.

Continued.

Table 36-1. Procedures for preparing stock solutions—cont'd

Antimicrobial agent	Manufacturer	Method of preparation
Nitrofurantoin	Norwich Pharmaceutical Co.	Weigh out 500 mg of drug. Dilute in 50 ml of polyethylene glycol (mw 300). Heat in 56 C water bath with shaking to dissolve. Solution contains 10,000 μg/ml.
Nalidixic acid	Winthrop Labs.	Weigh out 30 mg; add 2 ml of 1 N NaOH and allow to stand to dissolve (may require gentle heat). Dilute with sterile water (less 2 ml) to desired concentration.

the growth of pneumococci and other streptococci without the addition of serum or blood is preferable, since the addition of such enrichment adds another variable to the test and may influence the results. Trypticase soy broth, or preferably Mueller-Hinton broth* is recommended for sensitivity tests, with the following exceptions: microaerophilic streptococci, which are isolated from 10% of patients with subacute bacterial endocarditis, obligately anaerobic cocci, *Bacteroides* species, and clostridia should be tested in Brucella broth† enriched with $NaHCO_3$, hemin, and vitamin K_1; some strains may require additional enrichment factors to support good growth. In the case of all fastidious organisms, the growth requirements should be determined **before** sensitivity tests are performed, in order that the broth medium supporting the most luxuriant and rapid growth may be selected for the procedure. Organisms such as *Haemophilus* species should be tested in Mueller-Hinton broth containing 1% rabbit blood. The blood may be added to the broth before it is distributed into the test tubes, or it may be added with the inoculum.

Procedure for preparing serial dilutions and determining susceptibility

1. Thaw the frozen stock solution of the drug(s) required and dilute 1:5 with

---

*This medium is low in tetracycline and sulfonamide inhibitors and shows good batch-to-batch consistency.

†Available from Baltimore Biological Laboratory, Cockeysville, Md., or Difco Laboratories, Detroit.

sterile Mueller-Hinton broth. This gives a **working solution** for each drug containing 200 μg or units.

2. Select 10 clear, sterile, cotton-plugged or capped 13- × 100-mm tubes and mark from 1 to 10.

3. Using aseptic technique, pipette 0.5 ml of dilution broth into tubes 2 through 10. Do this for each drug to be tested.

4. Add 0.5 ml of the working solution (200 μg/ml) of the drug into tubes 1 and 2. Mix contents of the second tube well and transfer 0.5 ml to tube 3. Mix well and transfer 0.5 ml to tube 4, continuing this procedure to tube 9. Discard 0.5 ml from tube 9; the tenth tube receives no drug and serves as the control. Use a **separate** pipet for each transfer to avoid any carry-over.

5. To all tubes add 0.5 ml of an inoculum containing approximately 10^5 to 10^6 organisms per milliliter. This may be prepared in most instances by making a 1:1,000 dilution in broth of an overnight (6-hour if a rapidly growing organism) broth culture of the organism to be tested. With slow-growing organisms, such as microaerophilic streptococci, it may be necessary to use cultures incubated as long as 48 hours. If numerous drugs are to be tested, prepare sufficient inoculum in a flask, for uniformity.

The final volume in each tube is 1 ml, and the concentration of drugs is from 100 μg or units to 0.39 μg

Table 36-2. Suggested setup for the tube dilution method

Tube	Diluent (medium) added (ml)	Antimicrobial agent added	Diluted culture added (ml)	Final drug concentration (units or μg/ml)
1	None	0.5 ml working solution	0.5	100
2	0.5	0.5 ml working solution	0.5	50
3	0.5	0.5 ml from tube 2	0.5	25
4	0.5	0.5 ml from tube 3	0.5	12.5
5	0.5	0.5 ml from tube 4	0.5	6.25
6	0.5	0.5 ml from tube 5	0.5	3.125
7	0.5	0.5 ml from tube 6	0.5	1.56
8	0.5	0.5 ml from tube 7	0.5	0.78
9	0.5	0.5 ml from tube 8*	0.5	0.39
10	0.5	None	0.5	Zero

*Discard 0.5 ml from tube 9.

or unit per milliliter, in twofold steps (Table 36-2).

For organisms considered highly susceptible to antimicrobial agents, such as streptococci, a lower range of dilutions may be employed by further diluting the working solution (200 μg or units per milliliter) 1:10. Thus, the final concentrations in Table 36-2 will be 10, 5, 2.5, 1.25, 0.63, 0.3, 0.15, 0.08 and 0.04 μg or units per milliliter.

6. Incubate tubes at 35 C and examine macroscopically for **evidence of growth.** The tubes should be incubated only as long as is necessary for the control tube to show turbid growth; usually 12 to 18 hours is optimal. As noted earlier, the lowest concentration of drug in the series showing no growth is taken as the MIC and is expressed as micrograms (or units) per milliliter. It is well to remember that in the serial dilution technique there is a possible error equivalent to one tube dilution, so the MIC values are not necessarily actual values.

One method that can be used to determine the MBC is to pipette 0.5 ml from each tube that shows no visible turbidity into 12 ml of infusion agar, mix, and make a pour plate. In addition, a colony count of the **initial** inoculum is prepared by making a pour plate of the 1:1,000 dilution (see step 5 above). By comparing colony counts after an appropriate incubation period of 48 to 72 hours, one may calculate the lowest concentration of the antimicrobial that provided 99.9% and 100% bactericidal activity. A quantitative loop may be used in lieu of the more cumbersome pour plate technique.

AGAR PLATE DILUTION METHOD FOR DETERMINING SUSCEPTIBILITY TO ANTIBIOTICS

The agar plate dilution method[1] is similar in principle to the tube dilution method, except that a solid medium is used. Mueller-Hinton agar is recommended and is prepared in 100-ml amounts. Some workers incorporate 5% blood or heated blood in the medium when using it for organisms that require enriched media, such as streptococci and *Haemophilus*. There appears to be no significant inactivation of the drugs by the addition of the blood.

Procedure for preparing serial dilutions

Prepare twofold serial dilutions of the stock antimicrobial agents, as described in the previous section, using at least ten times the volumes indicated. Stock solutions containing 1,000 μg/ml are most useful, since decimal dilutions are readily prepared from these. For example, to prepare a 10 μg/ml plate, add 1 ml of the stock solution to 100 ml of melted and cooled agar and then pour plates.

In an effort to simplify the agar dilution procedure, some laboratories prefer to increase the size of the dilution steps tested or to use only a limited number of concentrations per drug. Table 36-3 shows the concentrations of antimicrobials used at the UCLA Hospital and Clinics. Concentrations were selected on the basis of anticipated serum levels at both high and low doses of the drugs. Note that some of the concentrations shown in Table 36-3 reflect levels of the antimicrobials achieved in urine (for drugs used in urinary tract infections).

Preparation of plate dilutions

Melt and cool sufficient screw-capped flasks (or bottles) of agar medium for the number of plates to be prepared (about 20 ml of medium is required per 90-mm diameter plate) and allow to equilibrate in a water bath at 50 C before adding the drugs. Add the required amount of the various dilutions to each flask (or bottle), mix gently by inversion, and pour into plates.* Allow the agar to harden and store in the refrigerator at 5 C until ready for use, preferably within 24 hours (and not after 1 week) of preparation. Media containing unstable antibiotics, such as ampicillin, should be prepared twice weekly.

Inoculation of plates

The inoculum size should be adjusted to contain approximately 10^8 organisms per milliliter (equivalent to a McFarland standard of 1 or 2); this will ensure dense, nearly confluent growth on a control plate containing no antibiotic.

Spot inoculation of the plates is made with a 1-mm loop (approximately 0.001 ml), a capillary pipet, or, preferably, by the use of the inocula-replicator of Steers and co-workers.[30] With this device, each

---

*It is **not** recommended that the drug dilutions and culture medium be mixed directly in the plates; this may produce uneven distribution of the antibiotic in the agar.

Table 36-3. Concentrations of antimicrobials tested at UCLA Hospital and Clinics against bacterial isolates by the agar dilution method

Antimicrobial agent	Concentrations (μg/ml)															
	0.06	0.12	0.25	0.5	1	2	4	8	16	32	50	64	100	128	200	256
Amikacin						x	x	x	x	x						
Ampicillin						x		x	x							
Carbenicillin											x		x		x	
Cephalothin				x		x		x	x							
Chloramphenicol						x		x	x							
Clindamycin		x			x		x	x								
Erythromycin			x			x		x								
Gentamicin		x		x			x	x								
Kanamycin						x	x	x								
Nalidixic acid										x		x		x		
Nitrofurantoin												x		x		x
Oxacillin		x		x		x	x									
Penicillin	x			x		x			x	x						
Tetracycline					x	x	x									
Tobramycin			x	x		x	x									

single manipulation can release 36 different cultures from the prongs on a replicator head to the surface of a 100- × 15-mm square plastic plate (Falcon) containing agar to a depth of 3 mm. Each prong will deliver about 0.001 ml; 36 inoculations can be made simultaneously.

In using the Steers replicating device, it is recommended that three spaces on each plate be allocated for control organisms, for example, *Staphylococcus aureus* (American Type Culture Collection 25923), *Escherichia coli* (ATCC 25922), and *Pseudomonas aeruginosa* (ATCC 27853). Thus, 33 spaces will be available per plate for the testing of clinical isolates.

Organisms that spread, such as *Proteus*, may be contained by the use of 12- × 12-mm Raschig rings.*[37]

Incubation and reading of plates

Incubate the plates at 35 C for 16 to 18 hours and examine for the presence of growth. The lowest concentration of drug producing **complete inhibition of growth** is taken as the end point. A very fine growth or one or two visible colonies may occur when the Steers replicator is used; these may be disregarded in the reading of the test. Control cultures on drug-free media should always show confluent growth.

Readers interested in a more detailed account of this agar dilution procedure should consult the excellent chapter in *Laboratory Procedures in Clinical Microbiology*.[39]

STANDARDIZED DISK–AGAR DIFFUSION METHOD FOR DETERMINING SUSCEPTIBILITY TO ANTIBIOTICS

Perhaps the most useful, and certainly the most widely used, laboratory test for antimicrobial susceptibility is the antimicrobial disk–agar diffusion procedure, the so-called **disk method.** Its simplicity,

---

*Available from Scientific Glass Apparatus Co., Bloomfield, N.J.

speed of performance, economy, and reproducibility (under standardized conditions) make it ideally suited for the busy diagnostic laboratory when the more laborious dilution methods may not be practical.

As originally described by Bondi and associates,[4] filter paper disks are impregnated with various antimicrobial agents of specific concentrations and are carefully placed on an agar culture plate that has been inoculated with a culture of the bacterium to be tested. The plate is incubated overnight and observed the following morning for a **zone of growth inhibition** around the disk containing the agent to which the organism is **susceptible. A resistant** organism will grow up to the edge of the disk.

No attempt is made here to discuss the complex physicochemical reactions that take place during diffusion of the antibiotic into the agar gel or the dynamics of bacterial growth under these conditions. The reader is referred to publications by Ericsson[6,7] for details.

Over the years, numerous attempts have been made to standardize the disk procedure, including the work of Bauer, Kirby, and co-workers,[3] Ericsson,[7] the World Health Organization (WHO), the Food and Drug Administration,[9] and most recently the National Committee for Clinical Laboratory Standards (NCCLS). The NCCLS performance standards for antimicrobial disk susceptibility tests[21] are presented herein.

Numerous proficiency testing surveys, including a nationwide laboratory evaluation by the Center for Disease Control (CDC), of the disk procedure have revealed that (1) the procedure as practiced is **not** standardized and (2) there are numerous variables that may contribute to these discrepancies. Among those that have been identified are:

1. Selection and concentration of antimicrobial disks
2. Selection, volume, and age of plating medium

3. Storage and handling of disks
4. Methodology of testing
5. Criteria used for interpreting results.

Selection of antimicrobial disks

The following set of antimicrobial agents* is available for susceptibility testing:

Amikacin	10 µg
Ampicillin	10 µg
Carbenicillin	100 µg
Cephalothin	30 µg
Chloramphenicol	30 µg
Clindamycin	2 µg
Cotrimoxazole	1.25 µg trimethoprim 23.75 µg sulfamethoxazole
Erythromycin	15 µg
Gentamicin	10 µg
Kanamycin	30 µg
Methicillin	5 µg
Nafcillin	1 µg
Nalidixic acid	30 µg
Nitrofurantoin	300 µg
Oxacillin	1 µg
Penicillin G	10 U
Polymyxin B	300 U
Streptomycin	10 µg
Sulfonamides	300 µg
Tetracycline	30 µg
Tobramycin	10 µg
Vancomycin	30 µg

A basic set of disks for routine testing against commonly isolated microorganisms is listed on p. 394.

Selection of plating medium

Although an ideal medium has not yet been perfected for the disk test, the NCCLS Subcommittee considers Mueller-Hinton agar the best compromise for routine susceptibility testing, since it shows good batch-to-batch uniformity and is low in tetracycline and sulfonamide inhibitors. With the addition of 5% defibrinated sheep, horse, or other animal blood, it will support the growth of more fastidious pathogens (i.e., those that will not grow on nonenriched medium). When required, the blood-containing medium may be "chocolatized," for testing *Haemophilus* species.

Mueller-Hinton agar* is prepared according to the manufacturer's directions and should be immediately cooled in a 50 C water bath after removal from the autoclave. The cooled medium is then poured into sterile Petri plates (on a level, horizontal surface) to a uniform depth of 4 mm; this is equivalent to approximately 60 ml in a 140-mm (internal diameter) plate, or approximately 25 ml for 90-mm plates. After solidifying at room temperature, the plates may be used the same day or stored in a refrigerator at 2 to 8 C for **not more than 7 days,** unless some method is used to minimize water loss from evaporation, such as storage in polystyrene plastic bags. As a sterility control, several plates from each batch of blood-containing Mueller-Hinton agar should be incubated at 35 C for 24 hours or longer. These plates should not be used subsequently.

Each batch of Mueller-Hinton agar should be checked for pH when prepared; the pH should be **7.2 to 7.4** at room temperature. This may be tested by macerating a small amount of medium in a little distilled water or by allowing a little of the medium to gel around a pH meter electrode in a small beaker. If available, a surface electrode is desirable.

Just prior to use of the medium, the plates should be placed in a 35 C incubator with lids partly ajar, until excess surface moisture has evaporated. This usually requires about 20 minutes.

Storage and handling of disks

Antimicrobial disks are generally supplied in separate containers with a

---

*Available from Baltimore Biological Laboratory, Cockeysville, Md.; Difco Laboratories, Detroit; Pfizer Diagnostics, Flushing, N.Y.; and others.

*Available from Baltimore Biological Laboratory, Cockeysville, Md.; Difco Laboratories, Detroit; and others.

desiccant* and should be kept under refrigeration (from 4 to 5 C). Disks containing the penicillins (including ampicillin and carbenicillin) and the cephalosporin family of drugs should always be **kept frozen** (at less than −14 C) to maintain their potency; a small working supply may be refrigerated for up to **1 week.** For long-term storage, disks are best kept in the frozen state until needed.

As they are required, the unopened containers are removed from the refrigerator or freezer 1 or 2 hours before the disks are to be used and allowed to adjust to room temperature. This is done to minimize condensation resulting from warm air reaching the cold containers. If disk dispensers are utilized, they should be equipped with tight covers and supplied with a satisfactory desiccant; when not in use, they should also be refrigerated.

Manufacturer's expiration dates should be noted and listed; disks **must be discarded** on their expiration date.

Preparation of inoculum

Various workers have shown that when certified antimicrobial disks and a single standard culture medium are used, the greatest factor contributing to reproducibility of the disk test is control of the inoculum size.

The currently recommended method of preparing a standardized inoculum is as follows:

1. With a sterile wire loop, the tops of four or five isolated colonies of a similar morphologic type are transferred to a tube containing 4 to 5 ml of soybean-casein digest broth.†
2. The broth is incubated at 35 C until its turbidity exceeds that of the standard (described in step 3). This usu-

ally requires 2 to 8 hours' incubation.

3. The turbidity is then adjusted to a **barium sulfate standard** prepared by adding 0.5 ml of 1.175% w/v (0.048 M) barium chloride hydrate ($BaCl_2 \cdot 2 H_2O$) to 99.5 ml of 1% w/v (0.36 N) sulfuric acid. The standard is distributed in screw-capped tubes of the same size as those used in growing the broth culture, which is approximately 4 to 6 ml per tube. These are then tightly sealed and stored at room temperature in the dark. Fresh standards must be prepared at least once every 6 months, although these standards can remain stable for a much longer period when heat sealed and stored in the dark.[40]
4. The barium sulfate standard must be vigorously agitated on a Vortex shaker just before use. The turbidity of the broth culture is then adjusted visually by adding sterile saline or broth, using adequate light and comparing the tubes against a white background with a contrasting black line.

Inoculation of the test plates

Within 15 minutes of adjusting the density of the inoculum, a sterile cotton swab on a wooden applicator stick (plastic or wire sticks are not satisfactory) is dipped into the standardized bacterial suspension. The excess fluid is removed by rotating the swab with firm pressure against the inside of the tube above the fluid level. The swab is then used to streak the dried surface of a Mueller-Hinton plate in three different planes (by rotating the plate approximately 60° each time) to ensure an even distribution of the inoculum.

Replace the plate lids and allow the inoculated plates to remain on a flat and level surface undisturbed for 3 to 5 minutes (no longer than 15 minutes) to allow for absorption of excess moisture,

---

*Humidity—particularly high humidity—heat, and contamination are important deteriorating factors.[10]
†Trypticase soy broth, Baltimore Biological Laboratory, Cockeysville, Md.; tryptic soy broth, Difco Laboratories, Detroit; and others.

Table 36-4. Recommended schemes for antimicrobial disks*

Organism	AK 10 µg	AM 10 µg	CB 100 µg	CF 30 µg	CM 30 µg	CC 2 µg	EM 15 µg	GM 10 µg	DP 5 µg	NA 30 µg	NF 300 µg	PN 10 U	SM 10 µg	TO 10 µg	VM 30 µg
Gram-negative rods															
Escherichia coli	x	x	[x]	x	[x]			x		(x)	(x)		[x]	x	
Klebsiella-Enterobacter-Serratia	x	x		x	[x]			x		(x)	(x)		[x]	x	
Other enteric bacilli	x	x		x	[x]			x		(x)	(x)		[x]	x	
Proteus species	x	x			[x]			x		(x)	(x)		[x]	x	
Pseudomonas aeruginosa, P. sp.	x		x		[x]			x		(x)				x	
Other nonfermentative bacilli	x	x	x	x	[x]			x		(x)	(x)		[x]	x	
Gram-positive cocci															
Staphylococcus aureus				x	[x]	x	x	[x]	x			x			x
Streptococcus pyogenes (group A)							[x]								
Group D streptococci, including enterococci				x	[x]		x	[x]				x	[x]		x
Streptococcus pneumoniae				[x]	[x]		x								
Other streptococci (alpha, and so forth)		x		x	[x]							x			x
Miscellaneous groups															
Neisseria meningitidis		x		x	[x]							x			
Haemophilus influenzae		x		x	[x]							x	[x]		
Other facultative organisms	As indicated by identity of isolate														

*From the Section of Microbiology and the Infectious Disease Research Laboratory, Wilmington Medical Center, Wilmington, Del. Abbreviations: AK, amikacin; AM, ampicillin; CB, carbenicillin; CF, cephalothin; CM, chloramphenicol; CC, clindamycin; EM, erythromycin; GM, gentamicin; DP, methicillin; NA, nalidixic acid; NF, nitrofurantoin; PN, penicillin G; SM, streptomycin; TO, tobramycin; VM, vancomycin. x, Recommended for use with the species indicated; (x), recommended for use with urine isolates *only*; [x], recommended for use with blood culture isolates *only*. Cotrimoxazole (trimethoprim-sulfamethoxazole) may be included with urinary isolates, on request.

then apply the disks, as described in the following section.

Placement of disks

With alcohol-flamed, fine-pointed forceps (cooled before using) or a disk dispenser,* the selected disks are placed on the inoculated plate and pressed firmly into the agar with sterile forceps or needle to ensure **complete contact** with the agar. The disks are distributed evenly in such a manner as to be **no closer** than 15 mm from the edge of the Petri dish and so that no two disks are closer than 24 mm from center to center. Once a disk has been placed, it should not be moved, since some diffusion of the drug occurs almost instantaneously.

An alternative method, using an agar overlay, has been described by Barry and colleagues.[2] This method is useful only for rapidly-growing organisms, such as *Staphylococcus aureus,* the enteric bacilli, and *Pseudomonas aeruginosa.* This method also must be standardized to correspond with results obtained by the cotton swab–streak method already described.

The plates are inverted and placed in the 35 C incubator within 15 minutes after application of the disks. Incubation under increased CO_2 tension should **not** be practiced, since the interpretative zone sizes were developed under aerobic conditions; CO_2 incubation may significantly alter the zone sizes.

Table 36-4 is presented as a practical guide in the selection of disks for routine susceptibility testing of facultative organisms isolated in clinical practice. Although not identical to that recommended by the NCCLS,[21] it has proved of value to clinicians at the Wilmington Medical Center and has served as an aid in reducing the misuse or overuse of antimicrobial agents in a large medical complex.

Reading of results

After incubation the relative susceptibility of the organism to the antimicrobic is demonstrated by a clear zone of growth inhibition around the disk. This is the result of two processes: (1) diffusion of the drug and (2) growth of the bacteria. As the antimicrobic diffuses through the agar medium from the edge of the disk, its concentration progressively diminishes to a point where it is no longer inhibitory for the organism. The size of this area of suppressed growth, the **zone of inhibition,** is determined by the concentration of antimicrobic present in the area. Therefore, within the limitations of the test, the **diameter of the inhibition zone** denotes the **relative susceptibility** to a particular antimicrobic.

After 16 to 18 hours' incubation (rapid growers can be read in 6 to 8 hours[15]), each plate is examined and the diameters of the complete inhibition zones are noted and **measured,** using reflected light and sliding calipers, a ruler, or a template prepared for this purpose and held on the bottom of the plate.* The **end point,** measured to the nearest millimeter, should be taken as the area showing no visible growth that can be detected with the unaided eye. Faint growth or tiny colonies near the edge of the inhibition zones are ignored, as is the swarming that may occur in the inhibition zones with some strains of *Proteus vulgaris* and *P. mirabilis.* With sulfonamides, slight growth (with 80% or more inhibition) is disregarded, and the margin of heavy growth is measured to determine the zone diameter.

---

*Dispensers for both the 90- and 140-mm Petri plates are available from Baltimore Biological Laboratory, Cockeysville, Md.; Difco Laboratories, Detroit; Pfizer Diagnostics, Flushing, N.Y.; and others.

*Microbial growth should be almost or just confluent; if only isolated colonies are present, the inoculum was too light and the test must be repeated.

Table 36-5. Zone size interpretative standards*

Antimicrobic or chemotherapeutic agent	Disk potency	Inhibition zone diameter to nearest mm		
		Resistant	Intermediate	Sensitive
Amikacin	10 μg	11 or less	12-13	14 or more
Ampicillin[1]				
Enterobacteriaceae and enterococci	10 μg	11 or less	12-13	14 or more
Staphylococci		20 or less	21-28	29 or more
Haemophilus		19 or less	—	20 or more
Carbenicillin	100 μg			
Pseudomonas sp.		13 or less	14-16	17 or more
Proteus and *Escherichia coli*		17 or less	18-22	23 or more
Cephalothin[2]	30 μg	14 or less	15-17	18 or more
Chloramphenicol	30 μg	12 or less	13-17	18 or more
Clindamycin	2 μg	14 or less	15-16	17 or more
Colistin (polymyxin E)[3]	10 μg	8 or less	9-10	11 or more
Erythromycin	15 μg	13 or less	14-17	18 or more
Gentamicin	10 μg	12 or less	13-14	15 or more
Kanamycin	30 μg	13 or less	14-17	18 or more
Methicillin[4]	5 μg	9 or less[9]	10-13	14 or more
Nalidixic acid[5]	30 μg	13 or less	14-18	19 or more
Nitrofurantoin[5]	300 μg	14 or less	15-16	17 or more
Penicillin G[6]				
Staphylococci	10 U	20 or less	21-28	29 or more
Other organisms[7]	10 U	11 or less	12-21	22 or more
Polymyxin B[3]	300 U	8 or less	9-11	12 or more
Rifampin (when testing *Neisseria meningitidis* susceptibility only)	5 μg	24 or less	—	25 or more
Streptomycin	10 μg	11 or less	12-14	15 or more
Sulfonamides	250 or 300 μg	12 or less	13-16	17 or more
Tetracycline[8]	30 μg	14 or less	15-18	19 or more
Tobramycin	10 μg	11 or less	12-13	14 or more
Trimethoprim-sulfamethoxazole[5]	25 μg	10 or less	11-15	16 or more
Vancomycin	30 μg	9 or less	10-11	12 or more

*See references 9 and 21.

[1]The ampicillin disk is used for testing susceptibility to both ampicillin and hetacillin.

[2]Class disk for cephalothin, cephaloridine, cephalexin, cefazolin, cephacetrile, cephradine, and cephapirin.

[3]The polymyxins diffuse poorly in agar, and the accuracy of the diffusion method is less than with other antibiotics. Resistance is always significant, but some relatively resistant strains of *Klebsiella* and *Enterobacter* may give zones in the lower end of the sensitive range (up to 15 mm). When treatment of systemic infections due to susceptible strains is considered, it is wise to confirm the results of a diffusion test with a dilution method.

[4]The methicillin disk is used for testing susceptibility to all penicillinase-resistant penicillins: methicillin, cloxacillin, dicloxacillin, oxacillin, and nafcillin. Methicillin-resistant strains of *Staphylococcus aureus* are best detected at 30 C.

[5]Urinary tract infections only.

[6]Class disk for penicillin G, phenoxymethyl penicillin, and phenethicillin.

[7]This category includes some organisms, such as gram-negative bacilli, that may cause systemic infections treatable by high doses of penicillin G.

[8]The tetracycline disk is used for testing susceptibility to all the tetracyclines: chlortetracycline, demeclocycline, doxycycline, methacycline, oxytetracycline, rolitetracycline, minocycline, and tetracycline.

[9]Staphylococci that exhibit resistance to the methicillin disk should be reported as resistant to cephalosporins, regardless of zone size (Thornsberry, C., and Hawkins, T. M.: Agar disc diffusion susceptibility testing procedure, U.S. Department of Health, Education, and Welfare, Public Health Service, Atlanta, 1977, Center for Disease Control).

Large colonies growing within a zone of inhibition may actually be a different bacterial species (a mixed, rather than a pure, culture) and should be subcultured, reidentified, and retested.

Interpretation of zone sizes

The diameters of the zones of inhibition are interpreted by referring to Table 36-5, which represents the NCCLS subcommittee's present recommendations.

The term "susceptible" implies that an infection caused by the strain tested may be expected to respond favorably to the indicated antimicrobial for that type of infection and pathogen. "Resistant" strains, on the other hand, are not inhibited completely by therapeutic concentrations. "Intermediate" implies that strains may respond to unusually high concentrations of the agent, due either to high dosage or high levels achieved, as in the urinary tract. In other circumstances, intermediate results might warrant further testing if alternative agents are not available.

Limitations of the test

This modified Bauer-Kirby procedure has been standardized for testing rapidly growing bacteria, particularly members of the Enterobacteriaceae, *S. aureus*, and *Pseudomonas* species. For testing *Haemophilus*, Mueller-Hinton agar plates supplemented with 1% hemoglobin and 1% IsoVitaleX (BBL) is recommended. Prepare the inoculum by suspending growth from a 24-hour chocolate agar plate in Mueller-Hinton broth to the density of a 0.5 McFarland standard (Thornsberry, C.: Antimicrobial test for *Haemophilus influenzae*. In Technical Improvement Service No. 25, Chicago, 1976, American Society of Clinical Pathologists). *Streptococcus pyogenes* and *S. pneumoniae* are considered susceptible to penicillin G and are not routinely tested; however, in patients hypersensitive to penicillin, the isolate may be tested against erythromycin or clinda-

mycin. The recent development of resistance of *S. pneumoniae* to penicillin G is noted in Chapter 18.

Hoo and Drew[11] pointed out that nitrofurantoin disks gave inconsistent results related to differences in pH of different batches of disks. They indicated that nonantibiotic chemotherapeutic agent disks are not regulated by the FDA. Accordingly, problems might be anticipated with other agents, such as cotrimoxazole, sulfonamides, and nalidixic acid. Daily inclusion of standard test strains will alert laboratories to batches of disks that give erroneous results.

In general, fastidious organisms that require increased CO_2 tension or an anaerobic atmosphere, or whose growth rate is unusually slow, do not lend themselves to susceptibility testing by the standardized disk–agar diffusion method; agar plate or broth dilution test procedures are recommended. Likewise, testing of *Neisseria gonorrhoeae* by the described procedure is not recommended. Susceptibility testing of anaerobes is described in a subsequent section.

Quality control procedures

It is essential that some form of quality control procedure be carried out to ensure precision and accuracy of the test results. The NCCLS recommends that tests be monitored **daily** with stock cultures of *S. aureus* (ATCC 25923), *E. coli* (ATCC 25922), and *P. aeruginosa* (ATCC 27853) using disks representative of those to be used in testing of clinical isolates.[21] These cultures may be grown on soy-casein digest agar slants and stored under refrigeration (4 to 8 C) and should be subcultured to fresh slants every 2 weeks.

For testing, the cultures are inoculated to soy-casein digest broth tubes, which are incubated overnight and streaked to agar plates to obtain isolated colonies; these are then picked to broth and tested as described in the preceding sections.

The control strains may be used as long

Table 36-6. Control limits for monitoring precision and accuracy of inhibitory zone diameters (mm) obtained in groups of five separate observations

Antimicrobial agent	Disc content	Individual test control Zone diameter (mm)	Accuracy control Zone diameter (mm) Mean of 5 values	Precision control Range* of 5 values Maximum	Precision control Range* of 5 values Average†
E. coli (ATCC 25922)					
Ampicillin	10 μg	15 to 20	15.8 to 19.2	6	2.9
Carbenicillin	100 μg	24 to 29	25.0 to 28.0	7	3.5
Cephalothin	30 μg	18 to 23	18.8 to 22.2	6	2.9
Chloramphenicol	30 μg	21 to 27	22.0 to 26.0	7	3.5
Colistin	10 μg	11 to 15	11.7 to 14.3	4	2.3
Erythromycin	15 μg	8 to 14	9.0 to 13.0	7	3.5
Gentamicin	10 μg	19 to 26	20.2 to 24.8	8	4.1
Kanamycin	30 μg	17 to 25	18.3 to 23.7	9	4.7
Neomycin	30 μg	17 to 23	18.0 to 22.0	6	3.5
Polymyxin B	300 units	12 to 16	12.7 to 15.3	4	2.3
Streptomycin	10 μg	12 to 20	13.3 to 18.7	9	4.7
Tetracycline	30 μg	18 to 25	19.2 to 23.8	8	4.1
Tobramycin	10 μg	18 to 26	—	—	—
Trimethoprim-	1.25 μg				
sulfamethoxazole‡	23.75 μg	24 to 32	25.3 to 30.7	9	4.7
S. aureus (ATCC 25923)					
Ampicillin	10 μg	24 to 35	25.8 to 33.2	13	6.4
Cephalothin	30 μg	25 to 37	27.0 to 35.0	14	7.0
Chloramphenicol	30 μg	19 to 26	20.2 to 24.8	8	4.1
Clindamycin	2 μg	23 to 29	24.0 to 28.0	7	3.5
Erythromycin	15 μg	22 to 30	23.3 to 28.7	9	4.7
Gentamicin	10 μg	19 to 27	20.3 to 25.7	9	4.7
Kanamycin	30 μg	19 to 26	20.2 to 24.8	8	4.1
Methicillin	5 μg	17 to 22	17.8 to 21.2	6	2.9
Neomycin	30 μg	18 to 26	19.3 to 24.7	9	4.7
Penicillin G	10 units	26 to 37	27.8 to 35.2	13	6.4
Polymyxin B	300 units	7 to 13	—	—	—
Streptomycin	10 μg	14 to 22	15.3 to 20.7	9	4.7
Tetracycline	30 μg	19 to 28	20.5 to 26.5	11	5.2
Tobramycin	10 μg	19 to 29	—	—	—
Vancomycin	30 μg	15 to 19	15.7 to 18.3	4	2.3
Trimethoprim-	1.25 μg				
sulfamethoxazole‡	23.75 μg	24 to 32	25.0 to 31.0	7	3.5
P. aeruginosa (ATCC 27853)					
Carbenicillin	100 μg	20 to 24			
Gentamicin	10 μg	16 to 21			
Tobramycin	10 μg	19 to 25			

From Approved Standard: ASM-2, Performance standards for antimicrobial disc susceptibility tests, Villanova, Pa., 1976, The National Committee for Clinical Laboratory Standards. Reprinted with permission.

*Maximum value minus minimum value obtained in a series of five consecutive tests should not exceed the listed maximum limits, and the mean should fall within the range listed under "accuracy control."

†In a continuing series of ranges from consecutive groups of five tests each, the average range should approximate the listed value.

‡Cotrimoxazole.

as there is no significant change in the mean inhibition zone diameters not otherwise attributable to technical error. If such changes occur, fresh strains should be obtained from a reference laboratory or other reliable source. Individual values of zone diameters and permissible variation are indicated in Table 36-6, which represents a computation based on standard statistical methods and is described in NCCLS's *Performance Standards for Antimicrobial Disc Susceptibility Tests.*[21]

It should be emphasized that some lots of Mueller-Hinton agar may contain increased concentrations of Ca^{++} and Mg^{++}. Since *S. aureus* and *E. coli* are not affected by these ions, they will demonstrate no changes in zone sizes, but growth of *P. aeruginosa* is enhanced and this organism will therefore demonstrate **smaller** zones. Thus, the effect of increased concentrations of these cations would influence the interpretation of susceptibility and should be predetermined. The concentration of calcium and magnesium is also an important factor in broth dilution susceptibility testing of *P. aeruginosa*, since these ions may be present at low levels in Mueller-Hinton and other broths. On the other hand, most solid (agar) media have higher concentrations of these divalent cations and therefore yield more reliable results.

Rapid tests

Several studies[15,19] indicated that reading Kirby-Bauer disk susceptibility tests at 8 hours, instead of at the usual 18 to 20 hours, gave accurate readings at least 85% of the time and even earlier readings could often be made accurately. This is not recommended as a routine, but could be very useful on occasion in the management of seriously ill patients.

A semiautomated commercial system for rapidly determining MICs (3 to 4 hours) gave results comparable to those obtained by the agar dilution method, using ampicillin and *H. influenzae.*[23] Fully automated systems are also available. Of these, the Autobac has been used most widely. These systems are convenient and provide results much more quickly than is true for conventional methods, but do not always give reliable results with certain organisms and drugs (Mogyoros, M., et al.: Antimicrob. Ag. Chemother. **11**:750-752, 1977). A recent article (Tilton, R. C., and Isenberg, H. D., Antimicrob. Ag. Chemother. **11**:271-276, 1977) describes a very promising commercial multiwelled plastic strip containing prediluted antibiotics. A rapid capillary tube method for demonstrating beta-lactamase production by *H. influenzae* has been described.[34] This test, performed in 5 to 15 minutes with colonies from an agar plate, provides evidence of ampicillin resistance. This test may also be used with *S. aureus* and *N. gonorrhoeae* versus pencillin G (Thornsberry, C., Gavan, T. L., and Garlach, E. M.: New developments in antimicrobial agent susceptibility testing, Cumitech 6, Washington, D.C., 1977, American Society for Microbiology). Other rapid tests include the use of a cephalosporin, which is chromogenic when broken down by beta-lactamase; use of a biplate, one side of which contains 2 μg/ml of ampicillin (Barkin, R. M., et al.: Am. J. Clin. Pathol. **67**:100-103, 1977); and use of an iodometric paper strip (Jorgensen, J. H., et al.: Antimicrob. Ag. Chemother. **11**:1087-1088, 1977). The latter test takes only 1 minute and is also useful for detecting beta-lactamase–producing strains of *N. gonorrhoeae;* the strips are stable for 1 year.

In selected instances, limited to sites normally sterile but from which a single organism is likely to be recovered in the event of an infection (as in a positive blood culture), it may be feasible to do direct disk susceptibility testing on the primary culture plate.[41] Again, this should be done only in the event of a seriously ill patient and should be repeated using a standard procedure.

Susceptibility testing of anaerobes

Many anaerobes have predictable patterns of susceptibility to antimicrobial drugs[32]; thus, if the organism is well identified, one may usually predict its susceptibility pattern fairly accurately. Susceptibility testing will be required in serious infections, in infections failing to respond to what was thought to be appropriate therapy, and in patients who relapse after an initial response to therapy. In most cases of anaerobic infection, antimicrobial therapy must be started before the availability of definitive bacteriologic data and susceptibility testing.

Although a number of specialized types of antimicrobial susceptibility tests have been devised specifically for anaerobic bacteria, the conventional tests used for other organisms may be used under anaerobic conditions.[20,21] It is extremely important, however, to realize that the Bauer-Kirby technique was not standardized for anaerobes and that reliable results cannot be obtained with anaerobes by this method.[35] Standardized disk tests designed specifically for anaerobes are available, but different techniques must be used for organisms with different growth rates.[17,33] For testing occasional isolates, most laboratories will probably find a broth test most convenient. An abbreviated test can be used, with three tubes covering the range of concentrations likely to be achieved with specific agents that might be employed therapeutically. An attractive alternative introduced by Wilkins and Thiel[42] utilizes antimicrobial-impregnated disks in broth. The drug is eluted from the disks to provide the desired concentration in broth quickly and conveniently. However, their method requires prereduced broth and use of a gassing apparatus. Kurzynski and associates[16] have modified the broth disk procedure by substituting aerobically incubated thioglycollate broth; this simple procedure seems ideally suited for the small laboratory doing small numbers of susceptibility tests

on anaerobes. However, we would suggest the use of different concentrations in the case of certain drugs. The choices of 6 μg/ml for tetracycline, 100 μg/ml for carbenicillin, and 3 μg/ml for erythromycin are appropriate. For chloramphenicol, we would recommend 24 μg/ml; for penicillin, 16 units/ml; and for clindamycin, 12 μg/ml. Another simple approach would be to freeze tubes with double the desired concentration of drugs in broth; tubes could be taken out of the freezer as needed, thawed, and an equal amount of broth with inoculum added.

At this writing, a collaborative group recommends the agar dilution method through the NCCLS as the reference anaerobic method for susceptibility testing.*

As a guide to the microbiologist and clinician, Table 36-7 is presented. It is based largely on correlations of in vitro findings and evaluation of clinical effectiveness.

Other uses for antimicrobial susceptibility tests

Antimicrobic disks may be used for the purpose of **selectively isolating** various microorganisms. Following Vera's suggestions,[36] one of us (EGS) has employed the practice of placing disks containing penicillin (10 units), neomycin (30 μg), and bacitracin (10 units) on all **primary plates inoculated with a potentially mixed flora.** The zones of inhibition around the neomycin disks have been particularly useful in exposing colonies of group A beta-hemolytic streptococci, pneumococci, enterococci, and other streptococci. *Haemophilus influenzae* has been isolated with ease from within the zones surrounding the 10-unit bacitracin disk on blood agar plates inoculated with sputum, and from material obtained from

---

*Sutter, V. L., Barry, A. L., Martin, W. J., Rosenblatt, J. E., Dowell, V. R., Jr., Thornsberry, C., Wilkins, T. D., and Zabransky, R. J.: Abstr. Intersci. Conf. Antimicrob. Ag. Chemother. Abstr. 453, 1976.

Table 36-7. Susceptibility of anaerobes to antimicrobial agents

Bacterium	Chloram-phenicol	Clinda-mycin	Erythro-mycin	Linco-mycin	Metroni-dazole	Penicillin G	Tetra-cycline	Vanco-mycin
Microaerophilic and anaerobic cocci	+++	++ to +++	++ to +++	+++	++	+++ to ++++	++	+++
Bacteroides fragilis	+++	+++	+ to ++	+ to ++	+++	+	+ to ++	+
B. melanino-genicus	+++	+++	+++	+++	+++	+++*	++ to +++	+
Fusobacterium varium	+++	+ to ++	+	+ to ++	+++	+++*	++	+
Other *Fusobac-terium* sp.	+++	+++	+	+++	+++	++++	+++	+
Clostridium perfringens	+++	+++†	+++	++ to +++	+++	++++*	++ to +++	+++
Other *Clostrid-ium* sp.	+++	++	++ to +++	+	++ to +++	+++	++	++ to +++
Eubacterium and *Actinomyces*	+++	++ to +++	+++	++ to +++	+ to ++	++++	++ to +++	++ to +++

From Sutter, V. L., and Finegold, S. M. In Altman, P. L., and Katz, D. D., editors: Human health and disease, Bethesda, Md., 1977, Federation of American Societies for Experimental Biology, pp. 37, 38.
*A few strains are resistant.
†Rare strains are resistant.
Aminoglycosides, such as gentamicin and kanamycin, are generally quite inactive against the majority of anaerobes. The activity of erythromycin varies significantly according to the testing procedure. Metronidazole is not yet approved by the Food and Drug Administration for anaerobic infections. Penicillin G: Other penicillins and cephalosporins are frequently less active. Ampicillin, carbenicillin, and cephaloridine are roughly comparable to penicillin G on a weight basis, but the high blood levels safely achieved with carbenicillin make it effective against 95% of the strains of *Bacteroides fragilis.* Cefoxitin, a compound resistant to penicillinase (β-lactamase I) and cephalosporinase (β-lactamase II), also appears promising, but is still in the experimental stage. Tetracycline: Doxycycline and minocycline are more active than other tetracyclines, but susceptibility testing is indicated to ensure activity.
Symbols: ++++, drug of choice; +++, good activity; ++, moderate activity; +, poor or inconsistent activity. There is no difference in activity between drugs rated +++ and those rated ++++; the symbol ++++ indicates a drug with good activity, good pharmacologic characteristics, and low toxicity.

the throat and nasopharynx. Penicillin disks (10 units) have been helpful in unmasking colonies of coliform bacilli, pseudomonads and species of *Proteus*, *Candida albicans*, and others. A 10-unit penicillin disk also is useful in revealing colonies of *Bordetella pertussis* on Bordet-Gengou plates of nasopharyngeal cultures. Kanamycin disks (30 μg) have been reported to be helpful in separating *Bacteroides* and *Clostridium* species from other wound bacteria when the plates are incubated anaerobically.[36]

Unique susceptibility patterns may also be useful in identification of bacteria (Buck, G. E., et al.: J. Clin. Microbiol. **6:** 46-49, 1977; and Southern, P. M., Jr., and Bagby, M. K.: Am. J. Clin. Pathol. **67:** 187-189, 1977).

DETERMINATION OF BACTERICIDAL LEVEL OF SERUM DURING ANTIMICROBIC THERAPY

A **direct** method for determining the antibacterial potency of serum of patients receiving antimicrobial therapy was first described by Schlichter and associates.[27,28] In selected cases of infection, particularly bacterial endocarditis, the Schlichter test may prove useful.

Schlichter test procedure*

1. Subculture a recent isolate of the organism to infusion agar or blood agar slant and store in the refrigerator until the test is run, then sub-

*Modified by John A. Washington II, Head, Section of Clinical Microbiology, Mayo Clinic.[37]

culture to a tube of broth early on the day of the test.

2. Obtain the first blood specimen **before** therapy, if possible; this serves as a control. Then take blood samples at any desired interval, although it is recommended that the low point of the blood concentration curve be included if the patient is on intermittent dosage. Collect 10 ml of the patient's blood in a sterile tube. On receipt in the laboratory, the clot is separated and the serum is obtained by centrifugation. The serum is then transferred to a sterile, rubber-stoppered tube; it may be tested at that time (or within 2 to 3 hours if refrigerated) or frozen at −20 C, a temperature at which it remains stable for several days.

3. Prepare serial twofold dilutions of serum in 1-ml amounts in Mueller-Hinton broth, using eight sterile, gauze-stoppered Kahn tubes (**use a separate pipet for each dilution**). Soy-casein digest broth or others, such as brain-heart infusion or Levinthal's broth can be used for organisms not growing in Mueller-Hinton medium. The first tube contains only **undiluted** serum, and a ninth tube contains only broth and serves as a culture **control**. The serum dilutions range from undiluted through 1:128. Very sensitive organisms, such as alpha-hemolytic streptococci, may require dilutions up to 1:2,048.

4. To each tube of the series add 0.05 ml of a 1:1,000 dilution of a 6-hour broth culture of the organism isolated from the patient. Also prepare a pour plate using 1 ml of the inoculum.

5. Incubate the tubes at 35 C for 18 to 24 hours and examine. The **bacteriostatic** end point is taken as the highest dilution in which no visible growth occurs. Because of the inherent turbidity of some sera, it is rec-

ommended that subcultures be made from each tube to a sector of a blood agar plate. To determine **bactericidal** end points, transfer 0.05 ml from each tube showing no growth to a tube of thioglycollate medium. Mix and incubate at 35 C for 72 hours. The tube in the series that shows no growth in thioglycolate is taken as the end point. Good growth should be evident in the control tube.

6. In cases where the organism grows slowly, a loopful of an overnight broth culture may be used as the inoculum. With microaerophilic or anaerobic bacteria, the tubes should be incubated anaerobically.

7. Schlichter indicated that optimal antimicrobial dosage (either single or combined drugs) had been achieved when a bactericidal level of 1:2 (complete inhibition in the first two tubes) had been demonstrated; others, however, believe that a bactericidal level of 1:8 is desirable.

A standardized test for serum bactericidal activity has recently been described (Reller, L. B., and Stratton, C. W.: J. Infect. Dis. **136:**196-204, 1977).

ASSAY OF ANTIMICROBIAL AGENTS

In recent years much interest has been focused on the development of techniques for the assay of antimicrobial agents.[18,25,29,31,44] These permit a laboratory to rapidly and accurately determine the concentrations of these agents in various body fluids (serum, spinal fluid, urine, and so forth). Assays are useful in the following situations: (1) with drugs such as gentamicin, which do not give predictable levels; (2) when relatively toxic drugs are used, particularly in patients with impaired kidney or liver function; (3) when inactivation of drugs may occur; (4) when it is uncertain how well a drug will penetrate the blood-brain barrier; and (5) with new agents whose pharmacology is not yet well known. For example, the aminoglycoside gentamicin is widely used to treat serious and life-threatening infections. However, the therapeutic-toxic ratio is relatively low with this agent. Because of its potential for nephrotoxicity and ototoxicity, it is felt that many patients receive suboptimal doses of gentamicin and, therefore, inadequate blood levels are achieved.[13,14,43]

A number of methods are in current use for assaying antimicrobial agents in body fluids. Included are radioimmunoassay, enzymatic, agar diffusion, turbidimetric, inhibition of pH or redox change, chemical, and high-pressure-liquid chromatography. It is not within the scope of this text to discuss these procedures in detail. Interested readers should consult references 24, 26, and 39, or others, for this information.

REFERENCES

1. Anderson, T. G.: Testing of susceptibility to antimicrobial agents and assay of antimicrobial agents in body fluids. In Blair, J. E., Lennette, E. H., and Truant, J. P., editors: Manual of clinical microbiology, Washington, D.C., 1970, American Society for Microbiology.
2. Barry, A. L., Garcia, F., and Thrupp, L. D.: An improved single-disk method for testing the antibiotic susceptibility of rapidly-growing pathogens, Am. J. Clin. Pathol. **53:**149-158, 1970.
3. Bauer, A. W., Kirby, W. W. M., Sherris, J. C., and Turck, M.: Antibiotic susceptibility testing by a standardized single disc method, Am. J. Clin. Pathol. **45:**493-496, 1966.
4. Bondi, A., Spaulding, E. H., Smith, E. D., and Dietz, C. C.: A routine method for the rapid determination of susceptibility to penicillin and other antibiotics, Am. J. Med. Sci. **214:**221-225, 1947.
5. Branch, A., Starkey, D. H., and Power, E. E.: Diversifications in the tube dilution test for antibiotic sensitivity of microorganisms, Appl. Microbiol. **13:**469-472, 1965.
6. Ericsson, H.: Rational use of antibiotics in hospitals, Scand. J. Clin. Lab. Invest. **12:**1-59, 1960.
7. Ericsson, H.: The paper disc method in quantitative determination of bacterial sensitivity to antibiotics, Stockholm, 1961, Karolinska Sjukhuset.
8. Fink, F. C.: Special features of the tube dilution method of antibiotic susceptibility testing, Presented at the Interscience Conference on Antimicrobial Agents and Chemotherapy, American Society of Microbiologists, Chicago, 1962.
9. Food and Drug Administration: Standardized disc susceptibility test, Federal Register, **37**(191):20527-20529, September 30, 1972.
10. Griffith, L. J., and Mullins, C. G.: Drug resistance as influenced by inactivated sensitivity discs, Appl. Microbiol. **16:**656-658, 1968.
11. Hoo, R., and Drew, W. L.: Potential unreliability of nitrofurantoin disks in susceptibility testing, Antimicrob. Ag. Chemother. **5:**607-610, 1974.
12. Isenberg, H. D.: A comparison of nationwide microbial susceptibility testing using standardized discs, Health Lab. Sci. **1:**185-256, 1964.
13. Jackson, G., and Riff, L. J.: *Pseudomonas* bacteremia: pharmacologic and other bases for failure of treatment with gentamicin, J. Infect. Dis. **124** (Suppl.):S185-S191, 1971.
14. Kaye, D., Levison, M. E., and Labovitz, E. D.: The unpredictability of serum concentrations of gentamicin: pharmacokinetics of gentamicin in patients with normal and abnormal renal function, J. Infect. Dis. **130:**150-154, 1974.
15. Kluge, R. M.: Accuracy of Kirby-Bauer susceptibility tests read at 4, 8, and 12 hours of incubation: comparison with readings at 18 to 20 hours, Antimicrob. Ag. Chemother. **8:**139-145, 1975.
16. Kurzynski, T. A., Yrios, J. W., Helstad, A. G., and Field, C. R.: Aerobically incubated thioglycolate broth disk method for antibiotic susceptibility testing of anaerobes, Antimicrob. Ag. Chemother. **10:**727-732, 1976.
17. Kwok, Y-Y, Tally, F. P., Sutter, V. L., and Finegold, S. M.: Disk susceptibility testing of slow growing anaerobic bacteria, Antimicrob. Ag. Chemother. **7:**1-7, 1975.
18. Lewis, J. E., Nelson, J. C., and Elder, H. A.: Radioimmunoassay of an antibiotic: gentamicin, Nature (London) New Biol. **239:**214-216, 1972.
19. Liberman, D. F., and Robertson, R. G.: Evaluation of a rapid Bauer-Kirby antibiotic susceptibility determination, Antimicrob. Ag. Chemother. **7:**250-255, 1975.
20. Martin, W. J., Gardner, M., and Washington, J. A. II: In vitro antimicrobial susceptibility of anaerobic bacteria isolated from clinical specimens, Antimicrob. Ag. Chemother. **1:**148-158, 1972.
21. NCCLS Subcommittee on Antimicrobial Susceptibility Testing: Performance standards for antimicrobial disc susceptibility tests. Approved Standard: ASM-2, January, 1976, The National Committee for Clinical Laboratory Standards.
22. Ryan, K. J., and Sherris, J. C.: Antimicrobial susceptibility testing, Human Pathol. **7:**277-286, 1976.

23. Ryan, R., and Tilton, R. C.: Rapid method for determining the minimum inhibitory concentration of ampicillin for *Haemophilus influenzae,* Antimicrob. Ag. Chemother. **11:**114-117, 1977.

24. Sabath, L. D.: The assay of antimicrobial compounds, Human Pathol. **7:**287-295, 1976.

25. Sabath, L. D., Casey, J. I., and Rych, P. A. Rapid microassay of gentamicin, kanamycin, neomycin, streptomycin, and vancomycin in serum or plasma, J. Lab. Clin. Med. **78:**457-463, 1971.

26. Sabath, L. D., and Matsen, J. M. Assay of antimicrobial agents. In Lennette, E. H., Spaulding, E. H., and Truant, J. P., editors: Manual of clinical microbiology, ed. 2, Washington, D.C., 1974, American Society for Microbiology.

27. Schlichter, J. G., and MacLean, H.: A method of determining the effective therapeutic level in the treatment of subacute bacterial endocarditis with penicillin, Am. Heart J. **34:**209-211, 1947.

28. Schlichter, J. G., MacLean, H., and Milzer, A.: Effective penicillin therapy in subacute bacterial endocarditis and other chronic infections, Am. J. Med. Sci. **217:**600-608, 1949.

29. Smith, D. H., Van Otto, B., and Smith, A. L.: A rapid chemical assay for gentamicin, N. Engl. J. Med. **286:**583-586, 1972.

30. Steers, E., Foltz, E. L., and Graves, B. S.: An inocula replicating apparatus for routine testing of bacterial susceptibility to antibiotics, Antibiot. Chemother. (N.Y.) **9:**307-311, 1959.

31. Stevens, P., and Young, L. S.: Rapid assay of aminoglycosides by radioenzymatic techniques. In Rapid diagnostic techniques in clinical microbiology, Washington, D.C., 1975, American Society for Microbiology.

32. Sutter, V. L., and Finegold, S. M.: Susceptibility of anaerobic bacteria to 23 antimicrobial agents, Antimicrob. Ag. Chemother. **10:**736-752, 1976.

33. Sutter, V. L., Kwok, Y.-Y., and Finegold, S.M.: In vitro susceptibility testing of anaerobes; standardization of a single disc test. In Balows, A.,

editor: Current techniques of antibiotic susceptibility testing. Springfield, Ill., 1974, Charles C Thomas, Publisher.

34. Thornsberry, C., and Kirven, L. A.: Ampicillin resistance in *Haemophilus influenzae* as determined by a rapid test for beta-lactamase production, Antimicrob. Ag. Chemother. **6:**653-654, 1974.

35. Thornton, G. F., and Cramer, J. A.: Antibiotic susceptibility of Bacteroides species, Antimicrob. Ag. Chemother. **10:**509-513, 1970.

36. Vera, H. D.: Sensitivity plate tests. In BBL manual of products and laboratory procedures, Cockeysville, Md., 1968, Baltimore Biological Laboratory.

37. Washington, J. A. II: Personal communication (E.G.S.), 1973.

38. Washington, J. A. II: Antimicrobial susceptibility of Enterobacteriaceae and nonfermenting gram-negative bacilli, Mayo Clin. Proc. **44:**811-824, 1969.

39. Washington, J. A. II, editor: Laboratory procedures in clinical microbiology, Boston, 1974, Little, Brown and Co.

40. Washington, J. A. II, Warren, E., and Karlson, A. G.: Stability of barium sulfate turbidity standards, Appl. Microbiol. **24:**1013, 1972.

41. Waterworth, P. M., and Del Piano, M.: Dependability of sensitivity tests in primary culture, J. Clin. Pathol. **29:**179-184, 1976.

42. Wilkins, T. D., and Thiel, T.: Modified brothdisk method for testing the antibiotic susceptibility of anaerobic bacteria, Antimicrob. Ag. Chemother. **3:**350-356, 1973.

43. Winters, R. E., Litwack, K. D., and Hewitt, W. L.: Relation between dose and levels of gentamicin in blood, J. Infect. Dis. **124** (Suppl.): S90-S95, 1971.

44. Wold, J. S.: Rapid analysis of cefazolin in serum by high-pressure liquid chromatography, Antimicrob. Ag. Chemother. **11:**105-109, 1977.

PART VI

Serologic methods in diagnosis

CHAPTER 37

Serologic identification of microorganisms

GROUPING AND TYPING OF BETA-HEMOLYTIC STREPTOCOCCI BY PRECIPITIN TEST

Both the group and the type of a beta-hemolytic streptococcus may be determined by the Lancefield precipitin procedure using the same antigen. Typing should be carried out soon after isolation of the organism because the M type-specific protein substance, which determines the type, can be lost on laboratory cultivation. Fresh isolates that have been frozen can be typed successfully later.

LANCEFIELD PROCEDURE
Materials

1. Two sizes of capillary tubing are required: (a) 1.2 to 1.5 mm outside diameter for grouping, and (b) 0.7 to 1 mm outside diameter for typing.* Tube lengths of 7.5 cm should be used.
2. Wooden blocks 12 inches long containing Plasticine are recommended for holding the capillary tubes upright after they are filled with antigen and antiserum.
3. Todd-Hewitt broth at pH 7.8 to 8, dispensed in 40-ml amounts and inoculated 24 hours previously with the culture to be tested.
4. Beta-streptococcus group-specific and type-specific antiserum.†
5. N/5 hydrochloric acid—1 ml 12 N HCl plus 59 ml of 0.85% saline.

6. Buffer solution—N/5 sodium hydroxide in M/15 phosphate buffer solution at pH 7. To prepare, dissolve 1 g of anhydrous acid sodium phosphate in 100 ml of N/5 sodium hydroxide.
7. Phenol red—0.01%. To prepare, add 0.01 g of phenol red to 60 ml of alcohol and 40 ml of water.
8. Thymol blue—0.01%. To prepare, add 0.01 g of thymol blue to 60 ml of alcohol and 40 ml of water.

Preparation of antigen*

1. Grow organisms to be tested for 18 to 24 hours in 40 ml of Todd-Hewitt broth.
2. Centrifuge the culture for 30 minutes and **remove all of the supernatant.**
3. Resuspend sediment in 0.4 ml of N/5 hydrochloric acid pH 2.0 to 2.4.
4. Mix well with a wooden applicator stick.
5. Add 1 drop of thymol blue indicator. A **peach** color will result.
6. Transfer to a 15-ml conical bottom centrifuge tube and heat in a boiling water bath for 3 to 10 minutes, shaking occasionally.

---

*Kimble Glass Co., Vineland, N.J.
†Difco Laboratories, Detroit; Baltimore Biological Laboratory, Cockeysville, Md.; Lee Laboratories, Grayson, Ga.; and others.

---

*A method for preparing streptococcal extracts by use of a proteolytic enzyme from *Streptomyces albus* has been described by Maxted.[5] It appears to be more rapid and simpler than the Lancefield method but is not suitable for some group D strains. The prepared enzyme may be obtained commercially from Baltimore Biological Laboratory, Cockeysville, Md., or Difco Laboratories, Detroit.

7. Cool in a refrigerator or a cold water bath for 10 minutes.
8. Centrifuge at 3,000 rpm for 10 minutes.
9. Decant the clear supernatant fluid into a clean test tube and add 1 drop of phenol red indicator. The solution will be a distinct **yellow.**
10. Add buffer solution drop by drop until a pale **pink** color develops (pH 7.4 to 7.8). Do not make too strongly alkaline.
11. Centrifuge as described previously and remove the supernatant fluid. This fluid must be **crystal clear** for use as the antigen.
12. The extract may be stored at 4 C for several days to a week.

Another widely used method of preparing the antigen extract is the autoclave method of Rantz and Randall[7]:

1. Inoculate streptococci into 40 ml of Todd-Hewitt broth containing 0.8% glucose and incubate 24 to 48 hours.
2. Centrifuge to completely sediment the bacterial cells; discard the supernatant fluid.
3. Add 0.5 ml of 0.85% NaCl.
4. Transfer the sediment to sterile flocculation tubes and autoclave for 15 minutes at 121 C.
5. Centrifuge to complete sedimentation.
6. Transfer the supernatant fluid to sterile flocculation tubes. This fluid is ready for use as the antigen extract.

Test procedure

For grouping beta-hemolytic streptococci. Clean the outside of a 1.2- to 1.5-mm capillary tube with tissue paper or lens paper. It cannot be overemphasized that all capillary tubes, antigen extracts, and antisera must be **perfectly clean.** Antiserum that becomes turbid on storage should be clarified by centrifugation. Dip the capillary tube into group A streptococcus antiserum and permit the serum to rise one third the length of the tube,

equivalent to a 2-cm column. Wipe the outside of the tube to remove excess serum and to prevent adulteration of the antigen. Dip the capillary tube into the prepared antigen extract and draw up an equal amount. Wipe the outside of the tube with tissue and invert the tube until there is an air space both above and below the column of liquid. Place the tube upright in the Plasticine in the wooden block.

Repeat the procedure, using groups B, C, D, F, and G antisera, respectively. Immediately after the tests are set up, examine the capillary tubes with a hand lens. This is best done using a strong light in front of a black background. The tubes should be perfectly clear if the test has been correctly performed. Leave the tubes at room temperature and reexamine after 15 to 30 minutes. In the tube containing antigen and homologous antiserum, a **milky ring** will be formed at the interface of the reactants. A **positive** reaction such as this is sharp and definite and appears in 5 to 10 minutes. After an hour or two the ring disappears and a heavy white precipitate settles to the bottom of the liquid. The **negative** tubes should remain perfectly clear.

If the culture tested proves to be group A streptococcus and typing of the strain is desired, the same extract can be used as the antigen; it can be kept in a refrigerator for at least a week and still give satisfactory results.

For typing beta-hemolytic streptococci. Clean the outside of a 0.7- to 1-mm capillary tube with tissue or lens paper. Use the same technique as just described, substituting beta-hemolytic streptococcus **type-specific serum** for the group-specific serum, and insert the capillary tube upright in the Plasticine. Incubate the tests for 2 hours, refrigerate overnight, and read as described above. Sometimes it is possible to read the tubes after incubation; in other cases it is necessary to refrigerate them to obtain a reaction.

Cross-reactions may occur with the type-specific antiserum. In such cases these reactions can be largely eliminated by repeating the test, using the antisera in which precipitation has occurred and diluting each one half, one fourth, or one eighth with isotonic salt solution. Incubate for 2 hours, refrigerate overnight, and read as in the regular test.[8]

Attention also is directed to the typing of group A streptococci by the T-agglutination technique of Griffith.[3] The reproducibility of the procedure has been well established internationally. For example, studies have shown an agreement of 86% in the T-typing of 355 group A strains by two separate laboratories.[9] Moody and co-workers also were able to T-type 88% of over 1,300 strains referred to the Center for Disease Control (CDC), compared with 47% typed by the M-precipitin technique.[6] The methods used were described in the report.

Immunofluorescence and counterimmunoelectrophoresis are also useful for grouping streptococci (see Chapter 39).

QUELLUNG METHOD FOR TYPING PNEUMOCOCCI

Pneumococcus typing by the **quellung reaction** was formerly one of the most important procedures in a medical diagnostic laboratory. With the advent of antimicrobial therapy for the treatment of pneumococcal infections, serologic typing of pneumococci is no longer necessary as a guide to therapy. However, the quellung test remains one of the most rapid and satisfactory methods for the **direct identification** of the pneumococcal organism in clinical material. For this reason it is still an important laboratory procedure and will be discussed here.

Clinical materials

The quellung test is considered a useful procedure with the following clinical specimens:
1. Sputum in which organisms resembling pneumococci are readily demonstrated in direct smears.
2. Cerebrospinal fluid sediment that contains organisms suggestive of pneumococci (successful typing of the organisms definitely and immediately establishes their identity).
3. Suspect fresh pneumococcus colonies obtained from blood agar plates and suspended in a few drops of broth.
4. Positive blood-broth cultures, empyema fluid, and other specimens that show organisms resembling pneumococci on microscopic examination.

Test procedure

1. On a clean glass slide, spread a small loopful of the specimen in a thin film and **allow to air dry.** Make six of these preparations. If stained smears of the material reveal more than 15 to 25 organisms per oil-immersion field, **dilute** the specimen, since too many organisms tend to agglutinate in the antiserum, thereby making the capsular reaction difficult to observe.
2. Place a large loopful of typing serum on a coverglass and add a small loopful of 1% aqueous methylene blue stain. Invert a No. 1 square coverglass over one of the dried films made previously. When available, the typing sera are individually prepared for pneumococcus types **1** through **34,** inclusive. Typing is carried out initially with pools labeled A through F,* until a positive quellung reaction is obtained, and then with antisera for the specific types making up that pool. As many as three coverglasses may be used on one slide. These sera are produced by the CDC and are available primarily to public health laboratories and government

*Pools A through F are available from Difco Laboratories, Detroit.

research agencies. A polyvalent diagnostic antiserum, **Omniserum,** is also available.* This serum can give a capsular swelling reaction with any of the 84 recognized types of pneumococci. Also available from the same source are nine pools (A through I) covering these types and 46 monovalent antisera. All contain methylene blue stain.

3. Examine each preparation under the microscope, using the oil-immersion lens with **reduced** illumination. A **positive** reaction is indicated by the appearance of a well-defined and refractile capsule surrounding the blue-stained pneumococcus. There is some variation in the reaction among strains; type 3 pneumococci, for example, possess very large capsules. In general, however, it is the **sharpness** of the capsular outline rather than the size of the capsule that indicates a positive reaction. Capsules that are visible but without adequate swelling are considered a negative reaction. In some negative reactions a thin halo with a definite outline may confuse the inexperienced worker. By focusing above and below such organisms nothing will be observed; in positive reactions the capsules are readily visible in these focal planes. Capsular reactions usually take place in several minutes; negative preparations should be reexamined after 1 hour before discarding. Avoid letting the preparations dry out.

QUELLUNG TEST ON SPINAL FLUID FOR DIAGNOSIS OF HAEMOPHILUS INFLUENZAE MENINGITIS

A rapid diagnosis of meningitis caused by *Haemophilus influenzae* can be made by performing a quellung test on spinal fluid. Gram-stained smears should first be made of the spinal fluid and observed carefully for the presence of small gram-negative coccobacilli; if none are observed, the fluid should be centrifuged and the sediment gram-stained. If gram-negative bacilli resembling *H. influenzae* are seen on the smears from either source, quellung tests should be performed using the appropriate specimen. Since almost all cases of *Haemophilus* meningitis are caused by **serotype b** organisms, the use of homologous antiserum is recommended.* The technique is the same as that described for serotyping the pneumococci. If influenzae bacilli of serologic type b are present, a typical quellung reaction will be observed.

QUELLUNG TEST FOR IDENTIFICATION OF SEROTYPES OF KLEBSIELLA PNEUMONIAE

As originally studied by Julianelle,[4] the klebsiellae were serologically categorized into four types: A, B, C, and a heterogeneous group X. Seventy-three capsular types are now recognized. Types 1, 2, and 3 correspond to types A, B, and C of Julianelle. Specific antisera made against the 73 capsular serotypes may be used in the performance of the quellung test.† The technique is the same as that described for the typing of pneumoccci.

SEROLOGIC GROUPING OF NEISSERIA MENINGITIDIS

Numerous serologic groups of *Neisseria meningitidis* are currently recognized.[1] However, groups A, B, and C are the most prevalent and therefore the most significant to test for serologically. The serologic grouping of meningococci is primarily one of epidemiologic significance, but if such identification is de-

---

*From Dr. Erna Lund, Statenserum-Institut, Copenhagen, Denmark.

*Difco Laboratories, Detroit; Hyland Laboratories, Los Angeles; and Statenserum-Institut, Copenhagen, Denmark.
†Capsular antisera for most serotypes and pooled antisera are available from Difco Laboratories, Detroit, and Lee Laboratories, Grayson, Ga.

sired, the following agglutination test can be recommended:

1. Prepare twofold serial dilutions of the group-specific antisera.*
2. Place 0.1 ml of each serial dilution in a very small test tube.
3. Suspend organisms from a young (5- to 24-hour) culture on solid medium using physiologic saline containing 0.05% potassium cyanide to not only enhance smoothness of the bacterial suspension but also to effectively kill the organisms.
4. Filter the suspension by pipetting through a wisp of nonabsorbent cotton or centrifuge for 2 minutes at low speed to remove coarse particles. Clumps will not settle out spontaneously.
5. Dilute the antigen to approximate the No. 8 McFarland standard (see Chapter 43).
6. Add 0.1 ml of the antigen to each serum dilution and shake the rack for 3 minutes at room temperature.
7. Add 0.8 ml of physiologic saline to facilitate reading of the reactions.
8. Read for macroscopic clumping of the organisms.

Typing can also be done by the quellung reaction, using type-specific antisera, as previously described. No quellung reaction is obtained with group B because no morphologic capsule exists.

SLIDE AGGLUTINATION TEST IN SEROLOGIC IDENTIFICATION

Final identification of various members of the Enterobacteriaceae is dependent on serologic analysis. Such analysis is used primarily for the numerous serotypes that make up the genera *Salmonella* and *Shigella*. It may also be applied to other organisms. With the aid of commercially available diagnostic antisera, the identification of many of these bacteria

---

*These antisera may be obtained from the Central Public Health Laboratories, London, and from Difco Laboratories, Detroit.

becomes readily obtainable for most laboratories, provided experienced personnel perform and interpret the serologic tests. However, exact antigenic analysis requires the use of specifically absorbed typing sera, which are available primarily through various state health laboratories and at reference centers, such as the CDC, Atlanta, and the Laboratory Center for Disease Control, Ottawa. Recently many of these absorbed typing sera have become available from commercial suppliers.

Test procedure

Mark off a number of squares (¾ inch) on a perfectly clean glass slide with a grease pencil (i.e., Blaisdell, red 169T). Prepare a milky concentration of cells in saline in a tube from the growth on a TSI (or KIA) slant, from an agar slant, or from colonies on an agar plate. Since large numbers of viable cells are involved, extreme caution should be exercised in carrying out this procedure. Place 1 loopful or a small drop of each antiserum to be tested per square and leave 1 square blank. Add a drop of 0.85% saline to the blank square. With a Pasteur pipet add a small drop of the cell suspension to each square containing serum and to the blank square (as a control). Tilt the slide back and forth for 1 minute to mix, then observe for agglutination macroscopically. Agglutination is recognized by the **prompt** formation of fine granules or large aggregates. The control and any negative tests should remain homogeneous.

K antigens, which mask the heat-stable somatic (O) complex, are found in *Salmonella*, *Shigella*, and *Escherichia*. Should the cells fail to agglutinate in O antiserum, the suspension should be heated at 100 C for 10 to 30 minutes, cooled, and retested in the appropriate O antisera. Suspensions of live cells will agglutinate in antisera that contain K antibody, for example, *Salmonella* Vi and *E. coli* OB.

For a more thorough discussion of the serologic examination of members of the family Enterobacteriaceae, consult Edwards and Ewing.[2]

The following list of diagnostic antisera* is provided for the reader's convenience.

Salmonella diagnostic sera

Salmonella polyvalent (contains primarily antibody against O antigens 1, 2, 3, 4, 5, 6, 7, 8, 9, 10, 15, 19, Vi)

Salmonella Group A (contains primarily antibody against O antigens 1, 2, 12)

Salmonella Group B (contains primarily antibody against O antigens 4, 5, 12)

Salmonella Group C_1 (contains primarily antibody against O antigens 6, 7)

Salmonella Group C_2 (contains primarily antibody against O antigens 6, 8)

Salmonella Group D (contains primarily antibody against O antigens 9, 12)

Salmonella Group E (E_1, E_2) (contains primarily antibody against O antigens 3, 10, 15)

Salmonella Group F

Salmonella Group G

Salmonella Group H

Salmonella Group I

Salmonella Vi

Salmonella H antiserum a (contains primarily antibody against flagellar a antigen)

Salmonella H antiserum b (contains primarily antibody against flagellar b antigen)

Salmonella H antiserum c (contains primarily antibody against flagellar c antigen)

Salmonella H antiserum d (contains primarily antibody against flagellar d antigen)

Salmonella H antiserum i (contains primarily antibody against flagellar i antigen)

Salmonella H antiserum k (contains primarily antibody against flagellar k antigen)

Salmonella H antiserum y (contains primarily antibody against flagellar y antigen)

Salmonella H antisera 1, 2; 1, 5; 1, 6; 1, 7 (contains primarily antibody against flagellar 1, 2, 5, 6, and 7 antigens)

Shigella grouping sera

Shigella group A *(S. dysenteriae)*
Shigella group B *(S. flexneri)*

Shigella group C *(S. boydii)*
Shigella group D *(S. sonnei)*

OTHER TESTS

The staphylococcal coagglutination technique has been used for identification of groups A, B, C, and G beta-hemolytic streptococci (Rosner, R.: J. Clin. Microbiol. **6:**23-26, 1977; and Kirkegaard, M. K., and Field, C. R.: J. Clin. Microbiol. **6:**266-270, 1977). A rapid fluorescent antibody test for identification of group A streptococci has been described (Waters, C. A., and Makens, M. A.: J. Clin. Microbiol. **5:**255-256, 1977). Portas and co-workers (J. Lab. Clin. Med. **88:**339-344, 1976) have been able to identify group D streptococci rapidly by counterimmunoelectrophoresis.

REFERENCES

1. Buchanan, R. E., and Gibbons, N. E.: Bergey's manual of determinative bacteriology, ed. 8, Baltimore, 1974, The Williams & Wilkins Co.
2. Edwards, P. R., and Ewing, W. H.: Identification of Enterobacteriaceae, ed. 3, Minneapolis, 1972, Burgess Publishing Co.
3. Griffith, F.: The serological classification of *Streptococcus pyogenes*, J. Hyg. **34:**542-584, 1934.
4. Julianelle, L. A.: A biological classification of *Encapsulatus pneumoniae* (Friedländer's bacillus), J. Exp. Med. **44:**113, 1926.
5. Maxted, W. R.: Preparation of streptococcal extracts for Lancefield grouping, Lancet **2:**255-256, 1948.
6. Moody, M. D., Padula, J., Lizana, D., and Hall, C. T.: Epidemiologic characterization of group A streptococci by T-agglutination and M-precipitation tests in the public health laboratory, Health Lab. Sci. **2:**149-162, 1965.
7. Rantz, L. A., and Randall, E.: Use of autoclaved extracts of hemolytic streptococci for serological grouping, Stanford Med. Bull. **13:**290-291, 1955.
8. Swift, H. F., Wilson, A. T., and Lancefield, R. C.: Typing group A hemolytic streptococci by M precipitin reactions in capillary pipettes, J. Exp. Med. **78:**127-133, 1943.
9. Wilson, E., Zimmerman, R. A., and Moody, M. D.: Value of T-agglutination typing of group A streptococci in epidemiologic investigations, Health Lab. Sci. **5:**199-207, 1968.

---

*Lederle Laboratories, Pearl River, N.Y.; Baltimore Biological Laboratory, Cockeysville, Md.; Difco Laboratories, Detroit; Lee Laboratories, Grayson, Ga.

CHAPTER 38

Antigen-antibody determinations on patients' sera

DIAGNOSIS OF PNEUMONIA CAUSED BY MYCOPLASMA PNEUMONIAE (PRIMARY ATYPICAL PNEUMONIA)

The differential diagnosis of *Mycoplasma* pneumonia may be clinically difficult and usually necessitates laboratory confirmation. Serologic evidence can be obtained by the use of the cold hemagglutination test or the complement fixation test.

Cold hemagglutination test*

The development of cold hemagglutinins in the serum of patients with primary atypical pneumonia was first reported by Peterson and co-workers in 1943.[12] They observed that these antibodies (now considered macroglobulins) caused human erythrocytes to form visible clumps when incubated at 0 to 10 C but not at 37 C.

Cold agglutinins are found in normal sera in low titers (less than 1:16) but are present in titers of 1:40 to 1:2,048 in a high proportion of patients with *Mycoplasma pneumoniae* infection. However, relatively high titers may be found in other conditions as well. The titer rises during the course of the illness, usually reaching a maximum during the third or fourth week, followed by its rapid disappearance thereafter. The cold agglutinin response is generally related directly to the severity and duration of the illness, although mild cases may also develop a significant titer.[5]

In obtaining serum for the cold hemag-

glutination test, it is essential that the drawn blood **not** be refrigerated before separation of the serum. This could result in absorption by the red cells of most or all of the cold agglutinins present, resulting in their removal and thus a valueless test. However, one can elute the antibody from the cells by incubating the patient's clot for 30 minutes in a 37 C water bath. This is followed by immediate centrifugation and serum separation. It should also be noted that prolonged refrigeration of the serum will usually result in the disappearance of the cold agglutinins. Inactivation of the serum by heat to destroy complement, however, will not affect the titer.

Technique

Prepare twofold dilutions of the patient's serum as shown in Table 38-1. A 2% suspension of group O human red cells is used. The red cells are washed in 0.85% saline. Add the saline to the cells, centrifuge the specimen for 5 minutes at 2,000 rpm, decant the supernate, and repeat the process until the supernate is clear. The cells should be centrifuged at least three times. If the supernate is not clear after five washings, discard the suspension and obtain fresh cells. Prepare a 2% suspension by diluting 0.1 ml of packed cells with 5 ml of saline. Add 0.5 ml of the suspension to each tube, including the control tube. The final serum dilutions will range from 1:10 to 1:2,560, and the final concentration of red cells is 1%. After mixing the antigen and antiserum by shaking the rack of tubes, place tubes in a refrigerator (4 C) overnight. Read

---

*Bennett's excellent text[1] on clinical serology is recommended for additional reading.

Table 38-1. Protocol for single agglutination test by serial dilution system

Tube	1	2	3	4	5	6	7	8	9	10
Amount of saline (ml)	0.8	0.5	0.5	0.5	0.5	0.5	0.5	0.5	0.5	0.5
Amount of serum (ml)	0.2 (Mix)	0.5 of tube 1	0.5 of tube 2	0.5 of tube 3	0.5 of tube 4	0.5 of tube 5	0.5 of tube 6	0.5 of tube 7	0.5 of tube 8	(Discard 0.5 from tube 9) (Control)
Initial serum dilution	1:5	1:10	1:20	1:40	1:80	1:160	1:320	1:640	1:1280	
Amount of antigen (ml)	0.5	0.5	0.5	0.5	0.5	0.5	0.5	0.5	0.5	0.5
Final serum dilution	1:10	1:20	1:40	1:80	1:160	1:320	1:640	1:1280	1:2560	—

immediately for agglutination after refrigeration; do not allow tubes to stand at room temperature before reading. The titer is the highest dilution of serum showing a 1+ (least amount of visible clumping) or greater agglutination. After reading the test, place tubes in a 37 C water bath for 2 hours, then reread the test; agglutination due to cold agglutinins will disappear. A titer of 1:32 or greater is considered significant, although it should be pointed out that the demonstration of a fourfold increase in titer in **paired** (acute and convalescent) sera is of greater significance.

Complement fixation test

This test, preferably done with the lipid antigen of *M. pneumoniae*, is preferred over the cold agglutinin test, since it is specific. The procedure is detailed elsewhere.[8]

DETERMINATION OF ANTISTREPTOLYSIN O AND RELATED TITERS

A significant number of patients who have had a recent infection with group A streptococci develop an antibody response to streptolysin O, a specific hemolysin of these strains (and an occasional group C or G strain). This antibody will combine with and neutralize streptolysin O in vitro, thereby inhibiting its hemolytic activity on erythrocytes. By a parallel tube dilution procedure using the patient's serum and a prestandardized fixed amount of streptolysin O with a red blood cell indicator system, the level of antistreptolysin O can be measured. Since the occurrence of this antibody in the patient's serum is dependent on the production of the streptolysin O by the infecting streptococcus, a **rising** antistreptolysin O (ASO) titer aids in the diagnosis of rheumatic fever, acute hemorrhagic glomerulonephritis, and other complications of group A streptococcal infection.

Because the reagents required for the determination of the ASO titer are readily available commercially,* accompanied by complete directions, the test is not described in detail here. In brief, the test is set up by preparing a series of dilutions of the patient's serum to which is added a constant volume of streptolysin O reagent. After 15 to 45 minutes' incubation at 37 C, a constant volume of group O human or rabbit erythrocytes is added to each serial dilution and the test is reincubated. The last tube of the series showing **no hemolysis** is the ASO titer, which is expressed as the reciprocal of that dilution and given in **Todd units.** For exam-

---

*Difco Laboratories, Detroit; Baltimore Biological Laboratory, Cockeysville, Md.; Hyland Laboratories, Costa Mesa, Calif.; and others.

ple, if the highest dilution showing no hemolysis is 1:250, the ASO titer will be 250 Todd units. A titer of 300 Todd units is considered significant because most normal adults show titers of up to 200 Todd units. An elevated titer appears from 1 to 3 weeks after onset; a rising titer on repeated weekly specimens is helpful diagnostically.

In patients with acute rheumatic fever, streptococcal antibody tests are generally a more reliable indicator of recent streptococcal infection than are throat cultures. The ASO titer is elevated in 80% to 85% of patients with acute rheumatic fever. Since the rest of such patients have a normal ASO titer, a diagnosis of acute rheumatic fever cannot be ruled out on the basis of the ASO test alone. An additional test, such as the antideoxyribonuclease-B (ADN-B), will frequently show an elevated titer in the latter patients.

Furthermore, the ASO test is not as useful as others, such as the ADN-B, in suspected cases of acute glomerulonephritis if this disease follows streptococcal skin infection rather than pharyngitis. The ADN-B test is also useful for Sydenham's chorea and also will not give false-positive results, as may be seen with the ASO test with (1) bacterial growth in the serum specimen, (2) liver disease, and (3) oxidation of the antigen.

Both the ASO and the ADN-B tests have good reproducibility, test for antibody to antigens produced by most strains of group A streptococci, and utilize commercially available antigens. The ASO test is much better known. The ADN-B titer rise usually occurs later than that of the ASO. Elevated serum DNase levels, such as occur in acute hemorrhagic pancreatitis, can result in a false-negative ADN-B titer. Details for performing the ADN-B test are given by Klein.[9]

Several other procedures for testing antibody response to streptococcal infection are available, but either the reagents are not available commercially, the reproducibility of the test is not good, or there are other problems. A multiple antigen test, the Streptozyme test, is also available*; it is a 2-minute slide hemagglutination procedure.

DETERMINATION OF C-REACTIVE PROTEIN

C-reactive protein (CRP) is an abnormal alpha globulin that appears rapidly in the serum of patients who have an inflammatory condition of either infectious or noninfectious origin and is absent in serum from normal persons.

This protein has the capacity for precipitating the somatic C carbohydrate of pneumococci; its presence was first determined by mixing patient's serum with the purified pneumococcal C polysaccharide. It was subsequently demonstrated that by injecting animals with the C-reactive protein, a specific antibody reacting with the protein could be produced. It is this anti-CRP serum that is used as the sensitive reagent in this precipitin test. The test has proved useful in follow-up of patients with rheumatic fever, since CRP disappears when the inflammation subsides, reappearing only when the disease process becomes reactivated.

The reagents for the test, complete with directions, are available commercially.†

PRECIPITIN TEST ON CEREBROSPINAL FLUID

In this test, cerebrospinal fluid and a specific antibacterial serum are allowed to react in a capillary tube. This procedure has proved useful in determining the etiologic agent of meningitis when negative cultures are obtained. The technique[6] uses various commercial anti-

---

*Wampole Diagnostics, Stamford, Conn.; Beckman Instruments, Fullerton, Calif.
†Difco Laboratories, Detroit; Baltimore Biological Laboratory, Cockeysville, Md.; and others.

sera* in individual capillary tubes over-laid with spinal fluid. The mixtures are incubated for 2 hours at 37 C, refrigerated overnight, and read. A positive test is indicated by the formation of a **precipitin ring** at the interface of the reactants.

QUELLUNG TEST

The quellung test, described in earlier chapters, uses specific antisera to induce capsular swelling in such organisms as pneumococci and type B *H. influenzae*. The test may be performed on spinal fluid, sputum, and other body fluids.

COUNTERIMMUNOELECTROPHORESIS AND GEL DIFFUSION

Basically, counterimmunoelectrophoresis (CIE) is the Ouchterlony gel-diffusion technique with the addition of an electric current to expedite interaction between antigen and antibody. Reactions occur in 30 to 60 minutes. The test has been useful in detecting antigens of pneumococci, meningococci, and *Haemophilus* in spinal fluid; this provides for rapid specific diagnosis and may be positive in the absence of live (or dead) bacterial cells.[13] In meningitis, CIE may also be used to detect staphylococci (teichoic acid), *Klebsiella,* and *Pseudomonas aeruginosa.* Other body fluids, such as serum, urine, and empyema fluid, may be studied effectively in various types of infections. CIE may also be used to detect antibody to agents such as *Candida* and *Staphylococcus.*

Gel diffusion may sometimes be positive when CIE is negative. It is, of course, a slower procedure.

MISCELLANEOUS SEROLOGIC PROCEDURES

Antibody-coated staphylococci have been utilized to detect various bacterial cells or antigens, the reaction causing

*Difco Laboratories, Detroit; Baltimore Biological Laboratory, Cockeysville, Md.; and others.

specific agglutination of the staphylococci.[15]

Latex agglutination has been used to detect capsular antigen of *H. influenzae* and *Cryptococcus.*

Hemolysis in gel, with sheep red blood cells coated with antigen, has been utilized to detect antibody to *Chlamydia.*

An indirect fluorescent antibody test for Legionnaires' disease is available through CDC and some state health departments.

RADIOIMMUNOASSAY

Radioimmunoassay (RIA) is the most sensitive method for detecting hepatitis B surface antigen and the antibody to it. It has also been used to detect antibody to *S. aureus* (teichoic acid).

ENZYME-LINKED IMMUNOSORBENT ASSAY (ELISA)

This "double antibody sandwich" method is utilized to detect and measure antigen. The technique is as follows:
1. Antibody is adsorbed to the well of a microplate; then the plate is washed.
2. Test solution thought to contain antigen is added, incubated in the well, and the plate is washed again.
3. Enzyme-labeled specific antibody is added and the plate is again washed.
4. Specific enzyme substrate is added (chosen to provide a color change on degradation).
5. Color change is assessed visually or in a spectrophotometer; the amount of hydrolysis is proportional to the amount of antigen present.

A modification of this procedure (using enzyme-labeled antiglobulin) may be used for detection of antibody.

The procedure has been used for diagnosis of a variety of viral infections,[17] *M. pneumoniae* infection, and parasitic and bacterial infections in animals.[14]

SLIDE AGGLUTINATION TESTS

A simple, rapid, quantitative slide test for the detection of serum agglutinins that develop during certain febrile infections is considered a useful diagnostic procedure in many laboratories. This technique can be as informative as the tube agglutination procedure.

The antigens are standardized and include the following*:

> *Brucella abortus* antigen
> *Proteus* OX19, OXK and OX2
> *Salmonella* Group A (O antigens 1, 2, 12)
> *Salmonella* Group B (O antigens 4, 5, 12)
> *Salmonella* Group C (C$_1$ and C$_2$) (O antigens 6, 7, 8, Vi)
> *Salmonella* Group D (O antigens 1, 9, 12, Vi)
> *Salmonella* Group E (E$_1$, E$_2$, E$_3$, E$_4$) (O antigens 1, 3, 15, 19, 34)
> Paratyphoid A antigen (flagellar a)
> Paratyphoid B antigen (flagellar b, 1, 2)
> Paratyphoid C antigen (flagellar c, 1, 5)
> Typhoid H antigen (flagellar d)

Technique

1. Using a glass slide 9 × 14 inches ruled in 1½-inch squares,† deliver 0.08-, 0.04-, 0.02-, 0.01-, and 0.005-ml volumes of patient's serum with an 0.2-ml pipet graduated in 0.001 ml to the squares of one row, from left to right. Repeat this for as many rows as there are antigens to be used (usually six).
2. Shake the antigen vials so that the contents are well mixed. By means of the standardized dropper provided with each antigen vial, deliver 1 drop of antigen (0.03 ml) on each volume of serum in each row, from left to right. When this amount of antigen is mixed with the volumes of serum indicated, the result will be approximately equivalent to a dilution series of: 1:20, 1:40, 1:80, 1:160, and 1:320 in a tube test using diluted antigen. Further dilutions may be prepared by using a 1:10 dilution of serum in physiologic saline and incorporating the volumes described in step 1.
3. Using applicator sticks or toothpicks, mix the serum and antigen, proceeding in each row from **right to left** to minimize the carry-over of serum from the low to the high dilutions.
4. Rotate the slide over a surface that is illuminated for maximum visibility for a period of 3 minutes.
5. Read and record the degree of agglutination as follows:

4+	Complete agglutination
3+	75% agglutination
2+	50% agglutination
1+ or less	25% or less agglutination

6. The highest dilution of serum with 2+ agglutination is considered the end point, or **titer.** Therefore, if a serum specimen shows the following pattern:

Serum	Equiva-lent dilution	Antigen A	Antigen B	Antigen C
0.08 ml	1:20	4+	4+	3+
0.04 ml	1:40	4+	3+	2+
0.02 ml	1:80	4+	2+	2+
0.01 ml	1:160	2+	+/−	−
0.005 ml	1:320	−	−	−

report it as the following serum titers:

> Antigen A = 1:160
> Antigen B = 1:80
> Antigen C = 1:80

Interpretation of test results is similar to that for tube agglutination tests, described below. Such agglutinin titers are to be considered as **presumptive** evidence (rather than specific evidence) in the diagnosis of disease. Also, the possibility of nonspecific reactions must be considered.

TUBE AGGLUTINATION TESTS

As indicated previously, the examination of a patient's serum for the detection of agglutinins against various organisms is an important diagnostic procedure. In

---

*Febrile antigens are available from Lederle Laboratories, Pearl River, N.Y.; Baltimore Biological Laboratories, Cockeysville, Md.; Lee Laboratories, Grayson, Ga.; and others.

†Permanently ruled glass slides are available from Arthur H. Thomas Co., Philadelphia. These are the Perma Slides, 20-ring 14-mm I.D., No. 6690-M10, or equivalent.

many infections, such as **tularemia** and **brucellosis,** the agglutination test may be the only means of laboratory diagnosis available and, as such, may be of considerable value.

Since agglutinins of variable titer may occur in normal sera, a positive agglutination test on a single specimen of serum has little or no significance. For results to be meaningful, at least a **fourfold rise in titer** must be demonstrated during the course of the infection. This can be determined only by comparing the titers of two or more samples of serum, one during the acute phase and one during convalescence.

When serum is submitted to the laboratory for an agglutination test, the clinician should always provide pertinent case history data to give some indication to the laboratory personnel as to the possible etiology. In cases of suspected **typhoid fever,** the patient's serum should be set up against both the H and the O antigens of *Salmonella typhi.* Additional H and O antigens also may be used for the detection of agglutinins against other salmonellae, although the value of these is questionable. For the serologic diagnosis of **rickettsial infections,** antigens prepared from *Proteus* OX19, *Proteus* OX2, and *Proteus* OXK should be used. The **typhus fevers** usually give a higher titer with *Proteus* OX19, whereas the **spotted fevers** may give a higher titer with *Proteus* OX2 antigens. *Proteus* OXK aids in the diagnosis of tsutsugamushi fever.

Antigens

Standardized suspensions of most bacteria for use as antigens are available commercially.* Readers interested in details of preparation of antigens should refer to Edwards and Ewing,[4] particularly with regard to the Enterobacteriaceae.

———

*Excellent standardized antigens are available from Central Public Health Laboratories, London.

Table 38-2. Protocol for agglutination test by parallel dilution system

Preparation of master dilutions									
Tube	1	2	3	4	5	6	7	8	9
Amount of saline (ml)	8	5	5	5	5	5	5	5	5
Amount of serum (ml)	2	Mix and transfer 5 ml from tube 1 to tube 2. Continue mixing and transferring 5 ml serially through to tube 9, discarding 5 ml from tube 9.								
Serum dilution	1:5	1:10	1:20	1:40	1:80	1:160	1:320	1:640	1:1280
Volume of serum dilution (ml)	5	5	5	5	5	5	5	5	5

Procedure for parallel agglutination tests

Commencing with the highest dilution in tube 9 of the master dilution series, pipette 0.5 ml of the dilution into the corresponding agglutination tube and proceed in reverse until tube 1 is reached, as indicated, using the same pipet.

Agglutination tube	1	2	3	4	5	6	7	8	9	10
Amount of master dilution (ml)	0.5 of dilution 1	0.5 of dilution 2	0.5 of dilution 3	0.5 of dilution 4	0.5 of dilution 5	0.5 of dilution 6	0.5 of dilution 7	0.5 of dilution 8	0.5 of dilution 9	0.5 of saline
Amount of antigen (ml)	0.5	0.5	0.5	0.5	0.5	0.5	0.5	0.5	0.5	0.5
Final serum dilution	1:10	1:20	1:40	1:80	1:160	1:320	1:640	1:1280	1:2560	Control

Technique

Two methods of preparing serial dilutions are available. The first method should be used if a single test is required; the protocol is given in Table 38-1. When a number of tests are to be performed, such as in the Widal test for typhoid and paratyphoid fevers, the **parallel dilution system** is employed for accuracy and speed. The protocol for this test is shown in Table 38-2.

Single serial dilution test. Set up ten clean, visually clear agglutination tubes (10 × 100 mm) in a rack. Add physiologic saline (0.85%) as the diluent to the tubes in the amounts shown in Table 38-1. Next, add the patient's undiluted serum with a 1-ml serologic pipet to the first tube, mix, and transfer as indicated. Discard 0.5 ml from tube 9. Tube 10 serves as the control and contains only saline. Add 0.5 ml of the appropriate antigen suspension to each tube of the series.

Parallel dilution method. Set up nine clean 18- or 20-mm test tubes in a rack. Add saline to each tube, as indicated in Table 38-2. Add undiluted patient's serum to tube 1 as shown, and, after thorough mixing, serially transfer 5 ml from tube to tube until nine master dilutions have been established. With a 5-ml pipet, transfer 0.5 ml of the highest dilution (tube 9) to the corresponding agglutination tube in each of the test series set up in agglutination racks. By starting with the highest dilution and working backward, the same pipet may be used throughout the procedure with little concern for carrying over any appreciable amount of serum antibody. With the system shown, at least nine parallel series can be established for agglutination tests with different antigens. Adjustments may be made either way in the master series to accommodate the number of agglutination tests planned. The antigen supensions are added as

shown in Table 38-2. Tube 10 serves as the control for each test.

Since antigens used in agglutination tests are subject to variation, it is recommended that **control tests,** using positive antisera of known titer, be routinely carried out as a check on the agglutinability of the test antigens. Such tests are performed by the technique described for the single serial dilution test.

Period of incubation

Febrile antigens are available commercially, accompanied by complete directions. The times and temperatures used for the incubation of tube agglutination tests will vary from one manufacturer to another.

Reading agglutination tests

Observe every tube for clearing of the supernatant fluid and the amount and character of the sediment and agglutinated particles. The pattern of the sediment can be more accurately observed if the tubes are read against a black background using an indirect light source. In the **control** and **negative** test, the antigen will settle to the bottom of the tube as a small round disk with smooth edges. In the **positive** tubes the cells will settle out over a larger area and may even extend up the sides of the tube. The pattern of this sediment will be somewhat irregular and will vary with the extent of the agglutination. After examining the pattern of the sediment, shake the tube gently. H agglutinins produce large floccular aggregates, which are easily broken up, whereas O agglutinins produce granular or small flaky aggregates. Complete agglutination with complete clearing of the supernatant fluid indicates a 4+ reaction. Decreasing amounts of agglutination and increasing cloudiness of the supernatant fluid are read as 3+, 2+, and 1+ reactions.

Interpretation of results

It is almost impossible to assign positive or negative values to arbitrary titers

*Bailey, W. R.: Unpublished data.

in any agglutination test. Variable factors, such as past infection, vaccination, time at which the specimen was taken, and naturally occurring agglutinins all can influence the titer. As mentioned earlier, only a fourfold (two tube) or greater **rise in titer** over a period of time is usually significant. However, the following suggestions may prove helpful in interpreting results.

Negative results. Negative results may be due to either of the following: (1) the sample of blood was obtained before the appearance of agglutinins in the serum or (2) incorrect diagnosis (patient not having infection for which the tests were requested).

Negative results are of particular value for comparing titers that may occur with later samples of serum. Any two-tube increase in titer during the course of an infection is usually significant.

Positive results. Interpretation of positive results obviously will vary with the infection and is given only brief mention in this chapter.

1. In **typhoid fever:** (a) Indication of current infection—titer of 1:160 with O antigen—titer rising in subsequent sera; (b) indication of past infection, recent vaccination, or an anamnestic reaction—titer of 1:80 to 1:160 with H antigen only.
2. In **brucellosis:** A titer of 1:160 or greater usually suggests infection, past or present, and 1:320 or greater usually indicates acute brucellosis.
3. In **tularemia:** A titer of 1:80 or greater usually suggests definite infection.
4. In **Rocky Mountain** and **typhus fevers:** Variable results are obtained with *Proteus* OX19 and *Proteus* OX2 antigens. In general, typhus fever gives a higher titer with OX19, and Rocky Mountain spotted fever and tick-bite fever may give a higher titer with OX2. In both instances, diagnostic titers are high, over 1:320, and only a rising titer is conclusive. Agglutination with

Proteus OXK may be diagnostic for tsutsugamushi fever, but it also may be indicative of the spirochetal disease, relapsing fever.

SEROLOGIC TESTS FOR SYPHILIS

Perhaps no other infectious disease process relies so heavily on the serologic test for a diagnosis than syphilis. This situation exists primarily because the etiologic agent, *Treponema pallidum,* cannot be cultured in the laboratory. Although darkfield examination is excellent for demonstrating *T. pallidum* in specimens (particularly in the early stages of the disease), most laboratories do not have the capability or the experience to perform this procedure properly (see p. 207).

Well over 200 serologic tests for syphilis (STS) have been described, but only a few are in use today. All tests for syphilis depend on the antigen-antibody reaction and are performed on either blood (serum or plasma) or cerebrospinal fluid specimens.

Tests are classified according to the type of antigen used. For example, **nontreponemal** or reagin tests are performed with extracts from normal tissue or other sources. **Treponemal** tests, on the other hand, employ treponemes or treponemal extracts to detect antibody. Further details on syphilis serology and its interpretation will be found in three excellent references cited here.[7,10,18]

Nontreponemal antigen tests

These tests are not immunologically specific for syphilis and are not the most sensitive tests. However, their ease of performance and relatively low cost account for the wide use of these antigen tests, particularly the flocculation tests (below), as screening procedures. The flocculation tests have largely replaced the more cumbersome but less sensitive Wasserman complement fixation test.[2] Of the nontreponemal antigen tests, the Venereal Disease Research Laboratory (VDRL) or the rapid plasma reagin (RPR) flocculation tests, along with the Kolmer complement fixation test, are the most widely used. Two studies (Fowler, E., et al.: Can. J. Publ. Health **67**:482-484, 1976; and Dzuik, P. E., et al.: J. Clin. Microbiol. **5**:593-595, 1977) have shown that the RPR card test was quite reliable and that the VDRL was somewhat less specific. The automated reagin test was less sensitive than the others.

Reactive nontreponemal tests confirm the diagnosis in the presence of early or late lesion syphilis. They offer a diagnostic clue in latent, subclinical syphilis and are effective tools for detecting cases in epidemiologic investigations. Finally, they are superior to the treponemal tests for following the response to therapy. For a thorough discussion on their application, as well as detailed instructions on how to adequately perform these tests, the reader should consult other references.[2,7,10,11,16,18]

Results of **qualitative** tests for syphilis are customarily reported as reactive (or positive, or 4+), weakly reactive (or weakly positive, or 3+, 2+, or 1+), or nonreactive (negative). **Quantitative** results may be obtained by diluting the serum in geometrical progression to an end point. The titer is usually expressed as the highest dilution in which the test is fully reactive.

Excessive production of antibody (particularly in the secondary stage of syphilis) occasionally results in a prozone phenomenon due to antibody excess. This is true in both complement fixation and flocculation tests. Undiluted specimens will give a nonreactive or weakly reactive test result. Testing at higher dilutions, however, gives reactive test results.

Careful attention must be paid to each reactive or weakly reactive serologic result. Many cases of untreated late latent or late syphilis will give only weakly reactive results with undiluted serum. On the other hand, the titer is usually high (>1:16) in secondary syphilis. A high titer does not necessarily mean early syphilis (or even syphilis), but it is strong evidence for the presence of syphilis. Some of the highest titers recorded have been in late visceral or cutaneous syphilis or in nonsyphilitic diseases (e.g., hemolytic anemia or systemic lupus erythematosus).[16]

Treponemal antigen tests

It has been long recognized that the nontreponemal antigen tests are not entirely specific for *T. pallidum* infection. Therefore, antigens for testing also have been made from treponemes (i.e., treponemal antigens). These treponemal antigen tests are primarily used as confirmatory tests, particularly in patients in whom the clinical, historical, or epidemiologic evidence of syphilis is questionable. Three such tests are the *Treponema pallidum* immobilization (TPI) test, the fluorescent treponemal antibody absorption (FTA-ABS) test, and the *Treponema pallidum* hemagglutination test(s).

Although time consuming, technically demanding, and extremely expensive to perform, the TPI test is considered by many to be the standard by which all treponemal antigen tests are compared. The antigen for this test is the Nichols strain of *T. pallidum* and is harvested from testicular syphilomas in artificially infected rabbits. In the TPI test, these harvested live treponemes are combined with the patient's serum and complement. After appropriate incubation, serum that contains treponemal antibodies will cause immobilization of the treponemes. Since the TPI antibody develops more slowly, the test is not reactive until later in early syphilis, in contrast with the nontreponemal antigen tests; that is, in some primary syphilitic patients the TPI test may be nonreactive, whereas the VDRL test is reactive. The TPI test is the test of choice for spinal fluids, especially when reagin tests give nonreactive or equivocal results. This test is not readily available in most serology laboratories.

The most widely used of the treponemal antigen tests is the FTA-ABS test. The antigen in this test consists of nonviable *T. pallidum* (Nichols strain). It is allowed to dry on a glass slide, is fixed, and then is combined with serum that has been previously absorbed to remove nonspecific treponemal antibodies. If syphilis antibody is present, it will combine with the nonviable organisms. Fluorescein-labeled antihuman globulin is then added and reacts with the serum that is attached to the organisms. When observed under the fluorescent microscope, this reaction is readily visible (i.e., the organisms **fluoresce**). If no syphilis antibodies are present in the test serum, the treponemes will not fluoresce and therefore are nonvisible under the fluorescent microscope.

The FTA-ABS test becomes reactive earlier than the TPI test in early syphilis and is about 5% more sensitive than the TPI in late latent or late syphilis. The FTA-ABS test is now widely available and because of its increased sensitivity and specificity is the confirmatory test of choice. A reactive test confirms the presence of treponemal antibodies but does not indicate the stage or activity of infection.

For a more detailed account of these and other treponemal antigen tests, including specific test instructions, the reader should consult the *Manual of Clinical Immunology*[18] and other selected publications.[2,3,7,10,16]

Several *T. pallidum* hemagglutination tests (TPHA) are available. Red cells from one or another animal species and components of the Reiter treponeme are utilized. The TPHA promises advantages of economy, specificity, and sensitivity with less need for expensive equipment and technical skill than are required for the TPI or FTA-ABS tests. The TPHA test also lends itself to automation. It is somewhat less sensitive than the FTA-ABS test in early primary syphilis, but the tests are comparable in other stages of the disease. It is not clear yet how effectively the TPHA will distinguish between specific treponemal reactions and biologic false-positive reactions.

False-positive reactions

All normal sera may contain minute amounts of reagin. The sensitivity of nontreponemal tests is altered by varying the proportion of reagents, temperature, mixing time, and other physicochemical variables. For these reasons, about one fourth of all false-positive reactions represent technical errors or day-to-day variability in testing.

Repeatedly reactive nontreponemal tests accompanied by nonreactive treponemal tests (TPI or the more sensitive FTA-ABS) characterize the **false-positive** reactor. The duration of reagin reactivity arbitrarily determines whether the false-positive reaction is **acute** (less than 6 months) or **chronic** (6 months or longer). Though false-positive reactions have been called "biologic," many such reactions are associated with specific diseases or follow vaccination or immunization; therefore, this adjective should be discarded. The terms acute false-positive or chronic false-positive adequately describe what is observed.[16]

Acute false-positive reactions are found in persons suffering from many viral and bacterial infections or who have had certain vaccinations and immunizations. Pregnancy may also result in a false-positive test. Chronic false-positive reactions are usually less frequent than "technical" or acute false-positive reactions.

Lepromatous leprosy, heroin addiction, lupus erythematosus, and occasionally malaria are associated with chronic false-positive nontreponemal tests for syphilis, whereas the nonvenereal treponematoses (yaws, pinta, and bejel) characteristically give reactive nontreponemal tests as well as treponemal tests and are not serologically distinguishable. It should be pointed out that syphilis and systemic lupus erythematosus or syphilis and leprosy can occur together. In situations such as these, a reactive serologic test is not to be considered as a false-positive result.[16]

OTHER DETERMINATIONS

Other types of antigen-antibody determinations and other specific applications of such tests are noted throughout the book in the appropriate areas (e.g., chapters on virus infections, fungal infections, parasitic infections, and so forth).

REFERENCES

1. Bennett, C. W.: Clinical serology, rev. ed., Springfield, Ill., 1968, Charles C Thomas, Publisher.
2. Center for Disease Control: The laboratory aspects of syphilis, Atlanta, 1971, The Center.
3. Deacon, W. E., Lucas, J. B., and Price, E. J.: Fluorescent treponemal antibody-absorption (FTA-ABS) test for syphilis, J.A.M.A. **98**:624-628, 1966.
4. Edwards, P. R., and Ewing, W. H.: Identification of Enterobacteriaceae, ed. 3, Minneapolis, 1972, Burgess Publishing Co.
5. Hayflick, L., and Chanock, R. M.: *Mycoplasma* species of man, Bacteriol. Rev. **29**:185-221, 1965.
6. Isenberg, H. D.: Personal communication (E.G.S.).
7. Jaffe, H. W.: The laboratory diagnosis of syphilis, Ann. Intern. Med. **83**:846-850, 1975.
8. Kenny, G. E.: Serology of mycoplasmic infections. In Rose, N. R., and Friedman, H., editors: Manual of clinical immunology, Washington, D.C., 1976, American Society for Microbiology.
9. Klein, G. C.: Immune response to streptococcal infection. In Rose, N. R., and Friedman, H., editors: Manual of clinical immunology, Washington, D.C., 1976, American Society for Microbiology.
10. Miller, J. N.: Value and limitations of non-treponemal and treponemal tests in the laboratory diagnosis of syphilis, Clin. Obstet. Gynecol. **18**:191-203, 1975.
11. Nicholas, L., and Beerman, H.: Present day serodiagnosis of syphilis, Am. J. Med. Sci. **249**:466-483, 1965.
12. Peterson, O. L., Ham, T. H., and Finland, M.: Cold agglutinins (autohemagglutinins) in primary atypical pneumonias, Science **97**:167, 1943.
13. Rytel, M. W.: Counterimmunoelectrophoresis in diagnosis of infectious disease, Hospital Practice, Oct. 1975, pp. 75-82.
14. Saunders, G. C., and Clinard, E. H.: Rapid micromethod of screening for antibodies to disease agents using the indirect enzyme-labeled antibody test, J. Clin. Microbiol. **3**:604-608, 1976.
15. Suksanong, M., and Dajani, A. S.: Detection of *Haemophilus influenzae* type B antigens in body fluids, using specific antibody-coated staphylococci, J. Clin. Microbiol. **5**:81-85, 1977.
16. Syphilis, a synopsis, U.S. Department of Health, Education, and Welfare, National Communicable Disease Center, Public Health Service Publ. No. 1660, Atlanta, 1968, pp. 96-108.
17. Voller, A., Bidwell, D., and Bartlett, A.: Microplate enzyme immunoassays for the immunodiagnosis of virus infections. In Rose, N.R., and Friedman, H., editors: Manual of clinical immunology, Washington, D.C., 1976, American Society for Microbiology.
18. Wood, R. M.: Tests for syphilis. In Rose, N. R., and Friedman, H., editors: Manual of clinical immunology, Washington, D.C., 1976, American Society for Microbiology.

CHAPTER 39

Fluorescent antibody techniques in diagnostic microbiology

The fluorescent antibody (FA) method, more properly known as the **immuno-fluorescence method,** has reached such a level of scientific maturity that it is no longer a laboratory curiosity. It is a rapid, reproducible, and reliable aid in the hands of experienced workers for the identification of microorganisms or the antibodies they engender.

Space does not permit a complete review of the vast literature about FA; the interested reader is referred to the excellent monographs and technical manuals on the method, including those of Coons,[5] Cherry and associates,[3] Beutner,[1] and the various manufacturers of reagents, as well as appropriate chapters in the *Manual of Clinical Microbiology*[2] and *Laboratory Procedures in Clinical Microbiology.*[9]

The FA method is essentially a sophisticated technique for **demonstrating antigen-antibody reactions.** In this technique a film preparation or tissue section is treated with an appropriate serum containing an immune globulin (antibody) that has been labeled (conjugated) with a fluorescent dye, such as fluorescein isothiocyanate. This preparation is then examined against a dark background (darkfield) illuminated by a very bright light source rich in the near-ultraviolet spectrum. This light causes the antigen-antibody complex to become **fluorescent** and appear a bright glowing **yellow-green** against the dark background (when stained with fluorescein isothiocyanate).

The diagram in Fig. 39-1 represents the basic optical system utilized in fluorescence microscopy and consists of a high-pressure mercury vapor lamp (Osram

HBO 200) that provides light of a very high intensity. This is passed through Schott BG-14 and BG-22 heat-absorbing filters and excited in wave-lengths of 350 to 450 mμm by passage through a BG-12 filter, the range in which the fluorescein dye is most brilliant. Since this wavelength is in the ultraviolet spectrum, it must be removed by the insertion of a barrier filter in the eyepiece (usually Schott OG-1) for protection of the observer's eyes. This orange filter holds back wavelengths below 500 mμm and transmits visible wavelengths emitted by the specimen. Corning, Wratten, or other equivalent filter systems also may be used, and newer optics and lighting systems are available.

Two important developments in fluorescent microscopy are the introduction of the halogen lamp with the interference filter and the introduction of incident-light, or epi-illumination. See Cherry[2] for details.

The choice of the particular FA technique to be used depends largely on the information desired. Generally, the **direct** staining procedure is used in identification of an unknown antigen, such as group A streptococci obtained from throat swabs. The **indirect** method is employed primarily in the detection of antibody, as in the fluorescent treponemal antibody (FTA-ABS) test for syphilis. Other uses of these basic principles are described in the references previously cited.

To date, a number of FA diagnostic procedures have been evaluated, and thus a number of standardized commercially prepared conjugates have become

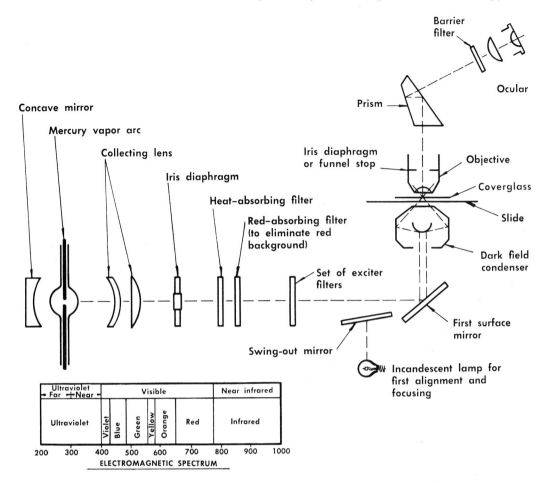

Fig. 39-1. Schematic representation of equipment for fluorescence microscopy. (Prepared with suggestions by Dr. Peter Bartals, E. Leitz, Inc., New York.)

available.* Fluorescent-antibody conjugates are stable for prolonged periods under various conditions.[6] These procedures are in regular use in many public health and hospital microbiology laboratories and include the following: rapid identification of group A streptococci; the rapid and specific diagnosis of rabies virus in tissue; identification of enteroviruses and respiratory viruses, herpes simplex, arboviruses, rickettsiae, and so forth, in isolates and clinical material; the

highly specific FTA-ABS test for syphilis; the rapid identification of *Neisseria gonorrhoeae* in exudates; and other tests mentioned in various sections of this text. For a critical review of the FA techniques used in diagnostic bacteriology, with an extensive bibliography, the reader is referred to the excellent monograph by Cherry and Moody.[4]

A list of FA tests useful in diagnostic bacteriology and mycology is presented in Tables 39-1 and 39-2. Indirect FA tests have also been useful in the diagnosis of chlamydial, viral, and parasitic infections.

There are numerous pitfalls in the application of immunofluorescence to diagnostic bacteriology; the procedures

---

*Baltimore Biological Laboratory, Cockeysville, Md.; Difco Laboratories, Detroit; Burroughs Wellcome Reagents Division, Greenville, N.C.; Clinical Sciences, Inc., Whippany, N.J.; and others.

Table 39-1. List of fluorescent antibody tests most useful in diagnostic bacteriology*

Test	Status
A. Group A streptococci. Direct test.	Most highly evaluated and extensively used of all FA tests.
B. Fluorescent antibody darkfield test (FADF).	Comparable to darkfield for demonstration of *T. pallidum.* As accurate and does not require motile treponemes of specific morphology.
C. Fluorescent treponemal antibody test (FTA-ABS). Indirect test.	Highly specific and sensitive for detection of true antibody to *T. pallidum.* Much easier to perform than treponemal immobilization test.
D. Identification of the gonococcus. Direct test. (An indirect FA technique has been developed for detecting gonococcal antibodies in the sera of females with uncomplicated disease.)[10]	A good test, particularly for rapid confirmation of a culture as *N. gonorrhoeae* (see pp. 145, 487).
E. Identification of the diphtheria bacillus. Direct test.	Good for use on nasopharyngeal specimens from patients suspected of having diphtheria. Not recommended for carrier surveys, contacts, and so forth. Specificity needs improvement.
F. *Bordetella pertussis.* Direct test.	A rapid, specific, and sensitive test as compared with conventional culture techniques.
G. Incitants of bacterial meningitis. Direct test on spinal fluids. 1. *Haemophilus influenzae* 2. *Neisseria meningitidis* 3. *Streptococcus pneumoniae*	A rapid, highly efficient test to apply to CSF sediments.
H. *Listeria monocytogenes.* Direct test.	Excellent for specific detection of organisms in formalin-fixed, paraffin embedded sections, CSF, impression smears of tissue, and so forth.
I. *Brucella* (three major species). Direct test.	Sensitive, rapid, and genus specific for detection of smooth strains of *Brucella* in tissues. No advantage over agglutination tests for detection of antibody.
J. *Yersinia pestis.* Direct test.	Excellent for rapid diagnosis of plague when used on blood, stomach contents of vectors, stored animal tissues, and phagocytic cell preparations. Direct or indirect staining of impression smears, frozen or freeze-dried material may be used. FA is method of choice for rapid and specific identification of plague bacillus in tissues and in culture.

*Courtesy William B. Cherry, Analytic Bacteriology Branch, Center for Disease Control, Atlanta. Modified.

Table 39-1. List of fluorescent antibody tests most useful in diagnostic bacteriology—cont'd

Test	Status
K. *Francisella tularensis.* Direct test.	Rapid, sensitive, and specific for detection of organisms in culture, impression smears, frozen sections, formalin-fixed, paraffin embedded sections of specimens from human cases, experimental animals, or from air.
L. *Bacillus anthracis.* Direct test.	Rapid and sensitive for detection of anthrax organisms in impression smears or formalin-fixed, paraffin-embedded tissue, also from human cutaneous vesicles, and tissues of experimental or naturally infected animals. Test not entirely specific.
M. Leptospirae.	Has been used successfully to detect leptospirae in urine sediments, frozen sections, impression smears, and in formalin-fixed frozen sections. Not reliable for serotyping.
N. *Salmonella typhi.* Direct test using sorbed Vi conjugate.	Approximately the same sensitivity and specificity as culture on bismuth sulfite agar but very rapid. Excellent when used for detection of chronic typhoid carriers. Culture is more reliable for use on acutely ill and convalescent patients.
O. *Shigella.* Direct test. S. *sonnei*	Sensitive and specific for screening fecal smears. Well evaluated.
S. *flexneri*	Appears promising for screening of fecal smears. Needs further evaluation.
P. Clostridia. Direct test.	Toxigenic types of *C. botulinum* can be differentiated with absorbed conjugates. Staining of smears from tissues or body fluids is very helpful in diagnosis of animal diseases due to clostridia.
Q. *Treponema pallidum.* Direct test.	Procedure may be applied to specimens, from lesions, transported either in capillary tubes or on dried slides and mailed to the laboratory.[8]
R. *Actinomyces israelii.* Direct test.	Specific identification for cultures and smears but not tissue sections.
S. *Actinomyces naeslundii.* Direct test.	Specific identification for cultures and smears but not tissue sections.
T. *Arachnia (Actinomyces) propionica.* Direct test.	Specific identification for cultures, smears, and tissue sections.

Table 39-2. Value and limitations of fluorescent antibody reagents in medical mycology*

Conjugate for	Procedure	Staining reaction with fungi in		
		Cultures	Tissue smears, pus, exudates	Tissue sections
Aspergillus species	Direct staining	—	Conjugate available for differentiating *Aspergillus* sp. from *Candida* sp. and phycomycetes[7]	Genus identification[7]
Blastomyces dermatitidis and *Paracoccidioides brasiliensis*	Direct staining	Specific identification of yeast form; no value for identification of mycelial form	Specific (tissue form) identification	Specific identification
Candida species	Direct staining	No specific reactivity for any *Candida* sp. but a good rapid screening agent		
Coccidioides immitis, tissue form	Direct staining	No value for identification of mycelial form	Specific identification	Specific identification
Cryptococcus neoformans	Direct staining	Specific identification; does not stain some strains	Specific identification	Specific identification
Histoplasma capsulatum	Direct staining in conjunction with *B. dermatitidis* conjugate	Specific identification; no value for identification of mycelial form	Specific identification	Not optimal
Sporothrix schenckii	Direct staining	Specific identification of yeast-form	Specific identification	Specific identification
Trichophyton sp.	Direct staining	Differentiation of *T. mentagrophytes* from *T. rubrum* only.		

*Courtesy Leo Kaufman, Chief, Fungus Immunology Section, Mycology Branch, Center for Disease Control, Atlanta. Modified.

must be carried out by workers well trained in a meticulous technique and with a strong background in serology and microscopy. The procedures must be monitored at all times by the inclusion of positive and negative controls; reagents must be standardized and their specificity must be checked regularly. In short, the FA method should not be used as a substitute for the conventional cultural procedures; rather, it should be utilized as an adjunct to these tests. Its great potential in the field of diagnostic microbiology is just now being realized.

REFERENCES

1. Beutner, E. H.: Immunofluorescent staining; the fluorescent antibody method, Bacteriol. Rev. **25**:49-76, 1961.
2. Cherry, W. B.: Immunofluorescence techniques. In Lennette, E. H., Spaulding, E. H., and Truant, J. P., editors: Manual of clinical microbiology, ed. 2, Washington, D.C., 1974, American Society for Microbiology.
3. Cherry, W. B., Goldman, M., Carski, T. R., and Moody, M. D.: Fluorescent antibody techniques in the diagnosis of communicable diseases, Public Health Service Publ. No. 729, Washington, D.C., 1960, U.S. Government Printing Office.
4. Cherry, W. B., and Moody, M. D.: Fluorescent antibody techniques in diagnostic bacteriology, Bacteriol. Rev. **29**:222-250, 1965.
5. Coons, A. H.: The diagnostic application of

fluorescent antibodies, Schweiz. Pathol. Bakt. **22:**700-723, 1959.

6. Green, J. H., Gray, S. B., Jr., and Harrell, W. K.: Stability of fluorescent antibody conjugates stored under various conditions, J. Clin. Microbiol. **3:**1-4, 1976.

7. Kaplan, W.: Direct fluorescent antibody tests for the diagnosis of mycotic diseases, Ann. Clin. Lab. Sci. **3:**25-29, 1973.

8. Kellogg, D. S., Jr.: The detection of Treponema pallidum by a rapid, direct fluorescent antibody darkfield (DFATP) procedure, Health Lab. Sci. **7:**34-41, 1970.

9. Washington, J. A. II, Martin, W. J., and Karlson, A. G.: Fluorescent antibody procedures. In Washington, J. A. II, editor: Laboratory procedures in clinical microbiology, Boston, 1974, Little, Brown and Co.

10. Welch, B. G., and O'Reilly, R. J.: An indirect fluorescent technique for study of uncomplicated gonorrhea, I. Methodology, J. Infect. Dis. **127:**69-76, 1973.

PART VII

Quality control

CHAPTER 40

Quality control in the microbiology laboratory

Although the concept of quality control has only recently come to the clinical microbiology laboratory, this is no excuse for its absence in any laboratory. With the present and continuing reexamination of health care delivery by the medical and allied health professions, it is incumbent on the microbiologist to ensure that the laboratory contributes to the highest quality of patient care by its expertise, its accuracy, its relevance and the prompt reporting of its findings to the attending physician. It is only by constant self-evaluation of the laboratory's performance through a quality control program that such a level of excellence can be developed and maintained.

It is not the function of this section to detail the methodology for carrying out a quality control program—these are presently available in excellent texts and publications[1-3,7,9,10,12-14]—but to suggest a basic format that can be tailored to individual situations or requirements.

CONTROL OF EQUIPMENT

Generally, equipment is simple to monitor; this consists chiefly of the daily, weekly, or monthly observation and recording of temperatures of incubators (optimal, 35 C), refrigerators, water baths, hot-air ovens, freezers, autoclaves (by temperature–time cycle charts), and so forth. A range of ± 1 C is generally acceptable for heating devices, with a greater range acceptable for refrigerating devices; both should be measured with thermometers that have been calibrated against a Bureau of Standards instru-

ment.* Autoclaves should also be checked at monthly intervals with biologic devices such as the Kilit ar Attest spore test ampules or strips.†

In addition to their temperature recordings, carbon dioxide incubators should also have their CO_2 concentration measured by the portable Fyrite CO_2 gas analyzer,‡ which determines the CO_2 level by absorption, with change in level of the absorbent in a graduated tube. Tank pressure also must be monitored, and an extra tank of gas should be kept on hand at all times.

Every laboratory handling material for mycobacteriologic, mycologic, and virologic examination should possess a **biologic safety hood,** which also requires regular recording of the negative pressure within the hood, the interval between the HEPA filter changes, the effective output of the UV Sterilamp (measured by a photometric cell§), and so forth. Directions for these tests are included in the hood operation manual or from the Biological Hazards Control Officer of the Center for Disease Control (CDC), Atlanta.

Other miscellaneous items requiring inspection, calibration, or replacement include the standard (0.01 and 0.001 ml) in-

---

*Thermometers should be in tubes of water wherever possible.
†Available from Baltimore Biological Laboratory, Cockeysville, Md.
‡Arthur H. Thomas, Co., Philadelphia, catalog No. 5566-C 10.
§Westinghouse SM-600 meter, Westinghouse Electric Corp., Lamp Division, Bloomfield, N.J.

oculating loops, chipped or cracked glassware, the presence of residual detergent in glassware; mechanical devices, such as centrifuges, vacuum pumps, or pipetting machines require frequent oiling and adjustments. Microscopes and balances should be cleaned semiannually; pH meters and other measuring devices, such as thermometers, graduated cylinders, pipets, and so forth, also should be calibrated periodically.

CONTROL OF CULTURE MEDIA

Most clinical bacteriology laboratories today utilize ready-made, commercially available solid and fluid culture media. These are, in general, of uniformly high quality and good batch-to-batch consistency and are readily obtainable. If, on the other hand, a laboratorian has the good fortune to include on the staff an experienced and dedicated media chef, he or she is assured of a product that is superior in many ways to that available in the marketplace. In this situation, however, strict attention to details of preparation must be adhered to, such as following exact directions of the manufacturer regarding methods of preparation and sterilization (overheating is the most frequent source of error), properly sealing and dating dehydrated media, and so forth. With certain media, such as decarboxylase media or Mueller-Hinton agar, it is also necessary to check the pH of the final product. In certain situations, as in anaerobic bacteriology, freshly made media of certain types will often be superior to purchased media.

The foregoing implies the use of a quality control program that is best carried out by **performance testing** for the desired reaction with a number of stock cultures of microorganisms of known stability. At the Wilmington Medical Center, a culture collection of 26 strains of bacteria and three strains of fungi is utilized in an ongoing program for the regular testing of culture media, stains, and reagents (Table 40-1). Each new batch of culture media is tested by the quality control technologist, while the stains and reagents are checked in use by the individual technologists performing the procedure. Records are kept on a daily basis and are eventually entered into a master log for monthly review by the microbiologist or laboratory director. A similar program exists at the UCLA Hospital and Clinics.

Stock cultures are preserved in the frozen state at approximately -70 C on glass beads, according to the method of Nagel and Kunz,[6] where they remain viable for many months. When required, a single glass bead is removed from its reservoir and placed in a tube of trypticase soy broth and incubated until growth is visible. This is then streaked on half of a blood agar plate to obtain isolated colonies and subsequently submitted for media evaluation, as indicated.

CONTROL OF REAGENTS AND ANTISERA

The various chemical reagents used in biochemical tests are best monitored, after preparation, by integrating into the daily laboratory workload through the use of a master schedule calendar.[12] Reagents should be dated on preparation and stored in light-proof, tightly stoppered bottles either at room temperature or in the refrigerator, as indicated.

Gram-stain reagents are best monitored by staining and examining slides prepared from a suspension of *E. coli* and *S. aureus* that has been grown in broth and sterilized by autoclaving; this procedure should be a part of each worker's **daily** staining schedule. Acid-fast and special staining reagents can be similarly checked with a suspension of the appropriate killed microorganisms.

Antisera that have been obtained from reliable sources are dated on receipt and diluted to the manufacturer's specifications. They are then tested with standard stock cultures for sensitivity and specificity with each new lot and every month thereafter. The manufacturer's directions

Table 40-1. Recommended cultures* for monitoring routine culture media, stains, and reagents†

Medium or test	Control organisms	Expected results
Arginine dihydrolase	*S. typhimurium*	Positive
	P. vulgaris	Negative
Bile esculin agar	*S. faecalis*	Growth, blackening
Blood culture bottles		
Biphasic (vented)	*H. influenzae*	Growth on subculture
Anaerobic (unvented)	*B. fragilis*	Growth on subculture
Brain-heart infusion agar with	*C. albicans*	Growth
antibiotics	*A. fumigatus*	Inhibited
Brucella agar with laked blood,	*Peptococcus* sp.	Growth (anaerobic)
K₁, and hemin		
Catalase test	*S. aureus*	Positive
	Group A streptococcus	Negative
Chocolate agar	*H. influenzae, S. pneumoniae*	Growth
Citrate, Simmons'	*K. pneumoniae*	Positive
	E. coli	Negative
Coagulase test (rabbit plasma)	*S. aureus*	Clot
	S. epidermidis	No clot
CNA agar	*S. faecalis*	Growth
	P. aeruginosa	No growth
CTA medium		
Glucose	*N. gonorrhoeae*	Acid (yellow)
Maltose	*N. gonorrhoeae*	No change
Sucrose	*N. gonorrhoeae*	No change
Lactose	*N. lactamicus*	Acid (yellow)
Differential carbohydrates	As indicated[13]	
DNase agar	*S. marcescens*	Positive
	E. coli	Negative
Egg yolk agar (Nagler)	*C. perfringens*	Positive (anaerobic)
	B. fragilis	Negative (anaerobic)
EMB agar	*E. coli*	Growth, metallic sheen
	S. typhimurium	Growth, lactose negative (colorless colony)
FA test, group A streptococci	Group A streptococcus	Positive
	Group B streptococcus	Negative
Germ tube test	*C. albicans*	Positive
GN broth	*S. typhimurium*	Growth on subculture
Gram stain	*S. aureus*	Gram positive
	E. coli	Gram negative
Hektoen agar	*S. typhimurium*	Growth, green colony with black center
	E. coli	Growth, orange colony (or no growth)
Hippurate hydrolysis	Group B streptococcus	Positive
	Group A streptococcus	Negative
Indole	*E. coli*	Positive
	K. pneumoniae	Negative
Kanamycin-vancomycin blood	*B. fragilis*	Growth (anaerobic)
agar	*E. coli*	Inhibited (anaerobic)
Kit for identification of Entero-	Directions from manufacturer	
bacteriaceae		

*These stable stocks have been collected from the Center for Disease Control reference strains, from the American Type Culture Collection, and from patient isolates confirmed by the Center for Disease Control. Certain ATCC strains of some of these organisms are available commercially on desiccated disks (Bact-Chek, Roche Diagnostics, Nutley, N.J.; Bactrol Disks,[8] Difco Laboratories, Detroit).

†Media are checked on each new batch; sterility tests are performed on 2% of a batch at 35 or 22 C. Stains and reagents are tested as used, but no more than once per day.

Continued.

Table 40-1. Recommended cultures for monitoring routine culture media, stains, and reagents—cont'd

Medium or test	Control organisms	Expected results
Lowenstein-Jensen medium	*M. tuberculosis* (H37-RA strain)	Characteristic colonies
Lysine decarboxylase	*S. typhimurium*	Positive
	P. vulgaris	Negative
Motility medium	*E. coli*	Positive
	K. pneumoniae	Negative
Mueller-Hinton agar	*S. aureus* (ATCC 25923)	Acceptable zone sizes (see
	E. coli (ATCC 25922)	Chapter 36)
	P. aeruginosa (ATCC 27853)	
Nitrate reduction	*P. aeruginosa*	Positive
	Acinetobacter calcoaceticus	Negative
OF media		
Glucose	*P. aeruginosa*	Oxidizer (yellow)
Maltose	*P. maltophilia*	Oxidizer (yellow)
Ornithine decarboxylase	*S. typhimurium*	Positive
	P. vulgaris	Negative
Optochin disk test	*S. pneumoniae*	> 18-mm zone
	S. mitis	≤ 18-mm zone
Oxidase reaction	*P. aeruginosa*	Positive (blue)
	E. coli	No color change
Rice agar with Tween 80	*C. albicans*	Characteristic chlamydospores
	C. tropicalis	No chlamydospores
Sabouraud dextrose agar with antibiotics	*C. albicans*	Growth
	A. fumigatus	Inhibited
Sheep blood agar	Group A streptococcus	Growth, beta hemolysis
	S. mitis	Growth, alpha hemolysis
Trypticase soy agar or broth	Group A streptococcus	Growth
Triple sugar iron agar	*S. typhimurium*	K/AG, H_2S
	E. coli	A/AG
	P. aeruginosa	K/no reaction
Urea agar (Christensen)	*P. vulgaris*	Positive (red)
	E. coli	Negative
XLD agar	*S. typhimurium*	Growth, pink-red with black center
	E. coli	Yellow (or no growth)
Ziehl-Neelsen, fluorochrome stain	*M. tuberculosis* (H37-RA)	Positive acid-fast
	E. coli	Negative acid-fast

Table 40-2. Criteria for rejection of requests for microbiologic tests*

Category	Criteria for rejection	Action
Identification	Discrepancy between patient identification on request form and on specimen container	Discuss with sender or return to sender for resolution. If discrepancy unresolved, specimen is processed but no report issued.
	No identification on container	Do not process unless request form wrapped around container. Write "Container not identified" on report form in such cases.
	Specimen source or type of culture ordered not on request form	Call physician or service for necessary information.
Specimen	Anaerobic culture request on	Cross anaerobic culture request off form.
	Sputum	Do not perform anaerobic cultures, except with microbiology staff approval. Explain reasoning to requesting physician.
	Midstream urine	
	Catheterized urine	
	Vaginal secretions	
	Prostatic secretions	
	Feces	
	Environmental material	
	Gastric washings	
	Bronchoscopic washings	
	Decubitus ulcer	
	Throat	
	Nose	
	Skin	
	Mouth	
	Ileostomy material†	
	Colostomy material†	
	Fistula†	
	Specimen identified by anatomic site only (e.g., chest, leg, etc.)	Request additional information.
	Anaerobic culture request on swab material (unless swab has been submitted in anaerobic transport tube or bag)	Return request form to floor immediately, stamped "This specimen is unacceptable for culture of anaerobic bacteria because it was not transported under anaerobic conditions. Please submit syringe aspirate injected into anaerobic transport tube, or, if necessary, a swab in proper transport setup." Do not perform anaerobic culture unless physician insists it is not feasible to obtain another sample.
	Anaeobic culture request on unacceptable material (specified above) in anaerobic transport tube	Return request form to floor immediately, stamped with "The above specimen was submitted in an anaerobic transport tube for anaerobic culture. However, since this material is generally contaminated with normal flora anaerobes, no anaerobic culture was performed. Please contact laboratory to discuss obtaining a suitable specimen."
	Improperly collected sputum (i.e., saliva)	Return request form to floor immediately, stamped with "Improper specimen; please resubmit." Explain problem.

*Adapted from criteria in use at the Mayo Clinic.
†Refers to intraluminal material.

Continued.

Table 40-2. Criteria for rejection of requests for microbiologic tests—cont'd

Category	Criteria for rejection	Action
	Material received in fixative (e.g., formalin)	Notify physician and request new specimen. Stamp "Specimen unsatisfactory; please submit new specimen" on request form and return.
	Gram stain smear of material from anus or rectum for gonococci	Cross Gram smear request off form.
	Dried out swab	Notify physician and request new specimen. Stamp "Specimen unsatisfactory—dried swab; please submit new specimen" on request form and return.
	Blood culture with request for Gram-stained smears, cultures for mycobacteria (TB), or viruses	Cross request off form.
	Foley catheter tips	Discard sample. Explain problem.
	24-hour urine or sputum collections for mycobacteria (TB) or fungi	Notify physician that 24-hour collections are unacceptable and request three single-voided urine specimens or three consecutive early morning freshly expectorated sputa.

	Urine
	  Held over 2 hours at room temperature
	  Improper container
	  Leaking container
		Return request form to floor immediately, stamped with "Improper specimen; please resubmit." Explain problem.

	Excess barium or oil in stool specimen for ova and parasites	Stamp "Exam for parasites unsatisfactory—oil (or barium) in excess" on report form and send out.
	Less than one swab per request for bacterial, mycobacterial (TB), and fungal cultures	Call physician to request additional material. If additional material cannot be obtained, ask physician to state priorities for culture.
	Multiple urine, stool, sputum, and **routine** throat specimens on the same day from the same source (AFB requests excluded)	One specimen will be processed. Stamp duplicate request forms with "Multiple specimens received; one will be processed. Notify lab within 24 hours if you wish the remaining specimens to be processed and explain circumstances."

for storage and test performance likewise should be explicitly followed. In a similar manner, bacterial antigens for agglutination tests should be monitored with known positive and negative antisera, if available.

PROCEDURE BOOK

The procedure book delineates the current standard procedures employed in the microbiology laboratory and should be comprehensive enough to include all techniques, from the simple preparation of reagents to the methods for identifying isolates of the common genera. It should be easily interpreted by the least experienced bench worker and should not include lengthy descriptions readily found in a standard microbiology text. It should be reviewed at regular intervals, obsolete material deleted and new methods inserted, and initialed and dated by the director of the laboratory. It should be the **working reference standard** for the entire staff.

QUALITY CONTROL IN SPECIMEN COLLECTION

Until recently, one of the least regulated areas of quality control in the microbiology laboratory has been the quality of the clinical specimen itself. Is it representative of the suspected infection site? Has it been collected correctly in a proper container? Were sites of normal flora avoided in obtaining the specimen, when this is important or feasible? Was it submitted promptly in a proper transport medium or vehicle, when this is important, and cultured without undue delay? These are questions that must be answered affirmatively so that the laboratory results are meaningful and not misleading to the attending physician.

The microbiologist must not be too lenient in acceptance of improperly collected specimens. Too frequently treatment has been initiated on the basis of a report of potential pathogens, including antibiograms, when in reality the isolates represent only colonization of the patient's oropharynx or skin. Examples of the criteria used for rejection of requests for microbiologic tests at the UCLA Hospital and Clinics are shown in Table 40-2.

Murray and Washington[5] have proposed screening all sputum specimens by Gram stain and accepting only those with fewer than 10 squamous epithelial cells per $100\times$ magnification field. These specimens correlated well, on culture, with transtracheal aspirates; the smaller number of epithelial cells indicated less contamination with oropharyngeal flora. This procedure has been followed successfully at the Wilmington Medical Center; approximately 75% of submitted specimens in a recent 3-month period were acceptable by these criteria. Van Scoy[11] has recently suggested, after reviewing the data of Murray and Washington, that a criterion of more than 25 polymorphonu-

clear leukocytes per high-power field, regardless of the number of squamous epithelial cells, would be more appropriate. This is discussed in more detail in Chapter 8.

While we have stressed the importance of obtaining the best possible specimen, it must be appreciated that the physician who has responsibility for the patient must have the **final** say on what is done. The microbiologist should serve as the physician's consultant and adviser. In the event that a physician repeatedly orders inappropriate and meaningless cultures (which may even be misleading) and thus also imposes an unnecessary burden on the laboratory, the microbiologist should sit down with the physician to discuss the situation in depth. If necessary, intercession should be sought from the hospital's infectious disease clinicians, clinical pathologists or laboratory medicine workers, or the hospital staff officers. Obviously, it is always better for the microbiologist and physician to settle things directly.

While it is well recognized that a bacterial count of 10^5 or greater in a clean-voided, promptly cultured urine specimen is associated with urinary tract infection, it should be noted that counts in the range of 10^3 to 10^5 per milliliter also may represent infection; in patients receiving therapy, even lower counts may represent residual infection. Such isolates should be speciated and their antibiotic susceptibilities determined.

Reference has been made to the prime importance of proper specimen collection in the isolation of anaerobic microorganisms (see Chapter 13). The use of "gassed out" tubes and other appropriate transport devices and proper anaerobic techniques is directly related to the successful isolation of these bacteria.

MISCELLANEOUS MEASURES

Unusual biochemical or other reactions should alert the microbiologist to problems in reagents or media or to the possibility of misidentification or a mixed culture. Thus, for example, an organism identified as a *Proteus,* which was found to be sensitive to polymyxin, should be rechecked, as should a group A streptococcus or pneumococcus resistant to penicillin.

The role of "**blind unknowns,**" that is, simulated clinical specimens (urine, wound, fluids, and so forth) seeded with known bacteria, is well established as an internal quality control procedure. Two such specimens may be introduced each week into the regular laboratory routine. They should be identified in such a way that the worker is not aware of the unknown. More important, a fictitious report should not be routed to the ward or physician indicated on the laboratory request slip. When the report is completed, the microbiologist should discuss the results with the technologist performing the culture, with recommendations for improvement, if indicated. Quality control procedures in antimicrobial susceptibility testing have already been described (see Chapter 36).

Proficiency test specimens, such as those disseminated from state laboratories, the CDC, the College of American Pathologists, and so forth, are not as valuable as quality control devices as are the internal "blind unknowns," in that realistic laboratory conditions are not applicable. The worker is aware of the source, there is usually a prolonged time limit, and the specimen is given a laboratory workup exceeding that normally carried out. However, the proficiency test unknowns can serve an exceedingly useful purpose as an ongoing educational program for the staff, particularly when accompanied by the excellent critiques published regularly by the Proficiency Testing Branch, Licensure and Development Division, Center for Disease Control.[4]

Other miscellaneous methods of quality control are self-evident: an ongoing educational program within the laboratory; weekly departmental conferences with the microbiologist regarding problems, changes, or personnel relations; encouragement and provision for staff members to attend medical conferences, workshops, and seminars; and so forth. It is also necessary to review constantly and correct the performances of night and weekend staff, particularly in plating procedures, relatively simple identification techniques, and especially in the performance and proper interpretation of a "stat" Gram stain.

In summary, good quality control rests essentially in the conscience of the individual worker, and it is the duty of the microbiologist, by leadership and knowledge, to encourage and nourish this desire for superiority and service at all times.

REFERENCES

1. Bartlett, R. C.: Medical microbiology; quality, cost and clinical relevance, New York, 1974, Wiley-Interscience, Division of John Wiley & Sons, Inc.
2. Bartlett, R. C., Irving, W. R., Jr., and Rutz, C.: Quality control in clinical microbiology, Chicago, 1970, Am. Soc. Clin. Pathol. Commiss. Contin. Educ.
3. Blazevic, D. J., Hall, C. T., and Wilson, M. E.: Practical quality control procedures for the clinical microbiology laboratory, Cumitech 3, Washington, D.C., 1976, American Society for Microbiology.
4. Hall, C. T., and Webb, C. D., Jr.: Proficiency testing—Bacteriology III and IV (July and Oct. 1972), Atlanta, 1973, Center for Disease Control.
5. Murray, P. R., and Washington, J. A. II: Microscopic and bacteriologic analysis of expectorated sputum, Mayo Clin. Proc. **50:**339-344, 1975.
6. Nagel, J. G., and Kunz, L. J.: Simplified storage and retrieval of stock cultures, Appl. Microbiol. **23:**837-839, 1972.
7. Prier, J. E., Bartola, J., and Friedman, H.: Quality control in microbiology, Baltimore, 1975, University Park Press.
8. Rhoden, D. L., Tomfohrde, K. M., Swenson, J. M., Smith, P. B., Balows, A., and Thornsberry, C.: Evaluation of the Bactrol disks, a set of quality control cultures, J. Clin. Microbiol. **1:**11-14, 1975.
9. Russell, R. L.: Quality control in the microbiology laboratory. In Lennette, E. H., Spaulding, E. H., and Truant, J. P., editors: Manual of clin-

ical microbiology, ed. 2, Washington, D.C., 1974, American Society for Microbiology.

10. Russell, R. L., Yoshimori, M. A., Rhodes, R. F., Reynolds, J. W., and Jennings, E. R.: A quality control program for clinical microbiology, Am. J. Clin. Pathol. **39:**489-494, 1969.

11. Van Scoy, R. E.: Bacterial sputum cultures: a clinician's viewpoint, Mayo Clin. Proc. **52:**39-41, 1977.

12. Vera, H. D.: Quality control in diagnostic microbiology, Health Lab. Sci. **8:**176-189, 1971.

13. Washington, J. A. II, editor: Laboratory procedures in clinical microbiology, Boston, 1974, Little, Brown & Co.

14. Woods, D., and Byers, J. F.: Quality control recording methods in microbiology, Am. J. Med. Technol. **30:**79-85, 1973.

PART VIII

Culture media, stains, reagents, and tests

CHAPTER 41

Formulas and preparation of culture media

The principles governing the selection, preparation, sterilization, and storage of laboratory culture media have been discussed in Chapter 1; thus the ensuing chapter will include the media to which reference is made throughout the text, their formulas, and any specific directions that may be needed. Complete directions for the preparation of media from the dehydrated products obtained commercially are provided by the manufacturers.*

1. Acetate agar

This agar medium is prepared in the same manner as Simmons' citrate agar (see No. 21) except that 0.25% sodium acetate is used in place of citrate. It is used to determine whether an organism is able to utilize acetate as its sole carbon source. Growth on the slant and development of a **blue** color indicate a positive test. A negative test is no growth or color change.

This test is useful in differentiating *Escherichia coli* from shigellae in that the latter do not utilize acetate.

2. Beef extract agar†

Add 2.5% agar to beef extract broth; dissolve by boiling; adjust to pH 7.6; tube and autoclave at 121 C for 15 minutes; and slant before solidifying.

The medium is used for pure culture preparation of *Candida* species prior to carrying out fermentation tests.

3. Beef extract broth*

Beef extract	3 g
Peptone	10 g
Sodium chloride	5 g
Distilled water	1,000 ml

Dissolve ingredients by boiling, adjust to pH 7.2, tube in 10-ml amounts in 18- × 150-mm tubes, and autoclave at 121 C for 15 minutes. For use as a fermentation base broth, add 100 ml of a 0.04% aqueous solution of bromthymol blue before tubing and sterilizing.

The medium is used for differentiation of *Candida* species by carbohydrate fermentation tests.

4. Bile esculin agar†

Beef extract	3.0 g
Peptone	5.0 g
Oxgall	40.0 g
Esculin	1.0 g
Ferric citrate	0.5 g
Agar	15.0 g
Distilled water	1,000 ml

a. Suspend 64 g of the dehydrated medium in water, heat to boiling to dissolve, and tube in screw-capped tubes.

b. Sterilize by autoclaving at 121 C for 15 minutes.

c. Cool to 55 C and add aseptically 50 ml of filter-sterilized horse serum (optional). Mix well.

d. Dispense in sterile tubes and cool in slanted position.

This medium is useful in the selective

---

*Von Riesen (J. Clin. Microbiol. **2:**554-555, 1975) has proposed a simple, convenient method of preparing media that are not in daily use.
†Baltimore Biological Laboratory, Cockeysville, Md.; Difco Laboratories, Detroit.

*Baltimore Biological Laboratory, Cockeysville, Md.; Difco Laboratories, Detroit.
†Pfizer Diagnostics, Flushing, N.Y.; Difco Laboratories, Detroit; Baltimore Biological Laboratory, Cockeysville, Md.

detection of group D streptococci. After inoculation, these organisms will form brownish black colonies, surrounded by a black zone. *Listeria monocytogenes* also reacts positively, as do some other organisms. Other forms of agar and broth media also are available.*

5. Bismuth sulfite agar†

Beef extract	5.0	g
Peptone	10.0	g
Dextrose	5.0	g
Disodium phosphate	4.0	g
Ferrous sulfate	0.3	g
Bismuth sulfite indicator	8.0	g
Agar	20.0	g
Brilliant green	0.025	g
Distilled water	1,000	ml

Bismuth sulfite agar medium is highly recommended for the isolation of *Salmonella typhi*, as well as other salmonellae. It may be used for both streak and pour plates. **Read's modification** of the medium (through the addition of 1% sodium chloride and 1% mannose, the elimination of the brilliant green, and adjustment of the final pH to 9.2) is recommended for the isolation of *Vibrio cholerae*.

6. Blood agar plates for streaking

Blood agar base medium	500 ml
Sterile blood	20 ml

Sterilize the base medium (trypticase soy agar [BBL] or tryptic soy agar [Difco] are recommended‡), cool to 48 or 50 C, and add 5% sterile defibrinated sheep, horse, or rabbit blood, aseptically. Rotate to mix thoroughly and pour into sterile Petri dishes in approximately 20-ml amounts. When agar is set, invert and incubate 1 or 2 plates per 100- to 200-ml batch over-

night to test for sterility. Discard test plates. Blood agar plates may be stored for 1 week at refrigerator temperature without deterioration, but they should be packaged to minimize water loss, which could amount to 7% per week if unprotected.

The pouring should be done carefully to avoid air bubbles. If bubbles appear in the poured plate, pass a Bunsen flame over the agar before it sets. This will cause them to break.

An alternative to pouring media from an Erlenmeyer flask is the use of a Kelly bottle with a release clamp on a piece of rubber tubing connected to a bell device at the other end. This assembly is mounted on a ring stand. Release of pressure on the clamp permits the heated media to flow evenly into the plate. An advantage of this system is that bubbles (when present) rise to the top of the liquid medium in the bottle prior to its being poured into the plate.

7. Blood agar pour plates
(Brown: Monograph No. 9, The Rockefeller Institute for Medical Research, 1919)

Heat 20-ml tubes of infusion agar in boiling water until the agar is thoroughly melted. Cool the agar to 48 C by standing the tubes in a container of warm water or a 48 C water bath. With a sterile pipet add about 1 ml of sterile blood to each tube. Inoculate the fluid blood agar with a proper dilution of culture or original material. Twirl the tube to mix the blood, inoculum, and agar, taking care not to form bubbles, and pour into a sterile Petri dish.

8. Blood cystine dextrose agar*
(Francis: J.A.M.A. **91**:1155, 1928; Rhamy: Am. J. Clin. Pathol. **3**:121, 1933)

Beef heart infusion	500 g
Proteose peptone	10 g

---

*Pfizer Diagnostics, Flushing, N.Y.; Difco Laboratories, Detroit; Baltimore Biological Laboratory, Cockeysville, Md.

†Baltimore Biological Laboratory, Cockeysville, Md.; Difco Laboratories, Detroit.

‡Brucella agar is an excellent base for blood agar for anaerobic bacteria (see No. 13 below). Columbia agar and Schaedler agar have also been recommended for this purpose.

---

*Available in dehydrated form from Difco Laboratories, Detroit, and Baltimore Biological Laboratory, Cockeysville, Md.

Glucose	10 g
Sodium chloride	5 g
Cystine	1 g
Agar	15 g
Distilled water	1,000 ml

Blood cystine dextrose agar medium may be used satisfactorily for the cultivation of *Francisella tularensis.*

To prepare this medium for the cultivation of *F. tularensis,* dissolve 16.8 g of dehydrated product in 300 ml of distilled water. Adjust the reaction to pH 7.3 and autoclave for 20 minutes at 15 pounds pressure. Cool to a temperature of 60 to 70 C, and add 18 ml of whole rabbit blood or dehydrated hemoglobin, mix well, and distribute into test tubes aseptically. Cool in slanting position.

9. Bordet-Gengou medium*

Potatoes, infusion from	125.0 g
Sodium chloride	5.5 g
Agar	20.0 g
Distilled water	1,000 ml

Sterilize at 15 pounds for 15 minutes. Cool to 50 C and add 15% to 20% sterile sheep, rabbit, or human blood aseptically.

The medium is recommended for cultivation of *Bordetella pertussis.*

10. Brain-heart infusion blood agar (BHIBA)†

Brain-heart infusion agar‡	26.0 g
Agar, powdered	2.5 g
Distilled water	500 ml
(Final pH approximately 7.4)	

Suspend the ingredients in the water and dissolve by boiling. Autoclave for 15 minutes at 121 C. Cool to about 45 C and add aseptically 30 ml of defibrinated animal blood. Mix well and distribute

aseptically in approximately 20-ml amounts in cotton-plugged and sterilized 25- × 150-mm Pyrex test tubes (without lips). Slant, allow to harden, and store in a refrigerator.

This medium is used for the isolation of fastidious fungi.

11. Brain-heart infusion broth*

Calf brains, infusion	200.0 g
Beef heart, infusion	250.0 g
Proteose peptone	10.0 g
Dextrose	2.0 g
Sodium chloride	5.0 g
Disodium phosphate	2.5 g
Distilled water	1,000 ml
(Reaction of medium pH 7.4)	

Recommended for cultivating the pneumococcus for the bile solubility test.

12. Brilliant green agar*

Yeast extract	3.0	g
Proteose peptone No. 3	10.0	g
Sodium chloride	5.0	g
Lactose	10.0	g
Sucrose	10.0	g
Phenol red	0.08	g
Brilliant green	0.0125	g
Agar	20.0	g
Distilled water	1,000	ml
(Final pH 6.9)		

Brilliant green agar is a highly selective medium recommended for the isolation of salmonellae other than *Salmonella typhi.* It is not recommended for the isolation of shigellae.

13. Brucella-vitamin K_1 blood agar (BRBA)

a. Vitamin K_1 solution (see Chapter 43)
b. Preparation of base:

Brucella agar†	43.0 g
Agar	2.5 g
Distilled water	1,000 ml

Sterilize by autoclaving at 121 C for 15 minutes; cool to 50 C.

---

*Available as dehydrated base from Difco Laboratories, Detroit; Baltimore Biological Laboratory, Cockeysville, Md.

†Brain-heart infusion agar, freshly poured and without added blood, is recommended for growth of *Actinomyces* and *Arachnia.*

‡Baltimore Biological Laboratory, Cockeysville, Md.; Difco Laboratories, Detroit.

---

*Available in dehydrated form from Difco Laboratories, Detroit, and Baltimore Biological Laboratory, Cockeysville, Md.

†BBL or Difco. The Pfizer product is not recommended for this purpose, as it contains citrate.

c. Add aseptically:

Defibrinated or laked sheep blood	50 ml
Vitamin K₁ solution	1 ml

Mix well and pour plates, using approximately 20 ml per plate. Store at room temperature.

The medium is used for the isolation and subculture of anaerobes.

14. Carbohydrate media for fermentation tests

When some carbohydrates are sterilized by heating in alkaline broth, they are more or less broken down into simpler carbohydrates. It has been found definitely advantageous to sterilize sugar solutions by filtration through Seitz or membrane filters and to add these aseptically to the broth base in the required amounts. Although several carbohydrates can withstand autoclave temperature and pressure, one is advised to sterilize the following by **filtration:** xylose, lactose, sucrose, arabinose, trehalose, rhamnose, and salicin. These may be prepared as 5% or 10% solutions, depending on solubility, then sterilized by filtration and added aseptically to the base containing an indicator, to give 0.5% to 1% final concentration.

Other less heat-susceptible carbohydrates may be added to the broth base containing indicator (bromcresol purple, for example) before autoclaving at 116 to 118 C (10 to 12 pounds pressure) for 15 minutes. Ten Broeck, in 1920, found that unheated serum contains an enzyme that hydrolyzes maltose to glucose. Serum to be added to maltose broth should therefore be heated for 1 hour at 60 C to inactivate the enzyme. Incubate the carbohydrate broth to test sterility.

15. Casein medium for separation of Nocardia and Streptomyces
(Center for Disease Control)

Prepare separately:

a. Skimmed milk (dehydrated or instant nonfat milk)	10 g

Distilled water	100 ml

Autoclave at 121 C for 20 minutes.

b. Distilled water	100 ml
Agar	2 g

Autoclave as above.

Cool both solutions to approximately 45 C, mix, and pour into sterile Petri dishes.

Test for hydrolysis. Streak or make point inoculations of each culture on casein plates, using half a plate for each organism. Incubate at 25 C (or 35 C if it does not grow at room temperature). Observe for clearing of casein in 7 and 14 days.

Nocardia asteroides does not hydrolyze casein.

N. brasiliensis and *Streptomyces* species hydrolyze casein.

16. Cetrimide agar*
(Lowbury and Collins: J. Clin. Pathol. **8:**47, 1955)

Peptone	20.0 g
Magnesium chloride	1.4 g
Potassium sulfate	10.0 g
Agar (dried)	13.6 g
Cetrimide†	0.3 g
Distilled water	1,000 ml
(Final pH 7.2±)	

Suspend powder in water, add 10 ml of glycerol, heat with frequent agitation, and boil for 1 minute.

Dispense in 5-ml amounts in 15- × 125-mm screw-capped tubes, autoclave at 118 to 121 C for 15 minutes, and slant to give a generous slant.

This medium is used for the selective isolation or identification of *Pseudomonas aeruginosa,* whose growth is not inhibited by the cetrimide. Other members of the genus (except *P. fluorescens*) and related nonfermentative organisms are inhibited.

---

*Pseudosel agar, Baltimore Biological Laboratory, Cockeysville, Md.
†Cetyl trimethyl ammonium bromide.

17. Chlamydospore agar

| Chlamydospore agar* | 18.5 | g |
| Distilled water | 500 | ml |

Suspend the dehydrated product in a 1-liter Erlenmeyer flask; dissolve by heating; and tube in approximately 15-ml amounts in screw-capped, 20- × 150-mm test tubes. Autoclave at 121 C for 15 minutes with caps loosened. When cool, tighten caps and store at room temperature. When needed, melt a tube in a water bath, pour into a sterile Petri plate, and allow to solidify.

18. Chocolate agar

Chocolate agar is agar to which blood or hemoglobin has been added and then heated until the medium becomes brown or chocolate in color. The recommended method of preparing this is as follows.

Suspend sufficient proteose No. 3 agar,† Eugonagar,‡ or GC agar base§ in distilled water to make a double-strength base. Mix thoroughly and heat to boiling for 1 minute, with frequent agitation; autoclave at 121 C for 15 minutes. At the same time, autoclave an equal volume of 2% hemoglobin,§ which is made by the gradual addition of distilled water to the dehydrated hemoglobin, to obtain a **smooth** suspension. Cool both solutions to approximately 50 C, add the supplement (Iso-VitaleX enrichment‡ or Supplement B or C† are recommended) and combine with bacteriologic precautions; then pour into sterile, disposable Petri plates. Best results will be obtained if plates are **freshly** prepared. One or two plates per 100- to 200-ml batch should be incubated at 35 C overnight to determine sterility, and later discarded.

---

*Available in dehydrated form from Difco Laboratories, Detroit; Baltimore Biological Laboratory, Cockeysville, Md.
†Difco Laboratories, Detroit.
‡Baltimore Biological Laboratories, Cockeysville, Md.
§Baltimore Biological Laboratory, Cockeysville, Md.; Difco Laboratories, Detroit.

Thayer-Martin agar (see No. 85) for the selective isolation of *Neisseria gonorrhoeae* and *N. meningitidis* may be prepared from the aforementioned chocolate agar by adding an antimicrobial agent inhibitor (V-C-N, BBL, or Difco). The use of Thayer-Martin medium results in a higher recovery of neisseriae from clinical specimens likely to be contaminated with other bacteria.

Chocolate agar slants may be prepared individually by the method just described, slanting the medium after it has been tubed in 5- to 10-ml amounts in sterile screw-capped tubes. To prepare a large number of slants, use a flask containing 50 to 100 ml of agar, proceeding as before.

19. Chopped meat glucose (CMG)

Ground beef, lean, fat-free	500	g
Distilled water	1,000	ml
Sodium hydroxide, 1 N	25	ml

a. Mix ingredients, bring to boil, and simmer with frequent stirring for 20 minutes.

b. Cool to room temperature, skim off fat, and filter through three layers of gauze; squeeze out gauze, retaining both meat particles and filtrate (filtered through coarse and fine paper).

c. Restore filtrate to 1 liter with distilled water, and add:

Trypticase	30	g
Yeast extract	5	g
Dipotassium phosphate	5	g
Resazurin solution (25 mg/100 ml water)	4	ml

d. Boil, cool, adjust to pH 7.8, and add 0.5% glucose and 0.5 g cystine.

e. Dispense 6- to 7-ml amounts into tubes containing meat particles (step b), 1 part meat to 4 or 5 parts broth, and autoclave at 121 C for 20 minutes.

This medium is recommended as a "backup" medium for the primary isolation of anaerobes and also for growing pure cultures for gas-liquid chromatographic analysis.* It may also be obtained as a dehydrated medium, cooked meat phytone.†

20. Chopped-meat medium*‡

To 1 pound of finely ground beef heart and other muscle (fat free) add 500 ml of boiling sodium hydroxide (N/15 to N/20); boil for 20 minutes. Cool, strain off fat, and filter through muslin. Adjust fluid to pH 7.5 and add 1% peptone. Add approximately 2 inches of meat and a small ball of steel wool to each tube and add enough broth to overlay the meat by about 1 inch. Heat tubes for 30 minutes in boiling water; sterilize by autoclaving at 121 C for 15 minutes. The medium should be boiled for a few minutes to drive off dissolved oxygen if not used the same day, unless the prereduced tubed media are used. Incubation in an anaerobic jar or under a petrolatum seal may be required for cultivation or maintenance of certain *Clostridium* species, except in the case of prereduced tubed media.

21. Citrate agar§
(Simmons: J. Infect. Dis. **39**:209, 1926)

Agar	20.0	g
Sodium chloride	5.0	g
Magnesium sulfate	0.2	g
Ammonium dihydrogen phosphate	1.0	g
Dipotassium phosphate	1.0	g
Sodium citrate	2.0	g
Bromthymol blue	0.08	g
Distilled water	1,000	ml

(Final pH 6.9±)

Citrate agar is used to determine the utilization of citrate as the sole carbon source.

22. Columbia CNA agar*

Polypeptone peptone	10.0	g
Biosate peptone	10.0	g
Myosate peptone	3.0	g
Corn starch	1.0	g
Sodium chloride	5.0	g
Agar (dried)	13.5	g
Colistin	10.0	mg
Nalidixic acid	15.0	mg
Distilled water	1,000	ml

(Final pH 7.3±)

a. Suspend 42.5 g of dehydrated medium in water, heat with frequent agitation, and boil for 1 minute.
b. Sterilize by autoclaving at 121 C for 15 minutes.
c. Cool to 50 C, add 5% defibrinated sheep blood and pour plates.

This medium is excellent for the selective growth of gram-positive cocci, particularly streptococci, when gram-negative bacilli, especially *Proteus* species, tend to overgrow on conventional blood agar plates. This medium may be inhibitory for certain strains of staphylococci.

23. Cornmeal agar*

Yellow cornmeal	125	g
Distilled water	3,000	ml

Heat cornmeal in water at 60 C for 1 hour, filter through paper, make up to volume, and add 50 g of agar. Expose to flowing steam for 1 hour, filter through absorbent cotton, dispense in tubes, and autoclave at 121 C for 30 minutes. Cornmeal agar suppresses vegetative growth of many fungi while stimulating sporulation.

With the addition of 1% Tween 80, it is useful in stimulating production of chlamydospores of *Candida albicans*.†

24. Cystine tellurite blood agar
(Frobisher: J. Infect. Dis. **60**:99, 1937)

a. Melt 100 ml of sterile 2% infusion

---

*Available in prereduced tubes from Scott Laboratories, Fiskeville, R.I.
†Baltimore Biological Laboratory, Cockeysville, Md.
‡Available in dehydrated form from the Baltimore Biological Laboratory, Cockeysville, Md.
§Baltimore Biological Laboratory, Cockeysville, Md.; Difco Laboratories, Detroit.

---

*Available in dehydrated form from the Baltimore Biological Laboratory, Cockeysville, Md., and others.
†Some workers find that "homemade" media give more consistent results.

agar in a flask and cool to 45 to 50 C. Care should be taken to maintain this temperature throughout the following steps in the preparation of the medium.

b. Add aseptically 15 ml of sterile 0.3% solution of potassium tellurite in distilled water. The tellurite solution may be sterilized by autoclaving.

c. Add aseptically 5 ml of sterile blood and mix well.

d. Add 3 to 5 mg of cystine. The dry powder is used and need not be sterilized. Since different lots of cystine vary, the optimal amount necessary to produce the best growth of *Corynebacterium diphtheriae* may vary from 3 to 5 mg/ 100 ml.

e. Thoroughly mix the medium and pour into sterile Petri dishes. Since the cystine does not go entirely into solution, shake the flask frequently while pouring the plates.

This medium is used for the isolation of *Corynebacterium diphtheriae*.

25. Cystine trypticase agar (CTA)*
(Vera: J. Bacteriol. **55:**531, 1948)

Cystine	0.5	g
Trypticase	20.0	g
Agar	3.5	g
Sodium chloride	5.0	g
Sodium sulfite	0.5	g
Phenol red	0.017	g
Distilled water	1,000	ml
(Final pH, 7.3±)		

CTA is an excellent all-purpose medium for the growth of pathogenic organisms. It can be used for the maintenance of cultures, including fastidious organisms (held at 25 C), the determination of motility, and with the addition of carbohydrates, for determining fermentation reactions of fastidious organisms, including *Neisseria*. Sterilize at 115 to 118 C (**not over** 12 pounds pressure) for 15 minutes.

---

*Available in dehydrated form from the Baltimore Biological Laboratory, Cockeysville, Md.

26. Decarboxylase test media
(Moeller: Acta Pathol. Microbiol. Scand. **36:**161, 1955)

Basal medium*:		
Peptone (Orthana special†)	5.0	g
Beef extract	5.0	mg
Bromcresol purple (1.6%)	0.625	ml
Cresol red (0.2%)	2.5	ml
Glucose	0.5	g
Pyridoxal	5.0	mg
Distilled water	1,000	ml
(Adjust to pH 6)		

Divide basal medium into four 250-ml amounts; one portion is tubed without addition of any of the amino acids (control). To another portion add 1% L-lysine dihydrochloride,‡ to a third add 1% L-arginine monohydrochloride, and to the fourth portion add 1% L-ornithine dihydrochloride. Adjust this last (ornithine) portion to pH 6 with N/1 NaOH before sterilization. The media are tubed in 3- to 4-ml amounts in small (13- × 100-mm) screw-capped tubes. Autoclave at 121 C for 10 minutes.

Inoculate all four tubes from an agar slant culture and overlay each tube with 4 to 5 mm of sterile mineral oil.§ If oil is not added, the reactions are not valid after 24 hours. Some workers recommend the addition of 0.3% agar to the basal medium in place of an oil overlay.

Incubate all four tubes at 35 C and read daily for not more than 4 days (most positive reactions with the enteric bacilli will occur in 1 to 2 days). **A positive** reaction is indicated by alkalinization of the medium with a change in color from yellow (due to the initial fermentation of glucose) to **violet** (due to the decarboxylation of the amino acid). Therefore, a **yellow** color after several days' incubation indicates a **negative** test, or the absence of the

---

*Baltimore Biological Laboratory, Cockeysville, Md.; Difco Laboratories, Detroit.

†Proteose peptone No. 3, 0.3%, has been found to be a suitable alternative.

‡Amino acids are available from Nutritional Biochemicals Corp., Cleveland, Ohio.

§Sterilize in test tubes at 121 C for 45 minutes.

enzymes decarboxylase or dihydrolase. All positive tests should be compared with the control tube, which remains yellow.

These media are used primarily for differentiating members of the Enterobacteriaceae.

27. Deoxyribonuclease (DNase) test medium*

Deoxyribonucleic acid	2 g
Phytone	5 g
Sodium chloride	5 g
Trypticase	15 g
Agar	15 g
Distilled water	1,000 ml

(Approximate final pH 7.3)

This medium may be sterilized in the autoclave at 15 pounds pressure for 15 minutes at 121 C, cooled, and poured into Petri dishes. The dry surface may be streaked in several places (½-inch streaks) with strains of staphylococci to be tested for DNase activity. Flooding of the plate after 24 hours of growth with N/1 hydrochloric acid will reveal **clear zones** around the growth of the DNase-positive strains. Some workers prefer flooding the plate with 0.1% toluidine blue; a bright **rose pink** color is produced around the growth of DNase-positive organisms.

28. Desoxycholate agar*
(Leifson: J. Pathol. Bacteriol. **40**:581, 1935)

Peptone	10.0	g
Lactose	10.0	g
Sodium citrate	1.0	g
Ferric citrate	1.0	g
Sodium chloride	5.0	g
Dipotassium phosphate	2.0	g
Sodium desoxycholate	1.0	g
Agar	16.0	g
Neutral red	0.033	g
Distilled water	1,000	ml

(Final pH 7.2±)

Desoxycholate agar is used for the isolation of gram-negative enteric bacilli and the differentiation of lactose-fermenting and nonlactose-fermenting species.

---

*Baltimore Biological Laboratory, Cockeysville, Md.; Difco Laboratories, Detroit.

29. Desoxycholate citrate agar*
(Leifson: J. Pathol. Bacteriol. **40**:581, 1935)

Meat, infusion from	350.0	g
Peptone	10.0	g
Lactose	10.0	g
Sodium citrate	20.0	g
Ferric citrate	1.0	g
Sodium desoxycholate	5.0	g
Agar	17.0	g
Neutral red	0.02	g
Distilled water	1,000	ml

(Final pH 7.3±)

Desoxycholate citrate agar is used for the isolation of salmonellae and shigellae from stool and other specimens.

30. Dextrose ascitic fluid semisolid agar for spinal fluid cultures

Phenol red broth base	4.0	g
Agar, powdered (weigh accurately)	0.5	g
Distilled water	250	ml

Suspend ingredients in a 1-liter Erlenmeyer flask, plug with cotton, and autoclave at 121 C for 15 minutes. At the same time sterilize a rack of 18- × 125-mm screw-capped test tubes and a wrapped Cornwall automatic pipet.† Cool medium and add the following aseptically:

Ascitic fluid‡	50	ml
Dextrose, sterile 20% solution	15	ml

Using the sterile pipet, tube in 5-ml amounts in screw-capped tubes, incubate overnight for sterility, and store at room temperature. This medium is poured over remaining spinal fluid sediment after smears and plate cultures have been made.

31. Drug-containing media for susceptibility testing of mycobacteria

See Chapter 31.

---

*Baltimore Biological Laboratory, Cockeysville, Md.; Difco Laboratories, Detroit.

†Beckton-Dickinson's Cornwall Pipetting Unit with 5-ml syringe.

‡Bacto-Ascitic Fluid, 10-ml ampules, Difco Laboratories, Detroit.

32. Egg yolk agar

Add 10 ml of sterile yolk emulsion* to 90 ml of melted and cooled blood agar base and pour the plates.

Proteose peptone No. 2	40.0 g
Sodium orthophosphate (Na$_2$HPO$_4$)	5.0 g
Potassium dihydrophosphate (KH$_2$PO$_4$)	1.0 g
Sodium chloride	2.0 g
Magnesium sulfate	0.1 g
Glucose	2.0 g
Hemin solution (see Chapter 43) (5 mg/ml)	1.0 ml
Agar	25.0 g
Distilled water	1,000 ml
(Adjust to pH 7.6)	

Autoclave at 121 C for 15 minutes. This medium is used for the isolation and identification of members of the genus *Clostridium* and certain other anaerobes.

33. Eosin-methylene blue agar (EMB)†

Peptone	10.0	g
Lactose	5.0	g
Sucrose	5.0	g
Dipotassium phosphate	2.0	g
Agar	13.5	g
Eosin Y	0.4	g
Methylene blue	0.065	g
Distilled water	1,000	ml

The agar concentration may be increased to 5% (use an additional 3.65 g of agar) to inhibit the spreading of *Proteus*. The Levine EMB agar, preferred by some investigators, does not contain sucrose.

Lactose-fermenting, gram-negative bacteria may be distinguished from nonlactose-fermenting types by their appearance. Colonies of *Escherichia coli* usually have a characteristic metallic sheen. If the sucrose-containing medium is used, *Proteus* colonies will also show this characteristic, provided they are inhibited from spreading by the higher agar concentration.

34. Fermentation broth for Listeria

Proteose peptone No. 3*	10.0 g
Beef extract	1.0 g

*From Colab Laboratories, Inc., Glenwood, Ill.
†Available in dehydrated form from Baltimore Biological Laboratory, Cockeysville, Md.; Difco Laboratories, Detroit.

Sodium chloride	5.0 g
Bromcresol purple	0.1 g
Distilled water	1,000 ml

Dissolve with the aid of heat and sterilize at 121 C for 20 minutes. Add aseptically 0.5% of the following carbohydrates: glucose, salicin, rhamnose, dulcitol, and raffinose.

These carbohydrate broths are helpful in the biochemical studies of *Listeria*.

35. Fermentation media for neisseriae

a. Suspend 4.3 g of dehydrated CTA medium (see no. 25) in 150 ml of distilled water and heat to boiling with frequent agitation.

b. Adjust to pH 7.4 to 7.6 with 1 N NaOH, dispense in 50-ml amounts in 250-ml Erlenmeyer flasks, and sterilize at not over 118 C (12 pounds) for 15 minutes.

c. Prepare 20% solutions of glucose, lactose, maltose, and sucrose in distilled water; tube and sterilize by membrane filtration.

d. Add 2.5 ml of each carbohydrate to separate flasks of the 50-ml cooled CTA medium.

e. Mix and dispense in 2.0-ml amounts in sterile screw-capped tubes (13 × 100 mm) and store in refrigerator at 4 C.

36. Fermentation medium for differentiating Staphylococcus and Micrococcus

(Facklam and Smith: Human Pathol. 7:187 1976)

Difco tryptone	1.0	%
Difco yeast extract	0.1	%
Bromcresol purple	0.004%	
Glucose or mannitol	1.0	%
Agar	0.22	%
(pH 7.0)		

Tube in 16- × 120-mm tubes to a depth of 7 to 8 cm. Before use, steam for 10 minutes, cool rapidly, and inoculate heavily by stabbing to the bottom. Incubate under anaerobic conditions.

*Difco Laboratories, Detroit.

37. Fildes enrichment* agar†

Sodium chloride (0.85% solution)	150 ml
Hydrochloric acid	6 ml
Defibrinated sheep blood	50 ml
Pepsin, granular	1 g

a. Mix well and heat in 56 C water bath for 4 hours, with occasional agitation.
b. Adjust to pH 7 with 20% NaOH (approximately 12 ml).
c. Readjust to pH 7.0 to 7.2 with HCl.
d. Add 0.25 ml chloroform, mix thoroughly, stopper tightly, and store in refrigerator at 4 C.
e. After heating 4 to 10 ml of this enrichment mixture in a sterile container in a 56 C water bath for 30 minutes to eliminate the chloroform, add to 200 ml of melted nutrient agar cooled to 56 C and pour into Petri dishes.

This is an excellent medium for the isolation of *Haemophilus influenzae.*

38. Fletcher's semisolid medium for Leptospira‡

a. Sterilize 1.76 liters of distilled water by autoclaving at 121 C for 30 minutes.
b. Cool to room temperature and add 240 ml of sterile normal rabbit serum.
c. Inactivate by incubation at 56 C for 40 minutes.
d. Add 120 ml of melted and cooled (not greater than 56 C) 2.5% meat extract agar at pH 7.4.
e. Dispense 5-ml amounts in sterile, 16- × 130-mm, screw-capped test tubes and 15-ml amounts in sterile, 25-ml, diaphragm-type, rubber-stoppered vaccine bottles.
f. Inactivate at 56 C for 60 minutes on 2 successive days.

39. Gelatin medium‡

To extract or infusion broth add 12%

---

*Fildes enrichment, itself, is also available commercially.
†Difco Laboratories, Detroit.
‡Difco Laboratories, Detroit; Baltimore Biological Laboratory, Cockeysville, Md.

gelatin. Heat in an Arnold sterilizer or in a double boiler until the gelatin is thoroughly dissolved. Adjust reaction to pH 7. Tube and autoclave at 121 C (15 pounds pressure) for 15 minutes. Sodium thioglycollate, 0.05%, may be added for cultivation of certain species of clostridia in an external aerobic environment.

40. Gelatin medium (dilute) for differentiation of Nocardia and Streptomyces
(Center for Disease Control)

Gelatin	4 g
Distilled water	1 liter
(Adjust to pH 7)	

Dispense in tubes, approximately 5 ml per tube. Autoclave at 121 C for 5 minutes. Inoculate with a small fragment of growth from a Sabouraud dextrose agar slant. Incubate at room temperature for 21 to 25 days. Examine for quantity and type of growth.

Nocardia asteroides exhibits no growth or very sparse, thin, flaky growth. *N. brasiliensis* shows good growth and round compact colonies. *Streptomyces* species show poor to good growth (stringy or flaky).

41. GN broth*
(Hajna: Pub. Health Lab. **13**:83, 1955)

Peptone	20.0 g
Glucose	1.0 g
D-Mannitol	2.0 g
Sodium citrate	5.0 g
Sodium desoxycholate	0.5 g
Dipotassium phosphate	4.0 g
Monopotassium phosphate	1.5 g
Sodium chloride	5.0 g
Distilled water	1,000 ml
(Final pH 7±)	

Dissolve the dehydrated medium in distilled water and autoclave at 116 C (10 pounds steam pressure) for 15 minutes or steam for 30 minutes at 100 C.

This broth medium is used as an en-

---

*Difco Laboratories, Detroit; Baltimore Biological Laboratory, Cockeysville, Md.

richment for isolating salmonellae and shigellae in fecal specimens.

42. Hektoen enteric agar*

Proteose peptone	12.0	g
Bile salts	9.0	g
Yeast extract	3.0	g
Lactose	12.0	g
Salicin	2.0	g
Sucrose	12.0	g
Sodium chloride	5.0	g
Sodium thiosulfate	5.0	g
Ferric ammonium citrate	1.5	g
Agar	14.0	g
Acid fuchsin	0.1	g
Bromthymol blue	0.065	g
Distilled water	1,000	ml

Suspend 76 g of dehydrated medium, if using the commercial product, in 1,000 ml distilled water. Heat to boiling and continue until the medium dissolves. **Do not autoclave.** Cool to 50 C and pour into Petri dishes.

This medium is useful for the isolation and differentiation of gram-negative enteric pathogens. The coliforms are usually salmon to orange in color, while the salmonellae and shigellae are bluish green.

43. Indole-nitrite medium

See No. 93, Trypticase nitrate broth.

44. Kanamycin-vancomycin blood agar (KVBA)

a. Preparation of base:

Trypticase soy agar	40.0	g
Agar	2.5	g
Distilled water	1,000	ml
Kanamycin base†	100	mg

Sterilize by autoclaving at 121 C for 15 minutes and cool to 50 C.

b. Add aseptically‡:

Defibrinated sheep blood	50.0	ml
Vancomycin§	7.5	mg

Vitamin K_1 solution (see Chapter 43)	1.0 ml

Mix well, pour plates (approximately 20 ml per plate), and store at room temperature.

This medium is extremely useful for primary inoculation of clinical specimens for selective isolation of anaerobes, particularly *Bacteroides,* but a nonselective medium (Brucella blood agar plate) should always be used as well.

44a. Kanamycin-vancomycin laked blood agar (KVLBA)

Same as KVBA, except that final concentration of kanamycin is 75 μg/ml and the blood is laked (hemolyzed) by freezing whole blood overnight and then thawing. This medium is particularly useful for isolating *Bacteroides melaninogenicus.*

45. Kligler's iron agar

See No. 90, Triple sugar iron agar—TSI agar.

46. Loeffler coagulated serum slants*
(Zentralbl. Bakt. **2:**105, 1887)

Add 3 volumes of beef, hog, or horse serum (collected as cleanly as possible) to 1 volume of glucose infusion broth (1% glucose). Dispense in tubes, slant, and inspissate for about 2 hours at 75 to 85 C on each of 3 successive days. The medium also may be sterilized in a slanted position in a horizontal autoclave. Close the autoclave tightly and, without allowing the air to escape, autoclave at 15 pounds pressure for 15 minutes to coagulate the medium. Then allow the air to escape slowly, and admit steam so that the pressure is maintained at 15 pounds. When an atmosphere of pure steam has been obtained, close the outlet valve tightly and sterilize the medium with

---

*Difco Laboratories, Detroit; Baltimore Biological Laboratory, Cockeysville, Md.
†Bristol Laboratories, Syracuse, N.Y.
‡Hemin may also be added: 1 ml of hemin solution (see Chapter 43).
§Eli Lilly and Co., Indianapolis.

---

*Available in tubed form from Baltimore Biological Laboratory, Cockeysville, Md., and Difco Laboratories, Detroit.

steam under 15 pounds pressure (121 C) for 15 or 20 minutes.

47. Lowenstein-Jensen medium*
(Holm and Lester: Public Health Rep. [abstr.] **62**:847, 1947)

Salt solution		
Monopotassium phosphate	2.4	g
Magnesium sulfate (7H$_2$O)	0.24	g
Magnesium citrate	0.6	g
Asparagin	3.6	g
Glycerol, reagent grade	12.0	ml
Distilled water	600	ml
Potato flour	30.0	g
Homogenized whole eggs	1,000	ml
Malachite green (2% aqueous)	20	ml

Add the potato flour to the flask of salt solution and autoclave the mixture at 121 C for 30 minutes. Clean fresh eggs, not more than 1 week old, by vigorous scrubbing in 5% soap solution and allow them to remain in it for 30 minutes. Place eggs in running cold water until water becomes clear. Immerse the washed eggs in 70% alcohol for 15 minutes, remove and break into a sterile flask; homogenize completely by shaking with glass beads. Then filter the homogenate through four layers of sterile gauze.

Add 1 liter of the homogenized egg suspension to the flask of cooled potato flour–salt solution. Add the malachite green to this emulsion and mix thoroughly. Dispense the medium aseptically with a sterile aspirator bottle in 6-ml amounts into sterile, screw-capped 150-mm glass tubes. Inspissate the tubes at 85 C for 50 minutes and check for sterility by incubating at 35 C for 48 hours. Store in a refrigerator, where the medium will keep for at least 1 month if the tubes are tightly sealed to prevent loss of moisture.

This medium is used for the cultivation of *Mycobacterium tuberculosis*.

48. Lysine-iron agar*
(Edwards and Fife: Appl. Microbiol. **9**:478, 1961)

Peptone	5.0	g
Yeast extract	3.0	g
Glucose	1.0	g
L-Lysine	10.0	g
Ferric ammonium citrate	0.5	g
Sodium thiosulfate	40.0	mg
Bromcresol purple	20.0	mg
Agar	15.0	g
Distilled water	1,000	ml
(Adjust to pH 6.7)		

Dispense 4-ml amounts into 13- × 100-mm tubes and sterilize at 121 C for 12 minutes. Slant tubes to obtain a deep butt and a short slant. This medium is inoculated by using a straight wire to stab the butt twice and to streak the slant. Incubation is at 35 C for 18 to 24 hours. If necessary, incubate for 48 hours.

This medium is useful in determining whether members of the family Enterobacteriaceae can decarboxylate or deaminate lysine. However, this medium is **not** to be considered as a substitute for the Moeller method. A **positive** reaction is indicated by an alkaline or **purple** reaction in the butt of the tube. The absence of lysine decarboxylase is indicated by an acidic or **yellow (negative)** reaction in the butt of the tube due to fermentation of glucose. Lysine deaminase is indicated by the formation of a **red** slant and is characteristic of the tribe Proteeae. It should be noted that H$_2$S-producing *Proteus* species do not blacken this medium. Furthermore, *P. morganii* does not consistently produce a red slant after 24 hours' incubation.

49. MacConkey agar*

Peptone	17.0	g
Proteose peptone	3.0	g
Lactose	10.0	g
Bile salts	1.5	g
Sodium chloride	5.0	g

---

*Prepared tubed media are available from Difco Laboratories, Detroit; Baltimore Biological Laboratory, Cockeysville, Md.

*Available in dehydrated form from Baltimore Biological Laboratory, Cockeysville, Md.; Difco Laboratories, Detroit.

Agar	13.5	g
Neutral red	0.03	g
Crystal violet	0.001	g
Distilled water	1,000	ml

The agar concentration may be increased to 5% (use an additional 3.65 g of agar) to inhibit the spreading of *Proteus*. The medium is inhibitory for gram-positive bacteria and differential rather than selective. Coliforms (lactose fermenting) produce red colonies on the medium while nonlactose fermenters produce colorless colonies.

50. Malonate (sodium) broth*
(Leifson: J. Bacteriol. **26:**329, 1933; Ewing et al.: Pub. Health Lab. **15:**153, 1957)

Yeast extract	1.0	g
Ammonium sulfate	2.0	g
Dipotassium sulfate	0.6	g
Monopotassium phosphate	0.4	g
Sodium chloride	2.0	g
Sodium malonate	3.0	g
Glucose	0.25	g
Bromthymol blue	0.025	g
Distilled water	1,000	ml
(Final pH 6.7±)		

Sterilize in small tubes in 3-ml amounts at 121 C for 15 minutes. Inoculate from TSI slants or broth cultures; incubate cultures at 35 C for 48 hours.

The medium is used to test for utilization of sodium malonate by members of the Enterobacteriaceae. A **positive** test is shown by a color change of the indicator from green to a **Prussian blue.**

51. Mannitol salt agar*

Beef extract	1.0	g
Proteose peptone No. 3	10.0	g
Sodium chloride	75.0	g
Mannitol	10.0	g
Agar	15.0	g
Phenol red	0.025	g
Distilled water	1,000	ml

Sterilize the medium at 15 pounds pressure for 15 minutes, cool to 48 C, and pour

into sterile Petri dishes. Use 15 to 20 ml per plate.

This medium is used for the selective isolation of pathogenic staphylococci, since many other bacteria are inhibited by the high salt concentration. Colonies of potentially pathogenic staphylococci are surrounded by a **yellow** halo, indicating mannitol fermentation.

52. McBride medium, modified
(J. Lab. Clin. Med. **55:**153, 1960)

Phenylethanol agar*	35.5	g
Glycine, anhydride	10.0	g
Lithium chloride	0.5	g
Distilled water	1,000	ml

Dissolve with the aid of heat and sterilize at 121 C for 20 minutes.

This medium is recommended for the cultivation of *Listeria.*

53. Methyl red–Voges-Proskauer medium (MR-VP) (Clark and Lubs medium)*

Buffered peptone	7	g
Glucose	5	g
Dipotassium phosphate	5	g
Distilled water	1,000	ml
(Final pH 6.9 ±)		

This medium is used for the methyl red test and the Voges-Proskauer test (production of acetylmethylcarbinol).

54. Middlebrook 7H10 agar with OADC enrichment

The preparation of 7H10 oleic acid–albumin agar as originally described is extremely tedious and complicated. Its preparation may be simplified by the use of a combination of six stock solutions prepared in advance, as described in the handbook *Tuberculosis Laboratory Methods of the Veterans Administration.*†

---

*Available in dehydrated form from Baltimore Biological Laboratory, Cockeysville, Md.; Difco Laboratories, Detroit.

---

*Available in dehydrated form from Baltimore Biological Laboratory, Cockeysville, Md.; Difco Laboratories, Detroit.
†Obtainable from the Superintendent of Documents, U.S. Government Printing Office, Washington, D.C.

However, for the average diagnostic microbiology laboratory, it is strongly recommended that **commercially prepared** 7H10–OADC complete medium be used whenever possible.* For purposes of orientation, the ingredients of each medium are listed, but the directions are intended solely for preparation from dehydrated medium and prepared enrichment.

a. Middlebrook 7H10 agar*

Ammonium sulfate	0.5	g
D-Glutamic acid	0.5	g
Sodium citrate	0.4	g
Disodium phosphate	1.5	g
Monopotassium phosphate	1.5	g
Ferric ammonium phosphate	0.04	g
Magnesium sulfate	0.05	g
Pyridoxine	0.001	g
Biotin	0.0005	g
Malachite green	0.001	g
Agar	15.0	g

b. Middlebrook OADC enrichment*

Oleic acid	0.5	g
Bovine albumin, fraction V	50.0	g
Glucose	20.0	g
Beef catalase	0.04	g
Sodium chloride	8.5	g
Distilled water	1,000	ml

To rehydrate the 7H10 agar, suspend 20 g in 1 liter of cold distilled water containing 0.5% reagent grade glycerol and heat to boiling to dissolve completely. Distribute in 180-ml amounts in flasks; sterilize by autoclaving at 121 C for 10 minutes; cool to 50 to 55 C; add 20 ml of OADC enrichment aseptically to each flask; and dispense in sterile plastic Petri dishes. Store at 4 C in the dark for not longer than 2 months in a plastic bag to prevent dehydration.

55. Milk media
a. Skimmed milk

Fresh, clean, skimmed cow's milk may be used. This is dispensed in tubes and sterilized either by tyndallization (flowing steam for 30 minutes on 3 successive days) or in the autoclave at 10 pounds pressure for 10 minutes.

Commercial dehydrated skimmed milk is also completely satisfactory. The concentration used is 10% in distilled water. The method of sterilization is as before.

b. Litmus milk

Skimmed milk powder	100 g
Litmus	5 g
Distilled water	1,000 ml

Autoclave at 10 pounds pressure for 10 minutes.

Reactions in litmus milk. Pink color of litmus indicates an acid reaction caused by fermentation of lactose. A purple or blue color (alkaline) indicates no fermentation of lactose. White color (reduction) results when litmus serves as an electron acceptor and is reduced to its leuco base. Coagulation (clot) is caused by precipitation of casein by the acid produced from lactose. Coagulation may also be caused by the conversion of casein to paracasein by the enzyme rennin.

Peptonization (dissolution of the clot) indicates digestion of the curd or milk proteins by proteolytic enzymes.

c. Milk medium with a reducing agent*

Skimmed milk powder	100	g
Peptone	10.0	g
Sodium thioglycollate	0.5	g
Litmus	5.0	g
Distilled water	1,000	ml

d. Methylene blue milk

Skimmed milk powder	10 g
Methylene blue (1% aqueous)	10 ml
Distilled water	90 ml

Methylene blue milk is used in the identification of the enterococci.

---

*Baltimore Biological Laboratory, Cockeysville, Md.; Difco Laboratories, Detroit.

---

*This medium has been found satisfactory for the cultivation of *Clostridium* species and will allow observation of their reactions in litmus milk. Autoclave at 10 pounds pressure for 10 minutes; stopper tightly.

56. Moeller cyanide broth*
(Acta Pathol. Microbiol. Scand. **34**:115, 1954)

Peptone (Orthana special†)	10.0	g
Sodium chloride	5.0	g
Monobasic potassium phosphate	0.225	g
Dibasic sodium phosphate (2 H₂O)	5.64	g
Distilled water	1,000	ml

(Adjust reaction to pH 7.6)

Autoclave the medium in flasks. Dissolve 0.5 g of potassium cyanide in 100 ml of **cold** sterile broth. Place 1 ml of the cyanide broth in 12- × 100-mm tubes. **Immediately** stopper with paraffined corks and store at 4 to 5 C. The medium remains stable for 2 to 3 weeks.

Inoculate medium with one small loopful of a 24-hour broth culture. Incubate at 35 C and observe daily for 2 days for **growth,** which is recognized by turbidity of the medium.

57. Motility test medium‡§

Beef extract	3	g
Gelysate peptone	10	g
Sodium chloride	5	g
Agar	4	g
Distilled water	1,000	ml

(Final pH 7.3±)

a. Suspend 22 g of dehydrated medium in water, add 0.05 g triphenyltetrazolium. Heat with frequent agitation; boil 1 minute to dissolve ingredients.
b. Dispense in screw-capped tubes and sterilize by autoclaving at 121 C for 15 minutes.
c. Tighten caps when cool; store at room temperature.

---

*The base medium is available from Baltimore Biological Laboratory, Cockeysville, Md.; Difco Laboratories, Detroit.
†Proteose peptone No. 3, 0.3%, has been found to serve as a satisfactory substitute.
‡Baltimore Biological Laboratory, Cockeysville, Md.
§Discussion adapted from Paik, G., and Suggs, M. T.: Reagents, stains, and miscellaneous procedures. In Lennette, E. H., Spaulding, E. H., and Truant, J. P., editors: Manual of clinical microbiology, ed. 2, Washington, D.C., 1974, American Society for Microbiology.

This medium is stabbed **once** and read after 1 to 2 days' incubation at 35 C. Motile organisms spread out from line of inoculation; nonmotile organisms grow only along stab.

If negative, follow with further incubation at 21 to 25 C for 5 days. For special purposes, such as enhancement of the motility and flagellar development in poorly motile cultures, it is often advisable to pass cultures first through a semisolid medium containing 0.2% agar tubed in Craigie tubes or in U tubes. Subsequent passages may be made in 0.4% agar medium.

Motility media containing concentrations higher than 0.3% produce gels through which many motile organisms cannot spread. Spreading in a semisolid medium is judged by macroscopic examination of the medium for a diffuse zone of growth emanating from the line of inoculation. Many aerobic pseudomonads fail to grow deep in semisolid medium in a test tube. Organisms possessing "paralyzed" flagella are nonmotile and cannot spread in the medium. Some filamentous organisms spread in or on semisolid medium but are nonmotile and nonflagellated. Although cultures may grow at 37 C or higher temperatures, the flagellar proteins of some organisms are not synthesized optimally at this temperature; hence, motility medium should be incubated at temperatures near 18 to 20 C. These observations require judicious interpretation of motility, and limit, to some extent, the reliability of using spreading in semisolid agar as the sole taxonomic criterion to delineate related species.

58. Motility test semisolid agar for Listeria

Bacto-tryptose*	10	g
Sodium chloride	5	g
Agar	5	g
Glucose	1	g
Distilled water	1,000	ml

---

*Difco Laboratories, Detroit.

Dissolve with the aid of heat, distribute in tubes, and sterilize at 121 C for 20 minutes.

Inoculate by stabbing with a 24-hour broth culture. Incubate tubes at 25 and 35 C and observe for growth away from the stab line (motile).

This medium is useful for demonstrating the motility of *Listeria*.

59. Mueller-Hinton agar*
(Proc. Soc. Exp. Biol. Med. **48**:330, 1941)

Beef, infusion from	300 g
Peptone	17.5 g
Starch	1.5 g
Agar	17.0 g
(Final pH 7.4±)†	

Suspend medium in distilled water, mix thoroughly, and heat with frequent agitation. Boil for about 1 minute and dispense and sterilize by autoclaving at 116 to 121 C (12 to 15 pounds steam pressure) for **not more than** 15 minutes, and cool by placing immediately in a 50 C water bath before pouring.

The medium is used primarily for the disk-agar diffusion method of testing for antimicrobial susceptibility of microorganisms. It has also been used to detect starch hydrolysis by *S. bovis* (Lee: J. Clin. Microbiol. **4**:312, 1976) and for culturing the Legionnaires' disease agent.

Five percent defibrinated animal blood may be added as enrichment; this may be chocolatized in the testing of *Haemophilus* species.

60. Mycoplasma (PPLO) isolation media
a. Mycoplasma isolation agar (PPLO agar)‡

Beef heart, infusion from, fresh tissue	50 g
Peptone	10 g
Sodium chloride	5 g

Agar	14 g
(Final pH 7.8)	

1. Suspend 34 g of dry medium in 1,000 ml distilled water, mix, and boil for 1 minute. Dispense in 70- to 75-ml amounts. Autoclave at 121 C for 15 minutes.
2. Cool to approximately 50 C and add aseptically:

Horse serum*	20 ml
Yeast extract (25%)	10 ml
(see preparation c)	
Penicillin G solution	2 ml
(100,000 units/ml)	

3. Mix gently, pour approximately 20 ml per plate into sterile Petri dishes (100 × 15 mm). The smaller (60 × 15 mm) dishes accommodate 8-ml amounts and are used for subculture.
4. If bacterial contamination is excessive, add 5 ml of a 1:2,000 dilution of thallium acetate; reduce fungal contamination by adding 0.7 ml of a 0.05% solution of amphotericin B.
5. Seal plates with parafilm and store in the refrigerator.

b. Mycoplasma isolation broth†

This fluid medium is identical to the above-mentioned agar medium with the agar omitted. It is prepared by dissolving 21 g of dry medium in 1,000 ml distilled water, dispensing in 70-ml volumes and sterilizing in the autoclave at 121 C for 15 minutes. After cooling to about 50 C, the horse serum, yeast extract, and penicillin are added as described previously.

c. Preparation of 25% yeast extract

1. Suspend 125 g of Fleischmann's pure dry yeast (Type 20-40)‡ in 500 ml distilled/deionized water in a 1-liter

---

*Baltimore Biological Laboratory, Cockeysville, Md.; Difco Laboratories, Detroit.
†Check by using a surface electrode, if available, after gelling.
‡Available from Difco Laboratories, Detroit, and Baltimore Biological Laboratory, Cockeysville, Md.

---

*Available from Baltimore Biological Laboratory, Cockeysville, Md.
†Available from Difco Laboratories, Detroit, and Baltimore Biological Laboratory, Cockeysville, Md.
‡Available from Standard Brands, Inc., New York.

beaker. Boil for 2 minutes with constant stirring.

2. Dispense in centrifuge tubes and refrigerate overnight to settle the yeast cells.
3. Centrifuge at 3,000 rpm for 30 minutes and carefully remove the supernatant.
4. Adjust to pH 8.0 with 1 N NaOH.
5. Dispense in 2.5-ml amounts in screw-capped tubes or small bottles; autoclave at 121 C for 15 minutes.
6. Store in frozen state at −20 C; use within 3 months. When thawing, the extract becomes quite turbid, but clears on standing at room temperature.

d. Biphasic isolation medium for M. pneumoniae

1. Prepare basal broth medium as described previously.
2. Add 1 ml of a 0.1% phenol red solution and 2 ml of a 0.1% methylene blue solution to 70 ml of broth.
3. Sterilize by autoclaving at 121 C for 15 minutes.
4. Cool to approximately 50 C and add aseptically the horse serum, yeast extract, and penicillin as described previously.
5. Add aseptically about 3 ml to tubes containing the isolation agar (2-ml amounts in screw-capped tubes) and seal tightly.

Growth of *M. pneumoniae* is indicated by a change in color of the indicator.

61. Nitrate broth for nitrate reduction test

Tryptone	5.0 g
Neopeptone	5.0 g
Distilled water	1,000 ml
Potassium nitrate (reagent grade)	1.0 g
Glucose	0.1 g

Before adding the potassium nitrate and glucose, boil the other ingredients in the water and adjust pH to 7.3 to 7.4. Dispense 5 ml per tube and sterilize at 15 pounds pressure for 15 minutes. The reagents and tests for nitrate reduction are listed in Chapter 43.

62. ONPG (beta-galactosidase) test

The ability of certain gram-negative bacilli to ferment lactose is a useful criterion for identifying certain members of the family Enterobacteriaceae. Since some coliform organisms, the so-called paracolon bacteria, ferment this carbohydrate slowly or not at all, they have been confused with the enteric pathogens.

Lactose fermentation is dependent on two enzymes: **permease,** which allows lactose to enter the bacterial cell, and **beta-galactosidase,** which splits lactose into glucose and galactose. The slow lactose fermenters are deficient in permease, and the demonstration of beta-galactosidase in such organisms permits more rapid identification of a lactose fermenter. This enzyme can be detected conveniently by the use of a tablet of ortho-nitrophenyl-beta-galactopyranoside (ONPG)* dissolved in distilled water and inoculated with a heavy suspension of the organism to be tested. The test is incubated at 35 C for 6 hours; hydrolysis of ONPG is detected by the liberation of ortho-nitrophenol, with its characteristic **yellow** color, often within 30 minutes. Thus a **positive** ONPG test indicates that the organism contains lactose-fermenting enzymes and may be classed as a lactose fermenter.

63. Oxidative-fermentative (OF) basal medium (Hugh and Leifson)†
(J. Bacteriol. **66:**24-26, 1953)

Trypticase or tryptone	2.0	g
Sodium chloride	5.0	g
Dipotassium phosphate	0.3	g
Agar (dried)	2.5	g
Bromthymol blue (3 ml of 1%	0.03	g
aqueous [not alcoholic] solution)		
Distilled water	1,000	ml
(Final pH 7.1 ± 0.2)		

Suspend the material in 1 liter of distilled water and heat to dissolve with frequent

---

*Key Scientific Products Co., Los Angeles.
†Available in dehydrated form from Baltimore Biological Laboratory, Cockeysville, Md.; Difco Laboratories, Detroit.

agitation; adjust reaction to a pH of approximately 7.1. Sterilize by autoclaving at 121 C for 15 minutes. To 100 ml of sterile base, add 10 ml sterile 10% glucose solution (10% lactose, mannitol, and sucrose also may be added to separate 100-ml portions if required). The completed medium should be **green**.

OF basal medium is used for differentiating organisms such as *Acinetobacter, Alcaligenes,* and *Pseudomonas* from members of the Enterobacteriaceae.

64. Peptic digest agar

Add 5% Fildes enrichment to trypticase soy agar base. This is a very good medium for demonstrating *H. influenzae;* colonies are translucent blue.

65. Phenol red broth base*

Trypticase	10.0	g
Sodium chloride	5.0	g
Phenol red	0.018	g
(Final pH 7.4±)		

Phenol red broth base can be used as a base for determining carbohydrate fermentation reactions. Carbohydrates can be added to the medium in 0.5% to 1% concentrations before dispensing and sterilization. Durham tubes should be inserted into the tubed medium for detection of gas formation.

66. Phenylalanine agar†
(Ewing et al.: Pub. Health Lab. **15**:153, 1957)

Yeast extract	3 g
DL-Phenylalanine	2 g
(or L-phenylalanine)	1 g
Disodium phosphate	1 g
Sodium chloride	5 g

Agar	12 g
Distilled water	1,000 ml

Dispense 3-ml amounts in small tubes and sterilize at 121 C for 10 minutes. Allow to solidify in a slanted position.

This medium is used to test for deamination of phenylalanine to phenylpyruvic acid by members of the Enterobacteriaceae.

After incubation of the culture for 18 to 24 hours at 35 C, allow 4 or 5 drops of fresh 10% ferric chloride solution to run down over the growth on the slant. If acid has been formed, a **green** color immediately develops on the slant and in the fluid at the base of the slant.

67. Phenylethyl alcohol agar* (Brewer and Lilley)
(J. Am. Pharm. Assoc. [Scient. Ed.] **42**:6-8, 1953)

Trypticase	15.0 g
Phytone	5.0 g
Sodium chloride	5.0 g
β-Phenylethyl alcohol	2.5 g
Agar	15.0 g
Distilled water	1,000 ml
(Final pH 7.3±)	

Suspend powder in water, mix thoroughly, heat with agitation, and boil 1 minute. Dispense in 16- × 150-mm screw-capped tubes. Sterilize at 118 to 121 C (12 to 15 pounds pressure) for 15 minutes. If desired, 5% blood may be added to the cooled medium (45 to 50 C) before pouring plates.

Phenylethyl alcohol agar is useful for the isolation of gram-positive cocci and the inhibition of gram-negative bacilli (particularly *Proteus)* when these are found in mixed culture.

68. Polymyxin staphylococcus medium
(Finegold and Sweeney: J. Bacteriol. **81**:636-641, 1961)

Nutrient agar	23.0 g
Tween 80	10.2 ml
Lecithin	0.7 g

---

*Available in dehydrated form from Baltimore Biological Laboratory, Cockeysville, Md. Similar media with bromcresol purple indicator can be obtained either from Baltimore Biological Laboratory or from Difco Laboratories, Detroit. If bromcresol purple is preferred to phenol red indicator, purple broth base is also available in dehydrated form from these laboratories.
†Difco Laboratories, Detroit; Baltimore Biological Laboratory, Cockeysville, Md.

---

*Difco Laboratories, Detroit; Baltimore Biological Laboratory, Cockeysville, Md.

Polymyxin*	75.0 mg
Distilled water	1,000 ml

Autoclave at 15 pounds pressure for 15 minutes. Pour plates.

Coagulase-negative staphylococci, micrococci, and most gram-negative rods are inhibited. In addition to *S. aureus*, *Proteus* will grow; the latter can be recognized by translucent colonies. *Proteus* does not swarm on this medium.

69. Potato-carrot agar

Carrots, peeled	20 g
Potatoes, peeled	20 g
Distilled water	1,000 ml

Wash potatoes and carrots, mash both, and place in distilled water for 1 hour. Boil for 5 minutes, filter through paper, make up to 1,000 ml of volume, and add 15 g of agar and 5 ml of Tween 80. Dispense in tubes, autoclave at 121 C for 20 minutes, slant, and cool. Potato-carrot agar is an excellent medium for demonstrating color characteristics of a fungal colony.

70. Potato-dextrose agar†

Potatoes, infusion from	200 g
Glucose (dextrose)	20 g
Agar	20 g
Distilled water	1,000 ml

Boil potatoes in water for 15 minutes, filter through cotton, and make up to volume with water. Add dry ingredients and dissolve agar with heat. No pH adjustment is required. Dispense as desired and autoclave at 121 C for 10 minutes. The agar is used to stimulate spore production of fungi.

71. Rice grain medium

White rice	8 g
Distilled water	25 ml

Place in a 125-ml Erlenmeyer flask and autoclave at 121 C for 15 minutes.

Rice grain medium is used for differentiation of *Microsporum* species. *M. canis* and *M. gypseum* grow and sporulate well in this medium; *M. audouini* grows poorly. Conidial formation is stimulated in some of the *Trichophyton* species.

72. Sabhi agar*
(Gorman: Am. J. Med. Technol. **33**:151, 1967)

Calf brain, infusion from	100	g
Beef heart, infusion from	125	g
Proteose peptone	5.0	g
Neopeptone	5.0	g
Glucose	21.0	g
Sodium chloride	2.5	g
Disodium phosphate	1.25	g
Agar	15.0	g
Distilled water	1,000	ml
(pH 7.0)		

a. Suspend 59 g in 1,000 ml distilled water; heat to boiling to completely dissolve medium.
b. Sterilize by autoclaving at 121 C for 15 minutes.
c. Cool to 50 to 55 C, add 1 ml sterile chloramphenicol solution (100 mg/ml).
d. Mix well, dispense in sterile, cotton-plugged 25- × 150-mm Pyrex test tubes.
e. Slant, allow to harden, and store in refrigerator.

Sabhi is an equal mixture of Sabouraud dextrose agar and brain-heart infusion agar and has proved useful for isolation of clinically significant fungi, particularly from body sites containing bacteria, such as sputum.

73. Sabouraud dextrose agar (SAB)

Sabouraud dextrose agar†	32.5 g

---

*75 mg = 75,000 µg or 750,000 units of activity.
†Available in prepared tubes from Baltimore Biological Laboratory, Cockeysville, Md.; Difco Laboratories, Detroit.

---

*Available from Difco Laboratories, Detroit (No. 0797).
†This medium and also media containing the antimicrobial agents are available from Baltimore Biological Laboratory, Cockeysville, Md.; Difco Laboratories, Detroit.

Agar, powdered	2.5	g
Distilled water	500	ml

Suspend ingredients in water; dissolve by heating to boiling; dispense in approximately 20-ml amounts in cotton-plugged, 25- × 150-mm Pyrex test tubes (without lips). If antimicrobial agents are to be added, this may be done after heating the medium and before autoclaving. The following amounts are recommended (Center for Disease Control):

Cycloheximide* (0.5 mg/ml)	250	mg
Chloramphenicol† (0.05 mg/ml)	25	mg

Add the chloramphenicol dissolved in 5 ml of 95% ethanol and the cycloheximide dissolved in 5 ml of acetone. Mix well and distribute into tubes or bottles as indicated. Autoclave at 118 C for **not longer** than 10 minutes. Slant, allow to harden, and store in refrigerator. This selective medium is used for the isolation of fungi when contaminating microorganisms may be present.

74. Sabouraud dextrose broth

Dextrose	40	g
Peptone	10	g
Distilled water	1,000	ml

Dissolve ingredients, dispense in 10-ml amounts in 18- × 150-mm tubes, and autoclave at 121 C for 10 minutes. Sabouraud dextrose broth is useful for differentiation of *Candida* species.

75. Salmonella-Shigella (SS) agar‡

Beef extract	5.0	g
Peptone	5.0	g
Lactose	10.0	g
Bile salts mixture	8.5	g
Sodium citrate	8.5	g
Sodium thiosulfate	8.5	g
Ferric citrate	1.0	g

Agar	13.5	g
Brilliant green	0.33	g
Neutral red	0.025	g
Distilled water	1,000	ml

Dissolve by boiling. **Do not autoclave.**

Salmonella-Shigella agar is used primarily as a selective medium for isolation of salmonellae and shigellae, while inhibiting coliform bacilli. It can also differentiate lactose-fermenting from non-lactose-fermenting strains.

76. Scott's modified Castañeda media and thioglycollate broth for blood culture

a. Agar bottle—slant

Trypticase soy agar*	40	g
Agar, granulated	15	g
Distilled water	1,000	ml

Heat in autoclave at 121 C for 5 minutes to melt agar, or heat on a hot plate with a magnetic stirring bar.

b. Agar bottle—broth

Trypticase soy broth*	30	g
Sodium polyanethol sulfonate (SPS)	5	ml
("Grobax," 5% sterile solution SPS)†		
Distilled water	1,000	ml

c. Thioglycollate medium (135 C)‡

	30	g
Sucrose	100	g
Sodium polyanethol sulfonate (SPS)	5	ml
Distilled water	1,000	ml

(pH of all media after autoclaving should be 7.2 ± 0.2)

1. Dispense agar medium (a) in the melted state in approximately 20-ml amounts into each of 50 clean, non-chipped 4-ounce, square, clear glass, screw-capped bottles (10 ml in 2-ounce bottles).§

---

*Cycloheximide is available in 4-g amounts from Upjohn Co., Kalamazoo, Mich., as Actidione.
†Chloramphenicol is available from Parke, Davis & Co., Detroit, as Chloromycetin.
‡Available in dehydrated form from Difco Laboratories, Detroit; Baltimore Biological Laboratory, Cockeysville, Md.

---

*Available in dehydrated form from Difco Laboratories, Detroit; Baltimore Biological Laboratory, Cockeysville, Md.
†Roche Diagnostics, Nutley, N.J. (No. 43000).
‡Baltimore Biological Lagoratory, Cockeysville, Md.
§Blood culture bottles (St. Louis Health Dept. type) with special screw caps and disposable rubber diaphragms, 4-ounce and 2-ounce sizes, can be obtained from Curtin Scientific Co., Rockville, Md., and other addresses.

2. Insert the disposable rubber diaphragms in the special screw caps, which are then applied loosely on the bottle tops.

3. Dispense thioglycollate medium (c) into 50 clean bottles, approximately 75 ml per bottle (30 ml in 2-ounce bottles) and apply screw caps as noted earlier.

4. Make up trypticase broth (b) in a 2-liter flask and plug. Along with this prepare a dispensing buret (a 300-ml Salvarsan tube* or satisfactory substitute) to the end of which are attached a 2-foot length of rubber tubing, a needle holder, and a 1½-inch, 21-gauge needle, in that order. Insert the attached needle, together with a portion of the rubber tubing, in the top of the buret and make fast by plugging, thus assuring a closed unit during sterilization.

5. Autoclave all media and equipment at 118 C to 121 C for 12 to 15 minutes.

6. As soon as the pressure reaches zero, open the autoclave, remove all bottles rapidly, and tightly fasten the screw caps. The hands must be protected from the hot bottles by asbestos gloves.

7. Place the bottles containing the agar on their sides on a cool table top to permit hardening of the agar layer. After about an hour, place bottles upright and sterilize tops with alcohol sponges.† (The agar does not become detached from the sides of the bottles.)

8. Clamp the sterile dispensing buret to a ring stand. Using aseptic technique, fill the buret with sterile trypticase broth.

9. To each agar slant bottle add 60 ml of broth (25 ml in 2-ounce bottles) by puncturing the rubber diaphragm with the sterile needle. The partial vacuum in each bottle will readily permit addition of this amount.

10. Incubate agar slant bottles at 35 C for 48 hours to ensure sterility. The thioglycollate bottles do not need incubation for a sterility check.

11. Inspect bottles, label, and store at room temperature (shelf life, 6 months).

77. Selenite-F enrichment medium*
(Leifson: Am. J. Hyg. **24:**423, 1936)

Sodium hydrogen selenite (anhydrous)	0.4%
Sodium phosphate (anhydrous)	1.0%
Peptone	0.5%
Lactose	0.4%
(Final pH 7)	

Dissolve ingredients in distilled water. Sterilize gently; 30 minutes in flowing steam in an autoclave is sufficient. It is important to note that the medium should **not** be autoclaved. This medium is used for the selective isolation of *Salmonella* and some strains of *Shigella*.

78. Sellers differential agar†
(Bact. Proc. 1963, p. 65.)

Yeast extract	1.0	g
Peptone	20.0	g
L-Arginine	1.0	g
D-Mannitol	2.0	g
Bromthymol blue	0.04	g
Phenol red	0.008	g
Sodium chloride	2.0	g
Sodium nitrate	1.0	g
Sodium nitrite	0.35	g
Magnesium sulfate	1.5	g
Dipotassium phosphate	1.0	g
Agar	15.0	g

To rehydrate the medium, suspend 45 g in 1,000 ml of cold distilled water and

---

*Arthur H. Thomas Co., Philadelphia.
†At the Wilmington Medical Center, the use of a loose application of a piece of steam autoclave tape (No. 1222-3M) to the bottle top, which is then fastened down after sterilization, obviates the need for disinfection of the diaphragm when filling with broth (step 8) or when collecting the blood culture.

---

*Available in dehydrated form from Baltimore Biological Laboratory, Cockeysville, Md.; Difco Laboratories, Detroit.
†Difco Laboratories, Detroit; Baltimore Biological Laboratory, Cockeysville, Md.

Table 41-1. Reactions produced on Sellers medium by nonfermentative gram-negative bacilli

Organism	Slant color	Butt color	Band color	Fluorescent slant	Nitrogen gas
Pseudomonas aeruginosa	Green	Blue or no change	Sometimes blue	Yellow-green	Produced
Acinetobacter calcoaceticus var. *anitratus*	Blue	No change	Yellow	Absent	Absent
A. calcoaceticus var. *lwoffii*	Blue	No change	Absent	Absent	Absent
Alcaligenes faecalis	Blue	Blue or no change	Absent	Absent	Produced

heat to boiling to dissolve the medium completely. Dispense into test tubes and stopper with cotton plugs or loosely fitting caps. Sterilize in the autoclave for 10 minutes at 15 pounds pressure (121 C). Allow tubes to cool in the slanted position to give approximately 1½-inch butts and 3-inch slants. Immediately **before** inoculating, add 2 large drops or 0.15 ml of a sterile 50% glucose solution to each tube by letting it run down the **side of the tube opposite the slant.** Inoculate tubes by deep stab into the butt and by streaking the slant. Incubate for 24 hours at 35 C. The final reaction of the medium will be pH 6.7 at 25 C.

Sellers differential agar is useful for differentiating and identifying nonfermentative gram-negative bacilli that produce an alkaline reaction on TSI (or KI) agar. The dehydrated medium is prepared according to the formula of Sellers and is recommended. This medium is particularly useful in differentiating *Pseudomonas aeruginosa, Acinetobacter calcoaceticus,* and *Alcaligenes faecalis* (Table 41-1).

79. Sodium chloride broth (6.5%)

Heart infusion broth*	100 ml
Sodium chloride	6 g

Brain-heart infusion broth contains 0.5%

---

*Available in dehydrated form from Baltimore Biological Laboratory, Cockeysville, Md.; Difco Laboratories, Detroit.

sodium chloride. Thus, by adding an additional 6%, the desired concentration is obtained. The medium is selective for enterococci and other salt-tolerant organisms and is useful in identification as well.

80. Starch agar medium

Bacto-agar	20 g
Bacto-peptone	5 g
Beef extract	3 g
Sodium chloride	5 g
Soluble starch	20 g
Distilled water	1,000 ml
(Final pH 7.2±)	

a. Dissolve agar in 300 ml of water with heat.

b. Dissolve beef extract and peptone in 200 ml of water.

c. Mix the solutions in steps a and b and make up to 1,000 ml volume.

d. To this mixture add the starch, dissolve, and autoclave at 121 C for 15 minutes.

The medium may be dispensed in tubes (15- to 20-ml amounts) or flasks convenient for pouring plates and should be stored in a refrigerator. For pouring of plates, melt the medium in tubes or flasks as required. **If poured plates of starch agar are refrigerated, the medium becomes opaque.**

Starch agar is useful in developing smooth cultures by streaking borderline rough strains on the surface of the medium. It also may be used for testing cultures for starch hydrolytic activity.

81. Stock culture and motility medium
(Hugh. In ASM Man. Clin. Microbiol. ed. 2, 1974, p. 263)

Casitone	10 g
Yeast extract	3 g
Sodium chloride	5 g
Agar	3 g
Distilled water	1,000 ml

a. Suspend ingredients in distilled water; heat to boiling to completely dissolve agar.
b. Dispense in 13- × 100-mm screw-capped tubes, 4 ml per tube.
c. Sterilize by autoclaving at 121 C for 15 minutes; store as butts.

This medium is used for motility testing by stabbing the inoculum once into the agar and incubating overnight at 35 C. Motility is indicated by growth spreading out from the line of stab.

The medium is also excellent for preserving stock cultures of non-fermenting and fermenting gram-negative rods (up to 6 months). After inoculation and overnight incubation, the caps are sealed with tape and the tubes stored in refrigerator.

82. Sucrose (5%) broth or agar

Prepare a 50% aqueous solution of sucrose, sterilize in an autoclave at 10 pounds pressure for 10 minutes, and store in a refrigerator.

To prepare broth, add aseptically 0.5 ml of the stock sucrose solution to 5 ml of tubed sterile infusion broth.

To prepare agar plates, add aseptically 1.5 ml of the stock sucrose solution to 15 ml of sterile melted agar. Mix well and pour into a sterile Petri dish.

83. Tellurite reduction test medium
Medium

a. Suspend 4.7 g of Middlebrook 7H9 dehydrated base* in 900 ml of distilled water and add 0.5 ml of Tween 80.
b. Autoclave at 121 C for 15 minutes, cool to 55 C, and add aseptically 100 ml ACD enrichment.*

---

c. Dispense aseptically in 5-ml amounts in 20- × 150-mm screw-capped tubes, check for sterility, and store in the refrigerator.

Tellurite solution

a. Dissolve 0.2 g of potassium tellurite in 100 ml of distilled water.
b. Dispense in 2- to 5-ml amounts and sterilize by autoclaving at 121 C for 10 minutes.

This medium is used to test the ability of certain mycobacteria of Runyon group III nonphotochromogens to reduce tellurite rapidly to the **black,** metallic tellurium.

84. Tetrathionate broth*

Proteose peptone†	5 g
Bile salts	1 g
Calcium carbonate	10 g
Sodium thiosulfate	30 g
Distilled water	1,000 ml

Dispense medium in 10-ml amounts and heat to boiling. **Before use,** add 0.2 ml of iodine solution (6 g of iodine crystals, 5 g of potassium iodide in 20 ml of water) to each tube.

This is a selective liquid enrichment medium for use in the isolation of *Salmonella,* except *S. typhi.* The sterile base **without** iodine may be stored indefinitely in a refrigerator.

85. Thayer-Martin agar (TM)

GC agar base* (double strength)	72 g
Agar	10 g
Distilled water	1,000 ml

a. Suspend dehydrated medium in water, mix well, and heat with agitation; boil for 1 minute.
b. Sterilize by autoclaving at 121 C for 15 minutes.
c. At the same time, autoclave a suspension of 20 g of dehydrated hemoglobin

---

*Available in dehydrated form from Difco Laboratories, Detroit; Baltimore Biological Laboratory, Cockeysville, Md.
†Difco Laboratories, Detroit.

*Difco Laboratories, Detroit; Baltimore Biological Laboratory, Cockeysville, Md.

in 1,000 ml water for 15 minutes (suspension must be **smooth** before sterilizing).

d. Cool both to 50 C, mix aseptically, then add 20 ml IsoVitaleX enrichment (BBL) and 20 ml V-C-N inhibitor (BBL); pour plates using 20 ml per plate. Store in refrigerator.

This medium is recommended for the isolation of *Neisseria gonorrhoeae* from all sites that might contain a mixed flora, as well as for the recovery of *N. meningitidis* from nasopharyngeal and throat cultures.

Modified Thayer-Martin medium (MTM) has 2% agar, 0.25% glucose, plus trimethoprim lactate, which is inhibitory to *Proteus*.

86. Thioglycollate medium without indicator* (THIO)
(Brewer: J. Bacteriol. **39:**10, 1940; and **46:**395, 1943)

Peptone	20.0	g
L-Cystine	0.25	g
Glucose	6.0	g
Sodium chloride	2.5	g
Sodium thioglycollate	0.5	g
Sodium sulfite	0.1	g
Agar	0.7	g
Distilled water	1,000	ml

(Final pH 7.2±)

Dispense medium in 15-ml amounts in 6- × ¾-inch test tubes, making a column of medium 7 cm high. Autoclave for 15 minutes at 121 C. **Store at room temperature.**

Enriched THIO is prepared by adding to the freshly prepared and autoclaved medium (or to previously prepared medium that has been boiled for 10 minutes and then cooled) vitamin K_1 solution (see Chapter 43), 0.1 μg/ml; sodium bicarbonate, 1 mg/ml; and hemin (see Chapter 43), 5 μg/ml. Rabbit or horse serum (10%) or Fildes enrichment (5%) may also be added.

87. Thionine or basic fuchsin agar
(Huddleson, et al.: Brucellosis in man and animals, New York, 1939, The Commonwealth Fund)

Trypticase soy agar may be used as a base for differential media containing thionine and basic fuchsin.

Prepare the dyes, thionine and basic fuchsin, in 0.1% stock solutions in sterile distilled water. These stock solutions may be stored indefinitely. Before adding to the media, heat the dye solutions in flowing steam in an autoclave for 20 minutes, shake well, and while still hot, add to melted agar. In trypticase soy agar the final concentration of the dye should be 1:100,000 (10 ml per liter of medium). Thoroughly mix the dyes (added individually) and the melted agar and pour immediately into Petri dishes, one set containing thionine and one containing basic fuchsin. Place plates in a 35 C incubator until the water of condensation disappears, at which time they are ready for use. Inoculate plates within 24 hours of preparation.

Streak the surface of plates with a heavy suspension of *Brucella* prepared from a 48- to 72-hour trypticase soy agar slant culture. It is advisable to streak plates in duplicate, incubating one set aerobically and the other in 10% carbon dioxide. Incubate plates for 72 hours and observe for **inhibition of growth** by either thionine or basic fuchsin or by both dyes.

88. Todd-Hewitt broth, modified*
(J. Pathol. Bacteriol. **35:**973, 1932)

Beef heart infusion	1,000	ml
Neopeptone	20	g

Adjust to pH 7 with normal sodium hydroxide and add:

---

*Available in dehydrated form from Baltimore Biological Laboratory, Cockeysville, Md. (No. 11720); Difco Laboratories, Detroit (No. 0430). It is also available with indicator. This may be enriched by the addition of 10% normal rabbit or horse serum when cool.

---

*Available in dehydrated form from Baltimore Biological Laboratory, Cockeysville, Md.; Difco Laboratories, Detroit.

Sodium chloride	2.0 g
Sodium bicarbonate	2.0 g
Disodium phosphate	0.4 g
Glucose	2.0 g

(Final pH 7.8±)

Mix chemicals in broth and bring to a slow boil. Boil for 15 minutes, filter through paper, dispense in tubes, and autoclave at 115 C for 10 minutes.

Modified Todd-Hewitt broth is used for growing streptococci for serologic identification.

89. Transport media
a. Buffered glycerol-saline base* (Sachs' modification)

Sodium chloride	4.2	g
Potassium dihydrophosphate (KH$_2$PO$_4$)	1.0	g
Potassium orthophosphate (K$_2$HPO$_4$)	3.1	g
Phenol red	0.003	g
Distilled water	700	ml
Glycerol	300	ml

(Adjust final pH to 7.2)

Dispense in 10-ml amounts in 30-ml screw-capped bottles. Sterilize for 10 minutes at 116 C.

This solution serves as an excellent stool specimen preservative and transport medium for fecal material.

b. Cary and Blair transport medium*

Disodium phosphate	1.1	g
Sodium chloride	5.0	g
Sodium thioglycollate	1.5	g
Agar	5.0	g
Distilled water	991	ml

Add ingredients to a chemically clean flask rinsed with Sorensen's 0.067 M buffer (pH 8.1). Heat with frequent agitation until the solution just becomes clear. Cool to 50 C. Add 9 ml of freshly prepared aqueous 1% CaCl$_2$ and adjust the pH to 8.4.

Distribute 7 ml into previously rinsed and sterilized 9-ml screw-capped vials. Steam for 15 minutes, cool, and tighten caps.

---

*Available from Baltimore Biological Laboratory, Cockeysville, Md.

c. Specimen preservative medium*†
(Hajna: Publ. Hlth. Lab. **13**:83, 1955)

Sodium deoxycholate	0.5	g
Yeast extract	1.0	g
Sodium chloride	5.0	g
Sodium citrate · 2 H$_2$O	5.0	g
Potassium dihydrophosphate (KH$_2$PO$_4$)	2.0	g
Magnesium sulfate · 7 H$_2$O	0.4	g
(NH$_4$)$_2$ HPO$_4$	4.0	g
Distilled water	700	ml

Dissolve ingredients by heating. Add 300 ml of glycerol, mix well, dispense into tubes (or vials), and sterilize at 116 C for 10 minutes. Final pH is 7.

This medium is useful as a stool specimen preservative and to transport fecal material.

d. Amies transport medium.†
(Amies: Can. J. Pub. Hlth. **58**:296-300, 1967).

1. Add 4 g of agar to 1 liter of distilled water, heat until dissolved, and while hot add:

Sodium chloride	3.0	g
Potassium chloride	0.2	g
Sodium thioglycollate	1.0	g
Disodium phosphate, anhydrous	1.15	g
(or disodium phosphate · 12 H$_2$O)	2.9	g
Monopotassium phosphate	0.2	g
Calcium chloride, 1% aqueous, freshly prepared	10.0	ml
Magnesium chloride · 6 H$_2$O, 1% aqueous	10.0	ml

(Final pH 7.3)

2. Stir until dissolved and then add 10 g of pharmaceutical neutral charcoal. Dispense 5 to 6 ml per 13- × 100-mm screw-capped tube (or vial), with frequent stirring to keep the charcoal in suspension. Avoid cooling or gelling.
3. Sterilize at 121 C for 20 minutes. **Prior** to solidification, invert tubes to distribute the charcoal evenly. Store in refrigerator.
4. It should be emphasized that prolonged heating in open flasks should be avoided, since the reducing agent (sodium thioglycollate) is volatile.

---

*Available from Baltimore Biological Laboratory, Cockeysville, Md.
†Available from Difco Laboratories, Detroit.

90. Triple sugar iron agar (TSI agar)*
(Hajna: J. Bacteriol. **49**:516, 1945)

Peptone	20.0	g
Sodium chloride	5.0	g
Lactose	10.0	g
Sucrose	10.0	g
Glucose	1.0	g
Ferrous ammonium sulfate	0.2	g
Sodium thiosulfate	0.2	g
Phenol red	0.025	g
Agar	13.0	g
Distilled water	1,000	ml

(Final pH 7.3±)

Triple sugar iron agar is used for determining carbohydrate fermentation and hydrogen sulfide production as a first step in the identification of gram-negative bacilli.

This medium is considered to be a modification of Kligler's iron agar. The only difference between the two is that sucrose is not included in Kligler's iron agar. It should be stressed that pH changes in the butt and in the slant of the medium must be recorded **only** after 18 to 24 hours of incubation.

91. Trypticase dextrose agar†

Trypticase	20.0	g
Dextrose	5.0	g
Agar	3.5	g
Bromthymol blue	0.01	g
Distilled water	1,000	ml

(Final pH 7.3±)

Trypticase dextrose agar can be used to determine motility and dextrose fermentation of aerobic and anaerobic organisms.

Prepare according to directions on label.

92. Trypticase lactose iron agar†

Trypticase	20.0	g
Lactose	10.0	g
Ferrous sulfate	0.2	g
Agar	3.5	g
Sodium sulfite	0.4	g

---

*Available in dehydrated form from Baltimore Biological Laboratory, Cockeysville, Md.; Difco Laboratories, Detroit.
†Available in dehydrated form from Baltimore Biological Laboratory, Cockeysville, Md.

Sodium thiosulfate	0.08	g
Phenol red	0.02	g
Distilled water	1,000	ml

(Final pH 7.3±)

This medium can be used for the determination of motility, lactose fermentation, and production of hydrogen sulfide by aerobes and anaerobes.

93. Trypticase nitrate broth*†

Trypticase	20	g
Disodium phosphate	2	g
Glucose	1	g
Agar	1	g
Potassium nitrate	1	g
Distilled water	1,000	ml

(pH 7.2)

This medium is used to demonstrate indole production and nitrate reduction by aerobes and anaerobes.

94. Trypticase soy agar*

Trypticase	15	g
Phytone	5	g
Sodium chloride	5	g
Agar	15	g
Distilled water	1,000	ml

(Final pH 7.3±)

Trypticase soy agar is an excellent **blood agar base** and can be used for the isolation and maintenance of all organisms except some with very special nutritional requirements.

95. Trypticase soy broth*

Trypticase	17.0	g
Phytone	3.0	g
Sodium chloride	5.0	g
Dipotassium phosphate	2.5	g
Glucose	2.5	g
Distilled water	1,000	ml

(Final pH 7.3±)

Trypticase soy broth is excellent for the rapid (6 to 8 hours) growth of most organisms and will support growth of pneumococci and streptococci without the addition of blood or serum. It will also support the growth of *Brucella*. However, fermentation of the glucose present will

---

*Available in dehydrated form from Baltimore Biological Laboratory, Cockeysville, Md.
†Indole nitrate medium (No. 11298).

cause a drop in pH, and acid-sensitive organisms, particularly pneumococci, may die in 18 to 24 hours.

96. Trypticase sucrose agar*

Trypticase	20.0	g
Sucrose	10.0	g
Agar	3.5	g
Phenol red	0.02	g
Distilled water	1,000	ml

(Final pH 7.2±)

This medium can be used to determine motility and sucrose fermentation by aerobes and anaerobes.

97. Tryptophane broth

Tryptophane broth is a popular medium for the detection of indole production. Trypticase (BBL) or tryptone (Difco) is recommended, in 1% aqueous solution. Follow label directions for preparation.

98. Urea agar—urease test medium†
(Christensen: J. Bacteriol. **52**:461, 1946)

Peptone	1.0	g
Glucose	1.0	g
Sodium chloride	5.0	g
Monopotassium phosphate	2.0	g
Phenol red	0.012	g
Agar	20.0	g
Distilled water	1,000	ml

(Final pH 6.8 to 6.9)

Prepare the agar base and sterilize in the autoclave at 121 C for 15 minutes in flasks containing 100- to 200-ml amounts. Store until needed. Prepare a 29% solution of urea. Sterilize by filtering through a sterile bacteriologic filter. Add the sterile urea solution in a final concentration of 10% to a flask of the agar base that has been melted and cooled to a temperature of 50 C. Mix well and distribute aseptically into sterile small tubes in amounts of 2 to 3 ml. Allow the medium to solidify in a slanting position in such a way as to obtain an agar butt of ½ inch and an agar slant of 1 inch.

Urea agar can be used to demonstrate **urease production** by species of *Proteus*. It will also detect the small amounts of urease produced by other enteric bacilli, thus differentiating them from urease-negative *Salmonella* and *Shigella*. It also may be used to detect urease production by *Cryptococcus* species.

99. Urease test broth*
(Rustigian and Stuart)

See p. 494, No. 59.

100. XLD agar*
(Taylor: Am. J. Clin. Pathol. **44**:471, 1965)
This medium may be prepared by using the dehydrated xylose lysine agar base* and adding the sodium thiosulfate, ferric ammonium citrate, and sodium desoxycholate (procedure recommended by some workers) or by utilizing the complete xylose lysine desoxycholate (XLD) agar.

Xylose	3.5	g
L-Lysine	5.0	g
Lactose	7.5	g
Sucrose	7.5	g
Sodium chloride	5.0	g
Yeast extract	3.0	g
Phenol red	0.08	g
Agar (dried)	13.5	g
Sodium desoxycholate	2.5	g
Sodium thiosulfate	6.8	g
Ferric ammonium citrate	0.8	g
Distilled water	1,000	ml

(Final pH 7.4±)

Suspend the medium in distilled water and heat with frequent agitation just to the boiling point. **Do not boil.** Transfer immediately to a 50 C water bath and pour plates as soon as the medium has cooled. The medium should be red-orange and clear, or nearly so. Excessive heating or prolonged holding at 50 C may cause precipitation, which could lead to some differences in colony morphology.

This medium is useful for the isolation of enteric pathogens, expecially shigellae.

---

*Available in dehydrated form from Baltimore Biological Laboratory, Cockeysville, Md.
†Available in dehydrated form from Baltimore Biological Laboratory, Cockeysville, Md.; Difco Laboratories, Detroit.

---

*Baltimore Biological Laboratory, Cockeysville, Md.; Difco Laboratories, Detroit.

CHAPTER 42

Staining formulas and procedures

STAINS

Although a number of the more important staining formulas and procedures are presented in this chapter, space does not permit a comprehensive review of the subject. Further details are available in the *Manual of Clinical Microbiology*.[2] The solubilities of the more widely used stains and dyes, in water and in alcohol, are shown in Table 42-1.

1. Acid-fast stain (see also No. 5, Fluorochrome stain)
Kinyoun carbolfuchsin method
(Kinyoun: Am. J. Pub. Health **5:**867, 1915)

Basic fuchsin	4 g
Phenol	8 ml
Alcohol (95%)	20 ml
Distilled water	100 ml

Dissolve the basic fuchsin in the alcohol and add the water slowly while shaking. Melt the phenol in a 56 C water bath and add 8 ml to the stain, using a pipet with a rubber bulb.

Stain the fixed smear for 3 to 5 minutes (no heat necessary) and continue as with Ziehl-Neelsen stain.

By the addition of a detergent or wetting agent the staining of acid-fast organisms may be accelerated (Muller and Chermock: J. Lab. Clin. Med. **30:**169, 1945). Tergitol No 7* may be used. Add 1 drop of Tergitol No. 7 to every 30 to 40 ml of the Kinyoun carbolfuchsin stain. Stain the smears for 1 minute, decolorize, and counterstain as described in the following section.

The technique and interpretation of the acid-fast stain are given in Chapter 3.

---

*Carbide and Carbon Chemical Corporation, New York.

Table 42-1. Solubility of stains*

Stain	Percent soluble at 26 C	
	In water	In 95% ethanol
Bismarck brown	1.36	1.08
Congo red	0	0.19
Crystal violet (chloride)	1.68	13.87
Eosin Y	44.2	2.18
Fuchsin, basic (chloride)	0.26	5.93
Malachite green (oxalate)	7.60	7.52
Methylene blue (chloride)	3.55	1.48
Neutral red (chloride)	5.64	2.45
Safranin O	5.45	3.41
Thionin	0.25	0.25

*Based on data from Conn, J. G.: Biological stains, Commission on Standardization of Biological Stains, Geneva, N.Y., 1928, W. F. Humphrey Press.

Ziehl-Neelsen method

a. Carbolfuchsin stain

Basic fuchsin	0.3 g
Ethanol (95%)	10.0 ml

Mix these with the following:

Phenol, melted crystals	5.0 ml
Distilled water	95.0 ml

b. Acid alcohol *

Hydrochloric acid, concentrated	3.0 ml
Ethanol (95%)	97.0 ml

c. Counterstain

Methylene blue	0.3 g
Distilled water	100 ml

Some workers may prefer 0.5% aqueous brilliant green or a saturated solution of picric acid as a counterstain; the latter is pale and does not selectively stain cellular material.

---

*Use a 1% aqueous solution of sulfuric acid or 0.5% acid alcohol (see No. 5) as a decolorizer when staining smears of suspected acid-fast *Nocardia*, such as *N. asteroides*.

a. Prepare a smear of appropriate thickness; dry and fix as described previously.
b. Place a strip of filter paper slightly smaller than the slide over the smear.
c. Flood the slide with carbolfuchsin stain; heat to steaming with a low Bunsen flame or electrically heated slide warmer. **Do not boil** and do not allow to dry out.
d. Allow to stand 5 minutes without further heating; then remove the paper and wash the slide in running water.
e. Decolorize to a faint pink with acid alcohol while continuously agitating the slide until no more stain comes off in the washings (approximately 1 minute for films of average thickness). Thoroughness in decolorization is essential to prevent the possibility of a false-positive reading.
f. Wash with water; counterstain with methylene blue for 20 to 30 seconds.
g. Wash with water, dry in air, and examine under the oil-immersion lens.

2. Auramine-rhodamine stain (see No. 5, Fluorochrome stain)

3. Capsule stain

The principles of capsule stains are discussed in Chapter 3.

Anthony method

a. Make a thin even smear of a culture in skimmed milk or litmus milk by spreading with a glass slide or an inoculating needle bent at a right angle. If it is not a milk culture, a loopful of the material may be mixed with a loopful of skimmed milk and then spread to give a uniform background.
b. Air dry. Do not fix with heat.
c. Stain with 1% aqueous crystal violet for 2 minutes.
d. Wash with a solution of 20% copper sulfate.
e. Air dry in a vertical position and examine under the oil-immersion lens.

The capsule is unstained against a purple background; the cells are deeply stained.

Hiss method

Mix a loopful of physiologic saline suspension of growth with a drop of normal serum on a glass slide. Allow the smear to air-dry and heat-fix. Flood the smear with crystal violet (1% aqueous solution). Steam the preparation gently for 1 minute and rinse with copper sulfate (20% aqueous solution). Capsules appear as faint blue halos around dark blue to purple cells.

India ink method*

In the India ink method, the capsule displaces the colloidal carbon particles of the ink and appears as a clear halo around the microorganism. The procedure is especially recommended for demonstrating the capsule of *Cryptococcus neoformans*.

a. To a small loopful of saline, water or broth on a clean slide, add a **minute** amount of growth from a young agar culture, using an inoculating needle. Spinal fluid may be used directly.
b. Mix well; then add a small loopful of India ink and immediately cover with a thin coverglass, allowing the fluid to spread as a thin film beneath the coverglass.
c. Examine immediately under the oil-immersion objective, reducing the light considerably by lowering the condenser. Capsules, when present, stand out as **clear halos** against a dark background.

Muir method

Muir mordant

Tannic acid, 20% aqueous solution	2 parts
Saturated aqueous solution of mercuric chloride	2 parts

---

*Not all India inks are suitable. Pelikan India ink made by Gunther Wagner of Hanover, Germany, is recommended; add about 0.3% tricresol as a preservative.

Saturated aqueous solution of   5 parts
   potassium alum

a. Prepare a thin even film of bacteria; allow to dry in air.
b. Cover the film with a piece of filter paper the size of the smear and flood the slide with Ziehl-Neelsen carbolfuchsin.
c. Heat to steaming with a low Bunsen flame for 30 seconds.
d. Rinse gently with 95% ethanol and then with water.
e. Add the mordant for 15 to 30 seconds; wash well with water.
f. Decolorize with ethanol to a faint pink; wash with water.
g. Counterstain with 0.3% methylene blue for 30 seconds.
h. Air dry and examine under the oil-immersion lens. The cells are stained red and the capsules blue.

4. Flagella stain
(Gray: J. Bacteriol. **12:**273, 1926)

Mordant

Potassium alum, saturated aqueous solution	5 ml
Tannic acid, 20% aqueous solution	2 ml
Mercuric chloride, saturated aqueous solution	2 ml

Mix and add 0.4 ml of a saturated alcoholic solution of basic fuchsin. Make up fresh mordant for use each day.

Gray method
a. Using a grease-free, well-cleaned slide that has been flamed and cooled, spread a drop of distilled water on the slide to cover an area of approximately 2 sq cm.
b. Select part of a colony from a young agar culture or take a small amount of growth from a slant with an inoculating needle and **touch gently** into the drop of water at several places on the slide; then gently rotate the slide.
c. Allow to **air dry. Do not heat.**
d. Add the mordant and allow to act for 10 minutes.
e. Wash gently with distilled water or clean tap water.

f. Add Ziehl-Neelsen carbolfuchsin and leave on for 5 to 10 minutes.
g. Wash with tap water, air dry, and examine under oil.

Recently a simplified version of the Leifson flagella stain was reported by Clark.[1] Employing scrupulously clean slides and organisms taken directly from 24- and 48-hour blood agar plates, this method proved to be a simple and reliable procedure. Readers interested in this procedure should consult the original paper.

5. Fluorochrome stain (Truant method)

By staining a smear with fluorescent dyes, such as auramine and rhodamine, and examining by fluorescence microscopy using an ultraviolet light source, acid-fast bacilli, when present, will appear to glow with a **yellow-orange** color. These are visible under lower magnifications of the microscope; thus, a stained smear can be examined in much less time than is required by conventional methods. Numerous modifications of the procedure have been introduced; that reported by Truant and co-workers[3] is recommended.

Auramine O (Allied Chemical Corp., C.I. No. 41,000)	1.5	g
Rhodamine B (Allied Chemical Corp., C.I. No. 749)	0.75	g
Glycerol	75.0	ml
Phenol	10.0	ml
Distilled water	50.0	ml

Combine solutions, mix well (magnetic stirring device for 24 hours or heat until warm and stir vigorously for 5 minutes), filter through glass wool, and store in a glass-stoppered bottle at 4 C. The stain is stable for several months under refrigeration.
a. Heat fix on a slide warmer* at 65 C for 2 hours or overnight.
b. Cover smear with auramine-rhodamine solution.

---

*Micro-slide staining and drying bath are available from Scientific Products, Division of American Hospital Supply Corp., McGaw Park, Ill.

c. Stain for 15 minutes at room temperature or at 35 C.
d. Rinse off with distilled water.
e. Decolorize with 0.5% hydrochloric acid in 70% ethanol for 2 to 3 minutes, then rinse thoroughly with distilled water.
f. Flood smear with counterstain, a 0.5% solution of potassium permanganate (filter and store in amber bottle), for 2 to 4 minutes (no longer—excessive exposure results in loss of brilliance).
g. Rinse with distilled water, dry, and examine.

Ultraviolet light source

The smears are examined under a binocular microscope using an ultraviolet light source. The Leitz, Zeiss, and Reichert fluorescent microscopy units are highly recommended and are equipped with Osram HBO 200 maximum pressure mercury vapor lamps as light sources, BG12 (3 or 4 mm) or C5113 (2 mm) violet exciter filters, and OG1 deep yellow barrier filters in the eyepieces. This recommended filter combination results in **bright yellow-orange** staining bacilli against a dark background; nonspecific background debris fluoresces a pale yellow, quite distinct from the yellow-orange bacilli.

It is suggested that a drop of immersion oil be placed on the darkfield condenser, the slide inserted, and the microscope first focused under bright light until a clear central area is seen on the slide, then switched over to the ultraviolet source. Smears may be rapidly examined under low- or high-power objective (25× or 40×) with a 10× eyepiece; after a little practice all smears may be examined at these magnifications or lower in a matter of seconds. Occasionally, it may be necessary to switch to the oil-immersion objective to confirm typical morphologic characteristics, such as beading or cording. It is recommended that microscopic examination be carried out in a darkened room for maximum efficiency.

Quartz-halogen illuminator*

Excellent demonstration of fluorochrome-stained mycobacteria is obtained by use of a microscope equipped with a quartz-halogen illuminator and the proper combinations of primary (exciter) and secondary (barrier) filters. This system provides the same benefits as ultraviolet apparatus with the added advantages that the **blue light apparatus** does not require a special dark room (although subdued lighting is recommended), does not require oil on the condenser or slide, is simpler and more economical, and does not present radiation hazards. The following recommendations based on use of the Zeiss RA 38 microscope and attachments are applicable to other equipment having comparable features.

The equipment needed is a standard binocular microscope with an illuminator containing a collector lens and a 12-V, 100-W quartz-halogen lamp or high intensity tungsten bulb, front-surface reflecting mirrors, brightfield condenser, low-power objectives (10× or 25×) for scanning and high-dry (63×) planachromat (flat field; if the high-dry objective is corrected for coverslip, then a cover slip must be placed, not mounted, over the smear), a 100× oil-immersion objective for more critical examination of acid-fast (fluorescent) bodies, and 10× compensating eyepieces. A turret, or intermediate tube with holder for secondary filters located in the tube body between the objectives and the eyepieces facilitates filter changing. The optics and light path must be precisely aligned to avoid loss of light intensity. The light source is adjusted for Koehler illumination by centering and focusing the lamp filament on the closed iris diaphragm of the condenser. Maximal intensity is obtained by making small adjustments of

---

*From Runyon, E. H., et al. In Lennette, E. H., Spaulding, E. H., and Truant, J. P., editors: Manual of clinical microbiology, ed. 2, Washington, D.C., 1974, American Society for Microbiology.

the condenser while viewing an aura-mine-stained mycobacterial smear.

Combinations of primary and secondary filters are selected to provide good contrast between a dark background and the fluorescing, yellow bacillus. However, the background must be sufficiently light that nonfluorescing debris can be seen for maintaining focus while scanning the slide. A BG 12 primary filter transmitting only wavelengths less than about 500 nm (peak 404 nm) in combination with secondary filters that transmit only wavelengths above 500 nm or 530 nm, as Zeiss No. 50 or No. 53, respectively, provide satisfactory demonstration of fluorescing mycobacteria. The particular combination of complementary exciter and barrier filters determines the color of the background. The greater the overlap of transmission curves, the lighter the background, and vice versa. Therefore, the user should have on hand BG 12 filters of various thicknesses (1.0, 1.5, 2, 3 mm; obtainable from Fish-Schurman Corp., 70 Portman Road, New Rochelle, N.Y. 10802) for neutral density purposes and secondary filters having transmission cutoffs at 500, 515, and 530 nm to determine which combinations provide optimal background-contrast qualities. The following exciter and barrier filter combinations have been found to be excellent for demonstrating fluorochrome-stained mycobacteria: 3-mm BG 12 and No. 50—light green background; 4-mm BG 12 and No. 50—dark green; 3-mm BG 12 and No. 53—light brown; 3.5-mm BG 12 and No. 53—dark brown. Another primary filter, the fluorescein isothiocyanate (FITC) interference filter, used in combination with a 3-mm BG 12 and a No. 50 or 53 barrier filter, results in excellent dark green or dark red-brown backgrounds, respectively. The FITC laminated to a BG 38 (to reduce red transmission) and in combination with a 1.5-mm BG 12 produces a reddish-tinged gray background with the No. 50 and a pleasant red field with the No. 53.

All positive smears should be confirmed with a Kinyoun or Ziehl-Neelsen stain. This may be done without removing the auramine-rhodamine stain. The reverse of this procedure is not satisfactory.

6. Giemsa stain for chlamydiae*

Giemsa stain is prepared by dissolving 0.5 g of powder in 33 ml of glycerol at 55 to 60 C for 1½ to 2 hours. To this is added 33 ml of absolute methanol, acetone-free. The solution is mixed thoroughly and allowed to sediment and then is stored at room temperature as stock. Dilutions of the stock stain are made with neutral distilled water or buffered water in a ratio of 1 part of stock Giemsa solution to 40 or 50 of diluent.

The smear is air-dried, fixed with absolute methanol for at least 5 minutes, and again dried. It is then covered with the diluted Giemsa stain, freshly prepared each day, for 1 hour. The slide is then rinsed rapidly in 95% ethyl alcohol to remove excess dye, dried, and examined for the presence of the typical basophilic intracytoplasmic inclusion body.

7. Giménez stain for chlamydiae*

Stock carbol fuchsin contains 100 ml of 10% (w/v) fuchsin in 95% ethyl alcohol, 250 ml of 4% (v/v) aqueous phenol, and 650 ml of distilled water. This stock solution should be held at 37 C for 48 hours before use; for the "working" solution, the stock is diluted 1:2.5 with phosphate buffer, pH 7.45 (3.5 ml of 0.2 M NaH_2PO_4, 15.5 ml of 0.2 M Na_2HPO_4, and 19 ml of distilled water). This working solution is immediately filtered and is filtered again before every stain. It is usable for 3 to 4 days.

Malachite green is used as 0.8% aqueous malachite green oxalate.

---

*From Hanna, L., et al. In Lennette, E. H., Spaulding, E. H., and Truant, J. P., editors: Manual of clinical microbiology, ed. 2, Washington, D.C., 1974, American Society for Microbiology.

The smear should be heat-fixed. The smear is stained as follows: stain with fuchsin for 1 to 2 minutes, wash with tap water; stain with malachite green for 6 to 9 seconds, wash with tap water; restain with malachite green for 6 to 9 seconds, wash thoroughly with tap water, and blot dry. In yolk sac smears, most elementary bodies will stain red against a greenish background.

8. Gram stain (Hucker modification)

Reagents
  a. *Stock crystal violet*

Crystal violet (85% dye)	20 g
Ethanol (95%)	100 ml

  b. *Stock oxalate solution*

Ammonium oxalate	1 g
Distilled water	100 ml

Working solution: Dilute the stock crystal violet solution 1:10 with distilled water and mix with 4 volumes of stock oxalate solution. Store in a glass-stoppered bottle.

  c. *Gram iodine solution*

Iodine crystals	1 g
Potassium iodide	2 g

Dissolve these completely in 5 ml of distilled water; then add:

Distilled water	240 ml
Sodium bicarbonate, 5% aqueous solution	60 ml

Mix well; store in an amber glass bottle.

  d. *Decolorizer*

Ethanol (95%)	250 ml
Acetone	250 ml

Mix; store in a glass-stoppered bottle.

  e. *Counterstain* (Stock safranin)

Safranin O	2.5 g
Ethanol (95%)	100 ml

Working solution: Dilute stock safranin 1:5 or 1:10 with distilled water; store in a glass-stoppered bottle.

The principles of the Gram stain are discussed in Chapter 3. The following procedure is for the **rapid** method:
a. Prepare a thin film of the material to be examined; dry and fix as previously described.

b. Flood the slide with crystal violet stain and allow to remain on slide for 10 seconds.
c. Pour off stain and wash off the remainder with the iodine solution.
d. Flood with iodine solution and allow to mordant for 10 seconds.
e. Rinse off with running water. Shake off excess.
f. Decolorize with alcohol-acetone solution or 95% alcohol (an alcohol-acetone solution may prove to be too rapid) until no further color flows from the slide. This usually takes from 10 to 20 seconds, depending on the thickness of the smear. Care should be taken not to overdecolorize the film, since this may result in an incorrect reading of the stain.
g. Counterstain with safranin for 10 seconds; then wash off with water.
h. Blot between clean sheets of bibulous paper and examine under oil immersion.

9. Metachromatic granule stain
Albert stain

The Albert stain is a differential stain and is recommended for its simplicity in staining *Corynebacterium diphtheriae.*
a. Prepare smear and fix with heat.
b. Flood the smear with Albert stain for 3 to 5 minutes.
c. Wash in tap water and drain off excess.
d. Flood with Gram's iodine. Allow to react for 1 minute.
e. Wash, blot dry, and examine.

Granules appear blue-black, the bands appear blue to blue-green, and the cytoplasm appears green.

Methylene blue stain

Methylene blue	0.3 g
Ethyl alcohol (95%)	30.0 ml

When dissolved add:

Distilled water	100 ml

Cover the fixed smear with staining solution and stain for 1 minute. Wash with water and blot dry with blotting paper.

Loeffler methylene blue stain, as formerly used, was prepared by adding alkali to the foregoing solution. Current commercial preparations of methylene blue do not require the addition of alkali. The older preparations contained acid impurities.

This is a simple stain. Prepare a smear of the organism and fix with heat. Flood smear with Loeffler methylene blue and allow to react for 1 minute. Wash and blot dry. The granules readily absorb the dye and appear deep blue in color. Overstaining lessens contrast.

10. PPLO (Mycoplasma) stain
(Dienes and Weinberger: Bacteriol. Rev. **15**:245, 1951)

Reagents

Methylene blue	2.5	g
Azure II	1.25	g
Maltose	10.0	g
Sodium carbonate	0.25	g
Distilled water	100	ml

Dissolve ingredients in the water.
a. Spread a drop of stain on a grease-free coverglass and allow to dry.
b. With a sterile scalpel, cut out a small block of agar medium containing a few *Mycoplasma* colonies and place, with colonies up, on a clean glass slide.
c. Lay coverglass, stain side down, carefully on the agar block, without rubbing.
d. Seal the preparation with a mixture of 3 parts petrolatum and 1 part of paraffin to prevent drying.
e. Examine under low-power objective of the microscope.
f. *Mycoplasma* colonies are quite distinct with dense blue-stained centers and light blue peripheries.

11. Relief staining (Dorner)

Reagents

Nigrosin	10	g
Distilled water	100	ml

Boil for 30 minutes, add 0.5 ml of formalin when cool, filter through paper, and store in 2-ml amounts in sterile corked tubes.

a. Place a loopful of the bacterial suspension on a grease-free slide and immediately add a loopful of the nigrosin solution.
b. Spread out in a thin film.
c. Dry slide in air or hasten drying with gentle heat.
d. Examine under oil-immersion lens.
e. Cells are unstained against the dark background.

12. Rickettsial stains
Castañeda stain

Reagents
Solution A

Potassium phosphate (KH_2PO_4) (1% aqueous)	100 ml
Sodium phosphate ($Na_2HPO_4 \cdot 12H_2O$) (25% aqueous)	100 ml

Mix and add 1 ml of formalin.

Solution B

Methyl alcohol	100 ml
Methylene blue	1 g

Mix 20 ml of solution A with 0.15 ml of solution B and add 1 ml of formalin.

Solution C (counterstain)

Safranin O (0.2% aqueous)	25 ml
Acetic acid (0.1%)	75 ml

a. Prepare a homogeneous film and dry in air.
b. Cover film with stain (mixture of solutions A and B).
c. Drain off stain. Do not wash.
d. Counterstain with safranin O (solution C) for 1 to 4 seconds.
e. Wash with tap water. Blot dry.
f. Examine under oil-immersion lens.
g. The rickettsiae stain blue, whereas the cellular elements stain red.

Giemsa method*

a. Prepare a homogeneous film on a clean glass slide. Allow to dry in air.
b. Flood with methyl alcohol for 1 minute.
c. Drain off alcohol and allow to dry.

---

*This stain may also be used for spirochetes, which stain blue.

d. Cover film with Giemsa stain (15 drops) and allow to react for 1 minute.
e. Add distilled water (30 drops) and continue staining for 5 minutes. Drain off.
f. Wash with distilled water.
g. Place slide on end and allow to dry in air.
h. Examine under oil-immersion lens.
i. The rickettsiae stain a bluish purple.

13. Spore stain
Dorner method

a. Make a heavy suspension of organisms in distilled water in a test tube and add an equal volume of freshly filtered carbolfuchsin.
b. Place tube in boiling water bath for 5 to 10 minutes.
c. Mix a loopful of the aforementioned combination with a loopful of a boiled and filtered 10% aqueous solution of nigrosin on a clean slide.
d. Spread out and dry film quickly with gentle heat.
e. Examine under oil. The spores stain red, and the bacterial cells are almost colorless against a dark gray background.

A modification of the Dorner method may be employed whereby a smear of the culture is prepared and fixed. The smear is then covered with a strip of filter paper, to which carbolfuchsin is added. The dye is heated to steaming for 5 to 7 minutes with a Bunsen burner, and the filter paper is removed. Wash with water, blot dry, and cover with a thin film of nigrosin using a second slide or a needle. The appearance of the cells will be as described previously.

Wirtz-Conklin method*

Flood the entire slide with 5% aqueous malachite green. Steam for 3 to 6 minutes and rinse under running tap water.

---

*From Paik, G., and Suggs, M. T. In Lennette, E. H., and Truant, J. P., editors: Manual of clinical microbiology, ed. 2, Washington, D.C., American Society for Microbiology.

Counterstain with 0.5% aqueous safranin for 30 seconds. Spores are seen as green spherules in red-stained rods or with red-stained debris.

14. Wayson stain for smears of pus

Dissolve 0.2 g of basic fuchsin and 0.75 g of methylene blue in 20 ml of absolute ethanol. Add the dye solution to 200 ml of a 5% solution of phenol in distilled water. Filter. Stain smears for a few seconds. Wash, blot, and dry. This stain is useful in detecting polar staining morphology.

15. Wright-Giemsa method for staining conjunctival scrapings

The Wright-Giemsa stain, along with a Gram stain of the scrapings, gives immediate information to the ophthalmologist regarding (a) conjunctivitis of bacterial origin, (b) inclusion body conjunctivitis and trachoma, or (c) eosinophilia of allergic conjunctivitis.

a. Two slide preparations of scrapings are made; one is stained by the Gram method and one by the Wright-Giemsa technique.
b. To carry out staining by the Wright-Giemsa technique, apply Wright stain to the slide for 1 minute. Add an equal volume of neutral distilled water and stain for 4 minutes.
c. Shake off stain; then apply dilute Giemsa stain (1 drop to 1 ml of neutral distilled water) and allow to stain for 15 minutes.
d. Shake off, decolorize lightly with ethanol, and air dry (do not blot).

MOUNTING FLUIDS
1. Chloral lactophenol

Chloral lactophenol, recommended by Dr. F. Blank of Temple University School of Medicine, is used in place of 10% potassium hydroxide, and in the same manner.

Chloral hydrate	2 parts
Phenol crystals	1 part
Lactic acid	1 part

Dissolve the ingredients by gentle heating over a steam bath.

2. Lactophenol cotton blue

Phenol crystals	20 g
Lactic acid	20 g
Glycerin	40 g
Distilled water	20 ml

Dissolve these ingredients by heating gently over a steam bath. Add 0.05 g of cotton blue dye (Poirrier's blue).

This may be used for yeasts as well as molds and serves as both a mounting fluid and a stain.

a. Place a drop of this fluid on a clean slide.
b. Place a small amount of culture in this drop. If the culture is on agar, remove a piece of the medium with the embedded growth.
c. Cover with a coverglass and press down gently to flatten.
d. Warm gently to remove air bubbles if necessary.
e. Examine under the microscope with high-dry or oil-immersion objectives.
f. Ring edges of coverglass with nail polish if a permanent mount is required.

3. Sodium hydroxide-glycerin

Glycerin	10 ml
Sodium hydroxide*	20 g
Distilled water	90 ml

This mounting fluid is used in moist preparations when examining clinical material for fungi.

---

*Potassium hydroxide may be substituted for sodium hydroxide.

REFERENCES

1. Clark, W. A.: A simplified Leifson flagella stain, J. Clin. Microbiol. **3:**632-634, 1976.
2. Lennette, E. H., Spaulding, E. H., and Truant, J. P., editors: Manual of clinical microbiology, ed. 2, Washington, D.C., 1974, American Society for Microbiology.
3. Truant, J. P., Brett, W. A., and Thomas, W., Jr.: Fluorescence microscopy of tubercle bacillus stained with auramine and rhodamine, Henry Ford Hosp. Med. Bull. **10:**287-296, 1962.

CHAPTER 43

Reagents and tests

1. Acetate utilization

See Chapter 41.

2. Arylsulfatase color standards

a. Prepare a stock solution of 0.1 g of phenolphthalein in 10 ml of ethyl alcohol.
b. Prepare a 2 N sodium carbonate solution by adding 10.6 g of sodium carbonate to 100 ml of distilled water.
c. Prepare standards according to the chart shown.
d. Mix each dilution and dispense in 2-ml quantities in 16- × 125-mm screw-capped tubes.
e. Add 6 drops of 2 N sodium carbonate to each tube.
f. Solutions without added sodium carbonate may be stored in the refrigerator for several months; tubes containing sodium carbonate will fade in 2 to 4 weeks and must be freshly prepared as indicated.
g. A more stable set of standards, valid for 6 months, can be prepared using M/15 Na_2HPO_4 as reagent and 0.1% phenol red as color indicator.*

---

*Vestal, A. L.: Procedures for the isolation and identification of mycobacteria, DHEW Publ. No. (CDC) 75-8230, 1975.

3. Arylsulfatase test reagent

a. *Stock substrate:*
Dissolve 2.6 g of tripotassium phenolphthalein* in 50 ml of distilled water (0.08 M); sterilize by filtration and store under refrigeration.
b. *Stock solution of substrate:*
Prepare a 0.001 M substrate solution by adding 2.5 ml of the 0.08 M stock substrate to 200 ml of Dubos Tween-albumin broth.† Dispense aseptically in 2-ml amounts in 16- × 125-ml screw-capped tubes.

4. Benzidine test solutions

(Deibel and Evans: J. Bacteriol. **79**:356, 1960)

Partially dissolve 1 gram of benzidine dihydrochloride or benzidine base‡ in 20 ml of glacial acetic acid. Add 30 ml of distilled water and heat the solution gently. Cool; then add 50 ml of 95% ethyl alcohol. The reagent is stable for at least 1 month at refrigerator temperature (a

---

*Nutritional Biochemicals Corp., Cleveland; L. Light & Co., Colinbrook, Bucks, England.
†Baltimore Biological Laboratory, Cockeysville, Md.; Difco Laboratories, Detroit.
‡This may not continue to be available, since it is considered to be carcinogenic.

Arylsulfatase color chart

Tube	Phenolphthalein	Distilled water	Amount	2 N Na_2CO_3	Reading
1	1 ml of stock	50 ml	2 ml	6 drops	5+
2	5 ml of tube 1	25 ml	2 ml	6 drops	4+
3	2 ml of tube 1	25 ml	2 ml	6 drops	3+
4	1.5 ml of tube 1	50 ml	2 ml	6 drops	2+
5	0.5 ml of tube 1	50 ml	2 ml	6 drops	1+
6	0.5 ml of tube 1	100 ml	2 ml	6 drops	±

slight yellow color does not affect the reagent's sensitivity).

Fresh 5% hydrogen peroxide solution is prepared each week by diluting 30% reagent grade hydrogen peroxide.

5. Bile solubility

See Chapter 18.

6. Bile test

Inoculate a tube of thioglycollate medium (BBL 135 C) containing 2% commercial dehydrated oxgall (equivalent to 20% bile) and 0.1% sodium desoxycholate, as well as a control tube without bile. Incubate and compare growth in the two tubes. Observe bile broth for inhibition (less growth than in control; not necessarily total inhibition), no inhibition, or stimulation of growth.

7. Buffer solutions
Buffered glycerol-saline solution
(Teague and Clurman, 1916; modified by Sachs, 1939).

Sodium chloride	4.2	g
Dipotassium phosphate, anhydrous	3.1	g
Monopotassium phosphate, anhydrous	1.0	g
Glycerol	300	ml
Distilled water	700	ml

Dispense in bottles with tightly fitting screw caps in approximately 10-ml amounts; autoclave for 15 minutes at 116 C. Add sufficient phenol red to give a distinct red color; if the solution becomes yellow (acid), it should be discarded. The solution is used for preserving fecal specimens.

Sorensen pH buffer solutions

Buffer solutions may be added to culture media to prevent a significant change in hydrogen ion concentration. Sorensen buffers, prepared from potassium and sodium phosphates, are readily prepared from the anhydrous salts, or they may be purchased from commercial sources.

Reagents
  Solution A
    M/15 Na_2PO_4

Dissolve 9.464 g of the anhydrous salt, previously dried at 130 C, in distilled water, to make 1 liter of solution.

  Solution B
    M/15 KH_2PO_4

Dissolve 9.073 g of the anhydrous salt, previously dried at 110 C, in distilled water to make 1 liter of solution. Mix solutions A and B as indicated.

pH	Solution A	Solution B
5.29	0.25 ml	9.75 ml
5.59	0.5 ml	9.5 ml
5.91	1 ml	9 ml
6.24	2 ml	8 ml
6.47	3 ml	7 ml
6.64	4 ml	6 ml
6.81	5 ml	5 ml
6.98	6 ml	4 ml
7.17	7 ml	3 ml
7.38	8 ml	2 ml
7.73	9 ml	1 ml
8.04	9.5 ml	0.5 ml

8. Catalase test

Use an 18- to 24-hour agar slant culture* incubated at 35 C. Pour 1 ml of a 3% solution of hydrogen peroxide over the growth and set the tube in an inclined position. The reaction is **positive** if there is a rapid ebullition of gas. Micrococci and staphylococci are catalase positive; streptococci and pneumococci are catalase negative; *Bacillus* species are catalase positive.

The test may also be carried out with a 24- to 48-hour culture in broth or thioglycollate medium (microaerophiles and anaerobes). Add approximately 1 ml of the hydrogen peroxide to the culture and observe for gas as before.

The test is **not** recommended for cultures grown on blood agar because of the catalase present in the red blood cells.

It may be done on egg yolk agar for anaerobes. Expose to air for at least 30 minutes before testing. Growth may be removed to a drop of H_2O_2 on a glass slide and observed for evolution of bubbles. If

---

*The agar slant should be inoculated quite heavily. An old slant culture will not give a proper test.

subcultures are to be made, this should be done prior to exposure of plate to air.

Recently, a color streak catalase test has been described (Hanker, J. S., and Rabin, A. N.: J. Clin. Microbiol. **2**:463-464, 1975).

9. Citrate utilization*
(Simmons: J. Infect. Dis. **39**:209-214, 1926)

Prepare Simmons' citrate according to manufacturer's directions.

a. Inoculate the surface of the slant lightly, using a saline suspension of the organism and a straight wire.
b. Incubate for 24 to 28 hours (maximum 7 days) at 35 C.
   Positive: growth, alkaline reaction.
Negative: no growth, no change of indicator.

10. Coagulase test
See Chapter 16.

11. Decarboxylase (lysine and ornithine) and dihydrolase (arginine)

See Chapter 41, Decarboxylase test media (No. 26) and Lysine iron agar (No. 48).

12. Deoxyribonuclease (DNase) test
See Chapter 41.

13. Digesting and decontaminating solutions for culturing of sputum for Mycobacterium tuberculosis N-acetyl-L-cysteine-alkali method
(Kubica et al.: Am. Rev. Respir. Dis. **89**: 284, 1964)

---

*Adapted from Blazevic, D. N., and Ederer, G. M.: Biochemical tests in diagnostic microbiology, New York, 1975, John Wiley & Sons, Inc.

a. Prepare the necessary volume of digestant as shown in Table 43-1. The solution is self-sterilizing but should be used within 24 hours, since it deteriorates on standing.
b. In a well-ventilated safety cabinet, transfer no more than 10 ml of sputum to a sterile 50-ml screw-capped, aerosol-free centrifuge tube. Smaller volumes may be used in smaller tubes, but in no case should the volume of sputum exceed one fifth of the volume of the tube.
c. Add an equivalent volume of the acetyl cysteine–sodium hydroxide digestant (Table 43-1) to the specimen; mix well in a Vortex mixer. Digestion is generally effected in 5 to 30 seconds. Avoid extreme agitation, which may inactivate the acetyl cysteine by oxidation. Proceed as described on p. 244.

Trisodium phosphate–benzalkonium chloride method

a. Dissolve 1,000 g of trisodium phosphate · $12H_2O$ in 4,000 ml of hot distilled water. Add 7.4 ml of 17% aqueous benzalkonium chloride concentrate.*
b. M/15 phosphate neutralizing buffer, pH 6.6

(1) Sodium monohydrogen phosphate (anhydrous)	9.47 g
Distilled water	1,000 ml
(2) Potassium dihydrophosphate	9.08 g
Distilled water	1,000 ml

---

*17% Zephiran chloride is available from Winthrop Laboratories, New York.

Table 43-1. Preparation of acetyl cysteine–sodium hydroxide digestant

Reagent	Volume of digestant needed				
	50 ml	**100**	**200**	**500**	**1,000**
1 N (4%) sodium hydroxide	25 ml	50	100	250	500
0.1 M (2.94%) sodium citrate · $2H_2O$	25 ml	50	100	250	500
N-acetyl-L-cysteine powder*	0.25 g	0.5 g	1 g	2.5 g	5 g

*Powdered N-acetyl-L-cysteine is available from Mead Johnson Laboratories, Evansville, Ind.; Sigma Chemical Co., St. Louis; Baltimore Biological Laboratory, Cockeysville, Md.

Mix 625 ml of (2) with 375 ml of (1). Check the reaction and adjust the pH to 6.6 if required; dispense in small volumes in appropriate containers and sterilize at 121 C for 15 minutes. This is used to wash and neutralize the sediment.

14. Egg yolk plate reactions
Lecithinase

A positive lecithinase reaction is indicated by an **opaque zone** in the medium around the colonies.

Lipase

A positive lipase reaction is indicated by an **iridescent** ("oil on water") sheen on the surface of the growth (observed under oblique light). This reaction may be delayed. Therefore, plates should be kept 1 week before being discarded as negative.

Nagler reaction

Prior to inoculating an egg yolk agar plate, swab one half of the plate with *Clostridium perfringens* type A antitoxin and allow it to dry. Streak the inoculum across both halves of the plate, starting on the half without antitoxin. Incubate anaerobically 24 to 48 hours and observe. Inhibition of lecithinase production on the half of the plate containing the antitoxin indicates a positive reaction. This antitoxin is not specific for *C. perfringens*, but is an α-toxin inhibitor. Other species that produce α-toxin (a lecithinase) will also give a positive Nagler reaction. These are *C. bifermentans*, *C. sordellii*, and *C. paraperfringens*.

15. Esculin hydrolysis—anaerobes

Inoculate a tube of esculin broth (heart infusion broth with 0.1% esculin and 0.1% agar), and after good growth is obtained, add a few drops of 1% ferric ammonium citrate solution. A positive reaction is indicated by the development of a **black** color. Alternatively, the tube may be observed under long-wave ultraviolet light (365 nm). Loss of fluorescence indicates a positive reaction.

16. Fermentation of carbohydrates— anaerobes

Tubes of Bacto CHO base broth (Difco) containing various carbohydrates are inoculated.* After good growth is obtained, pH is determined using a pH meter equipped with a long, thin electrode. Interpretation is as follows: pH 5.5 and below, acid; 5.6 to 6.0, weak acid; and above 6.0, negative, providing the pH in control broth is 6.2 or higher. If pH in control broth is 6.1 or less, lower the values for interpretation accordingly. Uninoculated tubes from each batch of medium should be incubated along with the inoculated tubes. Ordinarily the pH of such uninoculated tubes will be 6.2 to 6.4. Occasionally, carbohydrate broths, such as arabinose and xylose, will have a pH of 5.8 or 5.9. Therefore, a pH of less than 5.4 would be acid and 5.4 to 5.6 or 5.7 would be weak acid production.

17. Ferric ammonium citrate

Used as a 1% aqueous solution to test for esculin hydrolysis. Keep in dark bottle.

18. Fildes enrichment†

Used as a supplement for thioglycollate and other media. It is added just prior to using the medium.

19. Gastric mucin (5%)
(Strauss and Klegman: J. Infect. Dis. **88:**151, 1951)

Emulsify 5 g of gastric hog mucin (granular type)‡ in 95 ml of distilled water in a blender for 5 minutes. Autoclave for 15 minutes at 121 C. Cool to room temperature; adjust to pH 7.3 with sterile sodium hydroxide. Check for sterility and store in a refrigerator.

---

*Thioglycollate medium without dextrose or indicator, supplemented with vitamin K_1 (0.1 μg/ml), hemin (5 μg/ml), and sodium bicarbonate (1 mg/ml) can also be used as a base.
†Available from Baltimore Biological Laboratory, Cockeysville, Md. (No. 20810).
‡Wilson Laboratories, Chicago, Ill.

Mix equal parts of 5% gastric mucin and a fungus suspension; inject 1 ml intraperitoneally into the appropriate laboratory animal.

20. Gelatin liquefaction*
Method 1: stab method
(Edwards and Ewing: Identification of Enterobacteriaceae, ed. 3, Minneapolis, 1972, Burgess Publishing Co., p. 345; Lennette et al.: Manual of clinical microbiology, ed. 2, Washington, D.C., 1974, American Society for Microbiology, p. 912)

Prepare 12% gelatin in nutrient broth. Dispense into tubes as deeps. Autoclave at 121 C for 12 minutes.
a. Inoculate gelatin deeps by stabbing to the bottom of the tube. Incubate at 20 C to 22 C for 30 days.
b. To detect liquefaction, place tubes in refrigerator for 30 minutes. Remove and observe for liquefaction. Continue to incubate until liquefaction occurs or until the 30-day period is over.

Strong positive: liquefaction within 3 days. **Weak positive:** liquefaction after 3 days.

Method 2
(Frazier: J. Infect. Dis. **39:**302-309, 1926; Cowan and Steel: Manual for the identification of medical bacteria, New York, 1970, Cambridge University Press, p. 156)

Prepare 12% gelatin in nutrient broth. Autoclave at 121 C for 12 minutes; pour into plates.
a. Inoculate plate in one spot. Incubate at 30 C for 3 days.
b. Flood plate with mercuric chloride solution:

$HgCl_2$	12 g
Distilled water	80 ml
Concentrated HCl	16 ml

Positive: clear zone around growth.

Rapid method
(Blazevic et al.: Appl. Microbiol. **25:**107-110, 1973)
a. Inoculate 0.5 ml of saline with a heavy loopful of growth.
b. Insert a strip of exposed, undeveloped x-ray paper (approximately 1 × 1¼ inches) into the saline suspension.
c. Incubate in a heating block or water bath at 37 C. Observe at 1, 2, 3, 4, and 24 hours for removal of the green gelatin emulsion from the strip.

Positive: appearance of the transparent blue strip support. **Negative:** strip remains green.

21. Gelatin liquefaction—anaerobes (Thiogel medium, BBL)

Test after good growth is observed by refrigerating an inoculated gelatin tube along with an uninoculated tube of gelatin until the uninoculated tube has solidified (usually ½ to 1 hour). Remove tubes to room temperature and invert. A positive reaction is indicated if the inoculated tube fails to solidify. A weak reaction is indicated when the inoculated tube begins to become liquid in approximately one half the time required for the control (uninoculated) tube to liquefy.

22. Gluconate oxidation test
(Haynes: J. Gen. Microbiol. **5:**939, 1951)

Pseudomonas aeruginosa is able to oxidize glucose or gluconate to ketogluconate, which in turn is detected by the reduction of copper salts, as found in Benedict's solution. This test is helpful in identifying nonpigmented strains of *P. aeruginosa.*

The use of gluconate substrate tablets* is recommended for this test. A single tablet is added to 1 ml of distilled water, which is then heavily inoculated with the test organism. After 12 to 18 hours' incubation at 35 C, test the culture for reducing substances with Benedict's solu-

---

*Adapted from Blazevic, D. N., and Ederer, G. M.: Biochemical tests in diagnostic microbiology, New York, 1975, John Wiley & Sons, Inc.

---

*Key Scientific Products Co., Los Angeles.

tion.* A positive test is indicated by a color change from blue to green-yellow.

23. Growth tests—anaerobes

The growth of some anaerobic isolates will be enhanced by the addition of supplements, such as bile, Fildes enrichment, or Tween 80. After 48 to 72 hours' incubation, growth in the tubes containing supplements is compared with growth in the conventional medium (Bacto CHO base with glucose). If one of the supplements enhances growth, it should be added to each tube required for biochemical tests before inoculation.

24. Hemin solution

Used as a medium supplement in a final concentration of 5 μg/ml. To prepare, dissolve 0.5 g of hemin† in 10 ml of commercial ammonia water (or 1 N sodium hydroxide), bring volume to 100 ml with distilled water, and autoclave at 121 C for 15 minutes. Stock solution, 5 mg/ml.

25. Hippurate hydrolysis
(Ayers and Rupp: J. Infect. Dis. **30:**388, 1922)

26. Hippurate hydrolysis (rapid)‡
(Hwang and Ederer: J. Clin. Microbiol. **1:** 114-115, 1975)

Prepare a 1% aqueous solution of sodium hippurate and dispense in 0.4-ml aliquots. Cork and store at −20 C.
a. Thaw tubes of sodium hippurate substrate.
b. Emulsify several small colonies or one large colony of beta-hemolytic streptococci in a tube of substrate. The suspension should be very cloudy.
c. Inoculate positive and negative control organisms, group B and group A streptococci, respectively.

d. Incubate tubes in a heating block at 37 C for 2 hours.
e. Add approximately 0.2 ml (5 drops) of ninhydrin reagent (3.5 g ninhydrin in 100 ml of a 1:1 mixture of acetone and butanol) to each tube. **Do not shake tube.**
f. Continue incubation for 10 minutes. **Do not incubate longer than ½ hour: false positives could occur.**
g. Remove tubes and immediately record results.
Positive: deep purple. **Negative:** no change or a very faint purple.

27. Hydrogen sulfide production
Lead acetate paper test

Saturate filter paper strips (5 × 1 cm) with 5% lead acetate solution. Air dry, then autoclave at 15 pounds pressure for 15 minutes.

Inoculate a sulfur-containing liquid medium and insert a lead acetate strip between the plug and inner wall of tube and above the liquid. Hydrogen sulfide production is evidenced by the **blackening** of the lower portion of the strip.

A negative test may be checked by adding a small amount of 2 N hydrochloric acid to the tube and closing the tube as before. Any dissolved sulfide will be liberated and will combine with the lead in the strip to form the black lead sulfide.

Note: The lead acetate paper test may be positive when the butt reaction in TSI (or KI) agar is negative or only weakly positive. It is more sensitive.

Triple sugar iron (TSI) or Kligler's iron (KI) agar method

The butts of these media are stabbed with the culture. Hydrogen sulfide production is detected by the blackening of the butt (p. 153). Lead acetate paper may also be used by inserting a strip between the loosened cap and inner wall of the tube.

---

*A Clinitest tablet from Ames Co., Elkhart, Ind., may be substituted.
†Sigma Chemical Co., St. Louis.
*Adapted from Blazevic, D. N., and Ederer, G. M.: Biochemical tests in diagnostic microbiology, New York, 1975, John Wiley & Sons, Inc.

28. Immunofluorescence (direct) procedure for Neisseria gonorrhoeae

a. Prepare a **thin** film of a suspected colony from a Thayer-Martin plate (may be made up to 15 minutes after performing the oxidase test) on a slide containing a 6-mm diameter etched circle.

b. Thoroughly dry the film in air.

c. Overlay with adsorbed (with anti-meningococcus group B serum) fluorescein-labeled *N. gonorrhoeae* antiserum,* keeping within the 6-mm diameter circle.

d. Incubate the slide at 35 C in a moist chamber for 30 minutes (alternatively for 5 minutes at room temperature).

e. Rinse with pH 7.2 phosphate buffer, dry, mount in buffered glycerine with a coverglass, and examine under fluorescence microscopy. Gonococci appear as **yellow-green** diplococci of typical size and shape.

f. Include a positive urethral smear or one prepared from a known **fresh** isolate of *N. gonorrhoeae*, along with one of a boiled suspension of *Enterobacter cloacae*, as positive and nonspecific staining controls.

29. Indicators for anaerobiosis
Fildes and McIntosh indicator

Prepare the following solutions:

a. 6% aqueous glucose (add a small crystal of thymol as a preservative).

b. 0.1 N sodium hydroxide; 6 ml to 94 ml of distilled water.

c. Aqueous methylene blue (0.5%); 3 to 100 ml of distilled water.

For use, mix 1 ml of each solution in a test tube, boil the mixture until colorless, then place in a loaded anaerobic jar before sealing. A blue color at the end of incubation indicates that anaerobiosis was **not** achieved.

---

*Available from Difco Laboratories, Detroit (No. 2361).

Smith modified methylene blue indicator

(Smith: ASM meeting, Washington, D.C., 1964)

Mix thoroughly 1 pound of sodium bicarbonate (commercial grade is satisfactory) with 50 g of glucose and 20 mg of methylene blue. For use, add about 1 inch to a 16- × 100-mm test tube, half fill with tap water, invert to mix, and place within the anaerobic jar. The solution will slowly become colorless during incubation at 35 C; if more rapid decolorization is required, the solution may be heated to boiling and cooled rapidly immediately before placing in the jar. The indicator should be **colorless** at the end of incubation if anaerobiosis was maintained.

30. Indole tests
Ehrlich indole test

(Modification of Bohme: Zentralbl. Bakt., Orig. **40**:129, 1906)

> *Reagent*
> | | |
> |---|---|
> | Paradimethylaminobenzaldehyde | 2 g |
> | Ethyl alcohol (95%) | 190 ml |
> | Hydrochloric acid (concentrated) | 40 ml |

Add 1 ml of xylene to a 48-hour culture of organisms in tryptone or trypticase broth or other appropriate medium. Shake well and allow to stand for a few minutes until the solvent rises to the surface.

Gently add about 0.5 ml of the reagent down the sides of the tube so that it forms a ring between the medium and the solvent. If indole has been produced by the organisms, it will, being soluble in solvent, be concentrated in the solvent layer, and on addition of the reagent, a brilliant **red ring** will develop just below the solvent layer. If no indole is produced, no color will develop.

Kovacs indole test

> *Reagent*
> | | |
> |---|---|
> | Pure amyl or isoamyl alcohol | 150 ml |
> | Paradimethylaminobenzaldehyde | 10 g |
> | Concentrated hydrochloric acid (A.R.) | 50 ml |

Dissolve the aldehyde in the alcohol and add the acid slowly. Prepare in small quantities and store in the refrigerator when not in use.

Inoculate tryptophane broth and incubate for 48 hours at 35 C. Add 5 drops of Kovacs reagent. A **deep red** color indicates the presence of indole.

31. Indole spot test
(Vracko and Sherris: Am. J. Clin. Pathol. **39:** 429, 1963)

This test utilizes a Whatman No. 1 filter paper moistened with 1 to 1.5 ml of a 5% solution of *p*-dimethylamino-benzaldehyde in 10% aqueous HCl placed inside the cover of a Petri dish. Colonies to be tested are picked carefully with a sterile loop from overnight growth on a sheep blood agar plate and smeared on a small area of the moistened filter paper. Development of a **brown-red** or **purple-red** color within 20 seconds indicates the presence of indole.

It should be emphasized that this procedure is to be used only with pure cultures of colonies that produce a metallic sheen on eosin-methylene blue agar. It can also be used on pure cultures of swarming colonies of *Proteus* (*P. mirabilis* does not produce indole).

Another spot test of significance (Sutter and Carter, Am. J. Clin. Pathol. **58:**335-338, 1972) is useful for detection of indole produced by anaerobic bacteria. As above, growth obtained from a single, pure culture on a blood agar plate is smeared on filter paper that has been saturated with 1% *p*-dimethylamino-cinnamaldehyde in 10% (v/v) concentrated HCl. Immediate formation of a **blue** color around the growth indicates a **positive** reaction. Negative reactions give no color change or a pinkish color. Late color development should be ignored. The reagent should be stored in a dark bottle and refrigerated when not in use.

32. Kanamycin stock solution
Dissolve 1 g kanamycin (base activity) in 10 ml of sterile phosphate buffer, pH 8.0.

Final concentration, 100,000 μg/ml. Store in refrigerator for up to 1 year. Can be autoclaved.

33. Lecithinase
See No. 14, Egg yolk plate reactions.

34. Lipase
See No. 14, Egg yolk plate reactions.

35. Malonate utilization*
Rapid method
(Blazevic: Laboratory procedures in diagnostic microbiology, ed. 2, St. Paul, 1974, Telstar Products, Inc., p. 111)

Prepare modified malonate broth according to manufacturer's directions. Dispense in 0.5-ml aliquots.
a. Inoculate the broth with a heavy loopful of an overnight growth from triple sugar iron agar or sheep blood agar (do not use growth from MacConkey agar, as false-negatives may occur from this medium).
b. Incubate in a heating block at 37 C for 3 hours.

Positive: blue color. **Negative:** green or yellow color.

36. McFarland nephelometer standards
a. Set up ten test tubes or ampules of equal size and of good quality. Use new tubes that have been thoroughly cleaned and rinsed.
b. Prepare 1% chemically pure sulfuric acid.
c. Prepare 1% aqueous solution of chemically pure barium chloride.
d. Add the designated amounts of the two solutions to the tubes as shown in Table 43-2 to make a total of 10 ml per tube.
e. Seal the tubes or ampules. The suspended barium sulfate precipitate corresponds approximately to homogeneous *Escherichia coli* cell densities per

---

*Adapted from Blazevic, D. N., and Ederer, G. M.: Biochemical tests in diagnostic microbiology, New York, 1975, John Wiley & Sons, Inc.

Table 43-2. McFarland nephelometer standards

Tube number	1	2	3	4	5	6	7	8	9	10
Barium chloride (ml)	0.1	0.2	0.3	0.4	0.5	0.6	0.7	0.8	0.9	1
Sulfuric acid (ml)	9.9	9.8	9.7	9.6	9.5	9.4	9.3	9.2	9.1	9
Approx. cell density ($\times 10^8$/ml)	3	6	9	12	15	18	21	24	27	30

milliliter throughout the range of standards, as shown in Table 43-2.

37. Meat digestion—anaerobes

Test is read in chopped meat–glucose. A positive reaction is indicated by disintegration and gradual disappearance of meat particles, leaving a flocculent sediment in the tube.

38. Methyl red test

(Clark and Lubs.: J. Infect. Dis. **17:**160, 1915)

To 5 ml of culture in MR-VP broth, add 5 drops of methyl red solution. A positive reaction is indicated by a distinct **red** color, showing the presence of acid (pH less than 4.5). A negative reaction is indicated by a **yellow** color. *Escherichia coli* and other methyl red–positive organisms produce high acidity from the dextrose in this medium within 48 hours, which turns the indicator red.

The solution of methyl red is prepared by dissolving 0.1 g of the indicator in 300 ml of 95% alcohol and diluting to 500 ml with distilled water.

Rapid method*

(Barry et al.: Appl. Microbiol. **20:**866-870, 1970)

a. Inoculate 0.5 ml of MR-VP broth in a 13- × 100-mm test tube with one colony. Incubate at 35 C for 18 hours.
b. Add 1 drop of methyl red reagent.
   Positive: bright red color. **Negative:** yellow or orange.
   Methyl red reagent: Dissolve 0.5 g of

---

*Adapted from Blazevic, D. N., and Ederer, G. M.: Biochemical tests in diagnostic microbiology, New York, 1975, John Wiley & Sons, Inc.

methyl red in 300 ml of 95% ethanol. Add 200 ml of distilled water.

39. Micromethods

Micromethods are available commercially (API, Analytab Products, Inc.; Enterotube, Roche Diagnostics; R-B, Corning Medical; Minitek, BBL, and others) for both aerobic and facultative bacteria and anaerobes. Results with the Enterobacteriaceae are good, averaging better than 90% correlation with conventional procedures. Evaluation with the anaerobes is less complete; the selection of tests is still inadequate for these organisms.

40. Milk reactions

See Chapter 41, p. 458.

41. Motility

See Chapter 41, p. 459, Motility test medium. Alternatively for anaerobes, prepare a hanging drop slide from a 4- to 6-hour thioglycollate medium culture. Observe under high-dry magnification.

42. Nagler reaction

See No. 14, p. 484, Egg yolk plate reactions.

43. Nitrate reduction test

Reagents
  Solution A

Sulfanilic acid	8	g
Acetic acid (5 N)	1,000	ml

  Solution B

Alpha-naphthylamine	5	g
Acetic acid (5 N)	1,000	ml

Add 5 drops of each reagent to the tube. A **positive** test for nitrites is revealed by the development of a **red** color in 1 to 2 minutes.

Some organisms can reduce nitrate beyond the nitrite stage to nitrogen gas or ammonia. **A negative test for nitrite, therefore, should not necessarily be construed as a negative nitrate reduction test without first testing for the presence of unreduced nitrate.**

Add a very small amount of zinc dust to the broth medium, which has shown a negative reduction test with the foregoing reagents. The presence of unreduced nitrate is revealed by the development of a red color, thus confirming a negative nitrate reduction test. If no color develops, nitrate was reduced beyond nitrite (positive test), and the remaining portion of the indole-nitrate culture should be tested for ammonia by adding a few drops of Nessler's reagent. A deep orange indicates a positive reaction.

The test for nitrate reduction is carried out after 24 to 48 hours of incubation at 35 C.

Rapid method*

(Blazevic et al.: Appl. Microbiol. **25**:107-110, 1973; Schreckenberger and Blazevic: Appl. Microbiol. **28**:759-762, 1974)

a. Inoculate 0.5 ml of nitrate broth with a heavy loopful of growth from an appropriate medium (e.g., TSI, blood agar, chocolate agar).
b. Incubate in a heating block or water bath at 35 C.
c. Add 1 drop of solution A and one drop of solution B. Shake.
   Positive: red color.
   If test is **negative,** add a minute amount of zinc dust. Shake.
   Positive: absence of red color.

Disk method for anaerobes

A simplified disk test for nitrate reduction by anaerobes has been described re-

cently (Wideman, P. A., et al.: J. Clin. Microbiol. **5**:315-319, 1977).

44. Nitrate reduction test for mycobacteria (nitrite standards)

a. Prepare a M/100 sodium nitrite solution by dissolving 0.14 g of sodium nitrite in 200 ml of distilled water.
b. Carry out twofold serial dilutions of the foregoing in 2 ml-volumes in 13 marked tubes.
c. To tubes 6, 7, 8, 10, 11, and 13 add 1 drop of 1:2 dilution of hydrochloric acid, 2 drops of 0.2% aqueous solution of sulfanilamide, and 2 drops of 0.1% aqueous solution of N-naphthylethylenediamine dihydrochloride.
d. Nitrite standards **fade rapidly** and must be freshly prepared each time.

Tube	Dilution	Reading
6	1:64	5+
7	1:128	4+
8	1:256	3+
10	1:1024	2+
11	1:2048	1+
13	1:8192	+/−

45. ONPG test

See Chapter 41.

46. Oxgall (40%)

Dissolve 40 g oxgall in 100 ml of distilled water. Autoclave at 121 C for 15 minutes. Store in refrigerator.

47. Oxidase test for detecting colonies of Neisseria

(Gordon and McLeod: J. Pathol. Bacteriol. **31**:185, 1928)

For this oxidase test a 1% solution (0.1 g in 10 ml) of para-aminodimethylaniline monohydrochloride* is used. Add the dye to distilled water and let stand for 15 minutes before using or until a definite purple color has developed. The solution should be used immediately and then

---

*Adapted from Blazevic, D. N., and Ederer, G. M.: Biochemical tests in diagnostic microbiology, New York, 1975, John Wiley & Sons, Inc.

*N, N-Dimethyl-*p*-phenylenediamine monohydrochloride, from Eastman Organic Chemicals, Rochester, N.Y.

discarded, the dye being active only 1 to 2 hours after the solution has been prepared.

Place a few drops of the dye solution on portions of the plate containing colonies suspected of being *Neisseria.* Colonies producing indophenol oxidase become **pink,** progressing to maroon, dark **red,** and finally to **black.** Organisms from pink colonies are usually viable, but after the colonies have become black the organisms are dead. The dye does not interfere with the Gram-stain reaction.

A 1% aqueous solution of tetramethyl-para-phenylenediamine dihydrochloride may also be used (see No. 48). It is less toxic and gives the colonies a lavender color that eventually turns purple. The reagent sometimes colors the surrounding medium, and it costs considerably more than the dimethyl reagent.

Indophenol method*
(Ewing and Johnson: *Int. Bull. Bact. Nomencl. Taxon.,* **10:**223-230, 1960)

a. Inoculate a nutrient agar slant; incubate at 35 C for 18 to 24 hours, no longer. Also inoculate an agar slant with *Aeromonas* for a positive control.
b. Add 2 to 3 drops of reagent A and reagent B to the growth on the slant. Tilt the tube so that reagents are mixed and flow over the growth.

   Positive: blue color in the growth within 2 minutes. Weak or doubtful reactions that occur after 2 minutes should be ignored.

Reagent A
   Dissolve 1 g of alpha napthol in 100 ml 95% ethanol.

Reagent B
   Dissolve 1 g of *p*-aminodimethylaniline HCl (or oxalate) in 100 ml of distilled water. Reagent B should be stored in the refrigerator and should be freshly prepared each month.

---

*Adapted from Blazevic, D. N., and Ederer, G. M.: Biochemical tests in diagnostic microbiology, New York, 1975, John Wiley & Sons, Inc.

48. Oxidase test for Pseudomonas
(Kovacs: Nature **178:**703, 1956)

Reagents

Tetramethyl-para-phenylenediamine dihydrochloride	0.1 g
Distilled water	10 ml

Add the dye to the water and allow to stand for 15 minutes before using. Prepare freshly for use each time. The dye loses its activity after 2 hours.*

Place 2 or 3 drops of the reagent on a piece of Whatman No. 1 filter paper (6 sq cm). Remove a colony† from plate, or some growth from an agar slant, and smear with a loop on the reagent-saturated paper. A positive reaction (oxidase-positive), recognized by a **dark purple** color, develops in 5 to 10 seconds.

49. Phenylalanine deaminase‡

See also Chapter 41, No. 66.

Method 1
(Ewing et al.: Publ. Health Lab. **15:**153-167, 1957)

Prepare phenylalanine agar in long slants according to manufacturer's directions.

a. Inoculate the phenylalanine agar slant by streaking heavily. Incubate at 35 C for 4 hours or 18 to 24 hours.
b. Allow 4 to 5 drops of 10% (w/v) ferric chloride to run over the growth.

   Positive: dark green color in the fluid and slant surface.

Method 2
(Ederer et al.: Appl. Microbiol. **21:**545, 1971)

Prepare phenylalanine-urea (PU) medium:

---

*A stable reagent in a dropper is available commercially and is recommended (Cepti-seal oxidase test reagent, Marion Scientific Corp., Rockford, Ill.).
†Use a platinum wire loop—nichrome may cause a false-positive reaction.
‡Adapted from Blazevic, D. N., and Ederer, G. M.: Biochemical tests in diagnostic microbiology, New York, 1975, John Wiley & Sons, Inc.

Yeast extract	1.0	g
$(NH_4)_2SO_4$	2.0	g
NaCl	3.0	g
K_2HPO_4	1.2	g
KH_2PO_4	0.8	g
DL-Phenylalanine	5.0	g
(or L-phenylalanine)	2.5	g
Distilled water	975	ml

Dissolve all ingredients; autoclave at 121 C for 15 minutes. Aseptically add 25 ml of urea agar concentrate (Difco). Mix well, and aseptically dispense 0.5 ml into 13- × 100-mm sterile tubes. If medium is to be kept for a long time, store at −20 C.

a. Inoculate 0.5 ml of PU medium with a colony of organism to be tested. Incubate overnight at 35 C.
Positive urease: definite pink color. **Negative urease:** yellow or faint tinge of pink.
b. Add 1 to 2 drops of 1% (v/v) HCl to adjust the medium to an acid pH (yellow).
c. Add 2 drops of 10% (w/v) ferric chloride.
Positive phenylalanine deaminase: dark green color. **Negative phenylalanine deaminase:** yellow. The phenylalanine deaminase reaction must be read within 10 seconds after adding the ferric chloride, as the green color fades rapidly.

With this method *Proteus*, most *Klebsiella*, some *Enterobacter*, and *Yersinia* will be positive for urease. All other Enterobacteriaceae will be negative.

Rapid method

a. Inoculate 0.5 ml of PU medium with a large loopful growth from TSI or other solid medium.
b. Incubate in a heating block or water bath at 37 C for 2 hours.
c. Read urease and phenylalanine deaminase reactions as under Method 2.

50. Preservation of fungal cultures*

Obtain a good grade of **heavy** mineral oil†

---

*Recommended by the Mycology Branch of the Center for Disease Control, Atlanta.
†Available from Parke-Davis Co., Detroit.

and dispense in approximately 125-ml amounts in 250-ml Erlenmeyer flasks (cotton plugged). Autoclave at 121 C for 45 minutes. Using sterile technique, pour the oil over a small but actively growing fungus culture on a short slant of Sabouraud agar. It is essential that the oil cover not only the fungus colony but also the whole agar surface (about 1 inch above the top of the slant); otherwise the exposed agar or colony will act as a wick and in time cause the medium to dry out.

The stock culture is stored upright at room temperature and will remain viable without further attention for several years.

In transferring from oiled cultures, remove a bit of the fungus with a long inoculating needle, drain off the oil by touching it on the inside of the tube, and inoculate to a fresh slant. Rinse off the needle in xylol before flaming.

51. Resazurin

Used as Eh indicator in PRAS media. Dissolve one tablet (Allied Chemical #506) in 44 ml distilled water. Store stock solution at room temperature.

52. Ringer's solution

Sodium chloride	8.5	g
Potassium chloride	0.2	g
Calcium chloride · $2H_2O$	0.2	g
Sodium carbonate	0.01	g
Distilled water	1,000	ml
(pH 7)		

Sterilize by autoclaving at 121 C for 15 minutes. To use in dissolving calcium alginate swabs, prepare in one-quarter strength and add 1% sodium hexametaphosphate. Approximately 10 minutes of shaking usually suffices for complete dissolution of the swab.

53. Sodium bicarbonate

Used as a medium supplement for anaerobes in a final concentration of 1 mg/ml. To prepare, dissolve 2 g in 100 ml distilled water and filter sterilize. Stock solution, 20 mg/ml. For use, add 0.5 ml to 10 ml of medium.

54. Spores—anaerobes

Spores can be observed in stained preparations (Gram or spore stain) made from solid or broth medium. Some commonly encountered species, such as *C. perfringens* and *C. ramosum*, sporulate poorly and spores are rarely seen. Heat tests are often used if spores cannot be demonstrated in stained smears. Growth from chopped meat slant or other solid medium should be suspended in two tubes of starch broth, being careful not to touch loop to sides of tubes above the level of the medium; at the same time, subculture to BAP for anaerobic incubation to check viability. Place one tube of starch broth in a water bath at 80 C, with water level above the level of the medium in the tube. Place tube, containing an equal amount of water, and thermometer in water bath at the same time. Leave starch tube in bath for 10 minutes after tube of water has reached 80 C. Remove starch tube and incubate with the unheated tube. Observe for growth (up to 10 days). Growth in both tubes indicates a positive test. If growth is questionable on visual inspection, both tubes should be subcultured.

55. SPS disk test

Inhibition zones at least 14 mm in diameter about disks impregnated with 20 µl of 5% sodium polyanethol sulfonate (SPS)* on brucella blood agar provide good presumptive evidence that an anaerobic gram-positive coccus is *Peptostreptococcus anaerobius*.

56. Starch hydrolysis†
(Allen: J. Bacteriol. **3:**15-17, 1918)

Prepare a broth or agar (appropriate to the type of organism being tested) with 0.2% soluble starch added. Prepare either slants or plates from the agar.

---

*From Harleco, Philadelphia.

†Adapted from Blazevic, D. N., and Ederer, G. M.: Biochemical tests in diagnostic microbiology, New York, 1975, John Wiley & Sons, Inc.

a. Inoculate starch medium. Incubate in an appropriate atmosphere at 35 C overnight or until sufficient growth has occurred.
b. To a broth add a few drops of Gram's iodine. Read immediately.
   Positive: no change in color. **Negative:** blue color (may fade after a while).
c. For plates or slants, flood with Gram's iodine.
   Positive: medium is blue with colorless area around growth. **Negative:** medium is blue even around growth.

57. Tween 80 hydrolysis substrate

M/15 phosphate buffer, pH 7	100 ml
Tween 80	0.5 ml
Neutral red (0.1% aqueous solution)	2.0 ml

a. Mix the foregoing solutions, dispense in 16- × 125-mm screw-capped tubes in 2-ml amounts.
b. Sterilize by autoclaving at 121 C for 10 minutes.
c. Check for sterility by incubating overnight at 35 C. Final color is amber or straw.
d. Store in refrigerator in the dark for not more than 2 weeks.

This substrate is used to test the ability of certain strains of mycobacteria to rapidly degrade the Tween 80 to oleic acid, which is detected by a change in color of the indicator.

58. Urease—anaerobes (see also Chapter 41)

Scrape growth from egg yolk agar or other solid medium. Make a heavy suspension in 0.5 ml of sterile urea broth. Incubate and observe for up to 24 hours aerobically. A bright red color indicates a positive reaction. With a heavy inoculum, urease production usually will be evident within 15 to 30 minutes. If indicator has been reduced, add Nessler's reagent to determine ammonia production. Presence of ammonia indicates a positive reaction.

59. Urease test broth

Urea	20.0	g
Monopotassium phosphate	9.1	g
Disodium phosphate	9.5	g
Yeast extract	0.1	g
Phenol red	0.01	g

(Final pH 6.8±)

Use 3.87 g per 100 ml of distilled water. **Do not heat.** When the powder has dissolved, filter the medium through a sterile bacteriologic filter. Distribute the broth in 0.5- to 2-ml amounts in small sterile tubes. Large amounts may be used if desired, but reactions are slower. If a filter is not available, it is possible to sterilize the medium in an autoclave, if the tubes are not tightly packed and the steam pressure is held at 5 pounds for 7 minutes or at 8 pounds for 20 minutes.*

In addition, the medium generally gives reliable results without sterilization, if prepared and inoculated immediately.

Urease test broth is prepared according to the formula of Rustigian and Stuart.†‡ It may be used for identification of bacteria on the basis of urea utilization and is particularly recommended for the differentiation of members of the genus *Proteus* from *Salmonella* and *Shigella* in the diagnosis of enteric infections.

Urease test broth may be inoculated from TSI (or KI) agar, trypticase soy agar, or other agar slants having heavy growth. It is recommended that large inocula be employed when it is desirable to obtain results rapidly. Incubate at 35 C. Normally, the finished medium has a pale pink or pinkish yellow color and a neutral pH of about 6.8 to 7.0. In those cultures that attack urea, ammonia is formed during incubation and makes the reaction of the medium alkaline, with a **deep purple** or bluish red color. In the medium described by the above investigators no other organism of the family *Enterobacteriaceae* has been found that will give evidence of urease production.

For a **rapid** urease test, Ewing recommends inoculation of 3 ml of broth with three loopfuls of an agar slant culture. After shaking, tests are incubated in a 37 C water bath and read after 10, 60, and 120 minutes.*

Urease test broth may be used in the same manner for the detection of urease activity of such organisms as members of the genera *Brucella*, *Bacillus*, *Sarcina*, and *Mycobacterium*.† Incubation usually should be longer than for enteric bacilli.

Note: Both prepared broth and dehydrated base should be stored in the refrigerator. If the seal on the bottle has been broken, the bottle should preferably be stored with desiccant in a sealed container.

60. Vancomycin stock solution

Dissolve 75 mg of vancomycin base activity in 5 ml N/20 HCl; add 5 ml sterile distilled water. Final concentration, 7,500 μg/ml. Store in refrigerator for up to 1 month or in freezer (-20 C) for up to 1 year.

61. Vitamin K₁ solution

Used as a medium supplement in a final concentration of 0.1 μg/ml for liquid media and 10 μg/ml for agar media. To prepare, weigh out 0.2 g of vitamin K_1 (Nutritional Biochemical Corp., Cleveland) on a small piece of sterile aluminum foil and aseptically add to 20 ml of absolute ethanol in a sterile tube or bottle. Stock solution, 10 mg/ml. The stock solution can be further diluted for use in sterile distilled water. Store solutions in a refrigerator in a tightly closed container protected from light.

---

*McKay, Edwards, and Leonard: Am. J. Clin. Pathol. **17**:479, 1947

†Rustigian and Stuart: Proc. Soc. Exp. Biol. Med. **47**:108, 1941.

‡Stuart, Van Stratum, and Rustigian: J. Bacteriol. **49**:437, 1945.

*Ewing: Enterobacteriaceae, Washington, D.C., 1960, U.S. Public Health Service (Publ. No. 734).

†Gordon and Mihm: J. Gen. Microbiol. **21**:736, 1959.

62. Voges-Proskauer test for acetyl-methylcarbinol or acetoin
(Coblentz: Am. J. Pub. Health **33:**315, 1943)
Barritt's test

Reagents
> Alpha-naphthol (5%) in absolute ethyl alcohol
> Potassium hydroxide (40%) containing 0.3% creatine

a. Pipet 1 ml of a 48-hour culture grown in MR-VP broth into a clean Wassermann tube.
b. Add 0.6 ml of 5% alpha-naphthol in absolute ethyl alcohol.
c. Add 0.2 ml of 40% potassium hydroxide–creatine solution.
d. Shake well and allow to stand for 10 to 20 minutes. If acetylmethylcarbinol has been produced, a bright **orange-red** color will develop at the surface of the medium and will gradually extend throughout the broth. The development of a copperlike color in some tubes is not considered a positive reaction.

O'Meara's (modified) test

Reagents

KOH	40.0 g
Creatine	0.3 g
Distilled water	100 ml

Dissolve the KOH in the distilled water and add the creatine.

a. O'Meara's (modified) reagent should be added in equal volume to a 48-hour culture
b. Incubate mixture at 35 C or at room temperature.
c. Final readings are made after 4 hours. Tests should be aerated by shaking the tubes.
d. Positive reactions are indicated by the formation of a **red-pink** color. If equivocal results are obtained, repeat tests with broth cultures incubated at 25 C for 48 hours.
e. O'Meara's reagent should be prepared frequently and should be refrigerated when not in use. It should not be kept more than 2 to 3 weeks, since it deteriorates rapidly beyond this time.

Note: Both the methyl red (modified) and the Voges-Proskauer (modified) test can be run at 24 hours if the volume of the broth is reduced to 0.5 ml. The broth-reagent mixture should be frequently shaken. A positive result is the appearance of a red color, usually within 15 minutes.

REFERENCES

1. Blazevic, D. N., and Ederer, G. M.: Biochemical tests in diagnostic microbiology, New York, 1975, John Wiley & Sons, Inc.
2. Mac Faddin, J. F.: Biochemical tests for identification of medical bacteria, 1976, Baltimore, Williams and Wilkins Co.
3. Paik, G., and Suggs, M. T.: Reagents, stains, and miscellaneous procedures. Chapter 96. In: Lennette, E. H., Spaulding, E. H., Truant, J. P. (eds.), Manual of clinical microbiology, ed. 2, 1974, Washington, D. C., American Society for Microbiology.
4. Sutter, V. L., Vargo, V. L., and Finegold, S. M.: Wadsworth Anaerobic Bacteriology Manual, ed. 2, 1975, UCLA Extension Division, Los Angeles.

Glossary

abscess Localized collection of pus.

accolé Early ring form of *Plasmodium falciparum* found at margin of red cell.

acid-fast Characteristic of certain bacteria, such as mycobacteria, which involves resistance to decolorization by acids when stained by an aniline dye, such as carbol fuchsin.

aerobe, obligate Microorganism that lives and grows freely in air and cannot grow anaerobically.

aerogenic Producing gas (in contrast to anaerogenic—non-gas-producing).

aerotolerant Ability of an anaerobic microorganism to grow in air, usually poorly, especially after anaerobic isolation.

alopecia Baldness.

aminoglycosides Group of related antibiotics including streptomycin, kanamycin, neomycin, tobramycin, gentamicin, amikacin.

anaerobe, obligate Microorganism that grows only in complete or nearly complete absence of air or molecular oxygen.

anergy Absence of reaction to antigens or allergens.

antibiotic Substance produced by a microorganism, which inhibits or kills other microorganisms. A broad-spectrum antibiotic is therapeutically effective against a wide range of bacteria.

antibody Substance (immunoglobulin) formed in the blood or tissues, which interacts only with the antigen that induced its synthesis (e.g., agglutinin).

antimicrobial Chemical substance, either produced by a microorganism or by synthetic means, that is capable of killing or suppressing the growth of microorganisms.

analytic reagent (AR) Grade of chemical.

ascitic fluid Serous fluid in peritoneal cavity.

autotroph Organism that can utilize inorganic carbon sources (carbon dioxide).

auxotroph Differing from the wild strain (prototroph) by an additional nutritional requirement.

bacteremia Presence of viable organisms in the blood.

bacteriocins Antibioticlike substances produced by bacteria, which exert a lethal effect on other bacteria.

bacteriuria Presence of bacteria in urine.

benign tertian malaria Malaria caused by *Plasmodium vivax.*

biopsy Removal of tissue from a living body for diagnostic purposes (e.g., lymph node biopsy).

blackwater fever Condition in which the diagnostic symptom is passage of reddish or red-brown urine, which indicates massive intravascular hemolysis *(Plasmodium falciparum).*

bronchoscopy Examination of the bronchi through a bronchoscope, a tubular illuminated instrument introduced through the trachea (windpipe).

bubo Inflammatory enlargement of lymph node, usually in groin or axilla.

buffy coat Layer of white blood cells and platelets above red blood cell mass when blood is sedimented.

capneic incubation Incubation under increased carbon dioxide tension, as in a candle extinguishing jar (ca 3% CO_2).

capsule Gelatinous material surrounding bacterial cell wall, usually of polysaccharide nature.

caseation necrosis Tissue death with loss of cell outlines and a "cheeselike" amorphous material.

catalase Bacterial enzyme that breaks down peroxides with liberation of free oxygen.

catheter Flexible tubular (rubber or plastic) instrument used for withdrawing fluids from (or introducing fluids into) a body cavity or vessel (e.g., urinary bladder catheter).

cellulitis Inflammation of subcutaneous tissue.

Charcot-Leyden crystals Slender crystals shaped like a double pyramid with pointed ends, formed from the breakdown products of eosinophils and found in feces, sputum, and tissues; indicative of an immune response that may have parasitic or nonparasitic causes.

chemotherapeutic Chemical agent used in the treatment of infections (e.g., sulfonamides).

chromatography Method of chemical analysis by which a mixture of substances is separated by fractional extraction or adsorption or ion exchange on a porous solid.

chromogen Bacterial species whose colonial growth is pigmented (e.g., *Flavobacterium* sp.-yellow).

colony Macroscopically visible growth of a microorganism on a solid culture medium.

commensal Microorganism living on or in a host but causing the host no harm.

conjunctivitis Inflammation of the conjunctivae or membranes of the eye and eyelid.

definitive host Host in which the sexual reproduction of a parasite occurs.

DNase Deoxyribonuclease, an enzyme that depolymerizes DNA, an essential component of all living matter, which contains the genetic code.

dysentery Inflammation of the intestinal tract, particularly the colon, with frequent bloody stools (e.g., bacillary dysentery).

dysgonic Growing poorly (bacterial cultures).

ectoparasite Organism that lives on or within skin.

effusion Fluid escaping into a body space or tissue (e.g., pleural effusion).

elephantiasis Condition caused by inflammation and obstruction of the lymphatic system, resulting in hypertrophy and thickening of the surrounding tissues, usually involving the extremities and external genitalia (filariasis).

elution Process of extraction by means of a solvent.

empyema Accumulation of pus in a body cavity, particularly empyema of the thorax or chest.

496

endogenous Developing from within (i.e., within the body).

endoparasite Parasite that lives within the body.

enterotoxigenic Producing an enterotoxin (e.g., enterotoxigenic *E. coli*).

enterotoxin Toxin affecting the cells of the intestinal mucosa.

erythema Redness of the skin resulting from various causes.

erythrocytic cycle Developmental cycle of malarial parasites within red blood cells.

eugonic Growing luxuriantly (bacterial cultures).

exoerythrocytic cycle Portion of the malarial life cycle occurring in the vertebrate host in which sporozoites, introduced by infected mosquitoes, penetrate the parenchymal liver cells and undergo schizogony, producing merozoites, which then initiate the erythrocytic cycle.

exogenous From without (outside the body).

exudate Fluid that has passed out of blood vessels into adjacent tissues or spaces; high protein content.

facultative anaerobe Microorganism that grows under either anaerobic or aerobic conditions.

fermentation Anaerobic decomposition of carbohydrate.

fluorescent Emission of light by a substance (or a microscopic preparation) while acted on by radiant energy, such as ultraviolet rays, as in the immunofluorescent procedure.

fusiform Spindle shaped, as in the anaerobe *Fusobacterium nucleatum*.

gangrene Death of a part or tissue due to failure of blood supply, disease, or injury.

glabrous Smooth.

granuloma Aggregation and proliferation of macrophages to form small (usually microscopic) nodules.

hemagglutination Agglutination of red blood cells caused by certain antibodies, virus particles, or high molecular weight polysaccharides.

hematogenous Disseminated by the bloodstream.

herpes Inflammation of the skin characterized by clusters of small vesicles (e.g., herpes simplex).

heterotroph Organism that requires an organic carbon source.

hypertrophy Increased size of an organ due to enlargement of individual cells.

immunofluorescence Microscopic method of determining the presence or location of an antigen (or antibody) by demonstrating fluorescence when the preparation is exposed to a fluorescein-tagged antibody (or antigen) using ultraviolet radiation.

inclusion bodies Microscopic bodies, usually within body cells; thought to be virus particles in morphogenesis.

indigenous flora Normal or resident flora.

induced malaria Malaria infection acquired by parenteral inoculation (e.g., blood transfusion or sharing of needles by drug addicts).

induration Abnormal hardness of a tissue or part resulting from hyperemia or inflammation, as in a reactive tuberculin skin test.

infection Invasion by and multiplication of microorganisms in body tissue resulting in disease.

intermediate host Required host in the life cycle in which essential larval development must occur before a parasite is infective to its definitive host or to additional intermediate hosts.

intertrigo Erythematous skin eruption of adjacent skin parts.

intramuscular (-peritoneal, -venous) Within the muscle (peritoneum, vein), as in intramuscular injection.

in vitro Literally, within a glass (i.e., in a test tube, culture plate, or other nonliving material).

in vivo Within the living body.

involution forms Abnormally shaped bacterial cells occurring in an aging culture population.

lag phase Period of slow microbial growth, which occurs following inoculation of the culture medium.

Leishman-Donovan body Small, round intracellular form (called amastigote or leishmanial stage) of *Leishmania* sp. and *Trypanosoma cruzi*.

leukocytosis Elevated white blood cell count.

leukopenia Low white blood cell count.

logarithmic phase Period of maximal growth rate of a microorganism in a culture medium.

lysis Disintegration or dissolution of bacteria or cells.

malignant tertian malaria Malaria caused by *Plasmodium falciparum*.

meconium Pasty greenish mass in intestine of fetus; made up of mucus, desquamated cells, bile, and such.

meningitis Inflammation of the meninges, the membranes that cover the brain and spinal cord (e.g., bacterial meningitis).

merozoite Product of schizogonic cycle in malaria that will invade red blood cells.

microaerophile, obligate Microorganism that grows only under reduced oxygen tension and cannot grow aerobically or anaerobically.

microfilaria Embryos produced by filarial worms and found in the blood or tissues of individuals with filariasis.

mixed culture (pure culture) More than one organism growing in or on the same culture medium, as opposed to a single organism in pure culture.

mucopurulent Material containing both mucus and pus (e.g., mucopurulent sputum).

mycoses Diseases caused by fungi (e.g., dermatomycosis, fungal infection of the superficial skin).

myocarditis Inflammation of the heart muscle.

nares External openings of nose (i.e., nostrils).

nasopharyngeal Pertaining to the part of the pharynx above the level of the soft palate.

necrosis Pathologic death of a cell or group of cells.

neonatal First 4 weeks after birth.

neurotrophic Having a selective affinity for nerve tissue. Rabies is caused by a neurotrophic virus.

nonsporulating Does not produce spores.

operculated ova ova possessing a cap or lid.

otitis Inflammation of the ear from a variety of causes, including bacterial infection. Otitis media, inflammation of the middle ear.

parasite Organism that lives on or within and at the expense of another organism.

paronychia Purulent inflammation about margin of a nail.

paroxysm Rapid onset (or return) of symptoms; term usually applies to cyclic recurrence of malaria symptoms, which are chills, fever, and sweating.

pathogenic Producing disease.

pathologic Due to or involving a morbid condition, as a pathologic state.

penicillinase (beta-lactamase I) Enzyme produced by some bacterial species, which inactivates the antimicrobial activity of certain penicillins (e.g., penicillin G).

percutaneous Performed through the skin (e.g., percutaneous bladder aspiration).

pericarditis Inflammation of covering of heart.

plasma Fluid portion of blood; obtained by centrifuging anticoagulated blood.

pleomorphic Having more than one form, usually widely different forms, as in pleomorphic bacteria.

pleuropulmonary Pertaining to the lungs and pleura.

prodromal Early manifestations of a disease before specific symptoms become evident.

proglottid Segments of the tapeworm containing male and female reproductive systems; may be immature, mature, or gravid.

prognosis Forecast as to the possible outcome of a disease (e.g., a guarded prognosis).

prophylaxis Preventive treatment (e.g.., the use of drugs to prevent infection).

prototroph Naturally occurring or wild strain.

psychrophilic Cold loving (e.g., microorganisms that grow best at low [4 C] temperatures).

purulent Consisting of pus.

pus Product of inflammation, consisting of a thin fluid and many white blood cells; often bacteria, cellular debris, and such, are also present.

pyogenic Pus producing.

quartan malaria Malaria caused by *Plasmodium malariae.*

saprophytic Nonpathogenic.

schizogony Stage in the asexual cycle of the malaria parasite that takes place in the red blood cells of humans.

scolex (plural, scolices) Head portion of a tapeworm; may attach to the intestinal wall by suckers or hooklets.

septicemia (sepsis) Systemic disease associated with the presence of pathogenic microorganisms or their toxins in the blood.

serum Cell and fibrinogen-free fluid after blood clots.

somatic Pertaining to the body (of a cell) (e.g., the somatic antigens of *Salmonella* sp.).

spore Reproductive cell of bacteria, fungi, or protozoa; in bacteria may be inactive, resistant forms within the cell.

sporogony Stage in the sexual cycle in the malarial parasite that takes place in the mosquito.

sporozoite Slender, spindle-shaped organism that is the infective stage of the malarial parasite; it is inoculated into humans by an infected mosquito and is the result of the sexual cycle of the malarial parasite in the mosquito.

stab culture One in which the inoculation of a tube of solid medium is made by stabbing with a needle to encourage anaerobic growth in the bottom.

stat (Do) immediately.

stationary phase Stage in the growth cycle of a bacterial culture in which the vegetative cell population equals the dying population.

sterile (sterility) Free of living microorganisms; the state of being sterile.

strobila Entire chain of tapeworm proglottids, excluding the scolex and neck.

sulfur granule Small colony of organisms with surrounding clublike material. Yellow-brown. Resembles grain of sulfur.

suppuration Formation of pus.

suprapubic bladder aspiration Obtaining urine by direct needling of the full bladder through the abdominal wall above the pubic bone.

syndrome Set of symptoms occurring together (e.g., nephrotic syndrome).

synergism Combined effect of two or more agents that is greater than the sum of their individual effects.

synovial fluid Sterile viscid fluid secreted by the synovial membrane; found in joint cavities, bursae, and so forth.

therapy, antimicrobial Treatment of a patient for the purpose of combatting an infectious disease.

thermolabile Adversely affected by heat (as opposed to thermostable, not affected by heat).

transtracheal aspiration Passage of needle and plastic catheter through the trachea for obtaining lower respiratory tract secretions.

transudate Similar to exudate, but with low protein content.

trophozoite Feeding, motile stage of protozoa.

urethritis Inflammation of the urethra, the canal through which urine is discharged (e.g., gonococcal urethritis).

virulence Degree of pathogenicity or disease-producing ability of a microorganism.

viscera Internal organs, particularly of abdominal cavity.

xenodiagnosis Procedure involving the feeding of laboratory-reared triatomid bugs on patients suspected of having Chagas' disease; after several weeks the feces of the bugs are checked for intermediate stages of *Trypanosoma cruzi.*

zoonoses Diseases of lower animals transmissible to humans (e.g., tularemia).

Index